Lecture Notes in Artificial Intelligence 6856

Subseries of Lecture Notes in Computer Science

LNAI Series Editors

Randy Goebel
University of Alberta, Edmonton, Canada
Yuzuru Tanaka
Hokkaido University, Sapporo, Japan
Wolfgang Wahlster
DFKI and Saarland University, Saarbrücken, Germany

LNAI Founding Series Editor

Joerg Siekmann
DFKI and Saarland University, Saarbrücken, Germany

W0193061

Roderich Groß   Lyuba Alboul
Chris Melhuish   Mark Witkowski
Tony J. Prescott   Jacques Penders (Eds.)

Towards Autonomous Robotic Systems

12th Annual Conference, TAROS 2011
Sheffield, UK, August 31 – September 2, 2011
Proceedings

 Springer

Series Editors

Randy Goebel, University of Alberta, Edmonton, Canada
Jörg Siekmann, University of Saarland, Saarbrücken, Germany
Wolfgang Wahlster, DFKI and University of Saarland, Saarbrücken, Germany

Volume Editors

Roderich Groß
The University of Sheffield, Sheffield, UK, E-mail: r.gross@sheffield.ac.uk

Lyuba Alboul
Sheffield Hallam University, Sheffield, UK, E-mail: l.alboul@shu.ac.uk

Chris Melhuish
University of Bristol and University of the West of England, Bristol, UK
E-mail: chris.melhuish@brl.ac.uk

Mark Witkowski
Imperial College London, London, UK, E-mail: m.witkowski@imperial.ac.uk

Tony J. Prescott
The University of Sheffield, Sheffield, UK, E-mail: t.j.prescott@sheffield.ac.uk

Jacques Penders
Sheffield Hallam University, Sheffield, UK, E-mail: j.penders@shu.ac.uk

ISSN 0302-9743   e-ISSN 1611-3349
ISBN 978-3-642-23231-2   ISBN 978-3-642-23232-9 (eBook)
DOI 10.1007/978-3-642-23232-9
Springer Heidelberg Dordrecht London New York

Library of Congress Control Number: 2011934359

CR Subject Classification (1998): I.2.9, I.2, I.4, H.5

LNCS Sublibrary: SL 7 – Artificial Intelligence

Typesetting: Camera-ready by author, data conversion by Scientific Publishing Services, Chennai, India

Printed on acid-free paper

Springer is part of Springer Science+Business Media (www.springer.com)

Preface

These proceedings contain the papers presented at TAROS 2011, the 12th edition of the conference Towards Autonomous Robotic Systems, held in Sheffield, UK, from August 31 to September 2, 2011.

The TAROS series was initiated by Ulrich Nehmzow in 1997 in Manchester, when the conference was known as "Towards Intelligent Mobile Robots" (TIMR). In 1999, Chris Melhuish joined Ulrich to form a conference Steering Committee, which was also joined by Mark Witkowski in 2003; at that time the name of the conference was also changed to "Towards Autonomous Robotic Systems" to better reflect the full breadth of UK research activities in robotics. The Steering Committee has provided continuity of vision as the conference has evolved to become the UK's premier annual conference on autonomous robots, whilst also achieving an increasingly international attendance. Sadly, Ulrich Nehmzow died in 2010 after a long period of illness—a very great loss to UK robotics. To honor Ulrich's contributions throughout the years, the TAROS best paper award has therefore been named the "Ulrich Nehmzow Best Paper Award."

For the 2011 edition, TAROS received 94 paper submissions. We offered two submission categories—full-length papers, of which 74 submissions were received, and extended abstracts, of which we received 20 submissions. Most papers were reviewed by three members of our International Programme Committee. Of the full-length paper submissions, 32 were accepted as full-length papers, corresponding to an acceptance rate of 43%. Also included in this volume are 30 extended abstracts which were selected from the remaining 62 submissions. Of the full-length papers, 20 of the most highly ranked were selected for oral presentation at the conference. All other contributions, including extended abstracts, were presented in poster format. The publication of the collected papers from TAROS 2011 by Springer, in the *Lecture Notes in Artificial Intelligence* (LNAI) series, testifies to the high standard of the research efforts reported here. Following the conference, the journal *Robotics and Autonomous Systems* will publish extended and revised versions of some of the best papers presented at the meeting. This theme issue will also be dedicated to the memory of Ulrich Nehmzow.

TAROS was originally intended as a forum enabling PhD students to have their first conference experience. Whilst the conference has become larger and now attracts a wide range of papers, from both senior and junior scientists, the ethos of providing an event that encourages new researchers has been retained. TAROS 2011 was therefore a single-track event with many of the platform talks presented by postgraduates—for some their first public talk. The conference also featured three distinguished plenary presentations: "Approaches to Autonomous

Behavior with Simple Neural Networks" by Joseph Ayers, "Robot Navigating, Mapping and Understanding with Laser and Vision" by Paul Newman, and "Bioinspired Assistive Robotics for Children and Adults with Special Needs" by Yiannis Demiris.

Worldwide robotics is evolving quickly, and whilst the UK has a strong research community in robotics (evidenced, for instance, by our leading contribution to the EU Challenge 2 programme for robotics and cognitive systems), it is essential that this community becomes better organized and more effective in recognizing and exploiting commercial opportunities. In recent years the aims of TAROS have therefore expanded to include the goal of forging links with UK companies and commercial organizations that build or use robots. To encourage these interactions, TAROS 2011 hosted an Industry-Academia day, featuring a plenary talk from Simon Blackmore on agricultural robots, a platform discussion with representatives from UK and European funding organizations, and a public exhibition with live demonstrations of robotic systems. The Industry-Academia day also included the launch of a new forum for UK robotics researchers which was established with the support of the British Automation and Robot Association (BARA) and will be known as the BARA Academic Forum (BARA-AFR). The forum will provide a direct link between robotics researchers and the UK robotics industry facilitating technology transfer and improving the focus of robotics research toward commercial and industrial needs.

Finally, TAROS 2011 provided the opportunity for the local hosts—Sheffield Hallam University and the University of Sheffield—to launch their new city-wide initiative for robotics research. The Sheffield Centre for Robotics (SCentRo) will bring together five existing research groups across the two universities with the aim of enhancing inter-disciplinary research and raising the profile of robotics research in Sheffield.

We take this opportunity to thank the many people that were involved in making TAROS 2011 possible. On the organizational side this included Amir Naghsh, Alan Holloway, Georgios Chliveros, Charles Fox, Tony Dodd, Kevin Gurney, Gillian Hill, Jane Wright and Gill Ryder. We would also like to thank the authors who contributed their work, and the members of the International Programme Committee, and the additional referees, for their detailed and considered reviews. We are grateful to the four keynote speakers who shared with us their vision of the future.

Finally, we wish to thank the sponsors of TAROS 2011: Sheffield Hallam University, The University of Sheffield, The British Automation and Robot Association, The Institution of Engineering and Technology (sponsor of a public talk), Springer (Ulrich Nehmzow Best Paper Award), and the IEEE Robotics and Automation Society UKRI Chapter (co-sponsor).

We hope the reader will find this volume useful both as a reference to current research in autonomous robotic systems and as a starting point for their own future work. If you are inspired by this research we would encourage you to submit to future editions of TAROS!

June 2011   The TAROS 2011 Programme Committee

Jacques Penders
Tony Prescott
Roderich Groß
Lyuba Alboul

The TAROS Steering Committee

Chris Melhuish
Mark Witkowski

Organization

TAROS 2011 was organized jointly by Sheffield Hallam University and the University of Sheffield.

Conference Chairs

Jacques Penders	Sheffield Hallam University, UK
Tony Prescott	The University of Sheffield, UK

Organizing Committee

Roderich Groß (Technical Programme)	The University of Sheffield, UK
Lyuba Alboul (Technical Programme)	Sheffield Hallam University, UK
Alan Holloway (Industry-Academia Workshop and Exhibition)	Sheffield Hallam University, UK
Georgios Chliveros (Industry-Academia Workshop and Exhibition)	Sheffield Hallam University, UK
Charles Fox (Industry-Academia Workshop and Exhibition)	The University of Sheffield, UK
Amir Naghsh (Publication Chair)	Sheffield Hallam University, UK
Tony Dodd	The University of Sheffield, UK
Kevin Gurney	The University of Sheffield, UK

General Enquiries

Gillian Hill	Sheffield Hallam University, UK
Gill Ryder	University of Sheffield, UK

TAROS Steering Committee

Chris Melhuish	University of the West of England, UK
Mark Witkowski	Imperial College London, UK

Programme Committee

Francesco Amigoni	Politecnico di Milano, Italy
Christos Ampatzis	Research Executive Agency, Belgium
Iain Anderson	The University of Auckland, New Zealand
Brenna Argall	EPFL, Switzerland
Ronald C. Arkin	Georgia Institute of Technology, GA, USA
Lijin Aryananda	Universität Zürich, Switzerland
Joseph Ayers	Northeastern University, MA, USA
Tony Belpaeme	University of Plymouth, UK
Luc Berthouze	University of Sussex, UK
Simon Blackmore	Unibots Ltd., UK
Martin Brown	The University of Manchester, UK
Joanna Bryson	University of Bath, UK
Guido Bugmann	University of Plymouth, UK
Lola Canamero	University of Hertfordshire, UK
Andrea Carbone	Université Pierre et Marie Curie, France
Gregory Chirikjian	Johns Hopkins University, MD, USA
Anders Lyhne Christensen	Instituto de Telecomunicações & Instituto Universitário de Lisboa, Portugal
David Johan Christensen	Technical University of Denmark, Denmark
Mark Cutkosky	Stanford University, CA, USA
Torbjørn Semb Dahl	University of Wales, UK
Kerstin Dautenhahn	University of Hertfordshire, UK
Geert De Cubber	Royal Military Academy, Belgium
Yiannis Demiris	Imperial College London, UK
Peter Dominey	CNRS and INSERM U846, France
Stéphane Doncieux	Université Pierre et Marie Curie, France
Marco Dorigo	Université Libre de Bruxelles, Belgium
Toshio Fukuda	Nagoya University, Japan
Simon Garnier	Princeton University, NJ, USA
Antonios Gasteratos	Democritus University of Thrace, Greece
Dongbing Gu	University of Essex, UK
Verena Hafner	Humboldt-Universität zu Berlin, Germany
Heiko Hamann	Karl-Franzens-Universität Graz, Austria
William Harwin	University of Reading, UK
Ani Hsieh	Drexel University, PA, USA
Huosheng Hu	University of Essex, UK
Phil Husbands	University of Sussex, UK
Roberto Iglesias Rodríguez	Universidade de Santiago de Compostela, Spain
Yaochu Jin	University of Surrey, UK
Serge Kernbach	Universität Stuttgart, Germany
Annemarie Kokosy	ISEN, France
Maarja Kruusmaa	Tallinn University of Technology, Estonia
Haruhisa Kurokawa	AIST, Japan

Additional Referees

Alberto Albiol	Micha Hersch	Basilio Noris
Verónica Esther	Geoffrey Hollinger	Jose Nunez-Varela
Arriola Ríos	Kaijen Hsiao	Dmitry Oleynikov
Alper Bozkurt	Karl Iagnemma	Amit Kumar Pandey
Ladislau Bölöni	Hannes Kaufmann	Pei-Luen Patrick Rau
Joost Broekens	Jean-François Lalonde	James Roberts
Lucian Busoniu	Friedrich Lange	Raphael Rouveure
Sotirios Chatzis	Antonio Leite	Selma Šabanović
Liang Ding	Mohamed Marzouqi	Felix Stephan Schill
Nadia Garcia-Hernandez	Matteo Matteucci	Jeffrey Too Chuan Tan
Jim Gilbert	Annalisa Milella	Rich Walker
Fernando Gómez Bravo	Michael Milford	Wenwei Yu
Dan Grollman	Luis Yoichi Morales Saiki	Claudio Zito
Lars Hammarstrand	Jozsef Nemeth	
Just Herder	Juan Nieto	

Sponsoring Institutions

Sheffield Hallam University, Sheffield, UK
http://www.shu.ac.uk

The University of Sheffield, Sheffield, UK
http://www.shef.ac.uk

The British Automation and Robot Association (BARA), UK
http://www.bara.org.uk

The Institution of Engineering and Technology (IET), UK
http://www.theiet.org

Springer, London, UK (as a sponsor of the Ulrich Nehmzow Best Paper Award)
http://www.springer.com

IEEE Robotics and Automation Society UKRI Chapter
(as a technical co-sponsor)
http://ieee-ukri.org

Table of Contents

A Cricket-Controlled Robot Orienting towards a Sound Source.......... 1
  Jan Wessnitzer, Alexandros Asthenidis, Georgios Petrou, and
  Barbara Webb

A General Classifier of Whisker Data Using Stationary Naive Bayes:
Application to BIOTACT Robots 13
  Nathan F. Lepora, Charles W. Fox, Mat Evans, Ben Mitchinson,
  Asma Motiwala, J. Charlie Sullivan, Martin J. Pearson,
  Jason Welsby, Tony Pipe, Kevin Gurney, and Tony J. Prescott

A Navigation System for a High-Speed Professional Cleaning Robot 24
  Gorka Azkune, Mikel Astiz, Urko Esnaola, Unai Antero,
  Jose Vicente Sogorb, and Antonio Alonso

A Recursive Least Squares Solution for Recovering Robust Planar
Homographies 36
  Saad Ali Imran and Nabil Aouf

Airborne Ultrasonic Position and Velocity Measurement Using Two
Cycles of Linear-Period-Modulated Signal 46
  Shinya Saito, Minoru Kuribayashi Kurosawa, Yuichiro Orino, and
  Shinnosuke Hirata

An Eye Detection and Localization System for Natural Human and
Robot Interaction without Face Detection 54
  Xinguo Yu, Weicheng Han, Liyuan Li, Ji Yu Shi, and Gang Wang

An Implementation of a Biologically Inspired Model of Head Direction
Cells on a Robot 66
  Theocharis Kyriacou

Contextual Recognition of Robot Emotions.......................... 78
  Jiaming Zhang and Amanda J.C. Sharkey

Costs and Benefits of Behavioral Specialization 90
  Arne Brutschy, Nam-Luc Tran, Nadir Baiboun, Marco Frison,
  Giovanni Pini, Andrea Roli, Marco Dorigo, and Mauro Birattari

CrunchBot: A Mobile Whiskered Robot Platform 102
  Charles W. Fox, Mathew H. Evans, Nathan F. Lepora,
  Martin Pearson, Andy Ham, and Tony J. Prescott

Deformation-Based Tactile Feedback Using a Biologically-Inspired
Sensor and a Modified Display 114
  *Calum Roke, Chris Melhuish, Tony Pipe, David Drury, and
  Craig Chorley*

Design and Control of an Upper Limb Exoskeleton Robot
RehabRoby 125
  Fatih Ozkul and Duygun Erol Barkana

Distributed Motion Planning for Ground Objects Using a Network of
Robotic Ceiling Cameras 137
  *Andreagiovanni Reina, Gianni A. Di Caro, Frederick Ducatelle, and
  Luca M. Gambardella*

Evaluating the Effect of Robot Group Size on Relative Localisation
Precision 149
  Frank E. Schneider and Dennis Wildermuth

Instance-Based Reinforcement Learning Technique with a Meta-learning
Mechanism for Robust Multi-Robot Systems 161
  Toshiyuki Yasuda, Motohiro Wada, and Kazuhiro Ohkura

Locomotion Selection and Mechanical Design for a Mobile
Intra-abdominal Adhesion-Reliant Robot for Minimally Invasive
Surgery 173
  *Alfonso Montellano López, Mojtaba Khazravi, Robert Richardson,
  Abbas Dehghani, Rupesh Roshan, Tomasz Liskiewicz,
  Ardian Morina, David G. Jayne, and Anne Neville*

Mapping with Sparse Local Sensors and Strong Hierarchical Priors 183
  Charles W. Fox and Tony J. Prescott

Multi-rate Visual Servoing Based on Dual-Rate High Order Holds 195
  *J. Ernesto Solanes, Josep Tornero, Leopoldo Armesto, and
  Vicent Girbés*

Optimal Path Planning for Nonholonomic Robotic Systems via
Parametric Optimisation 207
  James Biggs

Probabilistic Logic Reasoning about Traffic Scenes 219
  Carlos R.C. Souza and Paulo E. Santos

Real-World Reinforcement Learning for Autonomous Humanoid Robot
Charging in a Home Environment 231
  Nicolás Navarro, Cornelius Weber, and Stefan Wermter

Robot Routing Approaches for Convoy Merging Maneuvers 241
Fernando Valdes, Roberto Iglesias, Felipe Espinosa,
Miguel A. Rodríguez, Pablo Quintia, and Carlos Santos

Sensing with Artificial Tactile Sensors: An Investigation of
Spatio-temporal Inference 253
Asma Motiwala, Charles W. Fox, Nathan F. Lepora, and
Tony J. Prescott

Short-Range Radar Perception in Outdoor Environments 265
Giulio Reina, James Underwood, and Graham Brooker

Smooth Kinematic Controller vs. Pure-Pursuit for Non-holonomic
Vehicles 277
Vicent Girbés, Leopoldo Armesto, Josep Tornero, and
J. Ernesto Solanes

Supervised Traversability Learning for Robot Navigation.............. 289
Ioannis Kostavelis, Lazaros Nalpantidis, and Antonios Gasteratos

Task Space Integral Sliding Mode Controller Implementation for 4DOF
of a Humanoid BERT II Arm with Posture Control 299
Said Ghani Khan, Jamaludin Jalani, Guido Herrmann,
Tony Pipe, and Chris Melhuish

Towards Autonomous Energy-Wise RObjects 311
Florian Vaussard, Michael Bonani, Philippe Rétornaz,
Alcherio Martinoli, and Francesco Mondada

Towards Safe Human-Robot Interaction............................ 323
Elena Corina Grigore, Kerstin Eder, Alexander Lenz,
Sergey Skachek, Anthony G. Pipe, and Chris Melhuish

Towards Temporal Verification of Emergent Behaviours in Swarm
Robotic Systems 336
Clare Dixon, Alan Winfield, and Michael Fisher

Walking Rover Trafficability - Presenting a Comprehensive Analysis
and Prediction Tool 348
Brian Yeomans and Chakravathini M. Saaj

What Can a Personal Robot Do for You? 360
Guido Bugmann and Simon N. Copleston

Extended Abstracts

A Systems Integration Approach to Creating Embodied Biomimetic
Models of Active Vision.. 372
Alex Cope, Jon Chambers, and Kevin Gurney

A Validation of Localisation Accuracy Improvements by the Combined
Use of GPS and GLONASS 374
    Dennis Wildermuth and Frank E. Schneider

Adaptive Particle Filter for Fault Detection and Isolation of Mobile
Robots 376
    Michał Zając

An Approach to Improving Attitude Estimation Based on Low-Cost
MEMS-IMU for Mobile Robot Navigation 378
    Lu Lou, Mark Neal, Frédéric Labrosse, and Juan Cao

Cooperative Multi-robot Box Pushing Inspired by Human Behaviour ... 380
    Jianing Chen and Roderich Groß

Cooperative Navigation and Integration of a Human into Multi-robot
System 382
    Joan Saez-Pose, Amir M. Naghsh, and Leo Nomdedeu

Coordination in Multi-tiered Robotic Search 384
    Paul Ward and Stephen Cameron

Covert Robotics: Improving Covertness with Escapability and
Non-Line-of-Sight Sensing 386
    *Tom Moore, Richard Ratmansky, Bob Chevalier, David Sharp,
    Vincent Baker, and Brian Satterfield*

Designing Electric Propulsion Systems for UAVs 388
    Mohamed Kara Mohamed, Sourav Patra, and Alexander Lanzon

Enhancing Self-similar Patterns by Asymmetric Artificial Potential
Functions in Partially Connected Swarms 390
    Giuliano Punzo, Derek Bennet, and Malcolm Macdonald

Evolving Modularity in Robot Behaviour Using Gene Expression
Programming 392
    Jonathan Mwaura and Ed Keedwell

Forming Nested 3D Structures Based on the Brazil Nut Effect 394
    Stephen Foster and Roderich Groß

Grasping of Deformable Objects Applied to Organic Produce 396
    Alon Ohev-Zion and Amir Shapiro

Learning to Grasp Information with Your Own Hands 398
    Dimitri Ognibene, Nicola Catenacci Volpi, and Giovanni Pezzulo

Long-Term Experiment Using an Adaptive Appearance-Based Map for
Visual Navigation by Mobile Robots............................... 400
    Feras Dayoub, Grzegorz Cielniak, and Tom Duckett

Occupancy Grid-Based SLAM Using a Mobile Robot with a Ring of
Eight Sonar Transducers 402
    George Terzakis and Sanja Dogramadzi

On the Analysis of Parameter Convergence for Temporal Difference
Learning of an Exemplar Balance Problem 404
    Martin Brown and Onder Tutsoy

Online Hazard Analysis for Autonomous Robots 406
    Roger Woodman, Alan F.T. Winfield, Chris Harper, and Mike Fraser

Results of the European Land Robot Trial and Their Usability for
Benchmarking Outdoor Robot Systems 408
    Frank E. Schneider and Dennis Wildermuth

Solutions for a Variable Compliance Gripper Design 410
    Maria Elena Giannaccini, Sanja Dogramadzi, and Tony Pipe

Study of Routing Algorithms Considering Real Time Restrictions Using
a Connectivity Function 412
    Magali Arellano-Vázquez, Héctor Benítez-Pérez, and
    Jorge L. Ortega-Arjona

Systematic Design of Flexible Magnetic Wall and Ceiling Climbing
Robot for Cargo Screening 414
    Yuanming Zhang and Tony Dodd

Tactile Afferent Simulation from Pressure Arrays..................... 416
    Rosana Matuk Herrera

The Interaction between Vortices and a Biomimetic Flexible Fin 418
    Jennifer Brown, Lily Chambers, Keri M. Collins, Otar Akanyeti,
    Francesco Visentin, Ryan Ladd, Paolo Fiorini, and William Megill

Toward an Ecological Approach to Interface Design for Teaching
Robots 420
    Guillaume Doisy, Joachim Meyer, and Yael Edan

Towards Adaptive Robotic Green Plants 422
    Janine Stocker, Aline Veillat, Stéphane Magnenat,
    Francis Colas, and Roland Siegwart

Using Image Depth Information for Fast Face Detection 424
    Sasa Bodiroza

Using Sequences of Knots as a Random Search 426
    C.A. Pina-Garcia and Dongbing Gu

Vision-Based Segregation Behaviours in a Swarm of Autonomous
Robots 428
    Michael J. Price and Roderich Groß

Visual-Inertial Motion Priors for Robust Monocular SLAM 430
    Usman Qayyum and Jonghyuk Kim

Author Index 433

A Cricket-Controlled Robot
Orienting towards a Sound Source

Jan Wessnitzer*, Alexandros Asthenidis*, Georgios Petrou, and Barbara Webb

Institute of Perception, Action, and Behaviour, School of Informatics,
University of Edinburgh, Edinburgh, UK
jwessnit@inf.ed.ac.uk

Abstract. We designed a closed-loop experimental setup that interfaces an insect with a robot for testing phonotaxis (sound recognition and localisation) behaviour in crickets. The experimental platform consists of a trackball mounted on a robot, so that a cricket walking on the trackball has its movements translated into corresponding movement of the robot. We describe the implementation of this system and compare the performance with previous cricket data and a neural model circuit on the same robot. Crickets are able to drive the robot towards the sound source, although they show substantially longer walking and stopping bouts than in more standard experimental setups. The potential and the current limitations of the robot setup together with alternative designs are discussed.

Keywords: phonotaxis, insect, hybrid robot control, trackball.

1   Introduction

Insect biology has been inspirational and beneficial to the design and implementation of autonomous robots in terms of body morphology and mechanics (e.g., [4]), neural models of behaviour (e.g., [12]), and control architectures (e.g., [16]) but robotics has also been useful at unravelling insect behaviour (e.g., social integration in cockroaches [5].)

Some experimental insect/machine prototypes have already been developed (e.g., for pheromone-oriented behaviour [8,7], and for obstacle avoidance behaviour [6]) and provide potential advantages for studying insect behaviour. Insects, because of their size, may be interfaced with small mobile robots, allowing the resulting behaviour to be compared with the insects' normal behaviour or the behaviour of a robot running a computational model of the underlying circuitry under the same experimental conditions. Recordings of neural activities using electrodes are then possible on the behaving animal. Moreover, detailed behavioural and neural behaviour can be gathered while the animal is in closed loop, receiving feedback from its actions, for example a change in the relative direction or strength of a sensory signal as it approaches it. Feedback effects

---

* Equal contribution.

R. Groß et al. (Eds.): TAROS 2011, LNAI 6856, pp. 1–12, 2011.
© Springer-Verlag Berlin Heidelberg 2011

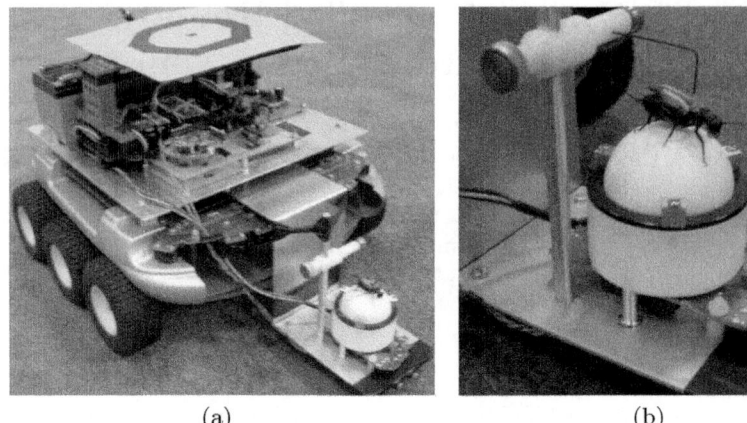

(a)   (b)

Fig. 1. (a) The complete system consists of a cricket walking on a trackball that is mounted on a robot base. A target on the robot is tracked by an overhead camera. Two microphones are located under the trackball. (b) Close-up of a cricket on the trackball.

between different behaviours and across sensory modalities can be explored further by systematically varying the sensorimotor contingencies between insect and robot movement.

However, despite a system designer's best efforts, the sensorimotor contingencies which an insect experiences (do insects have forward models? [15]) are likely to be different in an insect/machine system. Crickets, for example, may predict their own reafferent visual input during phonotactic turns and react differently if the feedback via the robot interface is larger or smaller than expected [10], or involves a delay. Thus, a fundamental question in building an insect/machine hybrid system is whether insects still behave and perform tasks successfully when in the unusual situation of being linked with a robot, and how the resulting behaviour compares to that in other experimental setups and conditions.

In this paper, we present an experiment in machine/insect hybrid control for phonotaxis. Female crickets are attracted by the males' calling song and this behaviour has been extensively studied, is relatively well understood at the neural and behavioural levels [13,3,11], and models have been developed and tested on robots [14,12]. Here, we designed a closed-loop experimental system to test whether crickets as part of a hybrid robotic system exhibit phonotaxis behaviour and we compare the resulting behaviour to other test scenarios for crickets and to a neural model of phonotaxis implemented on the same robot.

2   Materials and Methods

The experimental platform consists of a K-team Koala robot, an inexpensive custom-built trackball system based on optical mouse technology, and the cricket *Gryllus bimaculatus*, forming a closed-loop system. The next sections provide details of the implementation. Further details can be found in [1].

2.1 Hardware

Robot. The K-Team Koala robot was chosen as it has a robust structure and chassis hosting a range of sensors and user-configurable I/Os. Its 2 DC brushed servo motors with integrated encoders run relatively quietly (compared to other robotic platforms available in our laboratory) which was a concern during the design stages as noise disturbances could affect the cricket's performance. Its size (32cm long and 32cm wide) offers sufficient space and payload allowing a 12V battery and all circuitry to be mounted on the Koala.

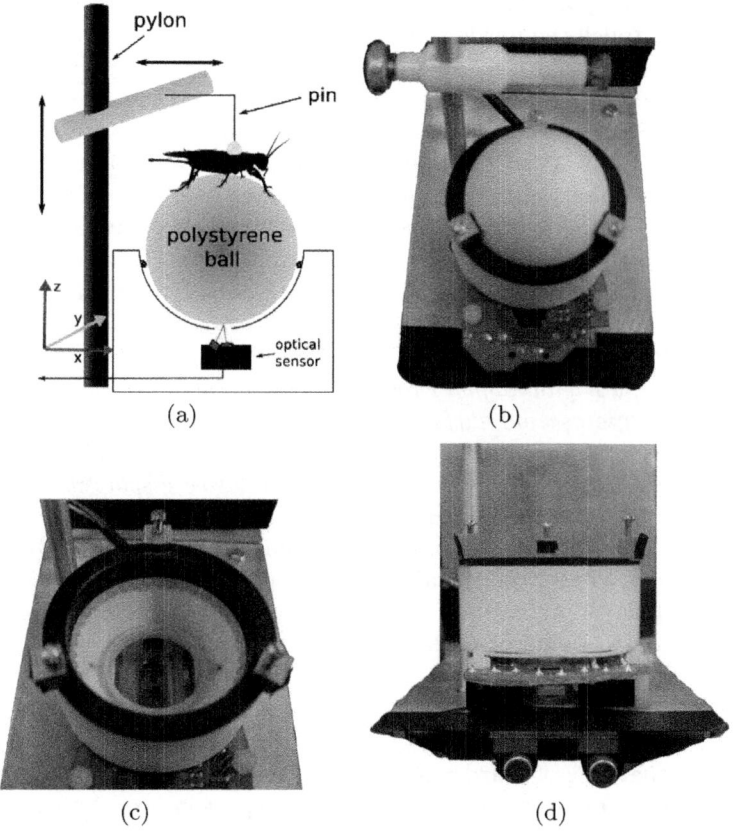

Fig. 2. (a) Schematic side view of the trackball system. The dome of the trackball platform from different viewpoints: (b) top view with polystyrene ball, (c) the inside of the dome and (d) front view showing the positions of the two microphones.

Trackball platform. The trackball platform (11cm long and 8 cm wide), as seen in figure 1(b), is made of aluminium and supports the hardware components of the trackball system: the optical mouse board, the dome holding a polystyrene ball, an adjustable mount for the cricket and two microphones. The platform is fixed at 2cm from the ground resulting in the cricket being held at a height of 8cm.

The mouse board is placed sensor side up facing the ball on the trackball. The forward movement axis of the Koala is aligned with the x axis of the mouse sensor. Due to the limited payload of the Koala, we decided against air flow to float the polystyrene ball (c.f., trackball system developed in [9]). The trackball's dome, shown on figures 2(b,c,d) holds a polystyrene ball of 20 mm radius (Graham Sweet Studios, Cardiff, UK). A light-weight and mouldable balsa filler (*J. Perkins*[1]) was used to smooth the surface of the polystyrene trackball. After the filler had dried, its surface was further smoothed using fine sandpaper. In order to decrease the friction between the ball's and the dome's surfaces, a ring of nylon balls are placed around the walls of the dome (figure 2(c)) such that the ball does not touch the inside walls but rests solely on the nylon balls which can move freely. Note that the friction between the polystyrene ball and the nylon balls is still relatively high when compared to other trackball systems (c.f., [9]). Three metallic prongs are placed around the rim of the dome to hold the ball, preventing it from being lifted and the nylon balls from falling. At the bottom of the dome, the lens of the mouse board is fixed at about 2.5mm from the ball in order to meet the specifications and working range of the mouse sensor [2].

Optical mouse sensor. The base of the trackball system is an ADNS-5020-EN optical mouse chip by Avago. Its theory of operation is described in its datasheet [2] as follows: *"The ADNS-5020-EN contains an Image Acquisition System (IAS), a Digital Signal Processor (DSP) and a three wire serial port. The IAS acquires microscopic surface images via the lens and illumination system. These images are processed by the DSP to determine the direction and distance of motion. The DSP calculates the Δx and Δy relative displacement values"*. The communication with the chip is established via a three wire serial peripheral interface (SPI). The chip's maximum running frequency is 1 MHz and it can detect high speed motion operating at either 500 (default) or 1000 counts per inch (cpi). This chip allows to read out the animals' walking responses on time scales relevant to signal propagation in the nervous system.

Ears circuit. For direct sensorimotor control (i.e., without the cricket in the control loop) the robot also carries an electronic circuit that simulates the behaviour of the cricket's eardrum [14]. Two microphones, are aligned under the trackball platform and spaced 18 millimetres apart in order to simulate the distance of cricket's ears as shown in figure 1. An analog circuit mimics the delays occurring in the cricket's tracheal tube that lead to phase cancellation and a strongly directional response, and outputs two values, one for the left and one for the right ear.

Microcontroller interface. The *mbed*[2] microcontroller was chosen to interconnect the different modules of the experimental platform. The *mbed* is based on the NXP LPC1768 with an ARM Cortex-M3 Core. Its running frequency is

---

[1] http://www.jperkinsdistribution.co.uk/
[2] http://mbed.org/handbook/mbed-NXP-LPC1768

96 MHz and it provides 512 KB Flash memory and 64 KB RAM. The *mbed* provides many interfaces, including SPI and Serial, and it is programmed in C/C++. The developed software is described in more detail in section 2.3. The *Koala* can be controlled by programs stored on its flash memory or by directly sending motor commands through serial communication. Thus, the *mbed* could be used to run all control code through a direct serial communication interface operating the robot's motors and/or to requesting sensor values (see section 2.3). In order to send the command to the RS-232 of the Koala, we used a *serial level shifter 03BBL274* by BitBox, a device for translating of the RS-232 levels (12V) to logic levels (3.3V). This level shifter (or one that is similar in function) is required to operate the *Koala* with our developed software described in section 2.3.

2.2   Wetware

Adult female *Gryllus bimaculatus* crickets were isolated after their final moult and maintained individually in small plastic cages at room temperature. To mount a cricket onto the platform, an insect pin is fixed to the cricket's thorax using wax [11]. A pylon placed further back from the centre of the trackball (figures 2(a) and 2(b)) holds the insect pin, allowing adjustments of the cricket's position along the $x-$ and $z-$ axes. Once a cricket has been positioned, a screw is fastened to hold the pin and insect in position.

2.3   Software

Control routines for the experimental platform and the neural network implementation were implemented in order to compare the resulting behaviours of the systems. The neural network only uses the *ears* board for sensory input whereas the routine for the experimental platform only uses the optical mouse sensor, i.e., the cricket's movements on the trackball. All software is available upon request.

Neural network. The neural circuitry underlying phonotaxis behaviour in the cricket has been the subject of much study. Here, we implement a simple neural network for phonotaxis consisting of six spiking neurons (see figure 3). This represents two pairs of identified neurons in the cricket prothoracic ganglion. Both are excited by auditory inputs, but the omega neuron (ON) pair perform cross-inhibition, such that the side with the stronger response will dominate. The auditory neuron (AN) pair act as low pass filters for the pattern of song, and adapting synapses between the AN and motor neuron (MN) pair act as a high-pass filter, so that the circuit as a whole acts as a bandpass filter for cricket song patterns. It otherwise operates by the same principle as a Braitenberg vehicle, turning to the side that has a stronger signal by decreasing the activation of the corresponding motor.

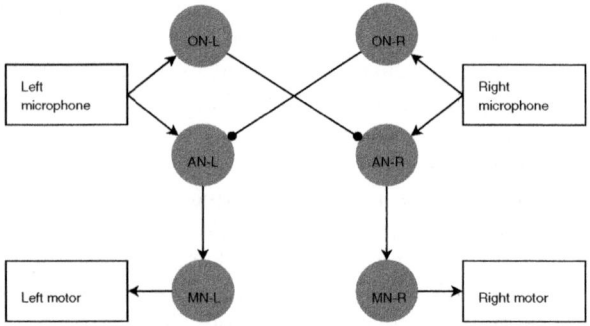

Fig. 3. Neural network for phonotaxis consisting of omega neurons (ON), ascending neurons (AN), and motor neurons (MN). Arrows are excitatory connections, filled circles are inhibitory.

Experimental platform. The Koala is driven by two differential wheel motors and we use simple differential drive kinematics to update the robot's movement when controlled by a cricket. The optical chip continuously reads and accumulates the trackball's displacement along the x and y axes. The control routine reads the registers holding the Δx and Δy displacement of the trackball every 100ms. Once read, these registers are reset and the control routine determines the motor speeds for the *Koala* to match this displacement. However, as the center of rotation of the trackball and the center of rotation of the Koala do not coincide, the values for the motor speeds were limited to maximum values in order to avoid large sudden movements of the robotic platform (see [1] for details).

C++ classes for the mbed were written allowing efficient communication with the optical mouse chip (Avago ADNS-5020-EN) and the K-team Koala robot and were published on the *mbed* website[3].

2.4   Experimental Methods

The system was evaluated by analysing the paths of the robot using an overhead 1.3 megapixel (Logitech) camera. A custom-built video tracking system written using OpenCV (Willowgarage[4]) and Qt (Nokia[5]) libraries was used in order to track the robot's path. The tracking algorithm is described in detail in [1].

The lights of the robot laboratory were kept off in order to simulate twilight lighting conditions. An artificial calling song (generated by a Matlab (Mathworks, Natick, MA, USA) script) at carrier frequency of 4.8 kHz with syllable duration of 21 ms (including 2ms rise and fall time), syllable period of 42 ms, chirp duration of 252 ms, and chirp period of 500ms was played from a speaker positioned in the centre of the lab. The robot was started from different positions at a distance of approximately 4 metres from the speaker.

---

[3] http://mbed.org/users/IPAB/programs/
[4] http://opencv.willowgarage.com/
[5] http://qt.nokia.com/

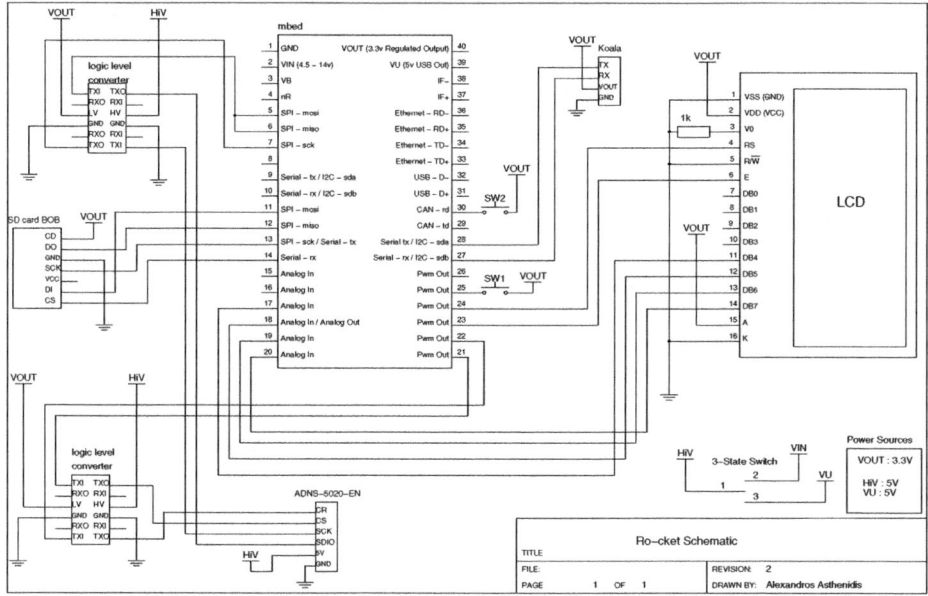

Fig. 4. Circuit diagram of the experimental platform. Note that most *mbed* pins can be used as standard digital I/Os.

3   Experiments

In the experiments, several crickets were mounted on top of the trackball system and were allowed to operate the platform either in the presence or absence of a calling song stimulus. When under control of the neural network, a calling song stimulus was always present (without sound, the robot would just drive forward in a straight line).

As seen in figure 6(a), the crickets successfully operate the system and are able to steer the robot towards the speaker. The robot's starting positions are on the left and the speaker position is marked with a red circle. For comparison, figure 6(c) shows the results of the robot under neural network control.

To control for other directional behaviours that may explain the results in figure 6(a) (for example, attractive visual stimuli behind the speaker), we repeated the experiments without calling song stimuli. As seen in figure 6(b), the paths were not directed towards the sound source (c.f., figure 8(d)). We cannot be sure what attracted the crickets to steer the Koala in that general direction. However, it should be noted that the cricket needs to be precisely aligned on top of the trackball since any misalignment of the cricket's position could bias the robot's movement to one side. Despite misalignment, a cricket performing phonotaxis adjusts its position continuously in order to reach the sound source.

We further compared the walking patterns to those of crickets on a trackball or in free-walking experimental setups. Our system's performance in terms of

Fig. 5. The robot is automatically tracked using an overhead camera and custom-built tracking software [1]

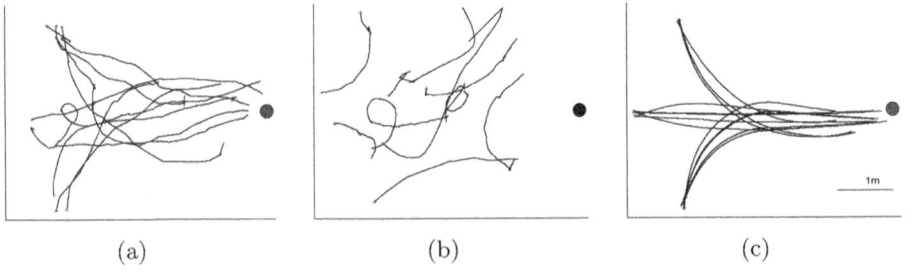

(a)   (b)   (c)

Fig. 6. (a) Paths of the robot under cricket control in the presence of cricket chirping. (b) Without calling song the crickets do not steer the Koala towards the sound source. (c) Resulting paths under control of the neural network, which can reliably localise the calling song.

translational velocity compares well with crickets on a trackball in open-loop conditions (c.f., figure 2 in [3]) although the robot seems a bit slower as the maximum velocity never exceeded 5 cm·s^{-1} as shown in figure 7. The robot's mean velocity was 1.322 cm·s^{-1} with a standard deviation 1.495. This is slower

Table 1. Behaviour statistics with and without calling song stimulus

	with	/ without
Total distance moved (m)	16.653	13.645
Velocity (cm/s)	1.322±1.495	1.083±1.441
Maximum velocity (cm/s)	4.943	4.667
Stop duration (s)	7.813±24.087	9.258±19.365
Activity duration (s)	6.801±9.105	6.717±8.320

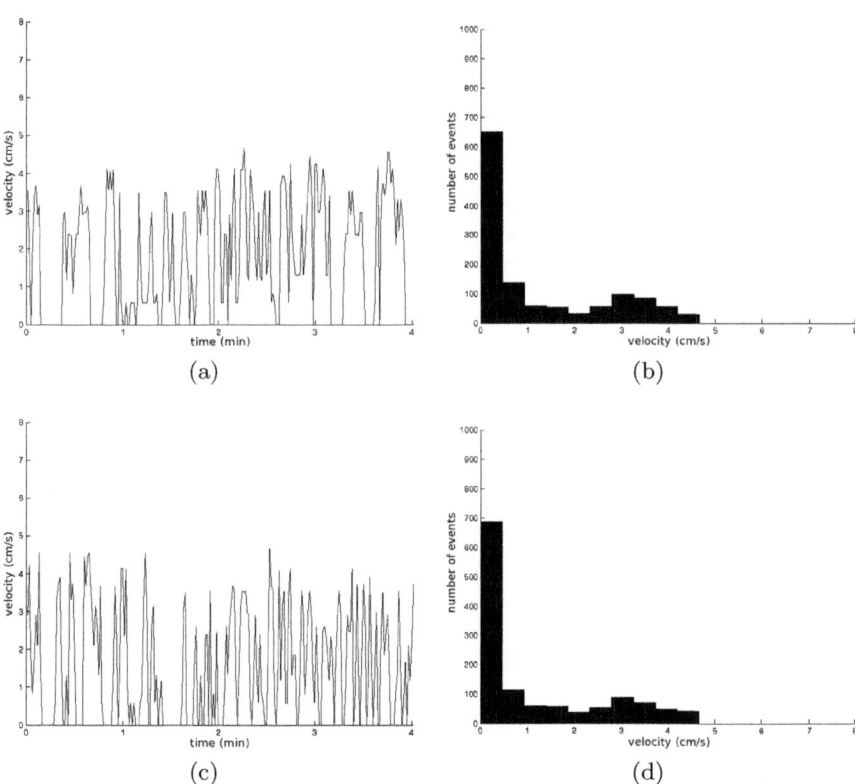

Fig. 7. Translational velocities of the cricket on the robot-mounted trackball. Upper plots, with calling song: a) velocity trace over four minutes; b) distribution of walking velocities over 20 minutes. Lower plots, without calling song: c) velocity trace over four minutes; d) distribution of walking velocities over 20 minutes.

than observed in [3] where the mean velocity was measured at 2.24cm·s^{-1}. The instantaneous rotational component fluctuated around 0 with a maximal deviation of about 100 degrees·s^{-1} as shown in figure 8 and is similar to what was found for crickets on a trackball [3]. The differences in speed might be explained by inaccuracies in the translation of trackball movement into robot motion and the higher friction in our trackball setup.

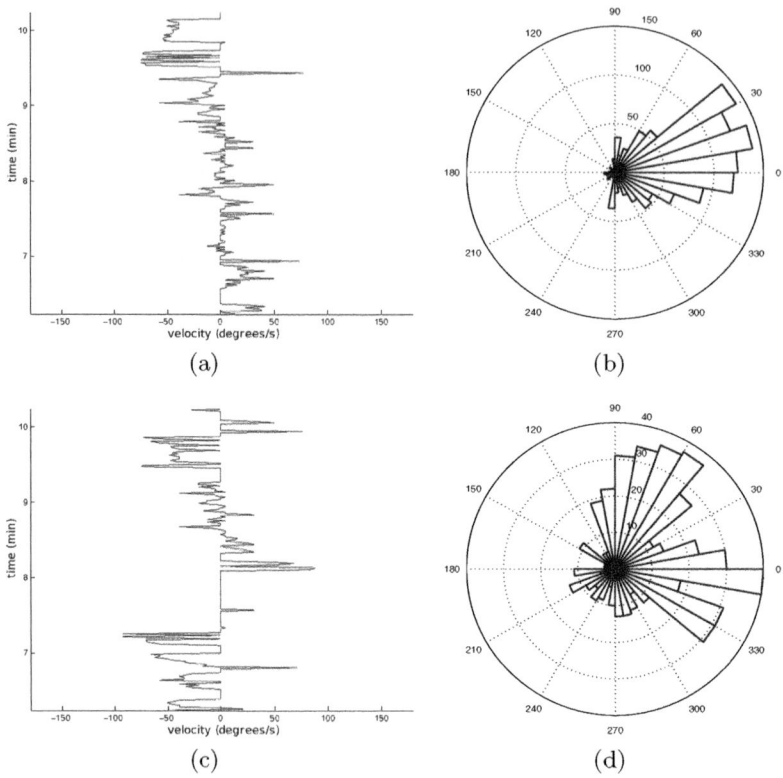

Fig. 8. Rotational velocities of the cricket on the robot-mounted trackball. Upper plots, with calling song: a) rotational velocity trace over four minutes; b) distribution of walking directions (sound source at 0 degrees) over 20 minutes. Lower plots, without calling song: c) rotational velocity trace over four minutes; d) distribution of walking directions over 20 minutes.

However, when analysing entire runs, significant differences have been observed in terms of activity and stop durations [13]. During phonotaxis, both mean activity duration and mean stop duration were much longer, 6.801s and 7.813s respectively, than observed in [13]. Similar results were found in the absence of a calling song stimulus, 6.717s and 9.258s. For comparison, on a walking compensator [13], crickets were found to move and stop on average for 1.38s and 1.12s respectively in the absence and 2.25s and 0.59s in the presence of a calling song stimulus. We can only speculate as to the reasons for these results. It is possible that insect/machine contingencies, such as the delay in translating trackball movement into robot movement or unexpected sensory feedback, causes crickets to panic resulting in longer bouts and demotivates them from moving resulting in longer pauses. A summary of the statistics is presented in table 1.

4   Discussion

We developed an insect/machine interface and tested our design on the well-studied phonotaxis behaviour of the cricket *Gryllus bimaculatus*. The cricket was capable of steering a mobile robot towards a sound source. The observed differences in walking behaviour may be down to the setup, e.g., open-loop [3] versus closed-loop in this paper, air cushion versus a ring of nylon balls holding the trackball (i.e., causing more friction). Sensory feedback is in all likelihood not as expected and slight delays between insect and robot motion may also cause these differences. Thus, immediate improvements should include the superposition of the trackball's and the robot's centres of rotation and the optimisation of insect/machine contingencies (e.g., shortening the delays between trackball and robot movements). With these refinements, the system could become a valuable tool for investigating cricket behaviour and evaluating models.

Acknowledgements. This work was supported by an EPSRC grant. We thank Hugh Cameron and Robert McGregor for technical support.

References

1. Asthenidis, A.: Cricket-Cyborg: an insect-controlled robot. Master's thesis, School of Informatics, University of Edinburgh, Edinburgh, UK (2010)
2. Avago Technologies: ADNS-5020-EN Optical Mouse Sensor. Datasheet
3. Böhm, H., Schildberger, K., Huber, F.: Visual and acoustic course control in the cricket *Gryllus-bimaculatus*. Journal of Experimental Biology 159, 235–248 (1991)
4. Finio, B., Wood, R.: Distributed power and control actuation in the thoracic mechanics of a robotic insect. Bioinspiration and Biomimetics 5, 045006 (2010)
5. Halloy, J., Sempo, G., Caprari, G., Rivault, C., Asadpour, M., Tache, F., Said, I., Durier, V., Canonge, S., Ame, J., Detrain, C., Correll, N., Martinoli, A., Mondada, F., Siegwart, R., Deneubourg, J.: Social integration of robots into groups of cockroaches to control self-organized choices. Science 318, 1155–1158 (2007)
6. Hertz, G.: http://www.conceptlab.com/roachbot/
7. Kanzaki, R., Nagasawa, S., Shimoyama, I.: Neural basis of odor-source searching behavior in insect brain systems evaluated with a mobile robot. Chemical Senses 30(suppl. 1), i285–i286 (2005)
8. Kuwana, Y., Nagasawa, S., Shimoyama, I., Kanzaki, R.: Synthesis of the pheromone-oriented behaviour of silkworm moths by a mobile robot with moth antennae as pheromone sensors. Biosensors and Bioelectronics 14, 195–202 (1999)
9. Lott, G., Rosen, M., Hoy, R.: An inexpensive sub-millisecond system for walking measurements of small animals based on optical computer mouse technology. Journal of Neuroscience Methods 161, 55–61 (2007)
10. Payne, M., Hedwig, B., Webb, B.: Multimodal predictive control in crickets. In: Doncieux, S., Girard, B., Guillot, A., Hallam, J., Meyer, J.-A., Mouret, J.-B. (eds.) SAB 2010. LNCS, vol. 6226, pp. 167–177. Springer, Heidelberg (2010)
11. Poulet, J., Hedwig, B.: Auditory orientation in crickets: pattern recognition controls reactive steering. PNAS 102, 15665–15669 (2005)
12. Reeve, R., Webb, B.: New neural circuits for robot phonotaxis. Philosophical Transactions of the Royal Society London A 361, 2245–2266 (2003)

13. Schmitz, B., Scharstein, H., Wendler, G.: Phonotaxis in Gryllus campestris l. I Mechanism of acoustic orientation in intact female crickets. Journal of Comparative Physiology A 148, 431–444 (1982)
14. Webb, B.: Using robots to model animals: a cricket test. Robotics and Autonomous Systems 16, 117–134 (1995)
15. Webb, B.: Neural mechanisms for prediction: do insects have forward models? Trends in Neuroscience 27(5), 278–282 (2004)
16. Wessnitzer, J., Webb, B.: Multimodal sensory integration in insects - towards insect brain control architectures. Bioinspiration and Biomimetics 1(3), 63–75 (2006)

A General Classifier of Whisker Data Using Stationary Naive Bayes: Application to BIOTACT Robots

Nathan F. Lepora[1], Charles W. Fox[1], Mat Evans[1], Ben Mitchinson[1], Asma Motiwala[1], J. Charlie Sullivan[2], Martin J. Pearson[2], Jason Welsby[2], Tony Pipe[2], Kevin Gurney[1], and Tony J. Prescott[1]

[1] Adaptive Behaviour Research Group, Department of Psychology,
University of Sheffield, Sheffield, UK
{n.lepora,c.fox,mat.evans,b.mitchinson,
a.motiwala,k.gurney,t.j.prescott}@sheffield.ac.uk
[2] Bristol Robotics Laboratory, Bristol, UK
{charlie.sullivan,martin.pearson,jason.welsby,tony.pipe}@brl.ac.uk

Abstract. A general problem in robotics is how to best utilize sensors to classify the robot's environment. The BIOTACT project (BIOmimetic Technology for vibrissal Active Touch) is a collaboration between biologists and engineers that has led to many distinctive robots with artificial whisker sensing capabilities. One problem is to construct classifiers that can recognize a wide range of whisker sensations rather than constructing different classifiers for specific features. In this article, we demonstrate that a stationary naive Bayes classifier can perform such a general classification by applying it to various robot experiments. This classifier could be a key component of a robot able to learn autonomously about novel environments, where classifier properties are not known in advance.

Keywords: BIOTACT, Active touch, Whiskers, Bayes' rule, Classifier.

1 Introduction

Robotics has much to learn from the ways that animals are constructed, control their bodies and utilize their sensory capabilities [1]. All of the robots described in this article result from a long-term collaboration between biologists and engineers that aims to further understanding of biological vibrissal (whisker) systems and determine the potential applications to engineered systems such as autonomous robots [14]. Research into these systems was undertaken as part of a European Framework 7 project termed BIOTACT (BIOmimetic Technology for vibrissal Active Touch) [2]. The vibrissal sensing technology used in these robots was inspired by many aspects of biological whiskers [13], including their morphology, control and sensory information processing.

This article focuses on how to process the sensory information from artificial vibrissal sensors to categorize and recognize the robot's nearby environment.

R. Groß et al. (Eds.): TAROS 2011, LNAI 6856, pp. 13–23, 2011.

This classification problem relies on characterizing stimuli from the environment that are similar every time they are encountered, for example the speed of a contacting object is typified by the peak whisker deflection it produces [4]. Classifiers of sensory data from artificial vibrissae have generally focussed on distinctive features of the whisker signal produced by the stimulus. Examples include peak deflection amplitude and duration for contact speed and radial distance [4], the profile of the whisker deflection power spectrum for texture [7,3,8], and the change in peak deflection across many whiskers for surface shape [9]. These classifiers perform well in the situations they are designed for, but are not accurate outside their operating range. For example, peak whisker deflection may not accurately characterize texture.

While it is possible to choose *specific* classifiers for robots sensing artificial environments, in more natural environments it would be useful to employ *generic* classifiers that apply across a broad range of stimuli. Traditionally this goal has not been a focus of classifier design for whiskered robots, because the first priority was to find a set of classifiers that allows the robot to interact with test environments. However, now that a fairly comprehensive library of classifiers is becoming available [4,7,3,8,9,6,5,10,15,16], such questions about the choice and flexibility of classifiers are becoming relevant.

In recent work, we found that a classifier based on generic probabilistic methods could recognize surface texture accurately by utilizing the probability distribution (histogram) of whisker deflections over a time window of data [10]. In principle, such a method could also characterize other stimuli by utilizing the overall statistical properties of whisker deflections to determine the salient aspects of each stimulus. To explore the hypothesis that a probabilistic classifier could apply across general types of stimuli, we demonstrate here that one method, stationary naive Bayes, applies well across a broad range of stimuli encountered by several robot platforms constructed for the BIOTACT project.

2   Classification with Stationary Naive Bayes

The probabilistic classifier used here is based on Bayes' rule, but makes two assumptions for simple application to time series of whisker data. First, the sensor measurements are assumed independent, or naive, and second the measurement distributions are assumed stationary in time. Naive Bayes is usually considered across different data dimensions rather than over time series with the stationary assumption, and so we refer to the present method as stationary naive Bayes.

This method relies on calculating the probability distributions of measured time series from the empirical frequencies with which values occur in classes of training data. The probability of a single measurement x being from a class is

$$P(x|C_l) = \frac{n_x(C_l)}{\sum_x n_x(C_l)}, \tag{1}$$

where $n_x(C_l)$ is the total number of times that the value x occurs in the time series for class C_l. The conditional probability $P(x|C_l)$ is commonly referred to as a likelihood of the sensor measurement x occurring.

Given a new set of test data, Bayes' rule states that the (posterior) probability $P(C_l | x_1, \cdots, x_n)$ for a time-series of measurements $x_1, \cdots, x_n$ being drawn from the training data for class C_l is proportional to the likelihood of those measurements $P(x_1, \cdots, x_n | C_l)$ estimated from that training data

$$P(C_l | x_1, \cdots, x_n) = \frac{P(x_1, \cdots, x_n | C_l) P(C_l)}{P(x_1, \cdots, x_n)}, \tag{2}$$

where $P(C_l)$ is the (prior) probability of the data being from class C_l and $P(x_1, \cdots, x_n)$ is the (marginal) probability of measuring $x_1, \cdots, x_n$ given no other information. A Bayesian classifier finds which class C has maximum *a posteriori* probability given the measurement time series

$$C = \arg\max_{C_l} P(C_l | x_1, \cdots, x_n) = \arg\max_{C_l} \frac{P(x_1, \cdots, x_n | C_l) P(C_l)}{P(x_1, \cdots, x_n)}, \tag{3}$$

where arg max refers to the argument (the class) that maximizes the posterior.

The following formalism uses that the marginals $P(x)$ are independent of the classes and can be ignored in the arg max operation. It will also be convenient to use the logarithm of the posterior probability to turn products of probabilities into sums. Then the classification in Eq. 3 is equivalent to finding

$$C = \arg\max_{C_l} \; \log P(x_1, \cdots, x_n | C_l) + \log P(C_l), \tag{4}$$

which is just the maximum over the log-likelihoods plus log priors of a series of measurements $x_1, \cdots, x_n$ being from each class of training data.

An important simplification occurs if the measurements are assumed statistically independent at each time across the data and the probability distributions from which these measurements are drawn are assumed stationary in time. Then the overall conditional probability of a series of measurements factorizes into a product of conditional probabilities for each individual measurement. Consequently, the classification can be rewritten as

$$C = \arg\max_{C_l} \sum_{i=1}^{n} \log P(x_i | C_l) + \log P(C_l), \tag{5}$$

Thus the most probable class is found from the maximum over the summed log-likelihoods for the time window, shifted by its log-prior. This equation defines the stationary naive Bayes classifier used in this paper, with 'naive' denoting the assumption of statistical independence over time and stationary denoting that the same likelihood distribution applies for all times.

The following examples are based around robot experiments in which similar amounts of data are collected for all classes. There is then no biasing towards any class from prior knowledge of its occurrence frequency. Therefore, we assume that the priors are equal and ignore them in the classification (Eq. 5). Although the classification is then equivalent to maximum likelihood, we retain the priors in the formalism to emphasize the Bayesian nature of the method in that in principle it applies also to situations with *a priori* class knowledge.

3   Experiment 1: Surface Texture Classification

Methods: Surface texture data was collected from whisker attached to an iRobot Roomba mobile robotic platform [3].

The whisker was mounted on the robot at a 45 degree angle to the direction of forwards motion with a downwards elevation sufficient to contact the floor (Fig. 1). The whisker sensor consisted of a flexible plastic whisker shaft mounted into a short, polyurethane rubber filled tube called a follicle case. Overall dimensions were approximately five-times a rat whisker (200mm long, 2mm diameter), with material properties matched to scaled-up biological whiskers [12].

Four surfaces were chosen for classification: two carpets of different roughnesses, a tarmac surface and a vinyl surface. Deflections of the whisker shaft were measured as 2-dimensional (X,Y) displacements at the base using a Hall effect sensor sampled at 2KHz. In these experiments, the robot moved in a stereotyped rotating manner, spinning either anticlockwise or clockwise (4 trials each of 16 seconds). Here the classification applies over both types of motion.

Results: The data resemble noisy time series interspersed with dead zones, jumps and systematic changes in the mean or variance (Figs 1A-D). The initial 8 seconds of each trial was used for training data, leaving the final 8 seconds for

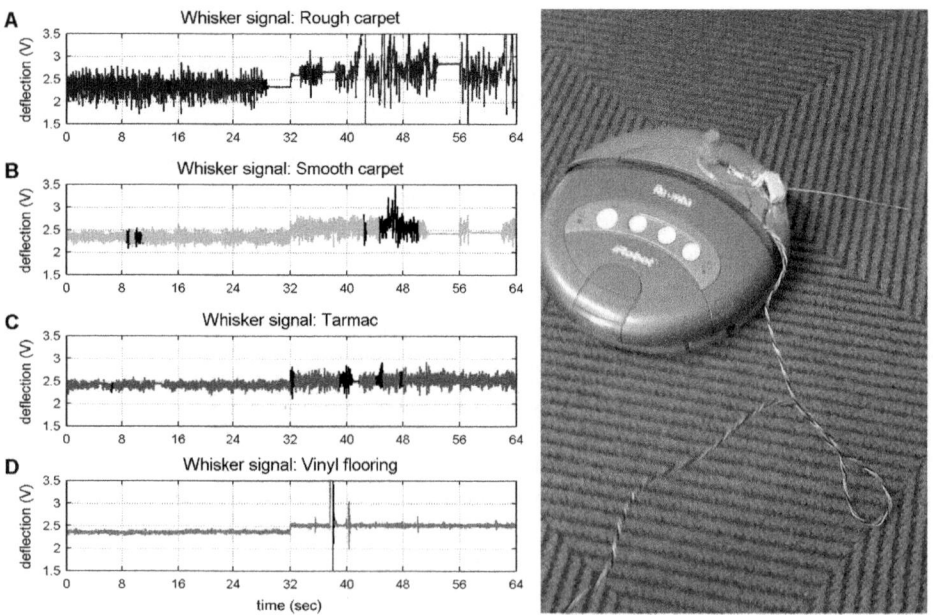

Fig. 1. Surface texture classification with the iRobot Roomba. The Y-deflection validation data is shown for each of the four textures. On each trace, the first four half-trials of 8 seconds are for anticlockwise motion and the latter four half-trials are for clockwise motion. Hits are shown in color and misses shaded in black.

validation testing. The texture likelihoods were then found from the histograms of the empirical frequencies for measurement values of the four textures (Eq. 1). In previous work, we used likelihood distributions that assumed statistical independence between horizontal (X) and vertical (Y) deflections of the whisker [10]. Here, we build on that study by utilizing a two-dimensional joint-likelihood for classification. This study also aims to present texture classification in the wider context of identifying other aspects of the environment, such as contact speed and distance or shape classification, as discussed later.

For validation, the data was separated into discrete segments of fixed temporal window size 500ms over which the texture is determined. The stationary naive Bayes classifier then considers the log-likelihood values for the four candidate textures from the conditional probability distributions at the same measurement value in the training data. As discussed above, two-dimensional likelihoods over pairs of X and Y deflection values were considered, binning measurements over a grid of resolution 10mV by 10mV. The log-likelihood values over the segments of validation data were then summed for each of the four textures (Eq. 5), with the maximal value giving the classified texture.

The correct classifications, or hits, for each of the four textures is shown on Figs 1A-D by the colored segments of data and the misses by black segments. The classification for all four textures was highly accurate (mean hit rates 89%, 88%, 91% and 99%). These results are apparently little different from those assuming independence between the X and Y directions (*c.f.* [10]). Thus in this situation, a saving in computational complexity can be made by moving to a classifier that naively treats the X and Y values as independent; however, we expect that in general the dependence between separate streams of information is important, and the determination of when it is remains an open research question.

In general, the stationary naive Bayes classifier was robust to changes in robot motion, since an identical classifier was used for both clockwise and anticlockwise rotations. This is perhaps surprising considering that such changes cause large systematic effects in the whisker deflection data (Fig. 1 at 32 seconds). It is also curious that the misses occurred mainly near jumps in the whisker deflection, corresponding to the whisker spuriously catching the floor surface, rather than directly confusing textures on normal data. Thus the accuracy on data without such artifacts would improve upon the 92% mean hit rate found here.

4 Experiment 2: Speed and Radial Distance Classification

Methods: An XY positioning robot moved a vertical bar onto a fixed, horizontal whisker to probe the whisker deflection dependency on contact speed and radial distance [4]. The data set consisted of four repeats of 2626 distinct contact speeds and distances along the whisker shaft.

The whisker sensor was similar to that in the preceding texture study, with slightly different dimensions (whisker length 185mm, tapering from 3mm to 0.5mm) and material properties again scaled up from biological rat whiskers [12]. Objects were moved onto the whisker and when contact was detected the robot

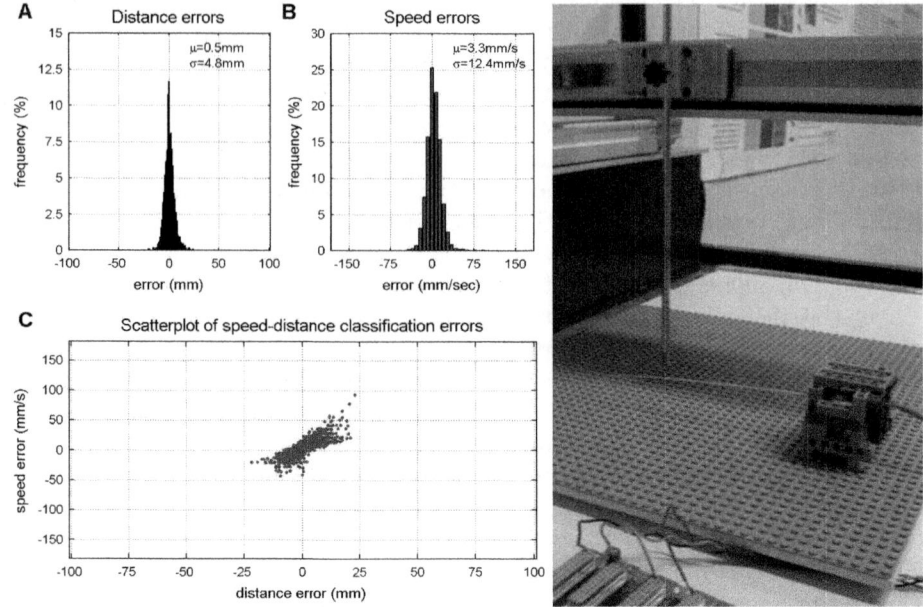

Fig. 2. Speed and radial distance classification with the XY table. Panels A and B show the contact speed errors and radial distance errors between the target value for the validation trial and the classified value from the stationary naive Bayes classifier. Panel C shows a scatter plot of these errors.

retracted the bar at its maximum speed after a short 300ms delay. Thus the contacts were actively feedback-controlled, as are whisker contacts observed in animal behaviour studies. Radial contact distance along the whisker shaft away from the base was sampled in 1mm steps from 80mm to 180mm and contact speed in 7.2mm/sec steps from 36mm/sec to 216mm/sec, giving $101 \times 26 = 2626$ distinct contacts. For more details we refer to [4].

Results: Whisker signals had a characteristic profile in which their X-deflection increased smoothly from rest until reaching their peak after a few hundred milliseconds, and then quickly returned towards baseline followed by damped oscillatory ringing of period $\sim$50ms [4, Fig. 3]. As observed in the original analysis, deflections of a similar peak amplitude can be produced for many contact speed and distance combinations (by compensating the increase in speed with a decrease in radial distance).

Traces from 100ms before the initial deflection to 650ms after were used for analysis. Empirically, it was found that whisker velocity traces gave a better localization of contact speed and distance than whisker deflection traces. Thus, the positional deflections were converted to velocities for the analysis (by taking their numerical derivative and Gaussian smoothing to reduce noise). For each contact combination, three trials were used for training and the remaining trial

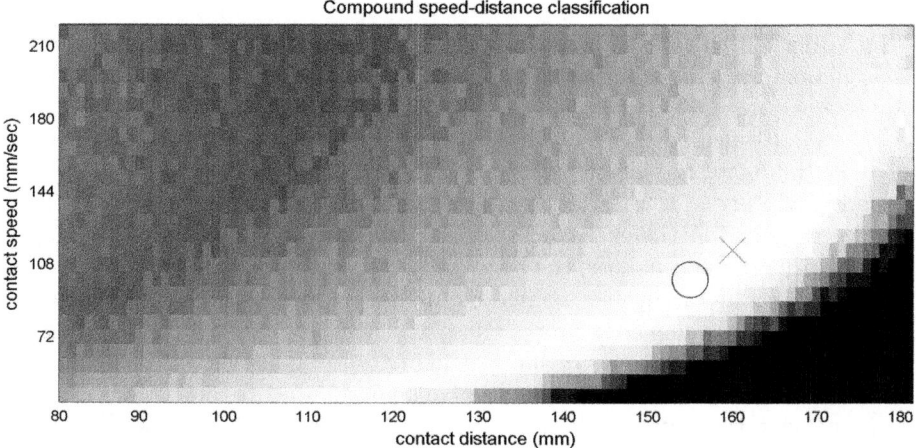

Fig. 3. Speed and radial distance classification with the XY table. The brightness of each pixel represents the magnitude of the log-posterior for each combination of speed and distance marked on the axes. The green cross represents the target speed and distance of the test data and the blue circle the classified values according to the maximal log-posterior (at the brightest pixel).

saved for validation. These three trials were concatenated and the likelihoods found from the histograms of the empirical frequencies for the time series values binned into 10mV/sec intervals (Eq. 1). This procedure resulted in 2626 distinct likelihood distributions, one for each speed-distance combination.

For validation, each likelihood distribution was used to construct a log-posterior from the test trial data (Eq. 5). Generally, the distribution of log-posteriors for each trial had an extended curving segment of high values close to the target speed and distance of the test trial, surrounded by a drop-off of values further away from these targets (example shown in Fig. 3). The maximum log-posterior gave the classified speed and radial distance for that test trial.

Histograms of the errors between the classified and target speeds and distances were approximately Gaussian with little systematic bias of their centers from zero error (Figs 2A,B). A two-dimensional scatter-plot of these errors was consistent with these observations, and also revealed correlations between the speed and distance errors (Fig. 2C) consistent with the prevailing direction of the high value region in the log-posterior distribution (*c.f.* Fig. 3, bright region). This direction in contact speed and radial distance coincided with the region of constant peak amplitude examined in the original study of feature-based classifiers [4]. Hence, the stationary naive Bayes classifier appears highly dependent on peak deflection amplitude, but must also use other aspects of the whisker signal to localize the contact speed and distance close to the target values. Overall, the spread of distance errors was around 5mm compared with an overall range of 100mm, and the spread of speed errors was around 12mm/sec compared with an overall range of 180mm/sec. Thus the overall accuracy was around 93-95%.

5   Experiment 3: Shape and Position Classification

Methods: The BIOTACT sensor (Fig. 4) was used to sense cylinders of various diameters placed horizontally at different positions along a platform and approached from a vertical direction.

The G1 (Generation 1) BIOTACT sensor consists of a truncated conical head that holds up to 24 whisker modules. The whisker modules are arranged in 6 radially symmetric rows of 4, oriented normally to the cone surface [17]. The head is mounted as the end-effector of a 7-degree-of-freedom robot arm (Fig. 4). For the present experiments, the head was fitted with a total of 4 whiskers in one row appropriate for sensing axially symmetric shapes such as cylinders aligned perpendicular to the whisker row. These whiskers were moved back-and-forth to repeatedly contact surfaces of interest akin to animal whisking behaviour [13], with individual whisker modules feedback modulated to make light, minimal impingement contacts on the surface [17]. Whiskers towards the front of the head were shorter than at the back (lengths 50mm, 80mm, 115mm and 160mm), and were designed with a taper from a 1.5mm base to a 0.25mm tip.

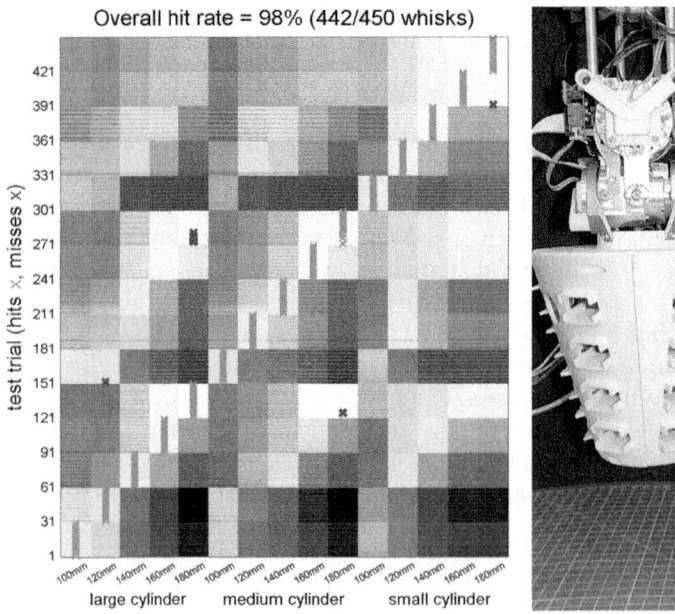
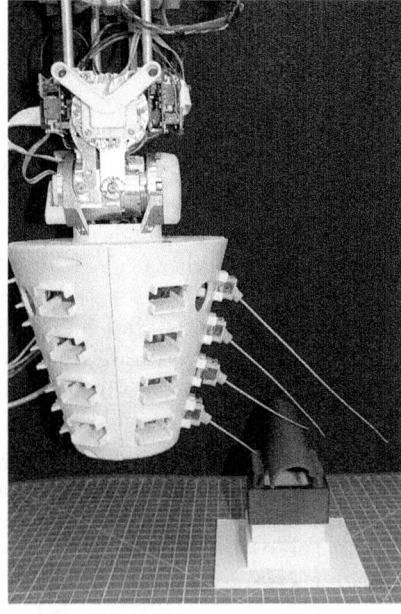

Fig. 4. Shape and position classification with the BIOTACT sensor. The x-axis represents the cylinder configuration that is being tested for (5 positions and 3 curvatures). The y-axis denotes the test whisk, which were taken in 30 whisks trials in order of the configurations shown on the x-axis. Green crosses represent test trials that were correctly classified and red crosses represent incorrect classifications. The shade of pixels represents the value of the log-posterior according to the test cylinder configurations, such that brighter pixels denote larger log-posteriors (with the classified configuration being the brightest for each test trial).

Three rigid plastic hemi-cylinders with radii of curvature 35mm, 25mm and 15mm were used as test objects. They were mounted with their curved surfaces lying upwards on bases of height 15mm, 25mm and 35mm to ensure their vertical dimensions were equal. Five different positions for each cylinder were used, with central axes offset 100mm, 120mm, 140mm, 160mm and 180mm from the central axis of the head cone of the BIOTACT sensor. The orientation of the BIOTACT head was such that the four whiskers contacted the cylinders along the horizontal and perpendicular to the cylinder axis (*i.e.* across the cylinder). When contacting a horizontal surface, the whisker tips are approximately 30mm apart, spanning a distance from 100mm to 190mm perpendicular to the head axis. The depth of the contacts was arranged to be equal for all trials and only contacting the curved sections of the test objects. Each of the 15 distinct trials was repeated twice with 30 whisks at 1Hz to give training and validation data sets.

Results: The four whisker contacts across the cylinder surfaces had a distinctive pattern depending upon the curvature and position of the cylinder. For the largest cylinder in a central position, four whiskers could contact simultaneously, while for the smallest cylinder only one or two whiskers could make contact.

Single whisk traces from all 4 whiskers were used for analysis, with each trace starting and finishing when the whisker was maximally retracted from the object. The 30 whisks from each training trial were concatenated and the likelihoods found from the histograms of the empirical frequencies for the time series values binned into 10mV/sec intervals (Eq. 1). This procedure resulted in 15 likelihood distributions, one for each of the 3 cylinder curvatures and 5 positions.

For validation, the classification accuracy over single test whisks were considered. Over the 15 test trials, this gave a total of 450 distinct test whisks. The stationary naive Bayes classifier considered the log-posterior values for the 15 candidate configurations of 3 curvatures and 5 positions, calculated by summing the associated log-likelihood values for the whisker deflections over each single test whisk (Eq. 5). Correct classifications, or hits, correspond to when the maximum log-posterior coincides with the test cylinder configuration.

Overall, the hit rate for this experiment was above 98% using single test whisks contacting a surface (Fig. 4; hits shown in green and misses in red). The most common mistake was confusing the largest and medium cylinders at their most distant position from the robot head, where only one or two whiskers could contact the object; even then, though, there were just 6 mistakes over 60 test whisks. Therefore, using four whiskers dabbing onto a surface, the BIOTACT sensor could reliably determine both the curvature and horizontal positioning of test cylinders. This was achievable with single whisks lasting less than a second. If instead the full trial of 30 whisks were considered, then all test configurations could be determined with a perfect classification rate of 100%.

6   Discussion

The main point of this article is that a classifier based on probabilistic methods can be a general classifier for artificial whisker sensors because it determines the salient aspects of the data on which to base the discrimination. Such a classifier contrasts with the more traditionally considered specific classifiers that are based on pre-chosen features of the signals, such as peak deflection or profile of the power spectrum. We see these two types of classifiers as complementary in their applications and use. Specific classifiers are appropriate for controlled environments in which many details are known in advance, whereas general classifiers should just 'work out of the box' and apply across a range of situations.

For this initial study, we used the simplest Bayesian classifier available for time series analysis, by assuming stationarity and naive independence over sensor measurements. Further work could investigate how relaxing these assumptions affects classification accuracy. That being said, all applications of stationary naive Bayes considered in this article gave highly accurate hit rates of 90% or greater over diverse stimuli and robots. Furthermore, the efficiency savings from computationally simple classifiers, compared with more sophisticated methods for Bayesian inference, are important for sensing in autonomous robots. Hence we expect that the classifier used here will be appropriate for many situations of practical interest for whiskered robots. That being said, future study should clarify when the classification is limited by the simplifying assumptions, for example when correlations in the time series degrade the performance.

One problem requiring classifiers with the general performance found here is to autonomously explore and learn about novel environments. Because novel stimuli are not known in advance, only general classifiers would be sufficiently versatile for each new situation that arises. Fortunately, the probabilistic classifier in this article, stationary naive Bayes, can also detect novelty [11]. In another study with a BIOTACT robot, SCRATCHbot, novelty detection was examined by training the classifier to recognize the whisker signals from a blank floor and then examining the classifier output when it passed a novel texture. The classifier output, corresponding to the log-probability of the data being from the floor, displayed an anomalous change that can be used to identify novel events. One future project would be to combine the novelty detecting properties of the stationary naive Bayes classifier with its classification abilities to investigate the autonomous learning of stimuli in unfamiliar environments.

A related question for future study is how to extend the standard training/testing protocol of presenting many stereotyped examples of a contact onto a stimulus and then testing with the same examples. In natural environments, the robot might be expected to be more varied in its movements while contacting an object. Or perhaps the motion of the robot would need to become stereotyped to achieve the most reliable classification? These questions relate to a central theme of BIOTACT on active sensing, in which the appropriate sensation for the task is actively selected by adapting the sensor in response to information received about the environment, and will be a focus of future work.

Acknowledgments. The authors thank members of ATLAS (Active Touch Laboratory at Sheffield), the Bristol Robotics Laboratory, and the BIOTACT

(BIOmimetic Technology for vibrissal ACtive Touch) consortium. This work was supported by EU Framework projects BIOTACT (ICT-215910), ICEA (IST-027819) and CSN (ICT-248986).

References

1. Bar-Cohen, Y.: Biomimetics - using nature to inspire human innovation. Bioinspiration & Biomimetics 1, 1 (2006)
2. BIOTACT consortium, http://www.biotact.org
3. Evans, M., Fox, C., Pearson, M., Prescott, T.: Spectral Template Based Classification of Robotic Whisker Sensor Signals in a Floor Texture Discrimination Task. In: Proceedings TAROS 2009, pp. 19–24 (2009)
4. Evans, M., Fox, C., Prescott, T.: Tactile discrimination using template classifiers: Towards a model of feature extraction in mammalian vibrissal systems. In: Doncieux, S., Girard, B., Guillot, A., Hallam, J., Meyer, J.-A., Mouret, J.-B. (eds.) SAB 2010. LNCS, vol. 6226, pp. 178–187. Springer, Heidelberg (2010)
5. Fend, M.: Whisker-based texture discrimination on a mobile robot. Advances in Artificial Life, 302–311 (2005)
6. Fend, M., Bovet, S., Yokoi, H., Pfeifer, R.: An active artificial whisker array for texture discrimination. In: Proc. IEEE/RSJ Int. Conf. Intel. Robots and Systems, IROS 2003, vol. 2 (2003)
7. Fox, C.W., Mitchinson, B., Pearson, M.J., Pipe, A.G., Prescott, T.J.: Contact type dependency of texture classification in a whiskered mobile robot. Autonomous Robots 26(4), 223–239 (2009)
8. Hipp, J., Arabzadeh, E., Zorzin, E., Conradt, J., Kayser, C., Diamond, M., Konig, P.: Texture signals in whisker vibrations. J. Neurophysiol. 95(3), 1792 (2006)
9. Kim, D., Moller, R.: Biomimetic whiskers for shape recognition. Robotics and Autonomous Systems 55(3), 229–243 (2007)
10. Lepora, N., Evans, M., Fox, C., Diamond, M., Gurney, K., Prescott, T.: Naive Bayes texture classification applied to whisker data from a moving robot. In: Proc. IEEE World Congress on Comp. Int., WCCI 2010 (2010)
11. Lepora, N., Pearson, M., Mitchinson, B., Evans, M., Fox, C., Pipe, A., Gurney, K., Prescott, T.: Naive Bayes novelty detection for a moving robot with whiskers. In: Proc. IEEE Int. Conf. on Robotics and Biomimetics, ROBIO 2010 (2010)
12. Pearson, M.J., Gilhespy, I., Melhuish, C., Mitchinson, B., Nibouche, M., Pipe, A.G., Prescott, T.J.: A biomimetic haptic sensor. International Journal of Advanced Robotic Systems 2(4), 335–343 (2005)
13. Prescott, T.: Vibrissal behavior and function. Scholarpedia (in press)
14. Prescott, T., Pearson, M., Mitchinson, B., Sullivan, J., Pipe, A.: Whisking with robots from rat vibrissae to biomimetic technology for active touch. IEEE Robotics and Automation Magazine 16(3), 42–50 (2009)
15. Solomon, J., Hartmann, M.: Biomechanics: Robotic whiskers used to sense features. Nature 443(7111), 525 (2006)
16. Solomon, J., Hartmann, M.: Artificial whiskers suitable for array implementation: Accounting for lateral slip and surface friction. IEEE Transactions on Robotics 24(5), 1157–1167 (2008)
17. Sullivan, J., Mitchinson, B., Pearson, M., Evans, M., Lepora, N., Fox, C., Melhuish, C., Prescott, T.: Tactile Discrimination using Active Whisker Sensors. IEEE Sensors 99, 1 (2011)

A Navigation System for a High-Speed Professional Cleaning Robot

Gorka Azkune[1], Mikel Astiz[1], Urko Esnaola[1], Unai Antero[1],
Jose Vicente Sogorb[2], and Antonio Alonso[2]

[1] Industrial Systems Unit, Tecnalia, Donostia, Spain
[2] Acciona R+D, Madrid, Spain
gorka.azcune@tecnalia.com

Abstract. This paper describes an approach to automate professional floor cleaning tasks based on a commercial platform. The described navigation system works in indoor environments where no extra infrastructure is needed and with no previous knowledge of it. A teach&reproduce strategy has been adopted for this purpose. During teaching, the robot maps its environment and the cleaning path. During reproduction, the robot uses a new motion planning algorithm to follow the taught path whilst avoiding obstacles suitably. The new motion planning algorithm is needed due to the special platform and operational requirements. The system presented here is focused on achieving human comparable performance and safety.

1 Introduction

A new approach for autonomous floor cleaning is presented in this paper. The goal is to achieve a high-speed robot capable of working in conventional environments, aware of obstacles. Many attempts have been made to solve this problem, but none has yet provided a suitable solution from an industrial point of view, in terms of safety, performance and cost-effectiveness. The navigation system presented in this paper addresses both challenges by combining the most appropriate methods in the literature with a new motion planning algorithm, adapted to the operational requirements of professional cleaning.

For practical reasons, the robot must work reliably with no previous knowledge of the environment. No map will be available, and no additional infrastructure is allowed. These constraints have supported the adoption of a teach&reproduce strategy: a human operator will teach how an area must be cleaned, and the robot will need to follow the same cleaning path as strictly as possible.

One of the most challenging problems consists of the physical constraints of these machines, particularly their weight (about 400Kg), their geometry (non cylindrical), and their working speed (above 1 m/s). Research efforts have been focused on obtaining a human-like behavior on both wide and narrow spaces, without compromising safety requirements. A new motion planning algorithm has been developed for this purpose, given that existing solutions do not satisfy

R. Groß et al. (Eds.): TAROS 2011, LNAI 6856, pp. 24–35, 2011.

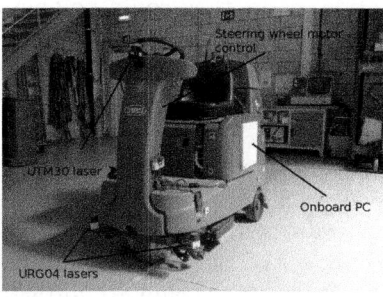

Fig. 1. The modified Tennant T7

all requirements. The main goal has been to follow taught paths with high precision, obstacle avoidance, and performing close-to-wall cleaning with minimum performance degradation.

The perception problem has been clearly separated from the navigation system, that is, the method to merge multiple sensor sources and build a consistent representation of the world being observed. Working on this representation allows different hardware configurations to be tested with minor software changes. In the presented experiments, three laser scanners have been mounted, as a standard proposal for conventional environments. For specific functional and cost requirements, this could be scaled up or down according to application needs.

This paper is organized as follows: in section 2 related work is analysed; section 3 gives an overview of the problem description and the implemented solution; section 4 explains in detail the new motion planning approach developed for the cleaning application; the results of the implemented solution can be seen in section 5 and to finish, section 6 contains the conclusions and future work to be done.

2   Related Work

Floor cleaning automation has been one of the application domains that has attracted the research interest of many roboticist in the past two decades. Many robotic solutions have been proposed, the most relevant of each time can be found in several surveys [13] [4] [7]. Floor cleaning machines can be divided in two categories depending on their use: (i) home cleaning machines and (ii) professional cleaning machines. The work presented in this article is oriented to professional cleaning.

During the nineties many industrial cleaning robots appeared in the robotics scene, some of which were developed up to commercial product state. For localization, some needed modifications of the environment in the form of landmarks (C100 of Robosoft, or Abilix 500 of Midi Robots) or magnetic field guides (Auror, Baror and CAB-X of Cybernetix). Others were able to localize themselves without requiring any modifications of the environment (RoboKent of ServusRobots, ST82 R Variotech of Hefter Cleantech/Siemens). Siemens developed a commercially available navigation system, named SINAS, oriented to cleaning robots [9].

Regarding to the operation mode, they could be operated in manual mode and in autonomous mode. For many of these solutions cleaning trajectories were programed in teach-in mode, where the robot learns the cleaning trajectories from a person guiding it (BR700 of Kärcher, C100 of Robosoft, Hako-Robomatic 80 of Hako/Anschütz).

Perception and localization techniques for robot navigation have now reached a mature state. These aspects have been important for the development of a new generation of industrial cleaning robots. Representative is DustClean, one of the robots developed under the European Community financed project Dust-Bot [11], a robot, in prototype stage, built to autonomously clean streets. The companies Subaru, with their floor cleaning robots Tondon and RFS1, and Intellibot, with their GEN-X cleaning robot product line [8], were able to employ their robots for real everyday operation. Another example can be found in the company Cognitive Robotics, which developed the CRB100 [1] system thought to convert ordinary mobile machines into robots. They have focused their activity on professional cleaning automation.

As far as navigation concerns, a lot of obstacle avoidance methods can be found in the literature. The most similar ones to our approach are probably the Vector Field Histogram [2] (VFH+) and the Dynamic Window Approach [6] (DWA). VFH+ uses polar histograms to process sensory input similar to our approach, while DWA forward simulates the trajectories of selectable motion commands to evaluate their suitability given an environment. However, both algorithms have a lot of parameters to configure, which makes harder the set up of a robot. Additionally, none of them could match the requirements of the cleaning process identified in this paper with a unique set of configuration parameters, which suggests the implementation of a specific algorithm, as shown in section 4.

3   Problem Description and System Overview

The problem being faced consists of reducing the costs of floor-cleaning by means of automation while maintaining speed, safety and process quality (70% of the cleaning costs are attributable to labor costs according to [7]). Several key requirements have been gathered from industrial partners: the need to avoid changes in the cleaning process, hardware solution based on a well-proven cleaning machine, the compatibility with traditional operator profiles, equivalent or improved cleaning performance, and above all, absolute safety. A deeper analysis has highlighted the need of the features listed below.

- The **solution must be human-comparable** in terms of driving style and speed. A predictable behavior is essential due to the presence of humans in the working space, which suggests a deterministic control strategy. Besides, the physical constraints of these machines represent a key challenge of the problem, mainly because of the geometry and the weight of the machine and the relatively high working speed (between 1 and 2 m/s). The solution is expected to deal with such constraints while keeping absolute safety conditions.

- The **navigation system must be independent of the environment** (no previous knowledge or environment map will be available), and must require no extra infrastructure. Typical working environments include both wide areas and narrow spaces with the presence of unknown obstacles. These factors should not influence considerably the performance of the robot in terms of precision and working time, compared to a human operator. In addition, the ability to clean close-to-wall spaces is strongly requested, given their relevance in the cleaning process.
- Other industrial requirements also play an important role, particularly **robustness, ease-of-use, and cost-effectiveness**.

The development presented in this paper has been strongly influenced by these drivers. One of the goals has been to test the solution in an operational environment, as a first step in the validation process of a pre-industrial prototype which is based on a common industrial cleanning machine, the T7 of Tennant. Sensor, motion and processing hardware has been incorporated to make the machine capable to perceive the environment and to move autonomously.

The geometry of the Tennant T7 has a rectangular footprint. Its dimensions are 1.52x0.82x1.27 (length x width x height) in meters. It is a car-like vehicle, with its associated mobility restrictions. The weight of the Tennant T7 is 265 kg. and has a water deposit of 110 liters. This makes a maximum total weight of 375 kg. when the tank is full. Two Hokuyo URG-04LX have been mounted at ankle-height at both sides of the machine. Additionally a Hokuyo UTM-30LX has been mounted on the machine at chest-height (see Fig. 1).

The Tennant T7 can be operated in manual mode, where the platform is controlled by a human operator and in autonomous mode, where the platform is controlled by the PC. One relay system controlled by the Technosoft IDM640 makes this possible. A safety watchdog system has been implemented through this motor controller. If no speed commands are received in 0.5 seconds, the controller switches to manual mode.

A general overview of the implemented solution presents several parts that will be roughly explained. The *perception for obstacle avoidance* merges sensor data coming from different sources in a probabilistic occupancy grid, which provides a consistent observation of the space surrounding the robot (see [3] and [12]). For *localization* purposes a Monte Carlo localization algorithm (see [5]) has been used, combining odometry and laser data to track the position of the robot. The *teaching by showing* module records the taught path using a sequence of odometry poses –position and orientation, without speed–, each of which will have an associated local map of the environment consisting of a set of geometric features, namely corners and planar walls, extracted from the laser scanner readings. During teaching, the machine will be driven manually, just like under ordinary operation. Finally, for the reproduction of the taught path, a new motion planning algorithm has been implemented. Section 4 contains a deep explanation of the mentioned algorithm.

4   Motion Planning

In the context of a teach&reproduce approach, the problem of motion planning consists in calculating the suitable motion commands given a taught path and some special constraints due to (i) the platform, (ii) the operation process and (iii) the user requirements.

Platform constraints cover geometric, kinematic and dynamic constraints of the Tennant T7. The most relevant ones are the rectangular footprint of the platform, the low angular speed of the steering wheel and the low braking rate due to the weight. **Operational constraints** refer to a set of navigation features that must be fulfilled while cleaning: navigation on slippery surfaces, fidelity to the taught path, reasonable speed during close-to-wall navigation, navigation through narrow spaces and avoidance of non-expected obstacles. **User constraints** refer mainly to safety and ease-of-use, which implies an intuitive, deterministic and easily configurable motion behavior.

The requirements of the problem described were found to demand a new approach for motion planning. The new approach presented here is founded in a two layer architecture: the first layer is a path follower that generates motion commands without considering any obstacle; the second layer is a new reactive obstacle avoidance algorithm. At a first stage, a standard PID controller for the steering angle has shown to be enough for path following. The steering angle generated by the path follower is used by the obstacle avoider to calculate new speed and steering angle commands that ensure a safe path through obstacles, fulfilling the platform, operational and user constraints enumerated above.

To understand how the motion planning algorithm deals with those requirements, three different solution sets are identified inside the space of all possible solutions. A solution set is a continuous range of steering angles θ and velocities v. Solution sets can be distinguished based on the following conditions: (i) **safety conditions** refer to the minimum and necessary requirements for navigation; if safety conditions are not guaranteed, the platform will no longer move, (ii) **navigable conditions** are those which guarantee conventional navigation requirements, whereas (iii) **desirable conditions** are those conditions that would like to be fulfilled under ideal circumstances. Three solution sets emerge from those conditions: the safe solution set, the navigable solution set and the desirable solution set. The so called solution sets are defined using very intuitive parameters. For example, a (v, θ) solution will be a navigable solution if the trajectory generated by (v, θ) has a collision distance which is greater than a certain value ($nav\_frontal\_clearance$), for a certain lateral clearance ($nav\_lateral\_clearance$). Both values can be configured and set by the user.

The key idea behind the algorithm is to use solution sets to evaluate the safety, navigable and desirable conditions, given any situation. The algorithm will try to find a solution inside a desirable solution set. But if it is not possible, requirements will be brought down to find solutions inside navigable solution sets or even safe solution sets. This way, the algorithm guarantees to find a solution which is as close as possible from the desirable conditions, providing safe navigation when required by environment conditions.

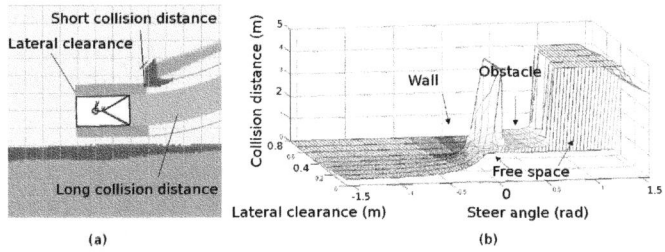

Fig. 2. (a) The occupancy grid generated by the perception component, where a curve trajectory has been drawn for two lateral clearances. (b) PCPH example for the occupancy grid of (a). Positive steer angles represent the left hand side of the platform. Valleys in the diagram represent small collision distances, hence obstacles, as shown by the arrows.

To be able to divide the whole solution space in the three sets defined above, a concrete data structure has been adopted called *platform clearance polar histogram* (PCPH). A PCPH is a data structure that represents the frontal clearance distance to obstacles (collision distance) for each precalculated trajectory and for each lateral clearance distance. So in other words, the idea of PCPH is to represent the collision distance for all the steering angles of the platform. However, this is done not only for the exact footprint of the platform but for different footprint inflations, resulting on a useful world representation, where lateral clearance can be handled easily.

To compute a PCPH an occupancy grid is required. As explained in section 3, sensor information is used to compute such an occupancy grid as the one shown in Fig. 2.a. Additionally, a grid of precalculated trajectories is generated at initialization time. This last grid stores the curve trajectories defined by each steer angle of the platform. So using the occupancy grid and the precalculated trajectories, a PCPH is calculated. Fig. 2.b shows a graphical representation of a PCPH.

One of the major challenges of PCPH is to calculate collision distances in real time at high rates. For obstacle avoidance to work efficiently, a minimum rate of 10 Hz is desirable. An efficient library has been implemented that can perform PCPH calculations at 20 Hz, with a grid-cell resolution of 5 cm, a maximum collision distance of 10 m and a steering angle resolution of 1 degree.

4.1   Steer Angle Selection

The process of steer angle selection is based on the segmentation of PCPH to find safe, navigable and desirable solution sets. Segmentation is the process of finding sets of steering angles that fulfill specific frontal and lateral clearance conditions. The pseudo-code for the algorithm is depicted in listing 1.1.

The pseudo-code shows how the algorithm, at its first stage, tries to find navigable sets. If there are no navigable sets, frontal and lateral clearance are iteratively decreased until a so called safe set is found or safety conditions are reached without any solution. In that case, the algorithm stops the robot, because safety requirements cannot be fulfilled in the environment. However, if the first stage is successful, i.e. if a navigable set exists, the algorithm continues selecting the best navigable set.

```
float selectSteerAngle(pcph, previous_set)
{
    if (!navigableSetSelection(nav_sets, nav_lateral_clearance,
        nav_frontal_clearance, pcph))
    {
        // There is no navigable set. Find safe sets
        return findSafeSteerAngle(nav_lateral_clearance,
            nav_frontal_clearance, safety_dist, pcph);
    }
    // Some navigable sets exist. Select the best one
    best_nav_set = selectBestNavSet(nav_sets, previous_set,
        target_steer_angle);
    // Increase frontal and lateral clearance to find desirable
        sets
    desirable_sets = selectDesirableSets(best_nav_set,
        previous_set, nav_lateral_clearance,
        nav_frontal_clearance, des_lateral_clearance,
        des_frontal_clearance);
    best_des_set = selectBestDesSet(desirable_sets,
        target_steer_angle);
    previous_set = best_des_set;
    return selectClosestSteerAngle(best_des_set,
        target_steer_angle);
}
```

Listing 1.1. Pseudo-code for steer angle selection

The selection of the best navigable set deserves further explanations. As it can be seen in the pseudo-code of listing 1.1, this is done by the function *selectBestNavSet*. The function uses the list of navigable sets, the set selected in the previous iteration of the algorithm and the target steer angle sent by the path follower. A two step process is run in the function:

(i) Initial selection: there is a key concept in this step: set matching. The algorithm stores in each iteration the set selected in the end of the whole process. In the next iteration, the stored set *previous_set* is used for the initial selection of navigable sets. The algorithm only considers those navigable sets which match *previous_set*. Matching is defined as the overlap of sets and it has been introduced to avoid oscillations in the trajectories of the platform. This solution has shown to be a very effective method in the tests, despite of its simplicity.

(ii) Best set selection: from all the sets that pass the initial selection, choose the one which is closest to target steer angle.

The process described above ensures that the selected set is navigable based on the frontal and lateral clearance parameters. Besides, it also removes oscillations

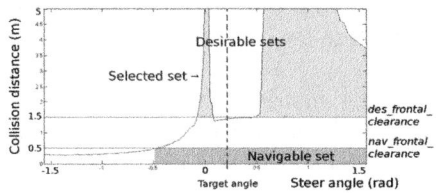

Fig. 3. Segmentation process for a PCPH slice corresponding to Fig. 2; previous selected set is supposed to overlap with both desirable sets

and selects a set that is close to the target steer angle, hence close to the path which is being followed by the path follower.

As navigability requirements are fulfilled, the algorithm attempts to find whether desirable conditions can also be fulfilled. For that purpose, *selectDesirableSets* function iteratively increases frontal and lateral clearance values. The criterion used to select a set is to match both, the selected navigable set and the last selected set. Hence, the function will increase clearance values until no sets can be found or desired clearance values are reached. The sets that survive this process will be called desirable sets. All of them will be as close as possible from fulfilling the desirability conditions and they will match the last selected set, preventing any oscillations.

Finally, the best set of all desirable sets is chosen inside the *selectBestDesSet* function. To select the best set as well as to select the best steer angle inside that set, a simple criterion has been implemented: choose the set and steer angle which are closest to the target steer angle. This simple criterion has shown to work very well keeping trajectories close to what the path follower commands. The most important concepts of the motion planning algorithm and their relations are illustrated in Fig. 3.

4.2   Speed Calculation

There are mainly four aspects that influence speed calculation: (i) the steering wheel latencies, (ii) the dynamic model of the platform, which specifies how the platform behaves under a speed command, (iii) frontal clearance and (iv) lateral clearance. Some assumptions have been done to simplify the dynamic model. The algorithm assumes that the platform acceleration is constant and that the latencies generated by the speed controller are also constant. With those assumptions, all the aspects commented above except (iv) are considered in equation 1:

$$v = \sqrt{2(col\_dist_{ref} - v_c t_L)a_b} \tag{1}$$

where v is the output speed calculated by the algorithm, v_c is the current speed of the platform, t_L is the estimated latency for the speed controller and a_b is the deceleration capacity of the platform. The variable col_dist_{ref} deserves further explanations, since it is the key to handle steering wheel latencies. The idea is to store in col_dist_{ref} the smallest collision distance found between current and selected steer angle and calculate the speed respect to that value. This strategy has shown to work well for the low angular speed of the wheel, which is around $\pi/4 \ rad/s$.

As said above, equation 1 handles all the aspects except lateral clearance, which is important to control the platform speed in narrow openings, close to walls or any other obstacle which is not in front of the platform. A simple and effective way to deal with it is to limit the maximum speed depending on the lateral clearance. Any relation between lateral clearance and maximum speed can be implemented. A linear relation has shown to work very well and it is very easy to configure.

5   Experimental Results

The experimental results section has been divided in two main sub-sections: the first one to evaluate the general performance of the system compared to human operators and the second one to show the performance of the new motion planning algorithm explained in section 4.

General performance of the cleaning machine has been measured by two parameters: (i) the comparison between human operator cleaning time and autonomous cleaning time and (ii) the error between taught path and reproduced path. For the solution to be acceptable from an industrial point of view, autonomous cleaning time and fidelity should be as close as possible from the taught ones. For the environment layout 3.5 meters wide corridors were used. The Tennant T7 is not well suited for narrower corridors due to its size and its poor maneuverability. Respect to the taught path, near optimal cleaning paths for corridors have been taught to the system, composed by long straights followed by 180 degree curves at the end of the corridor that cover the whole cleaning surface. Finally, for the experiments to be fair, the environment is exactly the same at reproduction time as in the teaching stage.

To measure the reproduction time against the teaching time, a total number of 50 paths were taught and reproduced. The shortest path was 50 meters long and the longest one 200 meters. After performing time measurements for each path teaching and reproducing, an average increment of 13.56% has been observed for reproduction time. However, the time increment measured for this kind of environments is remarkable. As a reference, the average speed of the machine while teaching was 1.3 m/s, which is a fast speed for cleaning. Fig. 4.a shows a graphic with all the tests performed.

The second parameter is the fidelity to the taught path. This can be measured as the reproduction error respect to the taught path. The teaching and reproduction tests done in Fig. 4.a have been also used for fidelity tests. After measuring the errors of all the paths reproduced, an average error of 0.09 m was found. The maximum error measured for the motion planning module was 0.32 m. Fig. 4.b shows the error measurements obtained during a reproduction of one of the taught paths. As it can be seen, up to the 85% of the measurements lie between [0.0, 0.2] meters, while only the 5% are bigger than 0.3 meters. These results are encouraging.

Motion planning performance: There are several aspects of motion planning that have to be tested for cleaning: mainly how the algorithm performs very close to walls, whether it avoids oscillations and its behavior in narrow spaces and crowded environments. As far as close to wall behavior concerns, a wall following path was taught to the platform, as close to the wall as possible. After teaching mode, the platform reproduced the taught path autonomously 50 times. The results show an average distance to wall of 0.13 meters (safety distance was set to 0.1 meters), while the average speed still keeps high: 0.7 m/s. Additionally, the resultant trajectory was smooth, very straight and free of oscillations.

As an example of a crowded environment, a slalom was prepared (Fig. 5). Wooden panels were set at 2.5 meters of distance (the Tennant T7 is 1.52 meters long). In a previous stage, the platform was taught a straight path (represented as the red dashed line in Fig. 5.b). The trajectory executed in reproduction mode can also be seen in Fig. 5.b. Localization poses were stored for visualization and analysis purposes. Fig. 5.b shows the superposition of the platform along the trajectory described in the environment of 5.a. The trajectories described by the platform are smooth and free of oscillations. Additionally, the average speed during the slalom was of 0.4 m/s, which, given the conditions, is a suitable speed.

Additionally, some trajectories were taught to enter narrow spaces and navigate trough them. Those situations are not very common for cleaning scenarios with such machines, but the algorithm showed good results. It could approach and cross narrow spaces of 1 meter wide (the Tennant T7 is 0.82meters wide), both in straight trajectories and curved trajectories.

To complete the tests of the motion planning algorithm, comparative tests were performed between the presented algorithm and DWA [6]. DWA was selected because it considers dynamic and geometric constraints of platforms (crucial for our application) and it is widely used. When DWA was configured to strictly follow the path, big obstacles could hardly ever be avoided (notice that tests were performed without any global planner). Besides, close to wall behavior was not satisfactory. The trajectories were not as smooth as desired and the

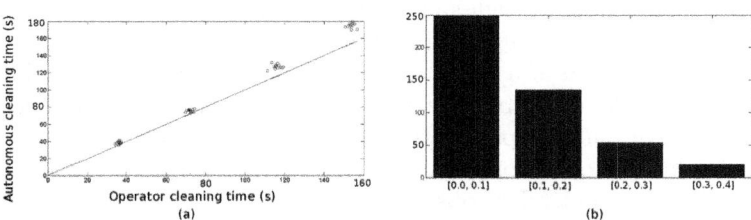

Fig. 4. (a) Comparison between operator cleaning time and autonomous cleaning time; black line represents equality between both variables. (b) Fidelity to the taught path. The histogram shows the number of pose measurements that fall into the error gaps depicted in the x axis. Nearly the 85% of the poses stored in a reproduction are closer than 0.1 meters to the taught pose.

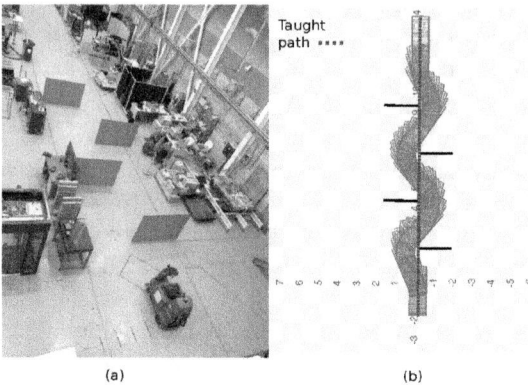

(a)   (b)

Fig. 5. Motion planning slalom test. (a) shows the environment used to perform the tests: wooden walls wich are located 2.5 meters far from each other. (b) shows the execution of the trajectory recorded in the real environment. Localization poses were stored for visualization and analysis purposes. Average lateral distance to wooden walls is 0.16 m.

average speed could not match our algorithm (0.4 m/s for DWA and 0.7 m/s for our algorithm). In general, it was almost impossible to find a set of parameters that could fulfill all the requirements of the cleaning process.

6   Conclusions and Future Work

An industrial autonomous floor cleaning machine has been presented in this paper. A teach&reproduce approach has been adopted to solve the problem, driven by the requirement that the machine should work in any environment with no extra infrastructure and no previous knowledge of the environment. Additionally the motion planning problem given platform restrictions and cleaning requirements has been addressed in depth. To overcome the problems arisen from those restrictions a new motion planning approach has been introduced, which has shown to work reliably at high speeds without trading off safety and enhancing the cleaning performance of the platform.

As shown in section 5, a lot of experiments have demonstrated that the approach presented here can provide a good solution for industrial cleaning, since autonomous reproduction parameters keep quite close in terms of performance to those of human operators. This human comparable performance and high speed motion does not affect obstacle avoidance. The platform behavior while avoiding obstacles is smooth and suitable.

Additionally, the system proposed in this paper can easily work on top of different hardware configurations. That means that the cost of the hardware can be reduced, for example replacing several laser sensors by ultrasound arrays. As perception is detached from motion planning, sensor replacement should not be a major problem. Indeed, some tests have been done, where ultrasound sensors were merged with a laser to compute an occupancy grid. The motion planning algorithm does not have to be changed to work on top of this configuration. Besides, the incorporation of a tilting laser can also be approached, to improve obstacle detection and achieve a higher security level. 3D perception is becoming a key feature for navigation applications (see [10]) and does not require any significant change in the system. Those improvements are believed to have important roles to definitely bring professional cleaning robots to market.

References

1. Cognitive robots (2010), http://www.c-robots.com/en/index.html
2. Borenstein, J., Koren, Y., Member, S.: The vector field histogram - fast obstacle avoidance for mobile robots. IEEE Journal of Robotics and Automation 7, 278–288 (1991)
3. Elfes, A.: Using occupancy grids for mobile robot perception and navigation. Computer 22, 46–57 (1989)
4. Elkmann, N., Hortig, J., Fritzsche, M.: Cleaning automation. In: Nof, S.Y. (ed.) Springer Handbook of Automation, pp. 1253–1264. Springer, Heidelberg (2009)
5. Fox, D.: Kld-sampling: Adaptive particle filters. In: Advances in Neural Information Processing Systems, vol. 14, pp. 713–720. MIT Press, Cambridge (2001)
6. Fox, D., Burgard, W., Thrun, S.: The dynamic window approach to collision avoidance. IEEE Robotics and Automation Magazine 4(1), 22–23 (1997)
7. Hägele, M.: World Robotics, Service Robots 2009 (2009)
8. Intellibot robotics (2010), http://www.intellibotrobotics.com/
9. Lawitzky, G.: A navigation system for cleaning robots. Autonomous Robots 9, 255–260 (2000), 10.1023/A:1008910917742
10. Marder-Eppstein, E., Berger, E., Foote, T., Gerkey, B.P., Konolige, K.: The office marathon: Robust navigation in an indoor office environment. In: International Conference on Robotics and Automation (May 2010)
11. Mazzolai, B., Mattoli, V., Laschi, C., Salvini, P., Ferri, G., Ciaravella, G., Dario, P.: Networked and cooperating robots for urban hygiene: the eu funded dustbot project. In: Proceedings of the Fifth International Conference on Ubiquitous Robots and Ambient Intelligence, pp. 447–452. MIT Press, Cambridge (2008)
12. Moravec, H.: Sensor fusion in certainty grids for mobile robots. AI Mag. 9(2), 61–74 (1988)
13. Prassler, E., Ritter, A., Schaeffer, C., Fiorini, P.: A short history of cleaning robots. Auton. Robots 9(3), 211–226 (2000)

A Recursive Least Squares Solution for Recovering Robust Planar Homographies

Saad Ali Imran and Nabil Aouf

Cranfield Defence and Security, Department of Informatics and Systems Engineering,
Sensors Group, Cranfield University, Shrivenham, UK
`s.a.imran@cranfield.ac.uk`

Abstract. Presented is a recursive least squares (**RLS**) solution for estimating planar homographies between overlapping images. The use of such a technique stems from its ability in dealing with corrupted and periodic measurements to provide the best solution. Furthermore, its capacity for providing reliable results for time varying parameter estimation also motivates its use in the context of real time cooperative image mosaicing where optimal transformation between mobile platforms is likely to change due to motion and varying ambient conditions and thus a way to tackle this problem real time is what is required. Additionally, and within the same context, a derived "match making" algorithm is introduced based on high curvature points (Harris points) and 3D intensity histograms which are in-turn matched using the L_2 and L_∞ and then compared to classical cross correlation(**CC**) techniques.

Experimental results show that for synthetic data heavily corrupted by noise the RLS does a decent job of finding an improved homography, provided that the initial estimate is good. Results from real image data show similar results where the homography estimate is improved upon by periodic measurements. The match making algorithm proposed fairs well compared to intensity vector techniques, with the L_∞ based method coming out on top.

1   Introduction

Cooperating agents are capable of achieving mission objectives more efficiently and with an added sense of security compared to individuals. With this added efficiency also comes an increase in the capabilities of the cooperating group in terms of the complexity of the missions they can handle. Consider the problem of live environmental surveillance where a single autonomous (or semi-autonmous agent) is relaying images back to an end user periodically. The information content of these images is constrained: the biggest restriction being the limited field of view (**FOV**), also known as the "Soda Straw Effect". An effective solution to overcome this constraint is the construction of image mosaics which fuse images together to provide a more information rich depiction of the environment. This is commonly done by installing a stereo imaging setup on mobile platforms. Now imagine a set of such mobile platforms sharing information together to construct

R. Groß et al. (Eds.): TAROS 2011, LNAI 6856, pp. 36–45, 2011.

larger image mosaics and thus providing a better depiction of the environment and so allowing the end user to infer decisions more confidently.

The process of constructing image mosaics is underlined by the problem of image registration, which in essence involves finding the best spatial, and in some cases intensity transformations, that align one image to another. Clearly put this means that a pixel location **(i,j)** in one image is mapped to a pixel **(k,l)** in another image via the transfer function τ, where $(\mathbf{k,l}) = \tau(\mathbf{i,j})$ .Typical spatial transformations include Rigid, Affine and Projective transformations, the latter being of importance here which maps straight lines to straight lines [**14**]. A planar projective transformation using homogenous coordinates is given by **Equation-1** or more briefly by **Equation-2**.

$$\begin{pmatrix} x'_1 \\ x'_2 \\ x'_3 \end{pmatrix} = \begin{bmatrix} h_{11} & h_{12} & h_{12} \\ h_{21} & h_{22} & h_{23} \\ h_{31} & h_{32} & h_{33} \end{bmatrix} \begin{pmatrix} x_1 \\ x_2 \\ x_3 \end{pmatrix} \tag{1}$$

$$\acute{x} = \mathbf{H}x \tag{2}$$

where $\mathbf{H}$ is a non-singular matrix, defined upto scale known as the homography. For a more in depth explanation on planar homographies refer to [**4**].

Estimating the parameters that constitute the homography is a challenging task. [**14**] and [**2**] entail various techniques for determining the optimal homography between two images. Among these are Fourier based techniques and non-parameteric techniques. The most ubiquitous though is that of exploiting distinct and common features between overlapping images to determine a global mapping. Such techniques are widely used in remote sensing as they tend to be faster when the distortion model parameters are not known *a priori*. Feature based techniques in general involve four steps to determine a homography: Feature extraction, Feature matching, Transform model estimation and Image transformation [**14**].

Estimating the homography matrix requires eight measurements as $\mathbf{H}$ has only eight degrees of freedom (defined up to scale). Thus a 2D projective homography requires at least four feature correspondences $\acute{x}_i \leftrightarrow x_i$ as each correspondence accounts for a couple of equations. It is however common practice to use more than four correspondences to overcome measurement noise and so estimating the homography matrix takes the form of a minimisation problem. Doing this for every time step leads to heavy computational requirements and maybe the solution isn't viable for real time operation. Thus an iterative solution where minimum computational effort is required for time varying parameter estimation is what is needed in the context of cooperative image mosaicing.

Real time image mosaicing on board single platforms has been previously attempted in [**10**] and [**11**] though for a sequence of images and for varying intentions, both using feature based techniques (the latter however employs a feature tracking technique as compared to extracting and matching features every time step). A similar approach is applied in [**5**] where image mosaics of image sequences is created using based on affine homography and 1D flow estimation. The problems with such techniques are that the motion between consecutive frames is required to be small otherwise the estimations tend to be quite bad. Real time video mosaicing is done in [9] based on a simplified SIFT algorithm for match making where computation time is reduced by decreasing the number of octaves used for scale space generation. Though the time taken to construct the mosaics is not included.

1.1   Overview

The following sections introduce a derived technique to establish fast and reliable feature correspondences followed by a novel proposal to recast the problem of homography estimation into the linear recursive form and subsequently using it for parameter estimation. Again the feature correspondence and the recursive estimation technique is put forward to try and deal with the requirements of real time image mosaicing on a cooperative level. The idea is to propose a basis for further work on this problem. The study finally concludes with a summary of the key findings. Implementation of the proposed algorithms is done in **Matlab**.

2   Feature Detection and Matching: A Derived Approach

To establish correspondences common features need to be detected and then consequently matched. The idea is to extract sufficient number of features and match them as quickly as possible and as accurately as possible. Various solutions to the feature extraction problem have been proposed over the years and a brief compilation is given in [12]. Among these approaches are the ones based on intensity gradients where changes in the pixel intensities above a defined threshold are taken as features. Moravec (1977) was the first to introduce such a technique which was later improved by Harris and Stephens [3]. Their technique is based on the second moment matrix which describes the gradient distribution in a local neighbourhood and whose eigenvalues define either a corner, an edge or a flat region. In order to avoid the computationally expensive task of evaluating the eigenvalues [3] introduced a "cornerness measure" given below.

$$cornerness = det\left(\mathbf{A}\right) - k \times trace\left(\mathbf{A}\right) \tag{3}$$

where $\mathbf{A}$ is the second moment matrix. The Harris corner detector is invariant to rotation as eigenvalues are rotationally invariant and is also invariant to a certain extent to illumination change.

After identifying the features they need to be distinctly described in order to be accurately matched. A thorough evaluation of feature descriptors is given [7]. Common feature descriptors range from simple intensity based vectors to 3D gradient based histograms. The latter approach was introduced by [6] and outperforms many if not all of its counterparts [7].

The feature correspondence technique proposed in this paper combines the Harris corner detector with the feature descriptor described in SIFT [6]. The premise behind this is to have a fast feature extractor with the Harris detector and combine it with a robust descriptor to increase speed and reliability. A comparison of matching techniques using simple intensity vectors and the gradient histogram is included. The intensity vectors are matched using the classical cross correlation (**CC**) technique and an adapted CC technique with added relaxation techniques to resolve matching ambiguities proposed in [**13**]. The feature descriptors are matched using the $\mathbf{L_2}$ ($\|\mathbf{D_i} - \mathbf{D_j}\|_2$) and $\mathbf{L_\infty}$ ($\|\mathbf{D_i} - \mathbf{D_j}\|_\infty$) norms as proposed in [**8**].

Figure-1 details the performance of each match making technique. Shown are the number of matched features, the number of outliers and the time taken to establish these correspondences. The test is done for image set no.1 shown in **Figure-3**. The thresholds for the match makers are chosen so as to have a similar number of matches in order to do a fair analysis. The threshold for the feature extractor is also the same for the various techniques and for the image analysed, produces an average of 205 features. Also to be noted is that the values for the time elapsed for the histogram based techniques is inclusive of the time taken to actually construct the descriptors. From the chart it can be seen that though the intensity histogram based techniques fair comparably with the amount of matches produced - as allowed be the individual thresholds - they do in relatively quick time, with the $\mathbf{L_\infty}$ based technique being the speediest. It does however produce slightly more outliers compared to its close counterparts - the $\mathbf{L_2}$ and CC without relaxation technique This in fact can be remedied by fine tuning the threshold, though a possibility exists where some good matches will also be discarded. The $\mathbf{L_\infty}$ based matching technique is computationally much less expensive compared to the others which accounts for its swiftness.

In order to quantify the capabilities of each match maker a performance index is introduced, given by $\mathbf{I_{Per}} = \frac{\mathbf{f_{Ratio}}}{\mathbf{et}}$ where $\mathbf{f_{ratio}} = \frac{correctly\ matched\ features}{average\ features}$. This index takes into account the ratio of good matches to average features and the time taken to establish these matches. **Table-1** details the performance indices from each match maker. Note that a score of one is the highest that can be achieved and defines the best performance.

Looking at the above table it can be seen that the intensity histogram based $\mathbf{L_\infty}$ match maker comes out on top. Further tuning the threshold will result in lesser outliers but also there is a chance good matches will be rejected. However for the increase in speed is sufficient and the requirement that only a limited number of matches are required is a good enough reason to consider this approach.

Table 1. Showing the performance index for the various correspondence establishing techniques used

	CC	CC_{Rel}	L_2	L_∞
$\mathbf{I_{Per}}$	0.44	0.11	0.45	0.53

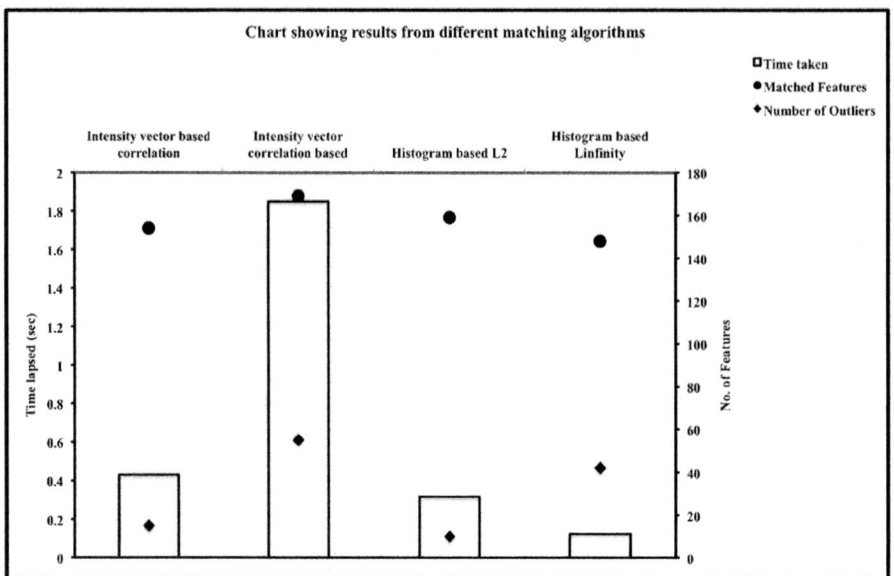

Fig. 1. First data set on the horizontal axis is for matches made using CC, second for CC+relaxation, third for L_2 based and fourth for L_∞ based. The left vertical axis shows time where as the right axis gives number of features.

3   A Recursive Solution for Estimating Homography

Solving for **H** can either be done using the "direct linear transform method" **(DLT)** for an over-determined solution using singular value decomposition **(SVD)** or using the inhomogenous solution described in [4]. Using the latter, the problem can be formulated into a linear least squares one. This technique involves imposing a condition $\mathbf{h_j} = \mathbf{1}$, which as the solution is up to scale can be done for the scale factor. Expanding and rearranging **Equation-1** the following equation is arrived at.

$$\begin{bmatrix} x_i & y_i & 1 & \mathbf{O} & -x_i\acute{x}i & -y_i\acute{x}i \\ \mathbf{O} & -x_i & -y_i & -1 & x_i\acute{y}_i & y_i\acute{y}_i \end{bmatrix} \mathbf{h} = \begin{pmatrix} \acute{x}_i \\ \acute{y}_i \end{pmatrix} \tag{4}$$

where **O** is $[\mathbf{0\ 0\ 0}]$ and **h** is a $\mathbf{8 \times 1}$ vector of parameters of **H**. The above equation can be written briefly as $\theta = \phi h$. For an over determined solution **i.e.** $i >$ eight, the optimal solution for **h** is given by

$$\mathbf{h}^* = \mathbf{inv}\left(\acute{\phi}\phi\right)\acute{\phi}\theta \tag{5}$$

which minimises the square error by finding the projection of θ in the column space of ϕh.

Incase of noisy measurements, which are expected, the solution for **h** using any method will not be exact. The beauty of putting a problem into the above form is that future measurements can be taken into account to update or enhance the estimate. Putting this into context of cooperative mosaicing, once initially a homography is established it can be updated every time by only a single measurement which in imaging terms translates to correspondences, hence in theory changing homographies can be tracked between platforms and at least for stereo setups if the camera positions are changed due to the terrain.

The recursive form of the least squares solution is given by the following three equations [1].

$$\hat{\mathbf{h}}_i = \hat{\mathbf{h}}_{i-1} + \mathbf{K}_i \left(\theta_i - \phi_i \hat{\mathbf{h}}_{i-1} \right) \tag{6}$$

$$\mathbf{K}_i = \frac{\mathbf{P}_{i-1}\phi_i}{\lambda + \phi'_i \mathbf{P}_{i-1}\phi_i} \tag{7}$$

$$\mathbf{P}_i = \frac{(\mathbf{I} + \mathbf{K}_i\theta_i)\,\mathbf{P}_{i-1}}{\lambda} \tag{8}$$

where $\mathbf{K}_i$ is the gain, λ is the forgetting factor and $\mathbf{P}_i$ bears similarity with the covariance matrix as found in the Kalman filter.

Results from an implementation of the above algorithm to synthetic data is given in **Figure-2**, where known points are transformed via a reference homography. These points are corrupted and assembled in **Equation-4** and initial estimate of **h** is then made using **Equation-5**. The RLS algorithm is initialised using this initial estimate and further corrupted measurements are taken into account every time step. The algorithm does a fairly good job of updating the estimate by dealing with measurement errors provided the initial estimate is a good one. Note that the error added in the periodic measurements was of magnitude 2.5.

Figure-3 shows the root means squared **(RMS)** residual transformation error for an initial estimation of the homography and for instances when new measurements have been taken into account for two different image data sets. Initial estimates for **H** are made using 5 feature correspondences. For image set no.1, it is visible that the RLS algorithm reduces the error eventually when more measurements are taken into account. Interesting to note is a bad initial estimate of the homography with good enough evenly spaced matches (**Figure-3** shows the spacing of the correspondences), the reason for which could be down to positional error where the matches do not exactly relate. The RLS shows some unpredictability in the sense that the error initially increases before decreasing. Data set no.2 shows further error analysis on another set of overlapping images: The initial estimate of the homography from initial correspondences is good and taking into account further measurements does initially reduce the error before diverging from zero. Here a way to counteract this problem is introduced where estimates for each iteration are compared to the previous one and if they are better then they are retained and if not then they are discarded and a new measurement is taken.

As additional analysis the best updated homography estimation from RLS is compared to results from an iterative minimisation of the initial estimate based

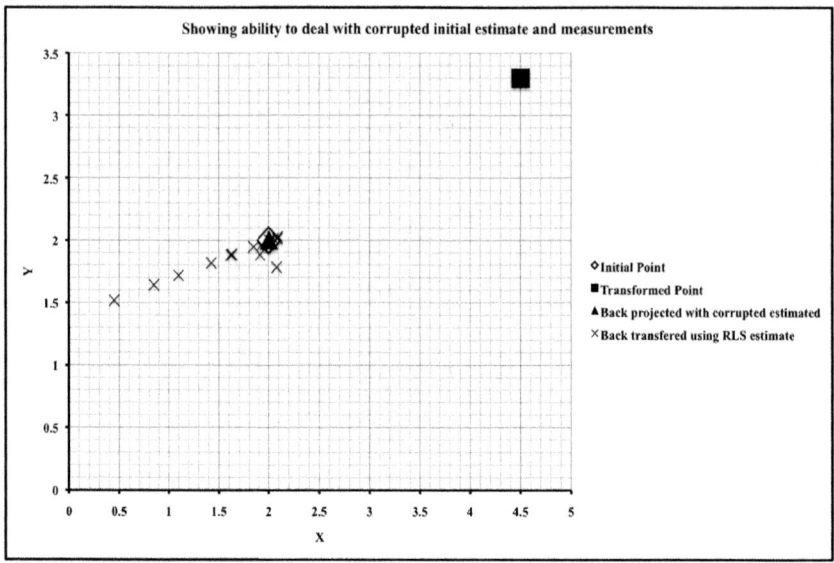

Fig. 2. Results from an implementation of the RLS algorithm to synthetic data. Note initial estimate is made using 5 corrupted measurement points transformed using the ground truth homography.

on the "Lavenberg Marquadt" algorithm **(LM)** [4]. The cost function to be minimised is the symmetric transfer error given by,

$$\sum_i d\left(x_i, \mathbf{H}^{-1}\acute{x}_i\right)^2 + d\left(\acute{x}_i, \mathbf{H}x_i\right)^2. \tag{9}$$

[4] provides a good account on various cost functions that can be used and states certain guidelines and drawbacks of using a minimisation technique. **Figure-4** give the RMS average residual errors for the initial estimate, the optimised estimate and the best estimate from the RLS for image set one and two from **Figure-3**. Note that the error given for the RLS is from the measurement where the error is minimal. The figure reveals that optimising **H** only slightly betters the initial estimate in terms of reducing the residual error. When compared with the the RLS mixed results are seen: for image set no.1(part a) the error is less compared to the minimised estimate whilst the error is slightly more for image set no.2 (part b). Note that the initial estimate for image set no.1 is quite bad and the optimisation technique does not significantly improve on the estimate where as for data set no.2 the initial estimate is rather good and the LM optimisation further improves on it. This implies that a good initial estimate is required to get the best out of such a technique. The RLS on the other hand does not require that the estimate be close to optimum: a corrupted estimate can be dealt with, though up to an extent as shown.

(a) Image data set no.1

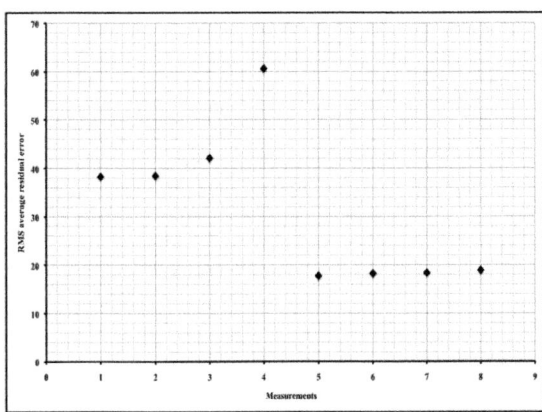

(b) Showing average residual RMS error for each measurement

(c) Image data set no.2

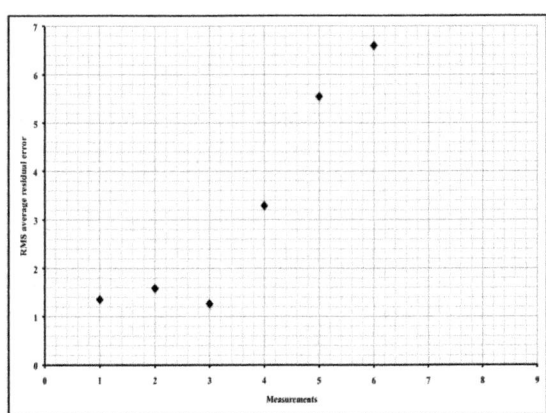

(d) Showing average residual RMS error for each measurement

Fig. 3. Results from an implementation of the RLS algorithm to real image data. The horizontal axis gives the number of iterations/measurements and the vertical axis gives the RMS average residual error.

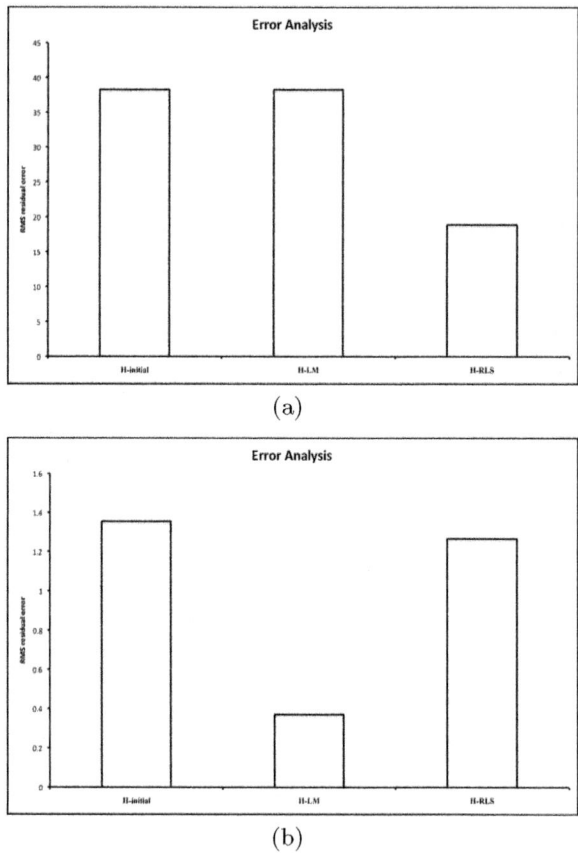

(a)

(b)

Fig. 4. Comparing initial, optimised and best RLS estimate. The first data set on the horizontal axis is the initial estimate, the second the optimised estimate using LM and the last is the best RLS estimate. The vertical axis gives the average RMS residual error. (part a is image set 1 part b is image set 2)

4   Summary

Presented is a recursive solution for robust planar homography estimation. The proposed methodology starts with a fast and reliable match making algorithm derived from the Harris feature detector and the 3D intensity based histograms proposed in the SIFT algorithm. These are then matched using the $\mathbf{L_2}$ and $\mathbf{L_\infty}$ and compared to classical CC techniques which show that the matching making technique is faster if not more reliable, in particular $\mathbf{L_\infty}$ based matching. Further tuning of the thresholding value though can result in even faster and more reliable matches, however the problem of losing good matches is likely to precipitate. Not a significant drawback as the minimum number of features requirement is adequately met. The RLS technique for homography update with incoming noisy measurements show fairly decent results: the experiments performed in real data do show a betterment of the initial estimate, though the algorithm is prone to diverging either before minimum error is reached or after it. A solution is to compare the iterative estimate to the previous one and if better then keep it and if vice versa then discard it. A partial analysis on optimising the initial estimate using the Lavenberg Marquadt algorithm revealed that an improvement was seen when the initial estimate was acceptable but not when the initial estimate was corrupted.

References

1. System identification: Theory for the user, 2nd edn. Prentice Hall PTR, Englewood Cliffs (1999)
2. Brown, L.: A survey of image registration techniques. ACM Computing Surveys 24, 325–376 (1992)
3. Haris, C., Stephens, M.: A combined corner and edge detector. In: Alvey Vision Conference, pp. 147–151 (1998)
4. Hartley, R., Zisserman, A.: Multiple view geometry, 2nd edn. Cambridge University Press, Cambridge (2003)
5. Hoshino, J., Kourogi, M.: Fast panoramic image mosaicing using 1d flow estimation. Real-Time Imaging 8, 95–103 (2002)
6. Lowe, D.: Distinctive image features from scale-invariant keypoints. International Journal for Computer Vision 2, 91–110 (2004)
7. Mikolajczyk, K., Shmid, C.: A performance evaluation of local descriptors. Pattern Analysis and Machine Intelligence 27 (2005)
8. Nemra, A., Aouf, N.: Robust feature extraction and correspondence for uav map building. In: 17th MED Conference (2009)
9. Ping, Y., Zheng, M., Anjie, G., Feng, Q.: Video image mosaics in real-time based on sift. In: Conference on Pervasive Computing Signal Processing and Applications (2002)
10. Santos, C., Stoeter, S., Rybski, P., Papanikolopoulos, N.: Mosaicking images: Panoramic imaging for miniature robots. IEEE Robotics and Automation Magazine, 62–68 (2004)
11. Saptharishi, M., Oliver, C., Diehl, C., Bhat, K., Dolan, J., Ollennu, A., Khosla, P.: Distributed surveillance and reconissaince using multiple autonomous atvs: Cyberscout. IEEE Robotics and Automation Magazine 18 (2002)
12. Schmid, C., Mohr, R., Bauckhage, C.: Evaluation of interest point detectors. Internation Journal of Computer Vision 37(151-172) (2000)
13. Zhang, Z., Deriche, R., Faugeras, O., Luong, Q.: A robust technique for matching two uncalibrated images through the recovery of the unknown epipolar geometry. Artificial Intelligence 78, 87–119 (1995)
14. Zitova, B., Flusser, J.: Image registration methods: a survey. Image and Vision Computing 21, 0–977 (2003)

Airborne Ultrasonic Position and Velocity Measurement Using Two Cycles of Linear-Period-Modulated Signal

Shinya Saito[1], Minoru Kuribayashi Kurosawa[1],
Yuichiro Orino[1], and Shinnosuke Hirata[2]

[1] Department of Information Processing, Interdisciplinary Graduate School of
Science and Engineering, Tokyo Institute of Technology, Tokyo, Japan
saitou.s.aa@m.titech.ac.jp, mkur@ip.titech.ac.jp,
orino.y.aa@m.titech.ac.jp
[2] The University of Electro-Communications, Tokyo, Japan
hirata@mce.uec.ac.jp

Abstract. Real-time position and velocity measurement of a moving object with high accuracy and resolution using an airborne ultrasonic wave is difficult due to the influence of the Doppler effect or the limit of the calculation cost of signal processing. The calculation cost had been reduced by single-bit processing and pulse compression using two cycles of linear-period-modulated (LPM) signal. In this paper, accuracy of the ultrasonic two-dimensional position and velocity vector measurement of the proposed method using two microphones is evaluated by experiments.

Keywords: ultrasonic position and velocity measurement, pulse compression, linear-period-modulation, single-bit signal.

1 Introduction

Acoustic sensing systems are used in many industrial applications due to advantages of acoustic sensors, their low-purchase cost, small size, and simple hardware. Method of airborne ultrasonic measurement are widely researched [4][5]. The pulse-echo method is one of the typical methods of ultrasonic distance measurement. The pulse-echo method is based on determination of the time-of-flight (TOF) of an echo reflected from an object [8]. Pulse compression has been introduced in the pulse-echo method for improvement of the signal-to-noise ratio (SNR) of the reflected echo and distance resolution [7].

A linear-frequency-modulated (LFM) signal is used in the pulse-echo method. The frequency of LFM signal linearly sweeps with time. A received signal is correlated with a reference signal which is the transmitted LFM signal. The TOF of the transmitted signal is estimated by the maximum peak in the cross-correlation function of received signal and reference signal. The signal processing for cross-correlation consists of huge iterations of multiplications and accumulations. Therefore, real-time ultrasonic measurement is difficult because of the high cost in digital signal process.

R. Groß et al. (Eds.): TAROS 2011, LNAI 6856, pp. 46–53, 2011.
© Springer-Verlag Berlin Heidelberg 2011

So a signal process method using a delta-sigma modulated single-bit digital signal has been proposed to reduce the calculation cost of cross-correlation. The proposed signal processing consists of a recursive cross-correlation by single-bit signals of and smoothing operation by a moving average filter. The calculation cost of cross correlation is reduced by the recursive cross-correlation operation of single-bit signals [2].

In the case of a moving object, the reflected signal is modulated due to the Doppler effect caused by the object motion. The linear shift of the signal period means the hyperbolic shift of the frequency. Therefore, a Doppler-shift LFM signal cannot be correlated with a reference LFM signal. So pulse compression using a linear-period-modulated (LPM) signal has been introduced for ultrasonic measurement of a moving object [6][1]. The signal period of the LPM signal linearly sweeps with time. Thus, a Doppler-shift LPM signal can be correlated with a reference LPM signal.

However, the cross-correlation function of the Doppler-shift LPM signal and a reference LPM signal is also modulated due to Doppler effect. The method of typical Doppler-shift compensation is high cost using the envelop but Doppler-shift compensation is required. A low-calculation-cost method of ultrasonic measurement by pulse compression using two cycles of LPM signal and Doppler-shift compensation has been already proposed [3].

Ultrasonic distance and velocity measurement have been already achieved by using one microphone[3]. In this paper, method of two-dimensional position and velocity vector measurement are considered by using two microphones, and accuracy of the two-dimensional position and velocity vector measurement are evaluated by experiments.

2   Cross Correlation by Single-Bit Processing

The proposed method of ultrasonic distance and velocity measurement by pulse compression using two cycles of LPM signal is illustrated in Fig. 1 [2]. In the proposed method, two cycles of LPM signal are transmitted by a loudspeaker. The received signal of one microphone is converted into a single-bit received signal by a delta-sigma modulator. The single reference LPM signal is converted into a single-bit reference signal of N samples by a digital comparator.

The cross-correlation function $c_1(t)$ of the received signal $x_1(t)$ and the reference signal $h_1(i)$ is expressed as

$$c_1(t) = \sum_{i=0}^{N-1} h_1(N-i) \cdot x_1(t-i) \tag{1}$$

The calculation of the cross-correlation operation of Eq. (1)requires huge numbers N of multiplica-tions and accumulations of single-bit samples.

The difference of the cross-correlation function, $C_1(t) - C_1(t-1)$, is expressed as

$$c_1(t) - c_1(t-1) = h_1(N) \cdot x_1(t) - h_1(1) \cdot x_1(t-N)$$

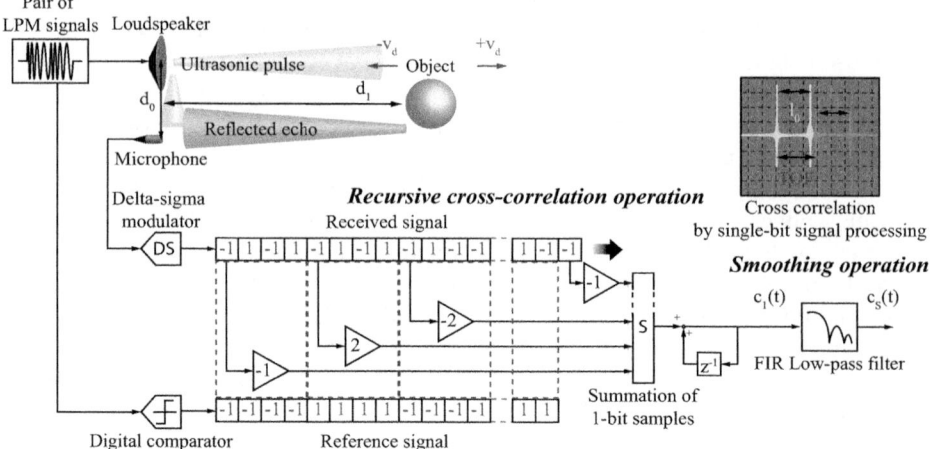

Fig. 1. The singal processing of ultrasonic position and velocity measurement by pulse compression using two cycles of LPM signal

$$+ \sum_{i=1}^{N-1} \{h_1(N-i) - h_1(N-i+1)\} \cdot x_1(t-i) \qquad (2)$$

The values of $h_1(1)$ and $h_1(N)$ are 1 and -1 respectively. Furthermore, $h_1(i)$ has several hundreded zero-cross points Z_i. The values of $h_1(N-1) - h_1(N-i+1)$ are expressed as

$$h_1(N-i) - h_1(N-i-1) = \begin{cases} 2, & \cdots N-i = Z_{2m-1}. \\ -2, & \cdots N-i = Z_{2m}. \\ 0, & \cdots \quad N-i \neq Z_i. \end{cases} \qquad (3)$$

where m is a natural number. The calculation of the recursive cross-correlation operation, which is performed by integrating the difference of the cross-correlation function, is expressed as

$$c_1(t) = c_1(t-1) - x_1(t-N) + 2 \cdot x_1(t-N+Z_1) \\ -2 \cdot x_1(t-N+Z_2) + \cdots - x_1(t) \qquad (4)$$

The calculation cost of the recursive cross-correlation operation is integration and summations of single-bit samples. The number of summations Z_i+2 only depends on the number of zero-cross points in the transmitted LPM signal.

3   The Method of Position and Velocity Measurement

3.1   Two-Dimensional Position Measurement

The TOF of an echo is estimated from the cross-correlation function. The arrangement of microphones, the loudspeaker, and the object is shown in Fig. 2. In

Fig. 2, d is a distance from the loudspeaker to the object, θ is an angle between the loudspeaker and the object, and L is a distance from the loudspeaker to microphones. The TOF of the microphone-1 and the microphone-2 have usually different values of TOF_1 and TOF_2 respectively. When d is much larger than $2L$, θ is

$$\theta = sin^{-1}\frac{(TOF_1 - TOF_2)c}{2L} \tag{5}$$

where c is the propagation velocity of an ultrasonic wave. Using the angle θ, the distance d is simply derivated by geometric calculation.

3.2   Two-Dimensional Velocity Vector Measurement

The method of the velocity vector measurement is illustrated in Fig. 3 and Fig. 4. In this method, the doppler velocity can be estimated from the signal length of the transmitted signal and the echo. The signal length difference is in proportion to the velocity of the object as shown in Fig. 3. The signal length is detected from the interval of the cross-correlation function peaks by two cycles LPM signals.

The measured velocities are the vector components of the ultrasonic propagation direction, which is v_1 and v_2 in Fig. 4, in the proposed method of ultrasonic measurement by pulse compression using two cycles of LPM signal. The measured velocity v_1 is a component of the object velocity v; the direction of v_1 is estimated from the geometrical relation between the distance d, the angle θ and the space L. Namely, from the measurement using the microphone-1, one velocity component of v is obtained correspondence to the direction of v_1. Similarly,

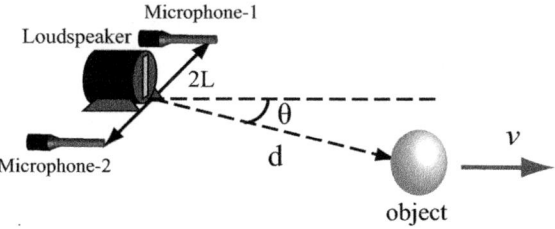

Fig. 2. The method of ultrasonic position measurement

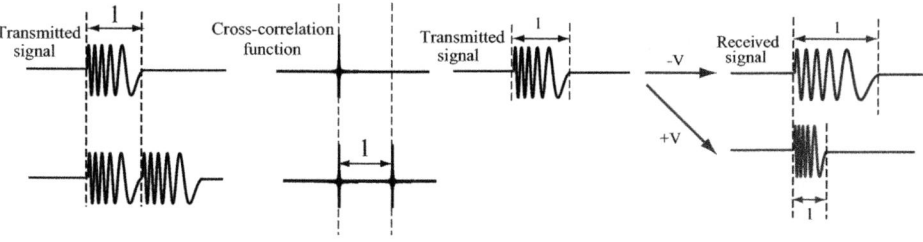

Fig. 3. The change of the signal length

another velocity component of the object velocity v, namely v_2, is acquired by using the microphone-2. Now, taken the vectors which is normal to the v_1 and v_2 respectively, the vectors intersect at one point. The velocity v is estimated from intersection point drawn in Fig. 4.

4   Experiment

4.1   Experimental Setup

The experimental setup for the ultrasonic two-dimensional position and velocity vector measurement is illustrated in Fig. 5. In the experiment, the period of the transmitted LPM signal linearly swept from 20 μs to 50 μs. The length of the transmitted LPM signal was 3.274 ms, the driving voltage of the loudspeaker was $2V_{p-p}$. The LPM signal was generated from the function generator and amplified by an amplifier. Two cycles of the LPM signal were transmitted by the loudspeaker, and the echo from the object was detected by two microphones. The distance from the loudspeaker to the microphones were 0.09 m. The propagation velocity of an ultrasonic wave in the air was approximately 348.8 m/s at 27.0°C. The received signals by the microphones were converted into the single-bit delta-sigma modulated signals. The sampling frequency of the delta-sigma modulator was 12.5 MHz.

The received signals were correlated with the single reference signal using MATLAB on the computer. The reference signal was simply converted into a single-bit signal by the digital comparator. The cross-correlation function of the received signal and the reference signal was obtained from a recursive cross-correlation operation of single-bit signals and a smoothing operation by a weighted moving average filter. The length of the filter was 141 taps.

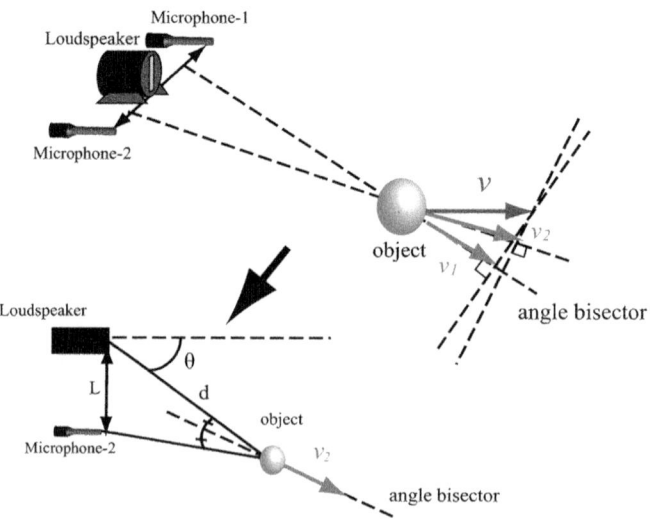

Fig. 4. The method of ultrasonic velocity measurement

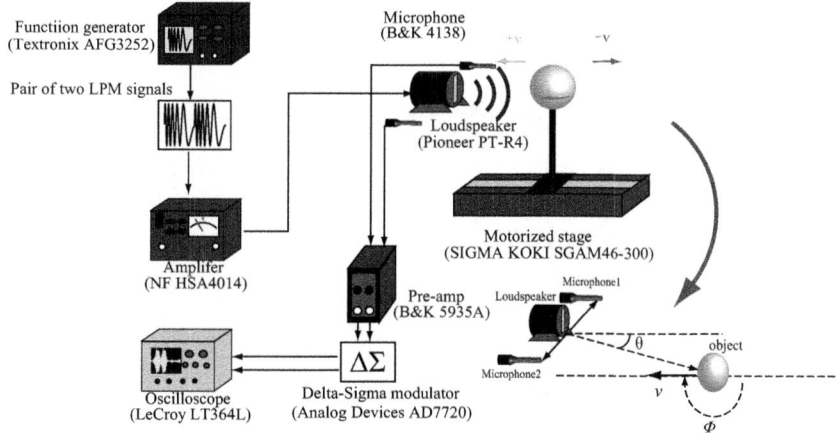

Fig. 5. Experimental setup of ultrasonic position and velocity measurement

Accuracy of the position and velocity measurement was evaluated by experiments. The object was a plastic ball whose diameter was 17 cm. The object distance d was 1.05 m and the angle θ was 20° to the object when two cycles of the LPM signal was transmitted. The velocity of the object was 0.4 m/s, the direction of movement was normal to a straight line that links the microphones and the loudspeaker, $\Phi = 180°$ in Fig. 5. The measurement was executed 150 times.

4.2   Experimental Result

The cross-correlation function was obtained by experiments as shown in Fig. 6. The first two peaks of cross-correlation function were caused by the waves which the microphone received from the loudspeaker directly. The second two peaks of cross-correlation function were caused by the echo from the object. The peak of the cross-correlation function was detected by limiting the time range of cross-correlation function and calculating the maximum point in the limited time range.

The probability distribution of the distance and the angle are illustrated in Fig. 7. Average and standard deviation of the distance and angle were 1.066 m and 73.8 μm, 19.5° and 0.09° respectively. The average of distance and angle were a little different from the setup value, but it is inevitable that errors are observed about 2 to 3 cm and 1 to 2° when the object's position was set. Given that, it can be said that high accuracy of the distance and angle by the proposed method is demonstrated by experiments of the ultrasonic two-dimensional position measurement.

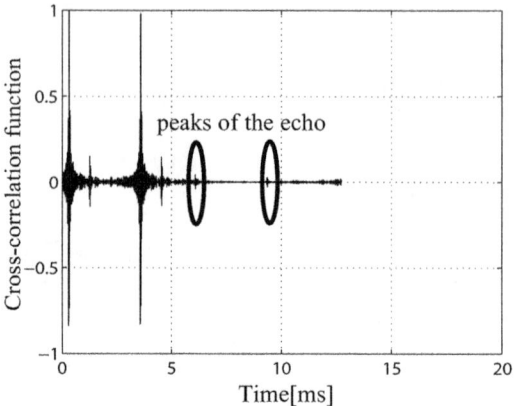

Fig. 6. Cross-correlation function

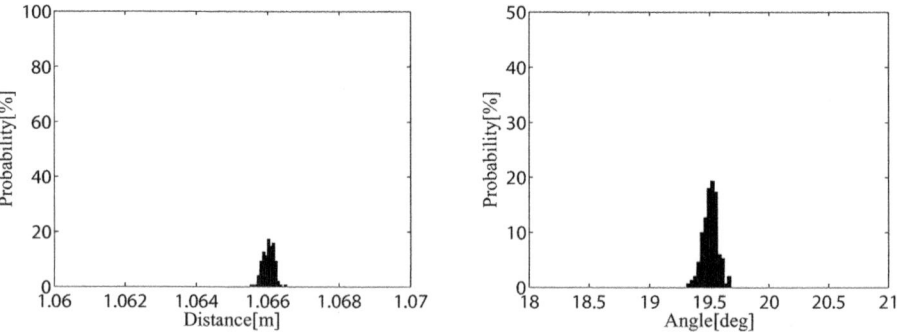

Fig. 7. The probablity distributions of the position, distance(left) and angle(right)

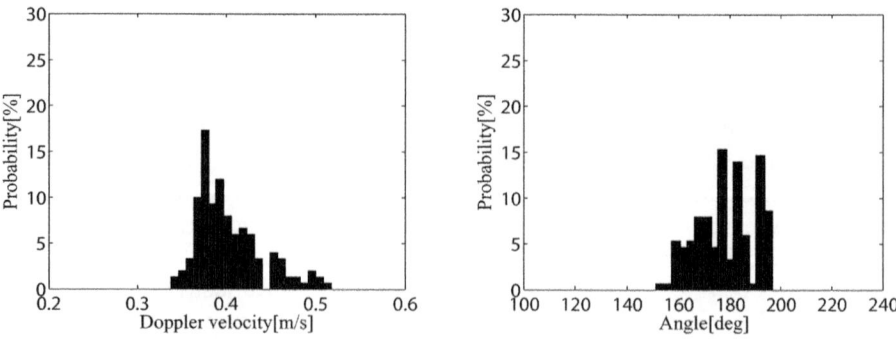

Fig. 8. The probablity distributions of the velocity(left) and direction angle(right)

The probability distribution of velocity is also illustrated in Fig. 8. Average and standard deviation of the velocity and the direction angle of movement were 0.403 m/s and 0.038 m/s, 178.3° and 10.9° respectively. Average of the velocity and direction of movement were near to setup value, but standard deviations of the velocity and the angle were a little large. The approximate value of the object's velocity vector can be estimated, but the measuring accuracy needs to be improved slightly.

5   Conclusion

Accuracy of airborne ultrasonic two-dimensional position and velocity vector measurement using two microphones with two cycles of LPM signal was examined by experiments. Experiments were executed using a recursive cross-correlation by single-bit signals and a low-calculation-cost method of Doppler-shift compensation. High accuracy regarding the ultrasonic position measurement was obtained. On the other hand, there was room for improvement regarding velocity vector measurement but the approximate value of the object's velocity can be estimated. Now a recursive cross-correlation is being implemented in FPGA because ultrasonic two-dimensional position and velocity measurement was not real-time in these experiments. For the future work, multi moving object in the area will be considered.

References

1. Altes, R.A., Skinner, D.P.: Sonar-velocity resolution with a linear-period-modulated pulse. J. Acoust. Soc. Am. 61(4), 1019–1030 (1977)
2. Hirata, S., Kurosawa, M.K., Katagiri, T.: Cross-correlation by single-bit signal processing for ultrasonic distance measurement. IEICE Trans. Fundam. E91(A), 1031–1037 (2008)
3. Hirata, S., Kurosawa, M.K., Katagiri, T.: Ultrasonic distance and velocity measurement by low-calculation-cost doppler-shift compensation and high-resolution doppler velocity estimation with wide measurement range. Acoustical Science and Technology 30(3), 220–223 (2009)
4. Jorg, K.W., Berg, M.: Sophisticated mobile robot sonar sensing with pseudo-random codes. Robotics and Autonomous Systems 25(3), 241–251 (1998)
5. Klahold, J., Rautenberg, J., Ruckert, U.: Continuous sonar sensing for mobile mini-robots. In: Proc. the 2002 IEEE International Conference on Robotics and Automation, vol. 1, pp. 323–328 (2002)
6. Kroszczynski, J.J.: Pulse compression by means of linear-period modulation. Proc. of the IEEE 57(7), 1260–1266 (1969)
7. Marioli, D., Narduzzi, C., Offelli, C., Petri, D., Sardini, E., Taroni, A.: Digital time-of-flight measurement for ultrasonic sensors. IEEE Trans. Instrumentation and Measurement 41(1), 93–97 (1992)
8. Marioli, D., Sardini, E., Taroni, A.: Ultrasonic distance measurement for linear and angular position control. IEEE Trans. Instrumentation and Measurement 37(4), 578–581 (1988)

An Eye Detection and Localization System for Natural Human and Robot Interaction without Face Detection

Xinguo Yu[1], Weicheng Han[2], Liyuan Li[1], Ji Yu Shi[1], and Gang Wang[1]

[1] Institute for Infocomm Research, Connexis, Singapore 138632
{xinguo,lyli,jyshi,gswang}@i2r.a-star.edu.sg
[2] Department of Electrical and Computer Engineering,
National University of Singapore, Singapore 117543
calayanrail@gmail.com

Abstract. There were many eye localization algorithms *depending on face detection* in the literature. Differently this paper presents a novel eye detection and localization system *not depending on face detection* for natural human and robot interaction using both stereo and visual cameras. To build a robust system we use stereo and visual cameras in synergy. The stereo camera is used to localize the head of the person to replace face detection. Then our eye identification algorithm detects and localizes two eyes inside head box. In eye detection step, our algorithm uses a HOG-moment (*Histogram Of Gradient*) feature to detect two eyes inside the head box. In eye localization step, we employ an iterative procedure to search the best location for eye pair. The experimental results show that the proposed eye detection and localization algorithm, not depending on face detection, has a similar robustness as the existing eye localization algorithms.

Keywords: Eye Detection and Localization, HOG, Face Detection, Disparity Image, Human and Robot Interaction.

1 Introduction

Eye detection and localization has been an important research problem in the past decades due to it is a key step in a variety of applications. In natural human and robot interaction eye identification is the key technology of eye contact between human and robot. The result of eye identification can facilitate to find out the direction of gaze. In the facial feature extraction two eyes can be located first thanks to their salience. And with the helpful innate geometrical constraint the accurate eye centers can facilitate the extraction of other facial features by providing good location estimation of other features [7-8]. Facial feature extraction is the essential step in multiple applications such as face tracking, face recognition, facial expression recognition, and human computer interface [22-23]. The facial feature extraction is to localize the facial components of interest including eyes, nose, and mouth and to estimate their scale. The accuracy of facial feature localization has big impact on the performance of the applications such as face recognition and facial expression recognition using locations of facial features as their input [22].

R. Groß et al. (Eds.): TAROS 2011, LNAI 6856, pp. 54–65, 2011.
© Springer-Verlag Berlin Heidelberg 2011

The algorithms of eye localization in the literature can be divided into active and passive two approaches. The active approach localize eyes from the images taken by near infrared (NIR) cameras [1-2], whereas the passive approach localizes eye from the images taken by visual cameras [3-22]. The principle of the former approach is red-eye effect in flash photographs, utilizing special IR illuminators and IR-sensitive CCD for imaging. In the indoor and relatively controlled conditions the spectral properties of the pupil under NIR illumination provide a very clean signal. Hence the algorithms in this approach are relatively simple and fast. And they can achieve high accuracy in eye localization when the required special conditions are met. The most significant conditions in this category are a relatively stable lighting condition, a camera set close to the subject, and open eyes [1-2].

Localizing two eyes from the visual images is more challenging than from the NIR images. In the literature there were more research efforts in this approach due to its wide variety of applications and its challenges and. These algorithms used various methods such as template matching [14], rule-based [17, 21], model-based [4, 16], feature based [5, 8, 12, 17], and hybrid [10]. In the template algorithms, multiple templates are created for searching eye pair [14]. Templates are used under certain order and rules. A number of templates are required to cope with the variety of the conditions of taking images and the variety of eye pairs. In the rule-based algorithms, several functions or transforms are used to compute several features and then rules are applied to these features to obtain the eye locations of two eyes. For instance, authors in [17, 21] used the projection functions to obtain the features, whereas authors in [3] used the radial symmetry transform to obtain the features. Authors in [8] proposed an algorithm based on the assumption that eye center is the center of isohphote curvature. However, the rules based on the prior knowledge are not easily decided. This method also has difficulty in finding features that can cope with the different conditions. In the model-based algorithms, single or multiple models are created to capture the eye characters. A model comprises of a set of equations including the eye location and some other variables of describing the considering region. The goal is to find the optimal solution of the model [4, 16]. In this method, the main challenge is that it is difficulty to define single or multiple models capable of capturing the variety of conditions. Finally, feature based algorithms used a trained learning machine to classify each region whether is an eye or an eye pair based on low-level feature. Some of examples of features include wavelets, Gabor feature, Haar-like feature, and HOG feature [8, 12]. HOG (Histogram Of Gradient) is one of effective methods in object detection and recognition. Monzo et al [12] used a HOG to model two eyes and a face together. The algorithm in [12] first detects the face region and then it employs Haar feature to obtain the eye candidates. The last step is to use HOG model to evaluate each pair of eyes and pick the best location of the pair of eyes.

Though there were many eye localization algorithms in the literature, they need some further work for building the system for real applications. The existing algorithms still have two issues. First, they take the face box as their input and assume that the face box is perfectly performed. The fact is that face detection still can not achieve the perfect performance, especially for occluded, makeup, camouflage faces. Second, they are not robust for the uncontrolled open environment. For example, they are not robust to locate the eyes under the scenario of natural human and robot interaction. In this paper, we develop a robust and fully automatic eye localization system based on the synergy of stereo and visual cameras. This system uses a procedure to replace *face detection* in the existing eye localization algorithms. Compared with face detection, this procedure has multiple merits. First, it can robustly locate the head for a wide range because the stereo can have a wide view. Second, it is much faster than face detection due to it is a shape analysis procedure. It is another critical merit for building the real system. Third, it still can locate the head when faces are of camouflage, makeup, or partial occluded. Fourth, it can know which head is closer to the robot.

The rest of the paper is organized as follows. Section 2 gives the overview of the proposed eye localization system. Section 3 and 4 describe the procedures of eye detection and eye localization respectively. Section 5 presents the evaluation of the algorithm. We conclude the paper in Section 6.

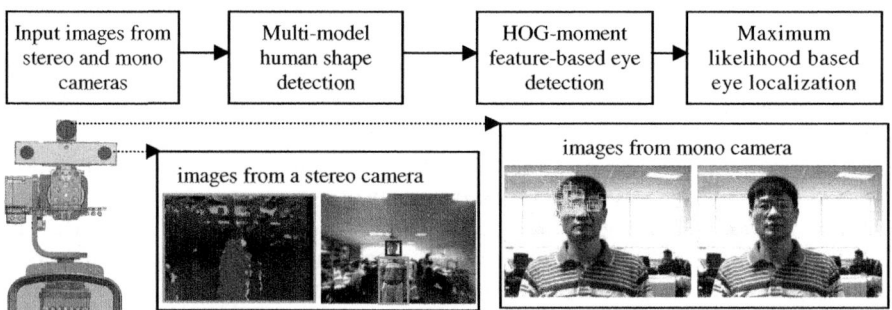

Fig. 1. Flowchart of our eye detection and localization system and the layout of stereo and mono cameras on our robot

2   Overview of the Eye Detection and Localization System

Here we give the overview of our eye detection and localization system, which is a robust and fully automatic eye localization system of using both stereo and mono cameras in synergy. As depicted in Fig 1, the system comprises of three main components: head localization, eye detection and eye localization. The component of head localization works on the images by the stereo camera. It first finds the person close to robot. Then it localizes person's head and converts the head box in the stereo-camera image into the one in mono-camera image. This component replaces the face detection in the existing eye localization algorithms in the literature [3-22]. The other two components actually are an eye detection and localization algorithm from the visual images with the known head box. Eye detection component is to obtain the eye candidates, which are the regions in certain size that probably contain an eye. Our eye candidate detection employs a scan procedure that scan the full image for all the regions. For each region, we use a SVM (Support Vector Machine) on the HOG-moment feature to evaluate whether it is an eye candidate. HOG-moment vector is the concatenation of the HOG vector and moment vector of this region. HOG (Histogram of Gradient) is a robust object detection technique and it especially has good performance in human detection. However, HOG can be fooled by some objects that have the similar edge distributions to eye. We complement HOG with moment vector.

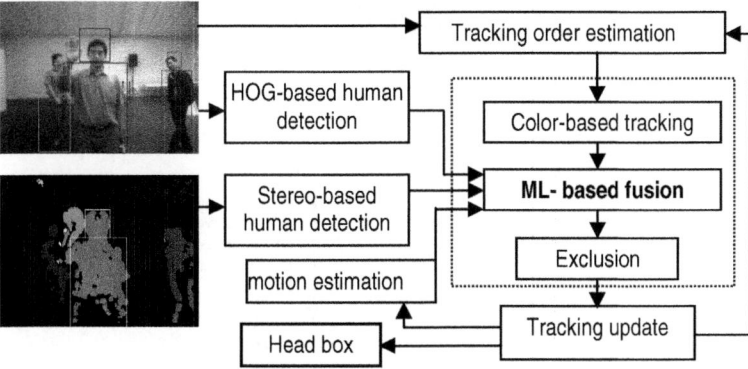

Fig. 2. The block diagram of multi-model human body and head detection and tracking

The succeeding eye localization component is to identify the eye pair from the eye candidates and find the accurate eye centers of two eyes. We first form a formula that calculates the likelihood that two points are left and right eye centers. Thus, our goal is to find two points that have a largest likelihood value. We design an iterative procedure to achieve this goal. The initial positions of two points are crucial for obtaining the best eye locations. We define a Bayesian formula to select the best pair from eye candidates. The measure used in this search is produced by integrating the appearance measure and the distance of two eyes.

3   Head Localization by Stereo Camera

3.1   Head Localization by Stereo Camera

The block diagram of our head localization component is shown in Fig 2. The input to this component is both disparity and color images from the stereo camera on the robot head. For each frame, the blocks in the diagram are executed as follows. First, the humans in the view are detected from the disparity and color images. Meanwhile, the positions of occluded humans are predicted based on their motion. To handle possible complex occlusions, multi-person tracking is performed sequentially from the closest one to the farthest (including the fully occluded ones) to approximate a globally optimal tracking process. The order is determined by the predicted 3D positions of all the tracked humans. Then, the blocks within the dotted box of the diagram in Fig 2 are executed to track humans one by one. For each human, a mean-shift tracking is first performed, and then the new position is located in the image by the ML-based fusion (*ML is the acronym of Maximum Likelihood*). Finally, the exclusion step is performed to suppress the visual features of the tracked human in both color and disparity images. This operation is to avoid other humans being trapped in the positions of those tracked humans. When the sequential multi-person tracking is completed, the system updates the 3D positions and appearance models of the tracked humans, as well as the initializations and terminations of the tracks. The results of

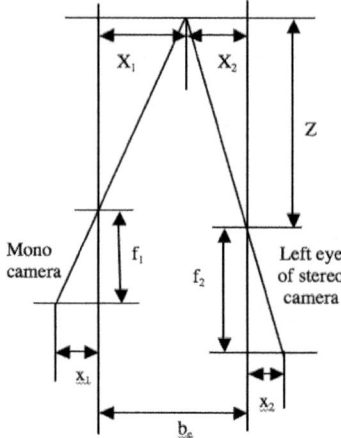

Fig. 3. The illustration of the relative relation of the variables of stereo and mono cameras

human tracking are the bounding boxes of human body and head. The following steps in this system just use the head box.

3.2 Head Location Conversion

To convert the head location in the stereo camera into the location in the mono camera we establish a mapping from mono camera to the left eye of the stereo camera. The variables $x_1, x_2, X_1, X_2, f_1, f_2, b_e, and\ Z$ are defined in the fig 3 and b_s is the baseline of the stereo camera. Then they have the following relation.

$$\begin{cases} \dfrac{x_1}{X_1} = \dfrac{f_1}{Z}, & \dfrac{x_2}{X_2} = \dfrac{f_2}{Z}, & \dfrac{x_2}{X_2} = \dfrac{f_2}{Z}, \\ x_2 = \dfrac{f_2}{Z} X_2 = \dfrac{f_2}{Z}(b_e - X_1) = -\dfrac{f_2}{f_1} x_1 + \dfrac{f_2 b_e}{Z}. \end{cases} \tag{1}$$

In stereo camera $Z = fbd^{-1}$, where d is the disparity value. Hence,

$$x_2 = -\frac{f_2}{f_1} x_1 + \frac{f_2 b_e}{f_2 b_s} d = -\frac{f_2}{f_1} x_1 + \frac{b_e}{b_s} d = k_1 x_1 + k_d d \tag{2}$$

where x_1 is the position from mono camera and x_2 is the estimated position in the color image from the left eye of the stereo camera. The coordinates should be computed with respect to the image centers, hence $x_1 = X_1 - X_{c1}$ and $x_2 = X_2 - X_{c2}$, where X_1 and X_2 are the positions in the image respect to the upper left center of the corresponding images, and X_{c1} and X_{c2} are the image centers of the corresponding images. By substituting the variables the model becomes

$$X_2 = k_1 X_1 + k_d d + C \tag{3}$$

where $C = X_{c2} - k_1 X_{c1}$ is a constant. The parameters for the model are $k_1, k_2,$ and C.

4   HOG-Moment Based Eye Detection

Eye detection is a procedure that scans all possible eye regions and evaluates each region whether it is an eye candidate using a SVM on HOG-moment feature. Given a grey image, it first estimates the eye dimension by the head size and forms five sizes of eye regions around the estimated eye dimension. For each size of region we scan the whole image to acquire the eye candidates.

4.1   HOG Feature

Histogram of oriented gradient (HOG) is an adaptation of Lowe's Scale Invariant Feature Transformation (SIFT) approach. A HOG feature is created by first computing the gradient magnitude and orientation at each image sample point in a window (*or region*) around an anchor point. The window is divided into a W x H cells. An orientation histogram for each cell is then formed by accumulating samples within the cell, weighted by gradient magnitude. Concatenating the histograms from all the cells forms the final HOG feature vector. A W x H cells window was used to scan the image with step length at S_w pixels in horizontal and S_h pixels in vertical.

The ratio between W and H is decided by object shape, which can be acquired by doing statistics on the annotated samples. And the values of W and H are chosen by considering accuracy and computation time. Each cell block in the window covers M x M pixels. Different values of M are used to detect the same target in different scales. In this way, HOG can handle the scale variance of object in images. In a cell block, the orientation of gradient of each pixel is classified into K bins. K bins evenly divide from 90 degree to -90 degree, i.e. each bin spans 180/K degrees. The value linking to a bin is in the interval [0.0, 1.0], which represents the ratio of pixels belongs to the bin. Thus, W x H x K bins are used for an anchor position. In this paper, W=4, H=3, K=9, S_w =W/2, S_h =H/2. The values of M are 16 + j*2 for j = 0 to 4 in this paper.

4.2   Moment Feature

Eye region has a distinctive gray value distribution pattern, *i.e.* the dark center iris, surrounding white sclera, then upper and lower eyelids, face skin, and eyebrow. This pattern is reflected by the spatial intensity pattern in gray eye images. This paper proposes a method to extract the moment feature that targets to capture this pattern. Let $B = (W, H)$ be the window of a normalized intensity image of an eye I(x) centered at $x_c = (x_c, y_c)$. Let us denote g_{iris} as the brightness value of the iris, which is selected as the minimum intensity value from the core center of 5 x 5 window centered at x_c. Then, the moments of up to the third order which characterize the spatial intensity variations related to the iris can be defined as

$$m_{ij} = \frac{1}{W^{i+1}H^{j+1}} \sum (x - x_c)^i (y - y_c)^j (I(x, y) - g_{iris}). \qquad (4)$$

We obtain ten items when we limit to $0 \le i, j \le 3$ and $i + j \le 3$. Besides the moment feature the spatial intensity distribution is also employed to characterize the eye graph pattern. We divide the window B into 3x3 grids, where the center block contains the iris and the other blocks cover the surrounding regions. The average of the intensities related to the iris brightness for each block is computed as

$$n_{kl} = \frac{9}{WH} \sum_{I(x,y) \in b_{kl}} (I(x,y) - g_{iris})$$ (5)

Where $k,l = 1, 2, 3$ and b_{kl} is the block at the kth row and lth column of the grid. These values characterize how the brightness spatially changes related to the iris. The average of each block can be computed from the integral image, so that the computational cost is very low when scanning the detection window over an image.

By combining (4) and (5) a vector of 19 dimensions can be obtained $v = (m_{00}, m_{10}, m_{01}, m_{20}, m_{02}, m_{11}, m_{03}, m_{30}, m_{12}, m_{21}, n_{11}, n_{12}, n_{13}, n_{21}, n_{22}, n_{23}, n_{31}, n_{32}, , n_{33})$. We still call it the moment feature, though it is not a pure moment feature in this paper.

5   Maximum Likelihood Based Eye Localization

5.1   Formulation

Given the eye candidates detected in eye detection step, the eye localization is to find the ideal positions of the left and right eyes (*i.e.*, $\mathbf{x}_l$ and $\mathbf{x}_r$ where $\mathbf{x} = (x, y)$) starting with these eye candidates. Let us denote the N detections as B= $\{B_i : i=1,2,...,N\}$ where B_i is the box of the ith detection centered at x_i. Assuming that the detections are independent each other, using Bayes' theorem, the likelihood of that the left eye is located at x_l can be expressed as

$$L(\mathbf{x}_l \mid \mathbf{B}) = \frac{\sum_{i=1}^{N} \pi_i^l P(\mathbf{x}_l \mid B_i) P(B_i)}{P(\mathbf{x}_l)} \propto \sum_{i=1}^{N} \pi_i^l P(\mathbf{x}_l \mid B_i) P(B_i)$$ (6)

where π_i^l denotes the association of the ith detection with the left eye, $P(\mathbf{x}_l \mid B_i)$ is the conditional probability of $\mathbf{x}_l$ being the ideal position of the left eye given the detection B_i and $P(B_i)$ is the confidence of the ith detection. Similarly, one can express the likelihood of the right eye position as

$$L(\mathbf{x}_r \mid \mathbf{B}) = \frac{\sum_{i=1}^{N} \pi_i^l P(\mathbf{x}_r \mid B_i) P(B_i)}{P(\mathbf{x}_r)} \propto \sum_{i=1}^{N} \pi_i^r P(\mathbf{x}_r \mid B_i) P(B_i)$$ (7)

When the probability measures follow simple Gibbs distributions, the conditional probabilities $P(\mathbf{x}_l \mid B_i)$ and $P(\mathbf{x}_r \mid B_i)$ can be written as

$$P(\mathbf{x}_l \mid B_i) = e^{-\frac{|\mathbf{x}_l - \mathbf{x}_i|^2}{\sigma_d^2}} \quad \text{and} \quad P(\mathbf{x}_r \mid B_i) = e^{-\frac{|\mathbf{x}_r - \mathbf{x}_i|^2}{\sigma_d^2}}$$ (8)

where σ_d is the spatial variance of the eye centers which is determined according to the statistics of human faces.

If $\mathbf{x}_l$ and $\mathbf{x}_r$ are the ideal positions of the two eyes, they should be separated at an interocular distance. Hence, the probability of that $\mathbf{x}_l$ and $\mathbf{x}_r$ represent a pair of eyes can be expressed as

$$P_{eye-pair}(\mathbf{x}_l,\mathbf{x}_r) = e^{-\frac{|(\mathbf{x}_r-\mathbf{x}_l)-\mathbf{d}_e|^2}{\sigma_e^2}} \tag{9}$$

where $\mathbf{d}_e = (d_x, d_y)$ is the distance vector between the two eyes, and σ_e denotes the variances of inter-ocular distances. They can be chosen according to the statistics of human faces. For an upright face, d_x equals the inter-ocular distance because $d_y \approx 0$. The vector $\mathbf{d}_e$ also indicates the pose of a front face.

Combining (6) to (9), the likelihood probability of the positions of the two eyes can be defined as

$$P_{eye}(\mathbf{x}_l,\mathbf{x}_r) = L(\mathbf{x}_l\,|\,\mathbf{B})L(\mathbf{x}_r\,|\,\mathbf{B})P_{eye-pair}(\mathbf{x}_l,\mathbf{x}_r) \propto$$

$$\left(\sum_{i=1}^{N}\pi_i^l e^{-\frac{|\mathbf{x}_l-\mathbf{x}_i|^2}{\sigma_d^2}}P(B_i)\right)\left(\sum_{i=1}^{N}\pi_i^r e^{-\frac{|\mathbf{x}_r-\mathbf{x}_i|^2}{\sigma_d^2}}P(B_i)\right)e^{-\frac{|(\mathbf{x}_r-\mathbf{x}_l)-\mathbf{d}_e|^2}{\sigma_e^2}} \tag{10}$$

The eye localization is to find the positions $\mathbf{x}_l$ and $\mathbf{x}_r$ that maximize the likelihood probability. Ideally, if $\mathbf{x}_l$ and $\mathbf{x}_r$ are the true positions of the left and right eyes in the face image, the likelihood reaches its maximum at the positions with $\dfrac{\partial P_{eye}(\mathbf{x}_l,\mathbf{x}_r)}{\partial \mathbf{x}_l} = 0$ and $\dfrac{\partial P_{eye}(\mathbf{x}_l,\mathbf{x}_r)}{\partial \mathbf{x}_r} = 0$. From (10), we can obtain

$$\begin{cases} \mathbf{x}_l = \dfrac{\sigma_e^2\sum_{i=1}^{N}\mathbf{x}_i\pi_i^l P(\mathbf{x}_l\,|\,B_i)P(B_i)}{(\sigma_d^2+\sigma_e^2)\,L(\mathbf{x}_l\,|\,\mathbf{B})} + \dfrac{\sigma_d^2(\mathbf{x}_r-\mathbf{d}_e)}{\sigma_d^2+\sigma_e^2} \\[3mm] \mathbf{x}_r = \dfrac{\sigma_e^2\sum_{i=1}^{N}\mathbf{x}_i\pi_i^l P(\mathbf{x}_r\,|\,B_i)P(B_i)}{(\sigma_d^2+\sigma_e^2)\,L(\mathbf{x}_r\,|\,\mathbf{B})} + \dfrac{\sigma_d^2(\mathbf{x}_l-\mathbf{d}_e)}{\sigma_d^2+\sigma_e^2} \end{cases} \tag{11}$$

Unfortunately, the system (11) does not form a close solution for maximizing the likelihood $P_{eye}(\mathbf{x}_l,\mathbf{x}_r)$ since both sides have $\mathbf{x}_l$ and $\mathbf{x}_r$. In this paper, we propose an iterative solution to this problem. It contains two steps. In the first step, a pair of good initialization positions are selected from the detections, and in the second step, an iterative algorithm is applied to refine the positions and scales of the two eyes.

5.2 Initialization

For iterative algorithms, good initialization is very important for success to real world problems. In this work, the initial eye positions are selected from the detections. Again, for each pair of two detections B_i and B_j, the probability that they are good candidates for the left and right eyes can be defined as

$$P_{eye}(i,j) = P(B_i)P(B_j)P_{eye-pair}(B_i,B_j) \tag{12}$$

where $P(B_i)$ is the confidence of the ith detection, and $P_{eye-pair}(B_i, B_j)$ is computed using (10) with d_x, the average of human interoccular distance and $d_y = 0$. The pair of maximum probability value is selected as the initial positions and scales of the two eyes.

5.3 Iterative Estimation

Let B_i and B_j be the selected initial positions of the left and right eyes. The initial center points of the left and right eyes can be denoted as $x_l = x_i$ and $x_r = x_j$. The initial scales of the two eyes can be represented as $B_l = B_i$ and $B_r = B_j$, where $B_i = (W_i, H_i)$ represents the width and height of the detection window. The distance vector between the two eyes characterizes the face pose. The initial face pose is assumed as upright face, hence, d_x is set as the average human inter-ocular distance and d_y is set as 0. From (11) and (12), an iterative algorithm for eye localization from detections can be defined. The updates in each step of the iterative algorithm can be expressed as

$$\begin{cases} \mathbf{x}_l^{t+1} = \dfrac{\sigma_e^2 \sum_{i=1}^{N} \mathbf{x}_i \pi_i^{l(t)} P(\mathbf{x}_l^t \mid B_i) P(B_i)}{(\sigma_d^2 + \sigma_e^2)\, L(\mathbf{x}_l^t \mid \mathbf{B})} + \dfrac{\sigma_d^2 (\mathbf{x}_r^t - \mathbf{d}_e^t)}{\sigma_d^2 + \sigma_e^2} \\[2ex] \mathbf{x}_r^{t+1} = \dfrac{\sigma_e^2 \sum_{i=1}^{N} \mathbf{x}_i \pi_i^{r(t)} P(\mathbf{x}_r^t \mid B_i) P(B_i)}{(\sigma_d^2 + \sigma_e^2)\, L(\mathbf{x}_r^t \mid \mathbf{B})} + \dfrac{\sigma_d^2 (\mathbf{x}_l^t - \mathbf{d}_e^t)}{\sigma_d^2 + \sigma_e^2} \\[2ex] \mathbf{d}_e^{t+1} = \mathbf{x}_r^t - \mathbf{x}_l^t \\[2ex] B_l^{t+1} = \dfrac{\sum_{i=1}^{N} B_i \pi_i^{l(t)} P(\mathbf{x}_l^t \mid B_i)\, P(B_i)}{L(\mathbf{x}_l^t \mid \mathbf{B})} \\[2ex] B_r^{t+1} = \dfrac{\sum_{i=1}^{N} B_i \pi_i^{l(t)} P(\mathbf{x}_r^t \mid B_i)\, P(B_i)}{L(\mathbf{x}_r^t \mid \mathbf{B})} \end{cases} \tag{13}$$

The update of the interocular distance vector $\mathbf{d}_e$ is derived from $\dfrac{\partial P_{eye}(\mathbf{x}_l, \mathbf{x}_r)}{\partial \mathbf{d}_e} = 0$, and the scales of the eye boxes are updated as the weighted average of the associated detection windows. Here, the association parameters π_i^l and π_i^r are computed based on the overlapping of the boxes. Let B_l^t be the estimated box of the left eye at the t iteration step, which is centered at $\mathbf{x}_l^t$. The association of B_l^t with the ith detection B_i is computed as $\pi_i^{l(t)} = \dfrac{|B_l^t \cap B_i|}{|B_l^t \cup B_i|}$, i.e., the numerator is the area of the intersection and the denominator is the area of the union of the two boxes. The iterative algorithm stops when both $|\mathbf{x}_l^{t+1} - \mathbf{x}_l^t| < \varepsilon$ and $|\mathbf{x}_r^{t+1} - \mathbf{x}_r^t| < \varepsilon$ hold, where ε is a small value.

With a good initialization, the iterative algorithm converges very quickly. Since the inter-ocular distance vector is involved in the updating in the iterations, the algorithm is able to adapt to a quite large variations of head poses. The updates of the eye scales can improve the estimations of the associations in the iterative steps.

6   Experimental Results

We conduct the experiments to compare the performance of the proposed algorithm with some existing algorithms on databases BioData, CVL, and Olivia. We also conduct the experiments on the effect of the particular methods of our algorithm.

6.1   Data and Evaluation Criterion

BioIData [24] is a frequently-used database for eye localization [3-5, 8-9, 11, 17-18, 21]. The dataset consists of 1521 gray images with a resolution of 384x286. Each one shows the frontal view of a face of one out of 23 different test persons.

The CVL [25] database contains 797 color images of 114 persons. Each person has 7 images of size 640x480 pixels: far left side view, $45°$ angle side view, serious expression frontal view, $135°$ angle side view, far right side view, and smile frontal view. The algorithm in [17] used the 335 frontal view face images from this database. We carry out the experiments on the same dataset 335 images.

The Olivia is the database that contains the 3000 images in the resolution 320x240 of ten persons. These images are recorded by our robot during the interaction between the robot and persons.

The aim of eye localization is to find the center of eye and the eye center of an eye is the center point of its pupil for open eyes and the middle point of two corners for closed eyes. In our evaluation, we adopt the measure proposed by Jesorskey et al. [11] including the variables. The error measure, defined as localization criterion,

$$e = \frac{\max(\| \mathbf{C}_l - \tilde{\mathbf{C}}_l \|, \quad \| \mathbf{C}_r - \tilde{\mathbf{C}}_r \|)}{\| \mathbf{C}_l - \mathbf{C}_r \|} \tag{14}$$

where $\mathbf{C}_l$ and $\mathbf{C}_r$ are the groundtruth positions and $\tilde{\mathbf{C}}_l$ and $\tilde{\mathbf{C}}_r$ are the detected eye centers of left and right eyes respectively in pixel. $\| \bullet \|$ is the Euclidean distance.

6.2   Experimental Results

To use HOG-moment feature to detect the eye candidates we need to train a SVM to judge whether a window containing an eye. The images used in our training comprise the ORL face database (AT & T) and 10% of BioID and CVL databases. For each image, we produce 2 positive samples for left eye and right eye respectively and 5 negative samples which are randomly selected windows from each image and are far enough from the correct eye region. We test our algorithm on BioID and CVL databases and compare our result with the result reported in [3-5, 8-9, 11, 17-18, 21]. For BioID database, our result is 98.55% with e<0.25. This result is not the best but in the best performance class. Our result is 90.14 with e<0.1. This result is better than all the results except the results in [5] and [11]. It is worth noticing that our algorithm does not depend on the face detection, whereas all the algorithms in [3-5, 8-9, 11, 17-18, 21] depend on the results of face detection and all assume that face detection achieved the perfect result, which is very difficulty to get. For 355 images selected from CVL, our algorithm achieve correct rate of 96.17% for e<0.25 and 90.56% for e<0.1 respectively. For e<0.25, our result 96.17% is worse than 99.7% reported in [17], whereas for e<0.1 our result 90.56% is much better than 80.9% of the result reported in [17].

Besides the experiments of comparison with the existing algorithms we also conduct some experiments to explore the performance of our algorithm under the different configurations. We compare the performance between with and without the face box and without HOG-moment features and the results are presented in Table 2. When we do not use the information of face region, our algorithm searches a larger area. When there are some objects in images that are very similar to eye in moment and appearance the algorithm would increase the rate of false alarm. However, such objects are very few in the databases used to do experiments in this paper.

Table 1. Comparison on the eye localization performance between our algorithm and the 8 algorithms in the literature

	#Ref	[3]	[4]	[5]	[8]	[9]	[11]	[17]	[18]	ours
BioID	0.25	96.00	96.1	98.00	93.00	91.8	99.9	99.46	98.49	98.55
	0.1	64.00	85.2	96.0	77.0	79.0	97.9	73.68	90.85	90.14
CVL	0.25							99.7		96.17
	0.1							80.9		90.56
Olivia	0.25									99.97
	0.1									91.65

Table 2. The performances of our algorithm when it is full one and no moment or no face box

Name	BioID	BioID	CVL	CVL	Olivia	Olivia
Threshold	0.25	0.1	0.25	0.1	0.25	0.1
Full Alg	98.55	90.14	96.17	90.56	99.97	91.65
No moment	94.61	80.34	92.63	89.38	98.33	61.58
No face box	84.42	77.71	96.17	90.56	99.94	90.10

7   Conclusions

We have presented a novel eye detection and localization system of using both stereo and mono cameras. This paper has multiple contributions in the technique development. First, it is an eye detection and localization system tested on the robot, not just an algorithm working on images. Second, it uses the stereo camera to obtain the head location in the image from mono camera. This procedure replaces the face detection in many other eye localization algorithms. Third, our algorithm is robust to the uncontrolled environment. This robustness is because of the reliable head box localization and the good distinguish ability of HOG-moment feature.

In the near future, we want to develop more robust eye localization algorithm by integrating more methods. In addition, we want to develop different eye locators and then use them in a scheme to achieve better performance of eye localization.

References

[1] Amir, A., Zimet, L., Sangiovanni-Vincentelli, A., Kao, S.: An embedded system for an eye-detection sensor. Comput. Vis. Image Underst. 98(1), 104–123 (2005)
[2] Haro, A., Flickner, M., Essa, I.: Detecting and tracking eyes by using their physiological properties, dynamics, and appearance. In: CVPR 2000, pp. 163–168 (2000)
[3] Bai, L., Shen, L., Wang, Y.: A novel eye location algorithm based on radial symmetry transform. In: ICPR 2006, vol. 3, pp. 511–514 (2006)

[4] Campadelli, P., Lanzarotti, R., Lipori, G.: Precise eye localization through a general-to-specific model definition. In: BMVC 2006, pp. 187–196 (2006)
[5] Cristinacce, D., Cootes, T., Scott, I.: A multi-stage approach to facial feature detection. In: BMVC 2004, pp. 277–286 (2004)
[6] Fasel, I., Fortenberry, B., Movellan, J.: A generative framework for real time object detection and classification. Computer Vision and Image Understanding 98, 182–210 (2005)
[7] Gernoth, T., Kricke, R., Grigat, R.-R.: Mouth localization for appearance-based lip motion analysis. WSEAS Transactions on Signal Processing 3(3), 275–281 (2007)
[8] Hamouz, M., Kittler, J., Kamarainen, J.-K., Paalanen, P., Kalviainen, H., Matas, J.: Feature-based affine-invariant localization of faces. In: PAMI 2005, vol. 27(9), pp. 1490–1495 (2005)
[9] Jesorsky, O., Kirchberg, K.J., Frischholz, R.W.: Robust face detection using the Hausdorff distance. In: Bigun, J., Smeraldi, F. (eds.) AVBPA 2001. LNCS, vol. 2091, pp. 90–95. Springer, Heidelberg (2001)
[10] Jin, L., Yuan, X., Satoh, S., Li, J., Xia, L.: A hybrid classifier for precise and robust eye detection. In: ICPR 2006, Hong Kong, vol. 4, pp. 731–735 (2006)
[11] Kroon, B., Maas, S., Boughorbel, S., Hanjalic, A.: Eye localization in low and standard definition content with application to face matching. Computer Vision and Image Understanding 113(8), 921–933 (2009)
[12] Monzo, D., Albiol, A., Sastre, J., Albiol, A.: Precise eye localization using HOG features, Machine Vision and Applications (May 2010), 10.1007/s00138-010-0273-0
[13] Niu, Z., Shan, S., Yan, S., Chen, X., Gao, W.: 2D cascaded AdaBoost for eye localization. In: ICPR 2006, vol. 2, pp. 1216–1219 (2006)
[14] Rurainsky, J., Eisert, P.: Eye center localization using adaptive templates. In: CVPR Workshops 2004, pp. 67–74 (2004)
[15] Song, J., Chia, Z., Liu, J.: A robust eye detection method using combined binary edge and intensity information. Pattern Recognition 39, 1110–1125 (2006)
[16] Tan, X., Song, F., Zhou, Z.-H., Chen, S.: Enhanced pictorial structures for precise eye localization under uncontrolled conditions. In: CVPR 2009, pp. 1621–1628 (2009)
[17] Türkan, M., Pardàs, M., Çetin, A.E.: Human eye localization using edge projections. In: Proceedings of 2nd International Conference on Computer Vision Theory and Applications (VISAPP 2007), Barcelona, Spain, vol. 1 (2007)
[18] Valenti, R., Gevers, T.: Accurate eye center location and tracking using isophote curvature. In: CVPR 2008, pp. 1–8 (2008)
[19] Wang, P., Green, M.B., Ji, Q., Wayman, J.: Automatic eye detection and its validation. In: CVPR Workshops, June 20-26, pp. 164–171 (2005)
[20] Wang, P., Ji, Q.: Multi-view face and eye detection using discriminant features. CVIU 105, 99–111 (2007)
[21] Zhou, Z.H., Geng, X.: Projection functions for eye detection. Pattern Recognition 37, 1049–1056 (2004)
[22] Shan, S., Chang, Y., Gao, W., Cao, B.: Curse of mis-alignment in face recognition: problem and a novel mis-alignment learning solution. In: Int'l Conf. Automatic Face and Gesture Recognition 2004, pp. 314–320 (2004)
[23] Viola, P., Jones, M.: Rapid object detection using a boosted cascade of simple features. In: CVPR 2001, vol. 1, pp. 511–518 (2001)
[24] BioID, http://www.humanscan.de/support/downloads/facedb.php
[25] CVL, http://www.lrv.fri.uni-lj.si/
[26] http://www.cl.cam.ac.uk/research/DTG/attarchive:pub/data/att_faces.tar.Z

An Implementation of a Biologically Inspired Model of Head Direction Cells on a Robot

Theocharis Kyriacou*

Research Institute for the Environment, Physical Sciences and
Applied Mathematics (EPSAM), Keele University, Keele, UK
t.kyriacou@cs.keele.ac.uk

Abstract. A biologically inspired model of head direction cells is presented and tested on a small mobile robot. Head direction cells (discovered in the brain of rats in 1984) encode the head orientation of their host
irrespective of the host's location in the environment. The head direction
system thus acts as a biological compass (though not a magnetic one)
for its host. Head direction cells are influenced in different ways by idiothetic (host-centred) and allothetic (not host-centred) cues. The model
presented here uses the visual, vestibular and kinesthetic inputs that are
simulated by robot sensors. Three test cases are presented that cover
different state combinations of the inputs. The test results are compared
with biological observations in previous literature.

Keywords: biologically inspired robot navigation, head direction cells.

1 Introduction: Biologicaly Inspired Robot Navigation

Biologically inspired navigation methods can perhaps be put in two broad categories. In the first, methods draw only from observations of animal behaviour
without considering the underlying cognitive mechanisms that play part in navigation. Tolman in 1948 (see [19]) was among the first who conducted experiments
with rats that allowed such observations but more recent work using rodents and
insects is presented for example in [2] and [22].

In contrast, "bottom-up" approaches to bio-inspired models of navigation
make use of knowledge obtained by observing the brain activity of animals while
they perform navigational tasks (see for example [11]). During such experiments,
the activity of a few brain cells can be recorded by means of microelectrodes.
This gives some clues as to how a navigational mechanism is implemented in the
brain. Bio-inspired models in this category make quite a lot of extrapolations
that try to fill in the gaps.

Three types of brain cells called *place cells* (see [13]), *head direction cells*
(see [18]) and *grid cells* (see [9]) have been discovered (mostly from experiments
conducted on rats) and are thought to play a significant role in animal navigation.

---

* The author would like to thank his colleagues Charles Day and John Butcher for
the useful discussions he had with them during the work presented in this paper.

R. Groß et al. (Eds.): TAROS 2011, LNAI 6856, pp. 66–77, 2011.

The work in this paper concentrates on head direction cells (or HD cells) and presents a model of these cells that is inspired by biological observations. A more detailed description of HD cells and an overview of previous work in modelling them is presented below.

1.1 Head Direction Cells

The most recent comprehensive review of neurophysiological observations related to HD cells is found in [23]. Here below, the main characteristics of the HD system are outlined.

Head direction cells were first discovered in 1984 by Ranck Jr. and more detailed findings on them were published in 1990 by Taube and colleagues (see [18]). An HD cell fires maximally when the animal's head points in a particular direction. This is called the *preferred head direction* of the particular cell. The HD system includes a population of HD cells with preferred head directions distributed through 360°. The activity of an HD cell does not depend on the location of the animal in the environment. The head direction cell system can thus be thought of as being a biological head compass (though not a magnetic one) that is influenced by several senses. When an animal is placed in a new environment the preferred head direction of each cell in the HD system quickly settles to an arbitrary value. The system maintains this alignment for the specific environment even if the animal is removed and re-introduced back to the same environment. This alignment will only be reset (thus the environment will be treated as a new, previously unseen one) if several weeks have passed before the animal is re-introduced in the environment ([18]). The visual sense is the major input that helps to align the HD system when the animal is introduced in a previously visited environment. Strong visual cues (for example large and prominent landmarks) can influence the preferred head direction of HD cells (see experiments and observations in [12] and [18]). Another major input to the HD system is the vestibular sense. This allows the animal to maintain a correct head direction for some time after visual input is removed (by switching off the lights for example). Apart from the most important two inputs to the HD system mentioned above, other cues also play a part in influencing the preferred head direction of HD cells. These include olfactory cues (see [8]) and cues that are involved in self-locomotion (motor, kinesthetic and proprioceptive) (see [17] for a review). Cues to the HD system are classified in two categories: they can be *allothetic*, i.e. not self-centred (for example visual and olfactory) or *idiothetic*, i.e. self-centred (for example vestibular and kinesthetic). When two or more inputs to the HD system are in conflict the preferred head direction of the HD cells depends mainly on two factors: (a) the relative influence to the HD system of each of the inputs in conflict and (b) the extent of the conflict (see example in [6]). Conflicts are extensively discussed in [23]. See also [17] for a more concise review.

1.2   Models of Head Direction Cells

Several attempts have been made to model the HD system. Redish and colleagues in [14] and Goodridge and Touretzky in [7] present anatomically faithful models but these only use the vestibular sense. In both cases the neural network weights are prescribed (i.e. not obtained by training). The models are however tested using real data obtained from experiments with rats. Models using both visual and vestibular inputs are presented in [21], [16] and [24]. In all three cases the model weights are also prescribed and tests are carried out in simulated environments. Also, only in [24] were input conflict situations simulated and compared with biological observations. Only a few examples exist in the literature of HD system models applied to real robotic agents. Of these, the most notable are presented in [1] and [4]. Both models incorporate visual and vestibular inputs but again, in both cases the models weights are prescribed.

In the work presented here, the head direction cell model by Stinger and colleagues in [16] is extended by incorporating in it the kinesthetic input (in addition to the visual and vestibular input). No previous work has been found that incorporates all three inputs to an HD cell model. Furthermore, the proposed model is trained and tested using real robot data. Conflict situations between the three inputs to the model are also presented and compared with biological observations.

Details of the model presented in this paper are given in section 2. The experimental procedure, testing and results are presented in section 3 and discussed in section 4. A summary and concluding remarks are presented in section 5.

2   Methods: The Model

The HD system model presented here is using a continuous attractor neural network (see introduction by Trappenberg in [3]). A continuous attractor neural network (CANN) is a network of interconnected nodes. In a fully connected network each node is connected via weighted connections to every other node including itself. A CANN is thus a form of recurrent network. The operation of the network is such that nodes in close association excite each other (via excitatory connections). The amount of excitation being proportional to the degree of association between the nodes. On the other hand, nodes that are less associated with each other are connected with inhibitory connections. The connection weights from one node to all other nodes are often prescribed by a Gaussian function with its tails below zero (the negative weight values for the inhibitory connections). A CANN gives rise to a self-sustained "hill" of excitation (the attractor) in the network. If the network is perfectly symmetrical about each node (both in connectivity and weight values to other nodes) the attractor will be stationary when the network has no external influences. External input stimuli that temporarily distort the network's symmetry (by biassing the activation of nodes) can cause the attractor to move.

In the case of the HD system model presented here the CANN network is comprised of 360 nodes (the HD cells) and it is fully connected. Each node is

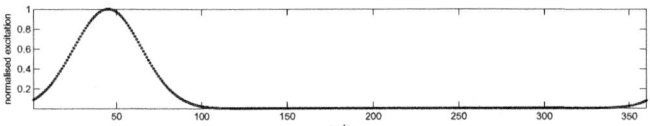

Fig. 1. The state of the HD system when it points at $45°$

associated with a preferred head direction and it is therefore most active when the subject (the host of the HD system) is facing in that direction. The nodes can be thought to be arranged in a circle in order of preferred head direction[1]. Figure 1 shows the state of the HD system when it points to $45°$ (from an arbitrary reference direction).

The state (i.e. the excitation of each node) of the network implementation presented here is a function of the previous state of the system and three inputs: visual, vestibular and the kinesthetic. Figure 2(left) is a partial diagram that shows how these inputs are connected to the HD cells. The reader is referred to the caption of the figure for a detailed explanation.

The activation h_i^{HD} at time t of a head direction cell i in the model presented here can be determined using the following differential equation:

$$\tau\frac{dh_i^{HD}(t)}{dt} = -h_i^{HD}(t) + \theta^{RC}\sum_j(w_{ij}^{RC} - w^{INH})r_j^{HD}(t) + \theta^{VIS}I_i^{VIS}(t)$$

$$+\theta^{VES}\sum_{jk}w_{ijk}^{VES}r_j^{HD}(t)r_k^{VES}(t) + \theta^{KIN}\sum_{jk}w_{ijk}^{KIN}r_j^{HD}(t)r_k^{KIN}(t) \tag{1}$$

where $r_i^{HD}(t)$ is the firing rate (excitation) of HD cell i given by the sigmoid function:

$$r_i^{HD}(t) = \frac{1}{1 + e^{-2\beta h_i^{HD}(t)}} \tag{2}$$

τ is the time constant of the system, w_{ij}^{RC} is the recurrent connection weight from HD cell j to HD cell i, I_i^{VIS} is the visual input to HD cell i, r_k^{VES} is the firing rate of vestibular sensor cell k, w_{ijk}^{VES} is the weight value of the connection from HD cell j to VCL cell i that is associated with the vestibular sensor cell k. Similarly, r_k^{KIN} is the firing rate of kinesthetic sensor cell k and w_{ijk}^{KIN} is the weight value of the connection from HD cell j to KCL cell i that is associated with the kinesthetic sensor cell k. The factors θ^{RC}, θ^{VIS}, θ^{VES} and θ^{KIN} control the influence of the recurrent, visual, vestibular and kinesthetic inputs respectively to the HD cells. Finally, w^{INH} is a constant negative offset to the recurrent connection weights that serves as a quick way to make the connection weights

---

[1] This arrangement however is not necessary. In fact, in the brain, HD cells have not been found to be arranged in any particular order that relates to their preferred head direction.

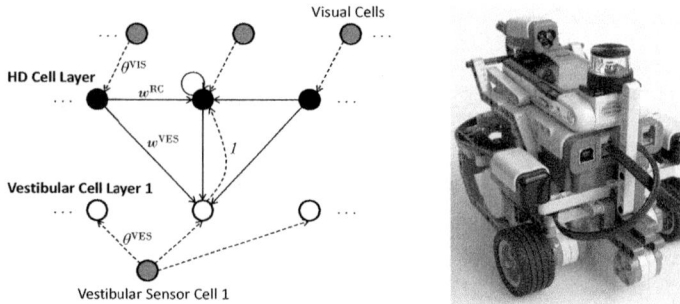

Fig. 2. Left: A partial diagram of the network of the HD model. The HD cell layer contains the CANN network nodes. These are the HD cells and they are fully interconnected. The visual input is a 360-element vector that is directly (one-to-one) connected to the HD layer. The excitation of the HD layer nodes is the output of the system and it indicates the direction in which the HD system is pointing at a given time. A layer labelled Vestibular Cell Layer 1 (VCL1) is also a 360-node layer that comprises of sigma-pi neurons. These are neurons that compute the sum of products of their inputs (see [10]). Each node in the VCL1 layer accepts input from all HD cells. In addition, each node in the VCL1 layer accepts input from a single cell (called the Vestibular Sensor Cell 1) that simulates the clockwise vestibular sense. Similarly, another layer (not shown for clarity) called Vestibular Cell Layer 2 also contains 360 sigma-pi neurons and it is connected to the HD layer in the same way as VCL1. This layer is associated with a cell that simulates the counter-clockwise vestibular sense (Vestibular Sensor Cell 2). In the same fashion, two more cell layers (KCL1 and KCL2 - also not shown) exist for the kinesthetic sense. In addition to the input from the HD cells each of these two layers accepts input from two cells correspondingly that simulate clockwise and counter-clockwise kinesthetic senses. Even though three HD cells are shown in the diagram above, only the connections to the middle one are explicitly drawn for clarity. Connections shown with solid lines are trained whereas those with dashed lines are not. Some of the weight connections are labelled but without their subscripts for clarity. **Right:** The LEGO® MINDSTORMS® NXT robot used for the experiments presented here. The robot is equipped with an on-board omnidirectional video camera (above the NXT brick) a gyroscopic sensor and an acceleration sensor (pictured to the left and right above the wheels). For locomotion, the robot uses two active wheels (seen at the front) and a dummy castor (not visible in the picture).

between distant cells negative (inhibitory). The weights w^{RC}, w^{VES} and w^{KIN} can be prescribed using Gaussian-like functions like for example in [21], [16] and [24] but this method is not biologically plausible and conveniently ignores the effects of noise to the weight values. Here, in order to train the network weights, data is collected from the robot as explained below.

3   Experimental Procedure and Results

The model described above was trained and tested using a LEGO® robot with an on-board omnidirectional video camera (see figure 2(right)). The robot is further

Fig. 3. Snapshot from the omni directional video camera on-board the robot

equipped with a gyroscopic sensor and an acceleration sensor. For locomotion, the robot uses two active wheels and a dummy castor. For the purposes of the work presented here, the visual input to the HD system was provided by processing the image from the omnidirectional video camera on the robot (see figure 3 for an example of a video image). The vestibular input was provided directly by the gyro sensor. One vestibular sensor cell (see figure 2(left)) was driven by the raw gyro signal and the other by the inverted version of the gyro signal. The two kinesthetic inputs were provided by differentiating the signals from the odometric sensors in the motors driving each wheel of the robot. The robot was controlled by a PC via USB connection in order to achieve maximum possible data transfer rates when reading the robot's sensors. The video during each recording session was independently recorded on the video camera and during post-processing it was time-corresponded with the robot's data. This setup allowed a 10Hz sampling rate in all data sources (gyro, motor position, video). Two sets of data were collected using the above setup. The first (training set) was used to train the network weights of the HD model and the second (the test set) was used to test the model.

3.1 Training the Model

The duration of the training set was 873 seconds (14.55 minutes). During this time the robot was programmed to continuously rotate in a random direction with a constant rotational speed of approximately 35 degrees/second. In order to obtain the expected output of the system at time t_n (i.e. training pattern p_n), initially, the direction of the robot (based on a world reference frame) was extracted from the video data by finding the maximum correlation (along the abscissa of the video image) between the video frame taken at t_n and the first captured video frame taken at t_0. As the robot was only rotating on the spot during the experiments described here, this provided a convenient way of establishing the world-based orientation of the robot. After that, p_n was created by translating a Gaussian function (of standard deviation σ) so that it would be centred at the the robot's orientation at t_n.

Training of the recurrent weights w^{RC} was achieved using the Hebbian learning rule:

$$\delta w_{ij}^{\mathrm{RC}} = k^{\mathrm{RC}} r_i^{\mathrm{HD}} r_j^{\mathrm{HD}} \tag{3}$$

where k^{RC} is the learning rate. Training of the vestibular weights was achieved using the following rule:

$$\delta w_{ijk}^{\text{VES}} = k^{\text{VES}} r_i^{\text{HD}} \bar{r}_j^{\text{HD}} r_k^{\text{VES}} \qquad (4)$$

where $\bar{r}^{\text{HD}}$ is a historical trace value of r^{HD} and is given by:

$$\bar{r}_j^{\text{HD}}(t + \delta t) = (1 - \eta) r^{\text{HD}}(t + \delta t) + \eta \bar{r}_j^{\text{HD}}(t) \qquad (5)$$

A similar rule to the one given by equation 4 was also used in order to train the kinesthetic weights w_{ijk}^{KIN}.

Note that equation 4 (through the use of equation 5) considers together current and past values of the state of the HD system. It is this feature that gives rise to idiothetic weight profiles that bias the CANN attractor to move with the right speed and in the right direction according to the idiothetic inputs. In equation 5, η is a parameter that dictates the influence of the current and previous state of the network in the trace rule (equation 4) and δt is the time delay between the two states considered.

It can be seen from the equations defining the HD model presented here that there are many parameters (τ, θ^{RC}, w^{INH}, δt, η, β, etc.) that define the behaviour of the system. Importantly, these parameters are not always orthogonal to each other. As in previous work, these parameters (listed in table 1) have been obtained here by trial-and-error. It should be stressed however that once determined, the same parameter values were used in all the test cases presented below.

3.2   Testing the Model

Test data was collected for 102.9 seconds (1.715 minutes). During this session the robot was programmed to continuously rotate on the spot and in the same direction with a constant rotational speed of approximately 35 degrees/second. The same test data set was used for the three cases presented below.

Test Case 1. For this test case the visual input was turned off for $30s < t < 60s$ and $t > 90s$. In addition, the vestibular and kinesthetic inputs were also turned off for $t < 10s$ and $t > 80s$. These input suppressions were done manually by overwriting the data during the "off" windows by zeroes. Figure 4 shows a set of plots with common time scale that display the inputs (plots 3-5) and output (plot 1) to and from the HD model during this test case. Plot 2 compares the actual head direction (obtained from video as described above) and the model-predicted head direction.

Table 1. The parameter values used in the HD system model presented

τ	0.5s	δt	0.1s	β	0.2
k^{RC}	0.01	θ^{RC}	3	η	0.9
k^{VES}	0.01	θ^{VES}	0.0195	σ	20°
k^{KIN}	0.01	θ^{KIN}	0.0025	w^{INH}	0.4
		θ^{VIS}	30		

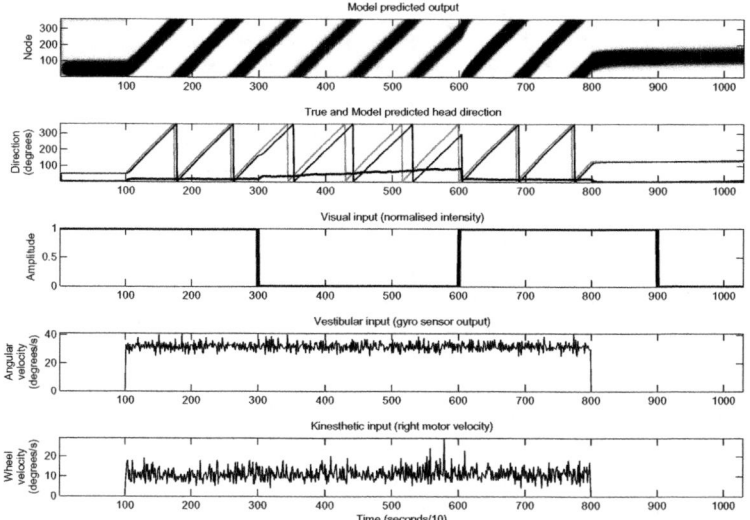

Fig. 4. Data describing test case 1. Plot 1: the model output. This is a surface plot viewed from the top. The black colour value is proportional to the excitation of each node. Plot 2: The robot's true head direction (grey line) and the model-predicted head direction (black line) with the absolute difference between them (wide black line). Note that the model-predicted head direction is the crest of the surface in plot 1. Plot 3: The normalised intensity of the visual input. When this is 0 the robot is in the dark (no visual input). Plot 4: The vestibular input (gyro sensor signal). Plot 5: The kinesthetic input (right motor velocity). Note that only the right motor velocity is plotted for clarity. The left motor velocity is very similar to this but in the opposite direction.

For $t < 10s$ the robot is stationary and only receives visual input ("lights on"). The attractor of the network is centred around $50°$. This is where the robot "thinks" it is facing and this is in agreement with its true orientation. For $10s < t < 30s$ the robot moves under its own volition while the lights are still on. The expected orientation follows closely the actual orientation of the robot over a bit more than two full revolutions. For $30s < t < 60s$ the robot keeps rotating but in the dark (no visual input). The output still follows the true orientation of the robot but there is a small drift (with the model output lagging). At $t = 60s$ visual input is re-introduced (lights turn on). The small discrepancy between the actual and model-predicted orientation is quickly rectified as the model uses the visual cue to reset its state. For $60s < t < 80s$ the model continues to follow the true orientation of the robot. At $t = 80s$ the robot stops moving (no idiothetic inputs) and 10 seconds later the lights are also turned off. For $t > 80s$ the model-predicted and actual orientations of the robot are stationary and in agreement.

Test Case 2. During this test case the visual and vestibular inputs are the same as in test case 1 but the kinesthetic input is set to zero throughout. Figure 5 only

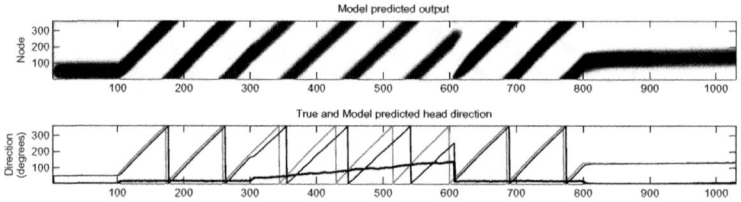

Fig. 5. Data describing test case 2. Plot 1: the model output. Plot 2: the true (grey line) and the model-predicted head direction (black line) with the absolute error (wide black line). Visual and vestibular inputs are the same as in plots 3 and 4 respectively in figure 4. The kinesthetic input is zero at all times and therefore not shown here.

shows the model-predicted output plot and the plot comparing the true and model-predicted head directions.

Since kinesthetic input is always zero here, a non-zero vestibular input implies that an external force is moving the robot. Initially the lights are on and the robot is first stationary ($t < 10s$) and then it is being moved ($10s < t < 30s$). The model follows the true orientation of the robot just like in test case 1 despite the conflict between the kinesthetic input and the other two senses (visual and vestibular)[2]. The same happens for $t > 60s$. The notable difference between this test and the previous one is visible in the interval $30s < t < 60s$. Here, it is as if the robot is being rotated in the dark. The drift between the model-predicted outpout and true orientation is greater than that in test case 1. This leads to a greater error at $t = 60s$ (when visual input is switched back on) but again this is quickly rectified by the system.

Test Case 3. In order to observe the effects of a persistent conflict between the visual and vestibular senses, during this test case the visual input was kept the same as before (see plot 3 in figure 4) but the vestibular input was (manually) inverted (see figure 6). The kinesthetic input was again kept at zero throughout this test case.

The most interesting observation here is during $10s < t < 30s$ and $60s < t < 80s$ when the visual and vestibular inputs are in persistent conflict. Even though the model mostly follows the true orientation of the robot, plot 1 in figure 6 is more informative of the state of the HD system during this test. The plot shows that there is a periodic effect during which the output is influenced more by the vestibular input than at other times. What happens when visual input is removed during $30s < t < 60s$ should not be surprising as the vestibular input is the only input to the system and therefore it solely drives the model-predicted orientation. Note that when the lights come on at $t = 60s$, it is only by coincidence that the expected head direction almost coincides with the actual head direction.

---

[2] Note that the word *conflict* in this text is used to imply any form disagreement between two inputs and not just one where two inputs oppose each other.

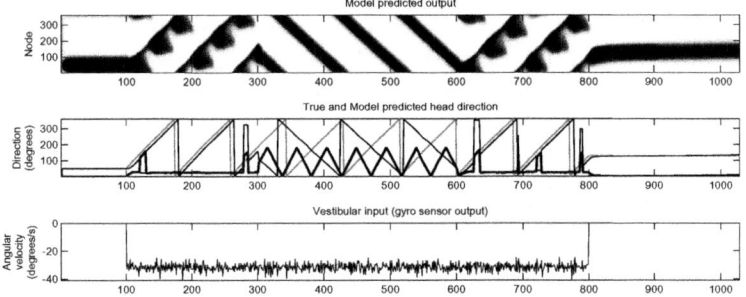

Fig. 6. Data describing test case 3. Plot 1: the model's output. Plot 2: true (grey line) and model-predicted head direction (black line) with the absolute error (wide black line). Plot 3: The vestibular input which is the inverted version of that shown in figure 4. Visual input is the same as in plots 3 of figure 4. The kinesthetic input is zero at all times and therefore not shown here.

4 Discussion

The qualitative observations in the test cases above are in agreement with biological observations made in experiments with animals (mainly rodents). More specifically it is shown that the visual input, when available, helps to set or reset the HD system to the true orientation ([17]). The visual input helps to maintain an absolute point of reference of the true orientation. When all inputs are in agreement the HD system follows the true orientation of the robot but when the visual input is turned off a slight drift is observed (see biological observations in [15]). The idiothetic inputs thus help to maintain a relative point of reference of the true orientation. In the absence of allothetic cues (i.e. visual input) a greater drift is observed when only vestibular input is used (i.e. the robot is being rotated by an external force) than when both idiothetic inputs are present (i.e. when the robot is self-rotating). Biological observations of this are presented in [5] who performed behavioural experiments with hamsters. When no input is present the HD system maintains a stable orientation. A very small drift of the output can be noticed in figures 4, 5 and 6 when all inputs are switched off ($t > 90s$). Observations of this drift and its time scales in rats are presented in [15]. It is speculated here that, in the case of a biological HD system, the synaptic weights cannot be perfectly symmetrical about each HD cell (noise, uneven number of connections etc). Similarly here, due to the training of the model using real (noisy) data from the robot the same effect is observed. Finally, when a persistent conflict is introduced between the visual and vestibular inputs, the output appears to follow the visual input but shows periodic tendency to follow the vestibular input. This effect is a consequence of the chosen values of the model parameters. The parameter values were not chosen to deliberately construct it but also, no attempts have been made to change or remove it by varying the model parameters. Conflicts between the visual and vestibular inputs to the HD system have been presented (after biological observations) in quite a few places (summarised in [17]). To the author's best knowledge however, the temporal effects of conflicts such as the one presented in test case 3 have not been documented before. It would be interesting therefore to recreate such conflict conditions with real animals in order to determine the validity of the model presented here.

5   Summary and Conclusion

A biologically inspired model of the head direction system was presented and implemented on a small mobile robot. The model takes three inputs (visual, vestibular and kinesthetic) that are among those which provide the most influencing cues to the biological HD system [17]. The model was trained and tested using real data obtained from the robot. Three test cases were conducted and the results obtained were compared with biological observations made in previous documented work.

According to Trullier and colleagues ([20]) biological navigation methods may not always produce the best, most mathematically optimal solution to a navigation problem but they are fast, flexible and adaptive. What is best and most mathematically optimal however depends on the employer of a particular navigational skill. We know little about even the simplest of organisms in nature and it could therefore really be that for a particular organism, their navigation strategy is the best in all respects. Modelling biological mechanisms of navigation helps us understand better the remarkably complex systems in nature. Besides the information value however, the understanding of how these mechanisms evolved, rather than just what they do and how they do it, may lead us to more generalised principles of designing artificial navigation systems that would be the best and most optimal for their intended application.

References

1. Arleo, A., Gerstner, W.: Spatial orientation in navigating agents: Modeling head-direction cells. Neurocomputing 38-40(1-4), 1059–1065 (2001)
2. Biegler, R., Morris, R.G.M.: Landmark stability is a prerequisite for spatial but not discrimination learning. Nature 361, 631–633 (1993)
3. de Castro, L.N., Von Zuben, F.J. (eds.): Recent Developments in Biologically Inspired Computing. Idea Group Publishing, USA (2005)
4. Degris, T., Lachèze, L., Boucheny, C., Arleo, A.: A spiking neuron model of head-direction cells for robot orientation. In: Proceedings of the Eighth International Conference on the Simulation of Adaptive Behavior, from Animals to Animats, pp. 255–263. MIT Press, Cambridge (2004)
5. Etienne, A.S., Maurer, R., Saucy, F.: Limitations in the assessment of path dependent information. Behaviour 106, 81–111 (1988)
6. Etienne, A.S., Maurer, R., Séguinot, V.: Path integration in mammals and its interaction with visual landmarks. Journal of Experimental Biology 199, 201–209 (1996)

7. Goodridge, J.P., David, Touretzky, D.S., Jeremy, P.: Modeling attractor deformation in the rodent head-direction system. Journal of Neurophysiology 83, 3402–3410 (2000)
8. Goodridge, J.P., Dudchenko, P.A., Worboys, K.A., Golob, E.J., Taube, J.S.: Cue control and head direction cells. Behavioral Neuroscience 112(4), 749–761 (1998)
9. Hafting, T., Fyhn, M., Molden, S., Moser, M.-B.B., Moser, E.I.: Microstructure of a spatial map in the entorhinal cortex. Nature 436(7052), 801–806 (2005)
10. Mel, B.W., Koch, C.: Sigma-pi learning: on radial basis functions and cortical associative learning. In: Touretzky, D.S. (ed.) Advances in Neural Information Processing Systems, vol. 2, pp. 474–481. Morgan Kaufmann Publishers Inc., San Francisco (1990)
11. Morris, R.G., Garrud, P., Rawlins, J.N., O'Keefe, J.: Place navigation impaired in rats with hippocampal lesions. Nature 297(5868), 681–683 (1982)
12. Muller, R.U., Kubie, J.L., Ranck, J.B.: Spatial firing patterns of hippocampal complex-spike cells in a fixed environment. Neuroscience 7(7), 1935–1950 (1987)
13. O'Keefe, J., Dostrovsky, J.: The hippocampus as a spatial map. preliminary evidence from unit activity in the freely-moving rat. Brain Research 34(1), 171–175 (1971)
14. Redish, A.D., Elga, A.N., Touretzky, D.S.: A coupled attractor model of the rodent head direction system. Network: Computation in Neural Systems 7(4), 671–685 (1996)
15. Mizumori, S.J., Williams, J.D.: Directionally selective mnemonic properties of neurons in the lateral dorsal nucleus of the thalamus of rats. Neuroscience 13(9), 4015–4028 (1993)
16. Stringer, S., Trappenberg, T., Rolls, E., de Araujo, I.: Self-organizing continuous attractor networks and path integration: one-dimensional models of head direction cells. Network: Computation in Neural Systems 13(2), 217–242 (2002)
17. Taube, J.S.: Head direction cells and the neurophysiological basis for a sense of direction. Progress Neurobioloy 55(3), 225–256 (1998)
18. Taube, J., Muller, R., Ranck Jr., J.: Head-direction cells recorded from the postsubiculum in freely moving rats. i. description and quantitative analysis. Neuroscience 10(2), 420–435 (1990)
19. Tolman, E.C., Ritchie, B.F., Kalish, D.: Studies in spatial learning. i. orientation and the short-cut. Journal of Experimental Psychology 36, 13–24 (1946)
20. Trullier, O., Wiener, S., Berthoz, A., Meyer, J.: Biologically-based artificial navigation systems: Review and prospects. Progress in Neurobiology 51, 483–544 (1997)
21. Skaggs, W.E., Knierim, J.J., Kudrimoti, H.S., McNaughton, B.L.: A model of the neural basis of the rat's sense of direction. Advances in Neural Information Processing Systems 7, 173–180 (1995)
22. Wehner, R., Menzel, R.: Do insects have cognitive maps? Annual Review of Neuroscience 13, 403–414 (1990)
23. Wiener, S.I., Taube, J.S. (eds.): Head direction cells and the neural mechanisms of spatial orientation. MIT Press, Cambridge (2005)
24. Zeidman, P., Bullinaria, J.A.: Neural models of head-direction cells. In: French, R.M., Thomas, E. (eds.) From Associations to Rules: Connectionist Models of Behavior and Cognition, pp. 165–177 (2008)

Contextual Recognition of Robot Emotions

Jiaming Zhang and Amanda J.C. Sharkey

Neurocomputing and Robotics Group, Department of Computer Science,
University of Sheffield, Sheffield, UK
acq08jz@sheffield.ac.uk, amanda@dcs.shef.ac.uk

Abstract. Researchers in emotional human-robot interaction (HRI) have often focused on the abilities of sociable emotional robots to express emotions themselves, and on the ability of people to recognize these. However, it has been shown that the recognition of human emotional expressions can be influenced by the surrounding context [17]. So far, no empirical research has been done to examine whether or not the recognition of robot emotions is similarly influenced. Two experiments are reported here that examine how a robot's simulated emotions were perceived by human observers, depending on what the surrounding context was. Evidence of an effect of surrounding context on user's perception of the synthetic robot emotions was obtained. Observers were better at recognizing the robot's expressions when they matched the emotional valence of accompanying pictures or recorded News announcements, than when they did not.

Keywords: robot, emotions, facial expressions, surrounding context, emotional congruence.

1 Introduction

Robots are beginning to be used in social situations where there is a need for them to seem to respond appropriately to humans. It will be easier to accept robot receptionists and robot guides that seem to respond to us with appropriately expressed facial emotions. There is a growing interest in human-robot interaction, but its researchers have tended to focus on the capabilities of sociable emotional robots themselves, for example, the ability to detect the affective states of users, the ability to express emotions through the use of synthetic facial expressions, speech and textual content, and the ability for imitating and social learning. MIT's Personal Robot Group has pioneered the development of sociable emotional robots, such as Kismet [4], Leonardo [5], and Nexi [1]. Meanwhile, there are some other sociable robots, such as Autom [13], ATR's Robovie [12], MAGGIE [10], Philips's iCat ("interactive cat") [11]. Some of these sociable robots are able to display emotional facial expressions.

The FACS (Facial Action Coding System) [6, 7] was initially developed as a set of guidelines for recognizing the facial expressions of humans. However, it has also been used to create the facial expressions of robots, such as MIT's Kismet [4] and Vrije Universiteit Brussel's Probo [9]. The FACS was found to be a reliable tool to create believable versions of six distinct and universal facial expressions (happiness, fear, surprise, anger, disgust/contempt, and sadness) for some emotional robots, even

R. Groß et al. (Eds.): TAROS 2011, LNAI 6856, pp. 78–89, 2011.
© Springer-Verlag Berlin Heidelberg 2011

though some emotions (fear, and happy+surprised) were harder to make convincing [4, 9]. Moreover, to generate dramatic but smooth shifts from one emotional state of a robot to another, Russell's circumplex model of affect containing two dimensions; valence and arousal [18] was used to construct an emotion space for Kismet and its new version [19] was for Probo.

The FACS system is based on the idea that there are certain emotions that can be universally recognized [6]. More recently, a growing body of empirical research has begun to demonstrate that the surrounding context can influence the perception and recognition of the facial emotional expressions of human beings. For example, Niedenthal et al. [17] found that both the human face and the surrounding context made important contributions to emotion judgments. In more detail, the way in which facial and context cues were combined (congruent or incongruent with each other), affected observers' judgments of facial expressions. When the surrounding emotional context did not match the facial expression, observers often reinterpreted either the facial expression (i.e., the person's face does not reveal the person's real feelings) or the meaning of the context (i.e., the situation does not have the usual meaning for the person).

Even though it seems that the principles of the FACS can be used to recognize both human faces and robot faces, there is good reason to suppose that the synthetic robot facial expressions [4, 9] are more ambiguous than human facial expressions [6, 7]. Contextual influences on the recognition of human emotional expressions are stronger when the expressions are ambiguous [17]. It seems likely therefore that the recognition of the simulated emotions shown by a robot will be influenced by a surrounding affective context. Although so far, to our knowledge, this question has not been empirically investigated.

A surrounding context can be either congruent or incongruent with the emotional valence of the simulated emotions shown by a robot. Emotional congruence has been shown to affect the accuracy in making judgements on the emotions displayed by computer simulated avatars [15, 16, 20].

Mower et al. [15, 16] used computer simulated avatars in an effort to determine how participants made emotional decisions when presented with both conflicting (e.g. angry avatar face, happy avatar voice) and congruent information (e.g. happy avatar face, happy avatar voice) in an animated display consisting of two channels: the facial expression and the vocal expression. Mower et al. [15, 16] found that: (1) when they were presented with a congruent combination of audio and visual information, users could differentiate between happy and angry emotional expressions to a greater degree than when they were presented with either of the two channels individually; (2) when faced with a conflicting emotional presentation, users predominantly attended to the vocal channel rather than the visual channel. In other words, they were more likely to see the avatar's expressed emotions as being expressed by means of its voice, than reflected in its facial expressions. Their findings indicate that the observer's recognition of facial expressions can be influenced by the surrounding context, and also that emotion conveyed by other modalities, in this case the voice, can override that expressed by the face.

Similar results were found by Hong et al. [20] when they paired neutral, happy and sad voices with neutral, happy and sad synthetic or real faces, and found that the emotional facial expressions were better recognized when they matched the context, than when it did not.

A robot's facial expressions can be viewed as a modality containing emotional information. At the same time, a surrounding context, such as recorded BBC News or selected affective pictures, can be treated as another modality. The two modalities may reflect congruent, or incongruent emotions. There is good reason to expect that the recognition of the emotional facial expressions of a robot will be affected by the emotional valence of a surrounding context that consists of either an auditory recording of positive/negative news, or a set of positive/negative pictures.

The study presented in this paper explores the influence of the surrounding emotional context (congruent or incongruent) on human perception of a robot's simulated emotions. We explored these questions with two experiments. In the first experiment, a robotic head displayed synthetic emotional expressions based on the principles of the FACS system. These expressions were either congruent or incongruent with the emotional valence of some recorded BBC News being played. In the second experiment, the same robotic head displayed synthetic emotional expressions that were either congruent or incongruent with the emotional valence (positive or negative) of a set of affective pictures shown to the subjects. In both experiments, the context and the presentation of emotional expressions occurred simultaneously.

2   Hypotheses

It seems that context effects can be found when people are making judgements about human emotional facial expressions, and that different judgements are made depending on whether the expressions and context are congruent, or incongruent [17]. A similar effect was also found when judgements were made about the expressions of an avatar: users were more able to recognize the happy and angry facial expressions of the avatar when confronted by a congruent audio-visual presentation than when they encountered a conflicting audio-visual presentation [15, 16]. In order to explore whether this is also the case when they make judgements about robot emotional facial expressions, we formulated one primary hypothesis as follows:

Hypothesis 1 (H1): When there is a surrounding emotional context, people will be better at recognizing robot emotions when that context is congruent with the emotional valence of the robot emotions than when the context is incongruent with the emotional valence of the robot emotions.

It was previously found in [15, 16] that users were influenced more by the vocal channel than by avatar's facial expression when making judgements about the avatar's emotional state. Contextual influence was found to increase when human facial expressions were ambiguous [17], and as suggested above, synthetic robot expressions are likely to be more ambiguous than human facial expressions. The surrounding context is therefore likely to influence the interpretation of the synthetic robot facial expressions, and even to dominate that interpretation. This reasoning leads to the following hypothesis:

Hypothesis 2 (H2): When subjects are presented with conflicting information from the robot's face and an accompanying emotional context, their perception of the robot emotions will be more influenced by the surrounding context than the robot itself.

3 Method

3.1 Interaction Design

We conducted two between-subjects experiments. The two experiments were based on a robot head known as CIM [2] (a believable emotional agent without vision and audio sensing capabilities) in the NRG lab, the University of Sheffield. The electronic interface with the CIM robot was built around the Sun SPOT (Sun Small Programmable Object Technology) system incorporating an I2C bus to connect the individual components to the dedicated microprocessor. This microprocessor was a dedicated java processor capable of running java programs. In the two experiments, the FACS (Facial Action Coding System) [7] was applied to set up the parameters of the servos (see Table 1) to make the robot head to produce a long sequence of some or all of the six distinct facial expressions (joy, fear, surprise, anger, disgust, and sadness).

Table 1. Servos set up for CIM

Facial expressions	Action set up	Servo Speed
Joy	Smiling Lips, Raised Checks	Relatively Fast
Sad	Crying Brows, Crying Lips, Raised Checks, Eyes Down	Slowest
Anger	Angry Brows, Eyes Narrowed, Tightened Lips	Fast
Fear	Raised Eye Brows, Mouth Opened	Fast
Disgust	Frown Eye Brows, Nose Wrinkled, Raised Upper Lip	Slow
Surprise	Raised Eye Brows, Eyes Widened, Eyes Up, Neck Backward	Fastest

Examples of simulated facial expressions shown by CIM (see Fig.1) are presented as follows:

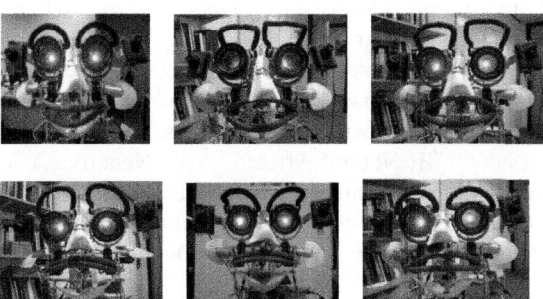

Fig. 1. Joy (top left), Surprise (top middle), Fear (top right), Sadness (bottom left), Anger (bottom middle), and Disgust (bottom right) of CIM.

The sequence of simulated facial expressions of the robot that the subjects watched while they were listening to the recorded BBC News in the first experiment or watching the affective pictures in the second experiment, could be divided into two types: Positive Affect, which mainly consisted of three different versions of joyful and surprised expressions (motions such as looking around and nodding were added in the gaps between these joyful and surprised expressions), and Negative Affect which mainly consisted of three different versions of sad, angry, and disgusted expressions (motions such as shaking and denying were added in the gaps between these sad, angry, and disgusted expressions). Each sequence of facial expressions was about three minutes long: the same time length as the recorded BBC News or time given to view the affective pictures.

In the first experiment, the source of the recorded News that the subjects listened to was BBC World News. Two types of News were used in this study, one was Positive/Neutral News, and the other one was Negative News. An example of a positive/neutral news item was a report that scientists had successfully used the Large Hadron Collider. An example of negative news was that collective international action was being called for to prevent the tiger population from dying out.

In the second experiment, the affective pictures were selected from the international affective picture system (IAPS), as the IAPS is currently used in experimental investigations of emotion and attention worldwide, providing experimental control in selection of emotional stimuli [3]. The selected affective pictures were either all pleasant or all unpleasant. Each type consisted of 32 slides of affective pictures (about 3-minute long) presented at intervals of 6 seconds.

3.2 Design and Results of the First Experiment

Warm up: The emotional robot head CIM showed six different static facial expressions to all subjects. Then subjects were asked to fill in a questionnaire about how they perceived the robot's simulated emotions.

Experiment: The subjects listened to a piece of recorded BBC News (positive/neutral, or negative) while being shown a 3-minutes sequence of facial expressions (positive or negative) of the emotional robot.

Responses: After listening to the recorded BBC News, each subject was asked to answer the following questions:

1. As a total impression, please select what kind of News you think you were listening to from the following given choices. _____
A: Positive News   B: Neutral News   C: Negative News
2. As a total impression, please select what kind of emotion (affect) you think the robot was feeling from the following given choices. _____
A: Positive Affect   B: Neutral Affect   C: Negative Affect
3. As a detailed impression, please select what emotions you think the robot was feeling from the following given choices. ___(you can choose more than one option)
A: Joy   B: Sadness C: Anger D: Fear E: Disgust   F: Surprise

Then, all subjects were asked to fill in one further questionnaire: the Brief Mood Introspection Scale (BMIS) [14].

Subjects were divided into four groups according to different combinations of BBC News (positive/neutral vs. negative) and robot expressions (positive vs. negative).

(Group 1: Positive/Neutral BBC News with Robot's Positive Affect; Group 2: Positive/Neutral BBC News with Robot's Negative Affect; Group 3: Negative BBC News with Robot's Positive Affect; Group 4: Negative BBC News with Robot's Negative Affect). Two of the groups thus received congruent emotional information from the news and the robot head (Group 1 and Group 4), whilst two received incongruent, or conflicting information (Group 2 and Group 3).

The 72 subjects (38 male and 34 female) with average age 26.82 who participated in this experiment, had various nationalities. There were 18 subjects in each group, which meant that there were 36 subjects in the Congruent Condition (group 1 and group 4) and 36 subjects in the Incongruent Condition (group 2 and group 3).

In the warming-up procedure, some of the robot's static expressions were more easily recognised than others (see Table 2). Most of the subjects (88%) could easily recognize the robot's sad expression, and a majority were able to identify the 'joy' expression (64%), and the surprise expression (65%). Anger was recognised correctly by 57% of the subjects, but most of the subjects (76%) confused the robot's fearful expression with surprised expression. Moreover, the disgust and the anger expressions of the robot were easily confused with each other, and some of the subjects (less than one third) found it difficult to tell the difference between the joy and the surprise expressions.

Table 2. Emotion-based evaluation of the six different facial expressions of CIM

% match	Joy	Sadness	Anger	Fear	Disgust	Surprise	% correct
Joy	64	0	0	3	5	28	64
Sadness	0	88	1	3	8	0	88
Anger	0	4	57	14	24	1	57
Fear	11	0	0	0	13	76	0
Disgust	0	4	45	11	40	0	40
Surprise	14	0	0	18	3	65	65

The responses to the questionnaire administered after viewing the robot were analysed. It was found that subjects were mostly good at identifying the emotional content of the News (see Table 3). A response was considered to be correct when the News was said to have been Positive or Neutral in the Positive/Neutral BBC News Condition, or Negative in the Negative BBC News Condition (neutral News was not counted in this condition).

Table 3. Subjects' perception of the BBC News

%match	Positive News	Neutral News	Negative News	%correct
Positive/Neutral BBC News Condition (group 1 and 2)	56	44	0	100
Negative BBC News Condition (group 3 and 4)	3	8	89	89

They were less good at reliably identifying the emotional expressions of the robot (see Table 4). A response was considered to be correct when the robot head was said to have shown Positive Affect in the Positive Affect Condition, or Negative Affect in the Negative Affect Condition (the neutral choice was counted as wrong in both conditions).

Table 4. Subjects' perception of the facial expressions of the robot

%match	Positive Affect	Neutral Affect	Negative Affect	% correct
Positive Affect Condition (group 1 and 3)	67	19	14	67
Negative Affect Condition (group 2 and 4)	22	11	67	67

However, although subjects were often not good at identifying the robot's expressions, they were much better when the robot's expressions were accompanied by a congruent context. A Chi-square test for independence (with Yates Continuity Correction) indicated a significant association between Information Style (congruent or conflicting) and Accuracy of recognizing robot's emotion. In other words, the accuracy of subjects' perception of robot's emotions was significantly different depending on whether the robot behaved congruently with the BBC News (a higher accuracy, 34/36 (94.4% correct, 5.6% incorrect)), or whether the robot behaved conflictingly with the BBC News (a much lower accuracy, 14/36 (38.9% correct, 61.1% incorrect)). χ^2 (1, n=72) =22.562, p<0.0001, correlation coefficient phi=0.589 (large effect). Consequently, H1 was supported.

The hypothesis (H2) that when subjects are presented with conflicting information, the surrounding context will have a stronger effect than the robot's expressions, was also tested. Two Chi-square tests for independence (with Yates Continuity Correction) indicated the BBC News modality had a stronger perceptual effect than the Robot Expression modality both in Condition 2 (18/18 accuracy for BBC News VS. 6/18 accuracy for Robot Emotions) and Condition 3 (15/18 accuracy for BBC News VS. 8/18 accuracy for Robot Emotions) when subjects were presented with conflicting information. In Condition 2, χ^2 (1, n=36) =18.000, p<0.0001, correlation coefficient phi=0.707 (large effect); and in Condition 3, χ^2 (1, n=36) =5.900, p=0.015, correlation coefficient phi=0.405 (mediate effect). Consequently, H2 was supported.

In addition, a Two-Way ANOVA (unrelated) in SPSS was conducted to explore the impact of the recorded BBC News and the synthetic robot emotions on subjects' moods, as measured by the BMIS Scale. The interaction effect between recorded BBC News and the synthetic robot emotions was not statistically significant, F(1,68)= 0.032, p=0.858; in addition, neither the main effect for recorded BBC News (F(1,68)= 0.013, p=0.910) nor the main effect for synthetic robot emotions (F(1,68)= 0.021, p=0.884) reached statistical significance. In conclusion, neither the recorded BBC News nor the synthetic robot emotions appeared to affect subjects' moods.

3.3 Design and Results of the Second Experiment

This experiment followed the same procedures used in the first experiment, except that a series of pictures was shown in place of the recorded news.

Warm up: Six different facial expressions of the emotional robot CIM and six static affective pictures on a computer screen selected from the IAPS were showed simultaneously to all subjects. Then, as in the first experiment, subjects were asked to fill in a questionnaire about how they perceived the robot's static facial expressions. This time, before the warming-up procedure, subjects were also asked to read the instructions for the Affect Grid used to measure current levels of pleasure and arousal [8], as it would be used later in the experiment.

Experiment: Subjects watched 32 slides of affective pictures while being simultaneously shown a 3-minute sequence of facial expressions (either positive or negative) of the emotional robot.

Responses: After the last picture was shown, subjects were asked to use the Affect Grid to indicate their current state, before they answered the following questions:

1. As a total impression, please select what kind of affective pictures you think you were viewing from the following given choices. _____

A: Pleasant Picture   B: Neutral Pictures   C: Unpleasant Pictures

2. As a total impression, please select what kind of emotion (affect) you think the robot was feeling from the following given choices._____

A: Positive Affect   B: Neutral Affect   C: Negative Affect

3. As a detailed impression, please select what emotions you think the robot was feeling from the following given choices. ____(you can choose more than one option)

A: Joy   B: Sadness   C: Anger   D: Caring   E: Disgust   F: Surprise

The same two sequences of facial expressions of the emotional robot were used in this experiment as in the first experiment.

Subjects were divided into four groups according to different combinations of affective pictures (positive vs. negative) and robot expressions (positive vs. negative) (Group 1: Positive Pictures with Positive Robot Affect; Group 2: Negative Pictures with Positive Robot Affect; Group 3: Positive Pictures with Negative Robot Affect; Group 4: Negative Pictures with Negative Robot Affect). As in the previous experiment, different groupings constitute the Congruent Condition (group 1 and group 4 together), and the Conflicting Condition (group 2 and group 3 together). 56 volunteers with an average age of 24.5 participated in this experiment. These 23 male and 33 female participants had various nationalities. There were 14 subjects in each group, resulting in 28 subjects in the Congruent Condition (group 1 and group 4), and 28 subjects in the Conflicting Condition (group 2 and group 3).

Table 5. Subjects' perception of the affective pictures

%match	Positive Pictures	Neutral Pictures	Negative Pictures	% correct
Positive Pictures Condition (group 1 and 3)	96	4	0	96
Negative Pictures Condition (group 2 and 4)	0	4	96	96

Table 6. Subjects' perception of the facial expressions of the robot

%match	Positive Affect	Neutral Affect	Negative Affect	% correct
Positive Affect Condition (group 1 and 2)	46	8	46	46
Negative Affect Condition (group 3 and 4)	14	11	75	75

The responses to the questionnaire administered after viewing the robot were analysed. It was found that subjects were mostly good at identifying the emotional content of the pictures (see Table 5).

They were less good at reliably identifying the emotional expressions of the robot (see Table 6).

However, although subjects were often not good at identifying the robot's expressions, they were better at identifying them when the robot's expressions were accompanied by a congruent context. A Chi-square test for independence (with Yates Continuity Correction) indicated significant association between Information Style (Conflicting Information or Congruent Information) and Accuracy of recognizing robot's emotions, in other words, subjects were less able to recognize the robot's emotional expressions, when the expressions conflicted with valence of the affective pictures (a lower accuracy, 10/28 (35.7% correct, 64.3% incorrect)) , than when the robot behaved congruently with the affective pictures (a much higher accuracy, 24/28 (85.7% correct, 14.3% incorrect)). χ^2 (1, n=56) =12.652, $p<0.0001$, correlation coefficient phi=0.512 (large effect). Consequently, H1 was again supported.

H2 was also tested. Two Chi-square tests for independence (with Yates Continuity Correction) indicated the Affective Pictures modality had stronger perceptual effect than the Robot Emotions modality in Condition 2 (14/14 accuracy for Affective Pictures VS. 2/14 accuracy for Robot Emotions) but not in Condition 3 (13/14 accuracy for Affective Pictures VS. 8/14 accuracy for Robot Emotions) when subjects were presented with conflicting information. In Condition 2, χ^2 (1, n=28) =17.646, $p<0.0001$ (smaller than 0.05), correlation coefficient phi=0.866 (large effect); and in Condition 3, χ^2 (1, n=28) =3.048, $p=0.081$ (bigger than 0.05), correlation coefficient phi=0.412 (medium effect). Consequently, H2 was only weakly supported.

In addition, an Independent-Samples T-Test in SPSS was conducted to compare the unpleasant-pleasant scores on the Affect Grid for participants who watched the unpleasant pictures and participants who watched the pleasant pictures. There was significant difference in scores for participants who watched the unpleasant pictures (M=-1.5357, SD=1.31887), and participants who watched the pleasant pictures (M=1.7143, SD=1.21281), with t(54)=9.598, p<0.0001 (two-tailed). The magnitude of the difference in means (means difference=3.25000, 95% CI: 2.57114 to 3.92886) was very large (eta squared=0.63, for 0.01=small effect, 0.06=moderate effect, and 0.14=large effect). There was evidence that the pleasant pictures made people feel happier than the unpleasant pictures.

4   Discussion

This study has examined how the surrounding emotional context (congruent, or incongruent) influenced users' perception of a robot's simulated emotional expressions. Hypothesis 1, that when there is a surrounding emotional context, people will be better at recognizing robot emotions when that context is congruent with the emotional valence of the robot emotions than when the context is incongruent with the emotional valence of the robot emotions, was supported in both of the experiments reported here. This suggests that, the recognition of robot emotions is strongly affected by the surrounding context, and such emotions are more likely to be recognised when they are congruent with that context. The second hypothesis, that when a robot's expressions are not appropriate given the context, subjects' judgements of those expressions will be more affected by the context than the expressions themselves was clearly validated in the first experiment, and weakly supported in the second. In summary, when confronted with congruent surrounding context (the recorded BBC News or the selected affective pictures), people were more able to recognize the robot emotions as intended than when faced by an incongruent surrounding context. In addition it was found that the recorded BBC News had a more dominant effect on judgements about the robot's expressions than the content of the expressions themselves. The affective pictures showed a weaker, but still dominant effect.

There are two possible explanations of the effect of the accompanying emotional context on judgements about the robot's emotional expressions. One is that the context affects the emotional state of the observer, which in turn affects their perception of the robot. Niedenthal et al [17] found that observers of human facial expressions were influenced by their own emotional state in their attributions of emotional states to the people they observed. The other is that the observer interprets the robot's expressions as though the robot was also witnessing the context, and responding to it. In this explanation, the observers' judgements are not determined by their own emotional state.

Which explanation is better supported by the experiments reported here? A significant association between subjects' moods and their perception of the synthetic robot emotions was only found in the second experiment. We suggest therefore that the effect of the surrounding context observed in the first experiment was not due to the observers' own emotional state coloring their perception of the robot, but resulted instead from witnessing the context and the robot at the same time, and interpreting the robot's expressions as if the robot was responding to the surrounding context. When the context was negative, the robot's expressions were seen to be negative, even when on objective criteria (the underlying FACS coding) they should have been seen as positive. The reverse was also true (the robot's expressions were seen as positive when the surrounding context was positive, even when there were objective reasons to expect the expression to be seen as a negative one). The interesting implication of this is that the robot's expressions seem to be being interpreted as if it were capable of understanding and responding to the current emotional situation. The same effect was found in the second experiment, although in this case the affective pictures were found to have influenced subjects' emotional state. In this experiment, judgements about the robot's expressions may have been influenced by both.

More research is needed to explore the relative effects of different kinds of surrounding context, and to distinguish between possible explanations of the contextual effects. Nonetheless, it remains the case that our results indicate that the recognition of a robot's emotional expressions can be affected by the situation in which they occur. These findings resemble the effects found in the recognition of human emotional expressions, where the surrounding context and the observer's emotional state can affect recognition. An interesting implication of these findings is that they suggest that one way of making a robot's emotional expressions seem more convincing, is to make sure that they are supported by, and match, the surrounding emotional context. It seems that the expressions on a robot's face, like the expressions on a human face, are interpreted in terms of the events that are happening around them.

References

1. Alonso, J.: Studying social cues in human robot interaction. In: Microsoft External Research Symposium, MIT Media Laboratory, Personal Robots (2009)
2. Bailey, A.: Synthetic Social Interactions with a Robot using the BASIC personality model. MSc thesis, University of Sheffield, Sheffield, England (2006)
3. Bradley, M.M., Lang, P.J.: The International Affective Picture System (IAPS) in the Study of Emotion and Attention. In: Coan, J.A., Allen, J.B. (eds.) The Handbook of Emotion Elicitation and Assessment, pp. 29–46 (2007)
4. Breazeal, C.L.: Designing Sociable Robots. A Bradford Book, The MIT Press (2002)
5. Brooks, A., Gray, J., Hoffman, G., Lockerd, A.T., Lee, H., Breazeal, C.: Robot's play: interactive games with sociable machines. Computers in Entertainment 2(3), 1–18 (2004)
6. Ekman, P., Friesen, W.: Facial Action Coding System. Consulting Psychologists Press, Palo Alto (1978)
7. Ekman, P., Friesen, W., Hager, J.: Facial Action Coding System. Research Nexus, Salt Lake City, Utah (2002)
8. Russell, J.A., Weiss, A., Mendelsohn, G.A.: Affect Grid: A Single-Item Scale of Pleasure and Arousal. Journal of Personality and Social Psychology 57(3), 493–502 (1989)
9. Goris, K., Saldien, J., Vanderniepen, I., Lefeber, D.: The Huggable Robot Probo, a Multi-disciplinary Research Platform. In: Proceedings of the EUROBOT Conference 2008, Heidelberg, Germany, pp. 63–68 (2008)
10. Gorostiza, J.F., Barber, R., Khamis, A.M., Malfaz, M., Pacheco, R., Rivas, R., Corrales, A., Delgado, E., Salichs, M.A.: Multimodal Human-Robot Interaction Framework for a Personal Robot. In: RO-MAN 2006: The 15th IEEE International Symposium on Robot and Human Interactive Communication, Hatfield, United Kingdom (September 2006)
11. Heerink, M., Kröse, B.J.A., Wielinga, B.J., Evers, V.: The Influence of a Robot's Social Abilities on Acceptance by Elderly Users. In: Proceedings RO-MAN, Hertfordshire (2006)
12. Kahn Jr., P.H., Freier, N.G., Kanda, T., Ishiguro, H., Ruckert, J.H., Severson, R.L., Kane, S.K.: Design Patterns for Sociality in Human-Robot Interaction. In: HRI 2008, Amsterdam, Netherlands, March 12-15 (2008)
13. Kidd, C.D., Breazeal, C.: Robots at Home: Understanding Long-Term Human-Robot Interaction. In: 2008 IEEE/RSJ International Conference on Intelligent Robots and Systems, Acropolis Convention Center, Nice, France, September 22-26 (2008)
14. Mayer, J.D., Gaschke, Y.N.: The experience and meta-experience of mood. Journal of Personality and Social Psychology 55, 102–111 (1988)
15. Mower, E., Lee, S., Matarific, M.J., Narayanan, S.: Human perception of synthetic character emotions in the presence of conflicting and congruent vocal and facial expressions. In: IEEE Int. Conf. Acoustics, Speech, and Signal Processing (ICASSP 2008), Las Vegas, NV, pp. 2201–2204 (2008)
16. Mower, E., Matarić, M.J., Narayanan, S.: Human perception of audio-visual synthetic character emotion expression in the presence of ambiguous and conflicting information. IEEE Transactions on Multimedia 11(5) (2009)

17. Niedenthal, P.M., Kruth-Gruber, S., Ric, F.: What Information Determines the Recognition of Emotion? In: The Psychology of Emotion: Interpersonal Experiential, and Cognitive Approaches. Principles of Social Psychology, pp. 136–144. Psychology Press, New York (2006)
18. Russell, J.: Reading emotions from and into faces: resurrecting a dimensional–contextual perspective. In: Russell, J., Fernandez-Dols, J. (eds.) The Psychology of Facial Expression, pp. 295–320. Cambridge University Press, Cambridge (1997)
19. Posner, J., Russell, J., Peterson, B.: The circumplex model of affect: an integrative approach to affective neuroscience, cognitive development, and psychopathology. Dev. Psychopathol. 17(03), 715–734 (2005)
20. Hong, P., Wen, Z., Huang, T.: Real-time speech driven expressive synthetic talking faces using neural networks. IEEE Transaction on Neural Networks (April 2002)

Costs and Benefits of Behavioral Specialization

Arne Brutschy[1], Nam-Luc Tran[1], Nadir Baiboun[1,2], Marco Frison[1,3],
Giovanni Pini[1], Andrea Roli[3,1], Marco Dorigo[1], and Mauro Birattari[1]

[1] IRIDIA, CoDE, Université Libre de Bruxelles, Brussels, Belgium
[2] ECAM, Institut Supérieur Industriel, Brussels, Belgium
[3] DEIS-Cesena, *Alma Mater Studiorum* Università di Bologna, Cesena, Italy
arne.brutschy@ulb.ac.be

Abstract. In this work, we study behavioral specialization in a swarm of autonomous robots. In the studied swarm, a robot working repeatedly on the same type of task improves in task performance due to learning. Robots may exploit this positive effect of learning by selecting with higher probability the tasks on which they have improved their performance. However, even though the exploitation of such performance-improving effects is clearly a benefit, specialization also entails certain costs. Using a task allocation strategy that allows the robots to behaviorally specialize, we study the trade-off between costs and benefits in simulation experiments. Additionally, we give a perspective on the impact of this trade-off in systems that use specialization.

Keywords: specialization, task allocation, swarm robotics, swarm intelligence, self-organization, division of labor.

1 Introduction

Division of labor is a concept that is common in the organization of large groups of individuals such as humans or social insects [1,5]. In division of labor, *"(a) each worker specializes on a subset of the complete repertoire of tasks performed by the colony, and (b) this subset varies across individual workers in the colony"* [1]. A common way to obtain division of labor is to let individuals adapt their behavior to focus on a subset of the tasks available—this is called *behavioral specialization* [13]. Behavioral specialization is known to increase the overall performance of an individual because of many different reasons. One of the most important of these reasons is learning: an individual grows more efficient in performing a task by repeating it multiple times [15]. A specialized individual can exploit this increased efficiency by behaviorally focusing on the tasks it is good at.

In this work, we study the costs and benefits of behavioral specialization in a swarm of autonomous robots. In the studied system, robots have to perform different types of tasks which are independent of each other. The robots can improve their performance on a task by repeatedly working on it (i.e., they learn). In order to maximize this performance improvement, robots should focus on a single task—first to learn this task to the full extent, and second to fully exploit the resulting performance improvement. We propose a simple self-organized strategy

R. Groß et al. (Eds.): TAROS 2011, LNAI 6856, pp. 90–101, 2011.

for task allocation that allows the robots to behaviorally specialize on a specific task, thereby maximizing the performance improvement available through learning.

However, even though the exploitation of such a performance improvement is clearly a benefit, specialization also entails certain costs [15]. An example for such a cost is the time a specialized robot spends searching for a suitable task. If the robot is specialized in a task that becomes less frequent in the environment, the robot has to spend more time searching for it. This search time increases the cost of specialization. A generalist, on the other hand, can accept the first task it encounters. As a result, specialists might be less efficient in environments where task types and distribution frequently change. We identify situations in which costs overcome benefits, and predict necessary changes in the system in order to avoid such disadvantages.

This paper is organized as follows. In Sec. 2 we review related works. In Sec. 3 we describe how we model leaning and specialization in our system and the task allocation strategy employed by the robots. In Sec. 4 we describe the experimental setup used for the study. In Sec. 5 we describe the experiments, and we report and discuss the results. In Sec. 6 we summarize the contributions of this work and present some directions for future research.

2 Related Work

Behavioral specialization has been observed in many species of animals [2,6]. Most works model division of labor and specialization in social insects (see [1] and [2] for an overview). These works focus almost exclusively on specialization as a means of increasing task performance by reducing external costs (e.g., travel times [15]). In these works, individuals repeatedly perform the same or a small subset of tasks, without improving in the performance of the individual tasks [6].

A way of improving task performance, other than by reducing external costs, is learning. Higher vertebrates are known to exploit this type of improvement by behavioral specialization, although it is disputed if it can be observed in social insects [5]. In robotics, improvements in task performance by learning are certainly possible (e.g., on-line adaptation of existing behaviors or learning techniques such as artificial neural networks). However, to the best of our knowledge, no works currently exist that exploit these improvements by using behavioral specialization. The general case of specialization decoupled from learning, on the other hand, has indeed been studied. Jones and Matarić studied a foraging problem, in which each individual specializes in foraging for one of two possible food types [8]. The study shows that the group behaviorally diverges depending on the ratio of food items present in the environment. Li et al. studied division of labor and specialization in an initially homogeneous swarm [10]. In their work, robots behaviorally diverge by assuming different roles in a stick-pulling experiment. The study confirms that specialization usually does not occur if there are more task types than individuals available (e.g., in small groups).

A frequently studied problem in relation to division of labor is the problem of foraging for energy. Commonly, two opposing behaviors exist (e.g., resting and foraging), that exhibit specific cost and benefits in terms of energy. The studied group has to adapt to optimize the collective energy level. Labella et al. found that, in their system, individuals effectively divide into active and passive foragers [9]. Liu et al. studied a similar system with four different strategies for foraging for energy, which exhibit also an effective division of labor in the swarm [11]. Recently, Ikemoto et al. proposed an adaptive mechanism for division of labor in a swarm of robots, which also divides into distinct groups that behaviorally specialize on a certain task [7].

3   Model of Specialization

In this section, we explain how we model the tasks and the environment, as well as the effect of learning on individual task performance. We consider an environment with different types of tasks, τ_i, which are independent of each other. The distribution of the task types is unknown and tasks appear at random locations. The goal of the swarm is to maximize its performance, measured as the number of completed tasks in a given period of time. The nature of the tasks is such that an individual repeatedly working on a task type will become more efficient for that task type. Thus, this individual will take less time to complete tasks of this type. However, if the individual switches to another type of task, it loses part of its performance improvement for the first task type. We will refer to these phenomena as *learning* and *forgetting*, respectively.

3.1   Learning

We model the learning effect mentioned above as an improvement in task performance of the individual. This improvement depends on the number of successive task completions by the individual. As found in natural systems, we model the improvement as rapidly increasing in the beginning, reaching a plateau with further repetitions [5]. We define the time it takes to complete a task τ_i as

$$w_i(n_i) = w_{std} - \frac{w_{std}}{k * (1 + e^{-n_i + c})} \quad \text{with} \quad 0 < n_i \leq n_{max} \quad (1)$$

where n_i is number of times a task τ_i was completed by the individual. n_i is incremented each time a task τ_i is performed, and decremented each time a different task is performed. The standard task completion time is called w_{std}; it applies to unlearned individuals. Notice that in this work we use the same w_{std} for all tasks in order to reduce the parameters of the system. c is a parameter that defines after how many repetitions of a task learning starts to show a meaningful effect. The factor k is used to vary the maximal time gain attainable through learning, reached after the individual successively completed n_{max} tasks of the same type. An individual reaching this state is called *specialist*. The resulting minimal task completion time is referred to as w_{min}. Fig. 1 shows a graphical representation of the learning model.

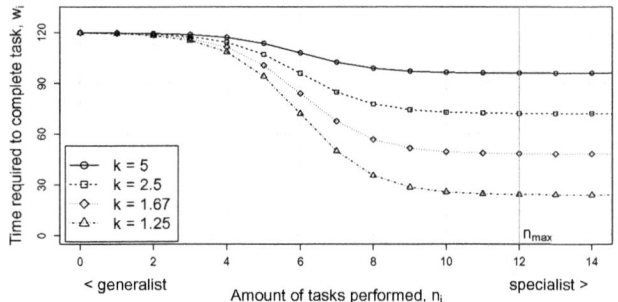

Fig. 1. The graph shows the effect of learning on the task completion time w_i for task τ_i. Learning applies when repeatedly working on the same type of task. Here, the standard task completion time in the unlearned state w_{std} is 120 s. The parameter k influences the time gain of learning. The values $k = \{1.25, 1.67, 2.5, 5\}$ shown here correspond to 20%, 40%, 60% and 80% of w_{std} at n_{max}, respectively.

3.2 Forgetting

In the model proposed, individuals either work on a task or search for a suitable task to work on. An individual that keeps searching gradually decreases the value of n_i, for each i, after traveling a distance d_f. This mechanism causes the individual to forget tasks previously completed, and thus to de-specialize while searching. Thus, individuals fall back to a *generalist* behavior over time. This also effectively keeps the individuals from remaining specialized on a task that is no longer available in the environment.

3.3 Task Allocation Strategy

The individuals of the swarm employ a simple stochastic strategy to decide if they start working on a task they encounter in the environment. The strategy is fully distributed and requires no communication between individuals, as it depends only on the individual's memory of previously completed tasks. We define the probability for an individual to start performing a task τ_i upon encountering it as a function of the number of completed tasks, n_i, as

$$p_i(n_i) = \frac{1}{1 + e^{-\alpha n_i}} \ , \tag{2}$$

with α being a control parameter for the steepness of the sigmoid probability curve. It influences the probability with which an individual accepts tasks of the same type: higher values of α will require a lower amount of tasks to be accomplished in order to reach the maximum probability. If the individual does not start to work on a task it encounters, it continues searching.

The function defined above leads to the specialization of the individual to a specific task as follows. By performing a task of type τ_i, an individual increases its probability of repeating this task type. This causes the individual to keep on repeating tasks of type τ_i (if available), thus becoming a specialist for this

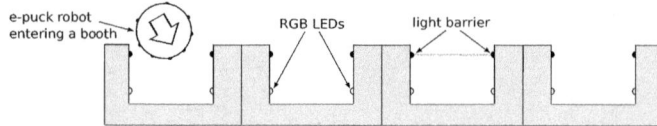

Fig. 2. Schematics of four booths arranged in an array. The light barriers detect a robot entering the corresponding booth. The color of the LEDs represent task type.

type of task. Conversely, if an individual remains inactive for a certain period of time, or performs other tasks than τ_i, the probability of accepting tasks of type τ_i upon encountering them decreases due to forgetting. As the probability of accepting a task decreases for all types of task while the individual searches, individuals slowly approach a generalist behavior if they remain searching for an extended period of time. We show later that by using this mechanism, the swarm will tend to organize itself in a way that the amount of individuals specialized in each task reflects the distribution of the tasks available in the envi ronment.

4   Experimental Setup

In this section, we describe the experimental framework we use to test the model presented in Sec. 3. We perform our experiments in simulation, using models of the e-puck robot. The e-puck is a small wheeled robot, designed as a research and educational tool for university students [12]. In the following we describe the way we abstract the tasks in the environment, the simulation framework we use for the experiments, and the metrics we use for evaluating the system.

4.1   Task Abstraction

In order to overcome the limited mechanical capabilities of the e-puck, we abstract the tasks that the robots can perform with a device developed in our laboratory [3]. The device is referred to as *booth*. Booths can be organized into arrays, as shown in Fig. 2. Each booth features a light barrier and two RGB LEDs. The LEDs can be perceived by a robot using its color camera. The robot can navigate to the booth and enter it depending on the information perceived. The presence of a robot can be detected by the booth using its light barrier. Upon the detection of a robot, the booth reacts by changing the color of its LEDs following a user defined logic.

In our experiments, each booth represents a task that can be executed. The type of the task is encoded by the LED color of the booth. Thus, a robot can perceive which type of task a booth represents. If a robot enters a booth, it is considered to work on the corresponding task. The booth acknowledges the robot's presence by temporarily turning the LEDs to red. The robot remains inside the booth for the time w_i required to complete the task. This time is, in general, different from robot to robot as it depends on the robot's internal state, regulated by the learning mechanisms explained in Sec. 3.3. After the robot has

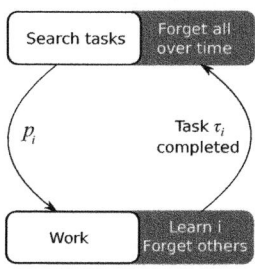
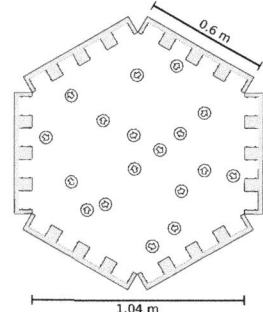

Fig. 3. Left: Finite state machine describing the behavior of the robots. Light rectangles represent actions executed by the robot, dark rectangles show the effect of learning and forgetting on the robot. A robot accepts a task it encounters with a probability p_i (see Eq. 2). Right: Representation of the arena with e-pucks at random initial positions. Tasks can appear at each of the booths located at the boundaries of the arena, with a total of 24 tasks simultaneously available in the environment. Task types are randomly distributed.

completed its task and left, the booth stochastically selects the color to light up next. By setting the probability with which a booth selects each color, we can modify the distribution of tasks types in the environment.

4.2   Simulation Tools

The work presented here has been carried out using the ARGoS simulation framework [14]. ARGoS is a discrete-time physics-based simulation framework developed within the Swarmanoid project.[1] ARGoS can simulate various robots at different levels of detail; the experiments presented in this work are carried out in a 2-dimensional kinematics-based simulation. ARGoS simulates the whole set of sensors and actuators available on the e-puck. For our experiments we employed the wheel actuators, the color camera for tasks detection, and the IR proximity sensors for obstacle avoidance. The booth, including its sensors and actuators, is also simulated in ARGoS.

In our experiments, tasks can be of two types, τ_g and τ_b, represented by green or blue LEDs, respectively. The robots can perceive tasks within a limited range using their camera, and recognize their type by their color. Fig. 3(left) shows the behavior of each robot. A robot randomly searches for a task to be performed. Upon the perception of a task (e.g., τ_g), a robot has an associated probability p_g of engaging in that task, defined by Eq. 2. Each robot is subject to the learning and forgetting mechanisms described in Sec. 3.

The environment consists of an obstacle free hexagonal arena, with six arrays forming the walls (see Fig. 3(right)). The tasks to be performed are represented by arrays at the boundaries of the arena. Each of the arrays consists of 4 booths, for a total of 24 tasks simultaneously available in the environment. The distance

---

[1] http://www.swarmanoid.org/

between an array and the one facing it on the opposite side of the arena is 1.04 m. This dimension is such that a robot that leaves a booth cannot directly see the tasks of the booth situated diametrically.

4.3   Specialization Measures

To quantify the degree of specialization in the swarm, we define two measures. A measure based on the internal probability of the robots to accept a certain task type, referred to as P *measure*, and a measure based on the frequency of task switches, referred to as F *measure*.

The P measure is based on the probability a robot has of accepting a task when encountering it. We say that a robot is specialized in a task of a certain type if its probability of accepting that type of task, as defined by Eq. 2, is greater or equal to 0.7. P is the number of robots specialized according to this definition. As the P measure depends on the internal probability of a robot to accept a task, it is only applicable to strategies that have such a probability.

The F measure is based on the frequency of switches between different task types by a robot. It has been developed by Gautrais et al. in their study of specialization in insect colonies [6]. The individual measure F_i consists of a value in the range $[-1, 1]$, representing the degree of specialization of a robot i. For a sequence of N_i tasks, it is computed as $F_i = 1 - (2D_i/N_i)$, where D_i is the total amount of switches between tasks of different types for robot i. F is the average over the values of F_i of all robots of the swarm. The value of F assumes 1, 0 and -1 for a fully specialized robot, systematic switching between task types and random task allocation respectively. Different from the P measure, it is independent of the underlying mechanism the robots use to select the tasks to work on.

5   Experiments

In this section of the paper we describe the experiments performed to study the proposed model, and present the results. We define to the ratio of tasks available in the environment as $\tau_b/(\tau_b + \tau_g)$, and refer to it as *task ratio*. It is set to 0.5 unless mentioned otherwise (equal probability of encountering one of the two types of tasks). The standard task completion time w_{std} is set to 120 s. The parameters of the learning function are set to $k = 1.25$ and $c = 1$, with $w_{min} = 24$. This results in an 80% time gain in task completion time at the maximal learning state, reached after $n_{max} = 12$ consecutive tasks of the same type. The forgetting distance d_f is set to 3 m. The swarm is composed of 18 e-pucks, randomly positioned in the arena at the beginning of each experimental run. The maximum speed of the robots is set to 3 cm/sec. The task acceptance probabilities p_g and p_b are initialized to 0.5; thus, all robots start as generalist. The value of α has been determined by trial and error in preliminary experiments and is set to 1. For each experimental condition we conduct 10 randomly seeded runs, for a duration of 10 000 simulated seconds each.

5.1   Basic Properties

In the first set of experiments we assess if the task allocation strategy presented in Sec. 3.3 successfully exploits the performance improvement available through learning. In the following, we refer to this strategy as *selective* strategy. We compare it to a reference strategy, the so called *greedy* strategy. Individuals using the greedy strategy accept, as the name indicates, any task they encounter. This allows us to quantify the performance improvement of specialization. Recall that the robots are subject to the learning and forgetting mechanisms, independently of the strategy they use for selecting their tasks. We measure the performance as the amount of tasks completed during 1000 seconds for each window of 1000 seconds after reaching the steady state ($t = 5000$ s). The data is normally distributed (see supplementary on-line material [4]); we therefore report the mean±SE. In case of the selective stra tegy the performance is 113.6±7.18, and in case of the greedy strategy it is 80.8±2.51. The performance of the two strategies is significantly different (Welch's t-test with $p = 5\%$).

Fig. 4 illustrates the degree at which the robots exploit learning. Two histograms show the task completion times for the greedy strategy (left) and the selective strategy (right). The histograms report data collected at the steady state ($t = 5000$ s). Only results for task τ_g are shown; analogous results are obtained for task τ_b. As the time spent working on a task decreases with learning, a higher frequency for low task completion times indicates an higher degree of learning. Fig. 4(right) shows that at the steady state, when employing the selective strategy, all robots complete their task at the minimal task completion time. This indicates that maximal learning for that type of task has been reached. Fig. 4(left) shows that tasks are performed mostly at high task completion times when using the greedy strategy. This indicates that there is poor or no learning in the swarm. As task allocatio n in the greedy strategy is random, some robots manage to improve their performance temporarily by learning, and thus complete their tasks in short time. Nevertheless, as this behavior is not persistent, the results show that a strategy that exploits learning through specialization outperforms a random task allocation strategy.

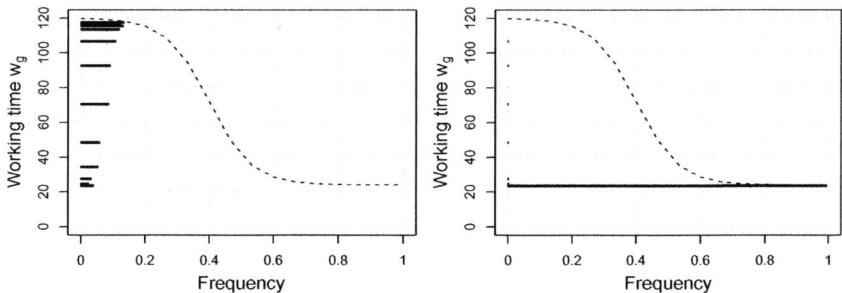

Fig. 4. Histogram for task completion time w_g, at steady state, for the greedy (left) and selective strategy (right). The learning curve (task completion time as function of the amount of tasks completed successively) is also superimposed.

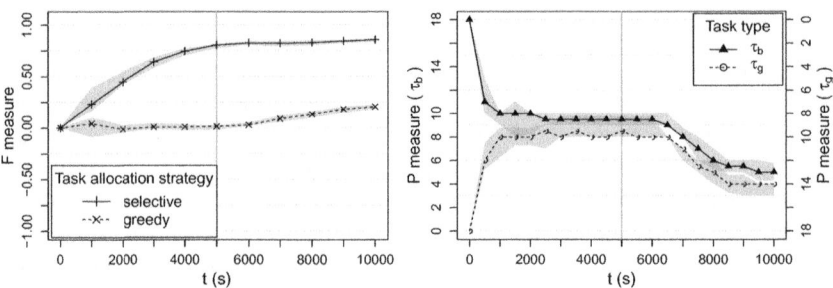

Fig. 5. Specialization of the robots of the swarm. Left: F measure for both strategies; right: P measure for the selective strategy only. The black line gives the median and the surrounding gray areas represents the interquartile range (IQR). The initial task ratio is 0.5, and is changed to 0.2 at $t = 5000\,\mathrm{s}$.

5.2   Adaptivity

In the second set of experiments we assess whether the task allocation strategy proposed is able to react to changes in the task ratio. The initial task ratio is 0.5; when the experiment reaches half its duration, the task ratio is changed to 0.2 in favor of τ_g. We evaluate the adaptivity by using the metrics F and P, as introduced in Section 4. For both measures we give the 25%,50% and 75% quantiles, as they are not normally distributed (see on-line supplementary material [4]). Fig. 5(left) reports the specialization measure F for both strategies. The F measure confirms that robots using the selective strategy specialize well, as its value is very close to 1 at the steady state (just before the change in the task ratio). This indicates a high degree of specialization in the group. The plot of the selective strategy shows that after the change, the swarm stays specialized on the predominant task. Different to this, the greedy strategy does not lead to a specialization of the swarm. The growth of F for the greedy strategy in the second part of the experiment reflects a higher amount of tasks of type τ_g in the environment.

Fig. 5(right) visualizes what happens after the change of the task ratio. The plot reports, for the selective strategy only, the number of robots specialized in the two types of task, using the probability-based measure P. The graph shows that before the change of the task ratio, half the swarm is specialized for each of the two task types. This matches the task ratio. At time $t = 5000\,\mathrm{s}$ the task ratio switches to 0.2 in favor of τ_g. After a period of approximately 1000 seconds, some of the robots de-specialize from τ_b and subsequently specialize in τ_g. When the steady state is reached, the number of robots specialized in τ_b/τ_g is 4/14, again matching the task ratio of the environment.

5.3   Costs and Benefits

In the third set of experiments we study whether specialization, which clearly has benefits in terms of task performance, entails costs that hinder the performance

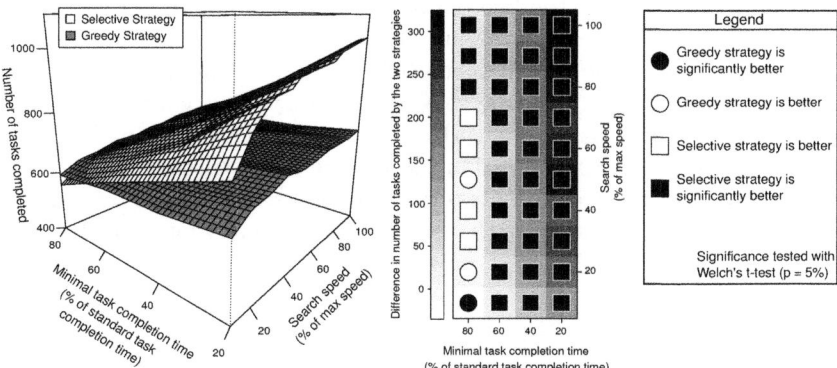

Fig. 6. Performance for different search speeds and task completion times at maximal learning. Left: observed mean of the number of completed tasks for the selective and greedy strategy (white and dark surface, respectively). SE < 2% for all values (not shown). Right: difference in number of tasks completed by the two strategies (gray shades), with indication of which strategy was better and whether the difference was statistically significant or not (see symbols in the legend).

of the swarm. The assumption is that a robot specialized for a certain task spends more time searching for it. Thus, specialists might be less efficient than generalists that accept the first task they encounter. In order to study the trade-off between costs and benefits, we vary the costs (search time of the robots) and benefits (the minimal task completion time in the learned state). We vary the search time of the robots by changing the speed they use while searching from 10% to 100% of the maximum speed, by steps of 10%. We vary the minimal task completion time w_{min} from a minimum of 20% to a maximum of 80% of the standard task completion time w_{std}, by steps of 20% ($k = \{5, 2.5, 1.67, 1.25\}$).

Figure 6(left) reports the mean of the number of completed tasks performed at the end of the experiment, using the selective strategy (white surface) and the greedy strategy (dark surface), for different values of search speed and minimal task completion time. The number of completed tasks is normally distributed with a SE < 2%; we therefore report only the observed mean (see on-line supplementary material [4]). Fig. 6(right) shows, for all values tested, the difference in number of tasks completed by the two strategies at the end of the experiment (gray shades), and if this difference is statistically significant or not (see symbols in the legend). The plot on the left shows that the greedy strategy is less affected by the two parameters. The change in minimal task completion time has no effect as the greedy strategy does not specialize and thus does not benefit from the effects of learning. Moreover, the performance of the greedy strategy is only slightly affected by the search speed because tasks are abundant in the environment and the robots using the greedy strategy accept every task they encounter. On the other hand, the performance of the selective strategy varies considerably in relation to the value of the two parameters, highlighting costs and benefits of specialization. The plot on the right shows that when task completion time is 80% of the w_{std} and the search speed 30% of the maximum speed, the greedy strategy performs better than the selective strategy. This confirms our assumption that robots specializing in a certain task are prone to loosing efficiency due to longer search times. Specialization is therefore not to be considered in terms of benefits only, as it is affected by external factors as task availability and spatial distribution, which might lower the benefits of specialization considerably.

6   Conclusions

In this work we focused on the topic of behavioral specialization in a swarm of robots. Behavioral specialization can often be observed in societies such as humans and social insects, as its advantages are many. Among others, it allows to exploit the improvements in task performance by learning. This means that if individuals improve their performance in a set of tasks by repeatedly carrying out those tasks, they should exploit this advantage by specializing in this set of tasks. Specialization can also entail costs: specialists may spend more time in searching for their tasks with respect to generalists.

In this work we proposed a simple model of a system in which individuals benefit from learning in terms of reduction of the time needed to perform tasks. We studied the case in which the robots can perform two types of tasks, available in the environment in a unknown ratio. We implemented a simple self-organized task allocation mechanism that leads to behavioral specialization of the individuals to one of the two tasks. We studied the system in simulation-based experiments, focusing on the response of the system to changes in the environment and on its behavior in different conditions in terms of costs and benefits of specialization. We identified cases in which costs of specialization overcome the benefits gained through learning. A task allocation strategy that does not use specialization is preferable in these cases. The results also suggest that specialization is not a good choice in highly dynamics environments, as specialists may not be able to adapt to changes fast enough. Future research will focus on the study of specialization in swarms of heterogeneous robots, where benefits and costs of specialization are linked to morphological differences between robots. Additionally, we plan to implement learning as actual improvements of behaviors instead of modeling it as an external parameter.

Acknowledgements. Marco Dorigo acknowledges support by the European Research Council through the ERC Advanced Grant "E-SWARM: Engineering Swarm Intelligence Systems" (contract 246939). Marco Dorigo, Mauro Birattari, and Arne Brutschy acknowledge support from the Belgian F.R.S.–FNRS. Marco Frison acknowledges support from "Seconda Facoltà di Ingegneria", Alma Mater Studiorum, Università di Bologna.

References

1. Beshers, S.N., Fewell, J.H.: Models of division of labor in social insects. Annual Review of Entomology 46, 413–440 (2001)
2. Bonabeau, E., Dorigo, M., Theraulaz, G.: Swarm Intelligence: From Natural to Artificial Systems. Oxford University Press, New York (1999)
3. Brutschy, A., Pini, G., Baiboun, N., Decugnière, A., Birattari, M.: The TAM: A device for task abstraction for the e-puck robot. Tech. Rep. TR/IRIDIA/2010-015, IRIDIA, Université Libre de Bruxelles, Brussels, Belgium (2010)
4. Brutschy, A., Tran, N.L., Baiboun, N., Frison, M., Pini, G., Roli, A., Dorigo, M., Birattari, M.: Costs and benefits of behavioral specialization – Online supplementary material (2011), http://iridia.supp/IridiaSupp2011-015/
5. Dornhaus, A.: Specialization does not predict individual efficiency in an ant. PLoS Biology 6(11), e285 (2008)
6. Gautrais, J., Theraulaz, G., Deneubourg, J.-L., Anderson, C.: Emergent polyethism as a consequence of increased colony size in insect societies. Journal of Theoretical Biology 215(3), 363–373 (2002)
7. Ikemoto, Y., Miura, T., Asama, H.: Adaptive division-of-labor control algorithm for multi-robot systems. Journal of Robotics and Mechatronics 22(4), 514–525 (2010)
8. Jones, C., Matarić, M.J.: Adaptive division of labor in large-scale minimalist multi-robot systems. In: 2003 IEEE/RSJ International Conference on Intelligent Robots and Systems (IROS 2003), pp. 1969–1974. IEEE Press, Piscataway (2003)
9. Labella, T.H., Dorigo, M., Deneubourg, J.-L.: Division of labour in a group of robots inspired by ants' foraging behaviour. ACM Transactions on Autonomous and Adaptive Systems 1(1), 4–25 (2006)
10. Li, L., Martinoli, A., Abu-Mostafa, Y.S.: Learning and measuring specialization in collaborative swarm systems. Adaptive Behavior 12(3-4), 199–212 (2004)
11. Liu, W., Winfield, A.F.T., Sa, J., Chen, J., Dou, L.: Towards energy optimization: Emergent task allocation in a swarm of foraging robots. Adaptive Behavior 15(3), 289–305 (2007)
12. Mondada, F., Bonani, M., Raemy, X., Pugh, J., Cianci, C., Klaptocz, A., Magnenat, S., Zufferey, J.C., Floreano, D., Martinoli, A.: The e-puck, a robot designed for education in engineering. In: Gonçalves, P.J.S., Torres, P.J.D., Alves, C.M.O. (eds.) Proceedings of the 9th Conference on Autonomous Robot Systems and Competitions, pp. 59–65. IPCB: Instituto Politècnico de Castelo Branco, Castelo Branco (2009)
13. Nitschke, G., Schut, M., Eiben, A.: Emergent specialization in biologically inspired collective behavior systems, pp. 100–140. IGI Publishing, New York (2007)
14. Pinciroli, C., Trianni, V., O'Grady, R., Pini, G., Brutschy, A., Brambilla, M., Mathews, N., Ferrante, E., Di Caro, G., Ducatelle, F., Stirling, T., Gutiérrez, A., Gambardella, L.M., Dorigo, M.: ARGoS: a pluggable, multi-physics engine simulator for heterogeneous swarm robotics. Tech. Rep. TR/IRIDIA/2011-009, IRIDIA, Université Libre de Bruxelles, Brussels, Belgium (2011)
15. Ratnieks, F.L.W., Anderson, C.: Task partitioning in insect societies. Insectes Sociaux 46(2), 95–108 (1999)

CrunchBot: A Mobile Whiskered Robot Platform*

Charles W. Fox, Mathew H. Evans, Nathan F. Lepora,
Martin Pearson, Andy Ham, and Tony J. Prescott

Active Touch Laboratory at Sheffield (ATL@S),
University of Sheffield, Sheffield, UK
charles.fox@sheffield.ac.uk

Abstract. CrunchBot is a robot platform for developing models of tactile perception and navigation. We present the architecture of Crunch-Bot, and show why tactile navigation is difficult. We give novel real-time performance results from components of a tactile navigation system and a description of how they may be integrated at a systems level. Components include floor surface classification, radial distance estimation and navigation. We show how tactile-only navigation differs fundamentally from navigation tasks using vision or laser sensors, in that the assumptions about the data preclude standard algorithms (such as extended Kalman Filters) and require brute-force methods.

1 Introduction

Touch-based navigation has two principal applications. Firstly, as a sole sensory system in environments where other types of sensors fail, such as smoky or dusty search-and-rescue sites, especially where covert (no signal emission) operation is required. Secondly, as a complement to other sensors such as vision, with which it can be fused or used as a 'last resort' during adverse conditions as in the sole sensor case.

This paper presents a new robot platform, CrunchBot (Fig. 1(a)) which draws together several strands of research towards systems for autonomous whisker-based tactile navigation. We and other authors have previously investigated individual components of such a system in isolation, including whiskered texture recognition [4],[11],[2], [8],[13], surface shape recognition [12],[9],[7],[3], and object recognition [6]. These components are often tested under ideal laboratory conditions or in individual mobile settings [16]. Here we present initial steps towards integrating them into a single platform, along with new results and observations on their performance 'in the wild' in a common arena environment. This integration report serves as a case study for other researchers wishing to work with similar platforms, listing the particular technologies used, the tools that link them together and providing new insights into the performance of such integrated systems.

---

* This work was supported by EU Framework project FP7-BIOTACT (ICT-215910).

R. Groß et al. (Eds.): TAROS 2011, LNAI 6856, pp. 102–113, 2011.

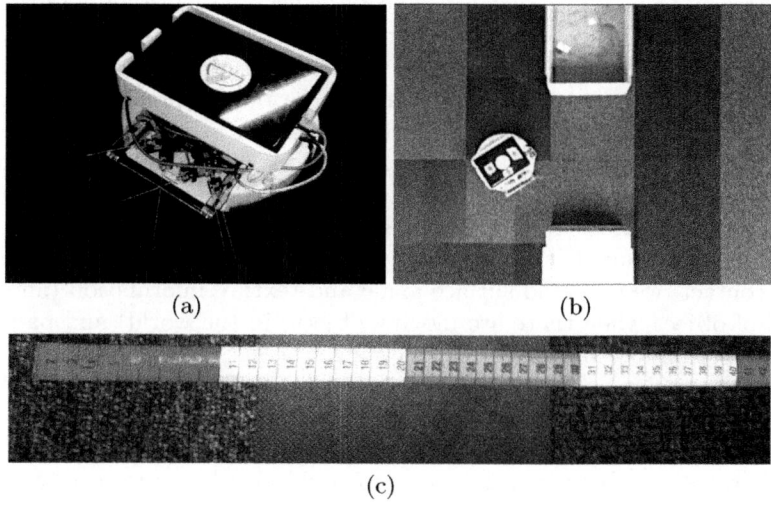

Fig. 1. (a) Crunchbot. (b) Overhead view of Crunchbot in the arena environment. Different carpet tile textures can be seen on the floor along with square obstacles. (c) Textured floor tiles used in the navigation task. From left to right; smooth carpet, vinyl and rough carpet.

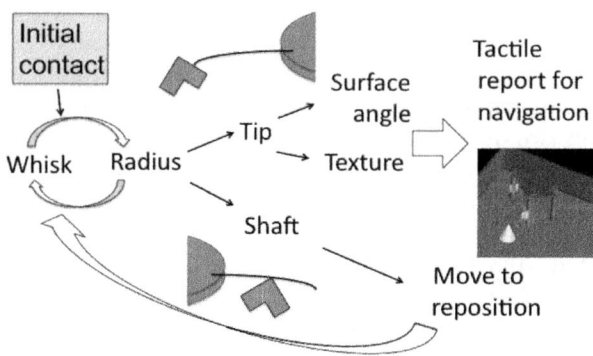

Fig. 2. Framework overview for whisker-based perception and navigation

Fig. 2 gives an overview of our general framework for perception and navigation with whiskers, which the present study works towards. When based on biological whiskers, whisker sensors have strain sensors at their base only. When a rat investigates an object it palpates the surface in a behaviour known as whisking [15],[1],[10]. It is thought that whisking is important for gathering the most reliable signals from whisker contacts. Whiskers can make two distinct types of contact with an object, contacting either at the tip or the shaft. Tip contacts are generally the most useful, because they provide a standardised, constrained setting (i.e. with the contact point at a known location at precisely the end of the whisker) from which surface properties such as orientation and texture can be identified [12],[4]. In contrast, shaft contacts are less informative; for example, the radial distance of an object can confuse attempts to classify surface orientation and texture [8]. Shaft contacts are rare in practice, occurring only when small objects enter the field of multi-whisker arrays between the whisker tip points. In the scheme used here, a radial distance estimator [3] is first used to make a decision of whether the contact is at the tip or the shaft. If it is a shaft contact, then the robot should use the radial distance information to move to another location that is likely to yield a more useful tip contact. Following a tip contact, we can read surface angle and texture information (and possibly speed of object when there are moving objects in the world) and pass them as an observation to a navigation or mapping system.

The remainder of this paper outlines an initial implementation of each of these stages, which for the first time are integrated into a complete system. The present navigation method is driven by texture and distance observations only. An intermediate object recognition step could also be performed, as described in a companion paper [6].

2   Methods

Whiskers. CrunchBot's six whiskers measure 160mm in length, 1.45mm diameter at the base tapering linearly to 0.3mm at the tip. They are built from nanocure25 using an Evisiontec rapid prototyping machine. A magnet is bonded to the base of the whisker and held in place by a plug of polyurethane approximately 0.75 mm above a Melexix 90333 tri-axis Hall effect sensor IC [14]. This sensor generates two outputs representing the direction of the magnetic field (in two axes) with respect to its calibrated resting angle. These two 16-bit values are sampled by a local dsPIC33f802 micro-controller which, in turn, is collected using an FPGA configured as a bridge to a USB 2.0 interface. Up to 28 whiskers can be connected to this FPGA bridge at one time. Using the vendor provided software driver and API (Cesys GmbH), a user can request the data from all whiskers at minimum intervals of $500\mu s$ (a maximum sample rate of 2kHz).

Robot platform. The whiskers are mounted in the cargo bay of an iRobot Create base (www.irobot.com), being positioned on an adjustable metal bar and rapid prototyped ball joint mountings. These mountings allow adjustment of the whiskers, which is particularly important for obtaining good floor contacts. We have also extended the cargo bay mounting to accommodate a netbook PC, which is used for local control of the robot. The netbook runs Ubuntu 10.10 on a single-core Intel Atom processor. A circular buffer in shared memory is used to make data from the Cesys driver available to other processes. The netbook hosts a Player server (playerstage.sourceforge.net) which provides high-level, networked API interfacing to the Create's serial port commands. Processes such as texture and shape recognition and basic motor control run on the netbook, reading the raw data from the fast circular shared memory buffer, and (usually, see Section 2 below) writing their results every 0.1s to a Python Pyro server (pyro.sourceforge.net) on the remote desktop. Differential and absolute odometry data from the Create is also sent to this server. Preliminary experiments in our lab show that the odometry of the Create, once loaded with the sensing and control hardware, is accurate to $< 5\%$ of any straight line or turn on the spot movements.

Navigation task. Fig. 1(b) shows the arena environment used in our tests. The arena is a 2.5m×2.5m square, surrounded by walls and paved with twenty five 0.5m×0.5m tiles. There are three types of tiles with different textures: vinyl, smooth carpet and rough carpet (see Fig. 1(c)). A few 0.5m×0.5m square obstacles are also placed over some carpet tiles. The current system implements a random exploratory behaviour of the arena, controlled by the netbook.

Previous work has shown that accurate object localisation with a whisker requires some measure of contact speed [3], or of the applied forces and bending moments at the base of the whisker [9],[12], values that are not always available in the mobile case as agent movement will affect these contacat properties. To address these points a 'body whisk' behaviour was included in the robot program. As the whiskers were not actuated the whole robot must rotate in a systematic way. Upon initial contact with an object the robot first reverses away a short distance before rotating at $\pi/24$ radians per second towards the object, then rotating at $\pi/24$ radians per second away from the object. This allows this whiskers to move over the surface of the contact object, collecting data about its shape. After the whisk the robot reverses again to clear the object, then rotates in a random direction and moves forward again.

The ultimate task is to infer the robot's location from noisy odometry and the whiskers. There is some subtlety in how odometry is reported. Inference running on the remote desktop may become computationally intensive, and its update cycle can fall behind the rate of reporting from the netbook. Markovian inference algorithms can simply discard any unused sensory observations, requiring only the latest observation, but the odometry must be *integrated* so that each inference step receives the total odometry occurring since the previous inference step. Thus multiple read outs may be required from the robot sensory systems, which were taken at 0.1s intervals.

Floor texture discrimination. The outer two of the six whiskers are angled downwards to make a light, brushing contact with the floor surface that CrunchBot is travelling over (Fig. 1). Classification software on the netbook then seeks to classify the whisker deflection signals into previously learnt classes (for example, vinyl or rough carpet), to infer which surface the robot is travelling over. In its current configuration, the signals from the two outer whiskers are classified individually, so that the robot can determine whether its left and right sides are on the same or different textures.

A stationary naive Bayes algorithm is used to infer the surface texture from the whisker contacts (see also Lepora *et al* (2011) in this conference volume). This algorithm has been demonstrated to be a reliable classifier of surface texture from whisker contacts on mobile robot platforms [13]. The classifier uses stored texture log-likelihoods $\log P(x|\mathrm{T})$ that represent the log-probability distributions of contact deflections in training data from each possible texture. The overall log likelihood over a temporal window of whisker deflection data is then obtained by integrating these the log-likelihoods over the individual samples, assuming naive sample independence and stationarity over time,

$$\log P(x_{n_s}, \cdots, x_{n_f} | \mathrm{T}_l) = \sum_{i=n_s}^{n_f} \log P(x_i | \mathrm{T}_l). \tag{1}$$

In previous studies of this classifier [13], these likelihoods were fed into Bayes rule to give the posterior probabilities of the probabilities for each texture having generated the window of data. Instead, the present method feeds the likelihoods directly into the navigation system for the robot, to be used to infer navigation information.

Another principal difference from our previous applications of stationary naive Bayes, is that the present implementation requires that the classification be done in real-time on board the robot. Moreover, the texture that the robot is sensing can change during the motion, as the robot moves from one tile to the next. For these reasons, only the most recent 100ms (200 sample) window of texture data was classified, to give good classification reliability while minimising possible boundary crossing. Fortunately, the algorithmic complexity of the classifier is low, so that classification times were much less than the inter-classification intervals of 100ms.

Object localisation. To determine whether an object has made contact with a whisker at the tip or the shaft, and to discriminate between contacts with the surfaces or corners of objects, object localisation was implemented. Previous work [3] has shown that peak deflection magnitude could be used as a feature for radial distance discrimination at a given speed. Whisker data was recorded during the 'body whisk' contact, and the maximum whisker deflection was measured. Deflection magnitude was taken as the Hall effect sensor output voltage at peak deflection, which is proportional to the bending moment. This feature f_1 can be defined as,

$$f_1 = \max_t \theta(t), \tag{2}$$

where $\theta(t)$ is the time-dependent deflection magnitude measured by the Hall effect sensor.

During the training phase a dataset was collected for each whisker, consisting of 5 contacts at each point along the whisker at 10mm intervals over a 50mm range from the tip of the whisker. Though the whisker is 160mm long, only 140mm is external to the 'follicle'. A model was then generated of the relationship between the deflection magnitude and the corresponding radial distance to contact by fitting a linear equation to the training data in MATLAB. To find an estimate of radial distance r,

$$r = a_1 f_1 + a_0, \tag{3}$$

was fitted to the data with a linear-in-the-parameters regression on the line, giving a least-squares fit for (a_0, a_1) for each whisker.

Navigation. A key question to address in tactile navigation and mapping is to what extent methods developed for other sensory modalities, such as laser scanners, are applicable to tactile sensors. A naive view of tactile navigation holds that touch sensors are just like laser scanners but with a very short range, and therefore standard approaches could be used.

Standard laser-driven methods include Extended Kalman Filtering (EKF) and Iterative Closest Point matching. In ICP, large point clouds are aligned between steps to compute odometry estimates. In EKF, unique or rare features are extracted from point clouds, and used as discrete landmarks. In both of these methods, the pose likelihood function varies smoothly over poses, because the lasers have a long range, and their values change gradually with pose. In particular, the assumption of this smoothness allows the EKF to approximate the likelihood function by a Gaussian, with parameters given by the local curvature around its prior mean. It is unclear in the tactile case whether the prior mean is closely related to the posterior mean, as the relationship depends on the likelihood smoothness.

Foveal grid. When these assumptions are violated, brute force computation is required, such as grid based or particle filtering methods. To examine what approach best matches the data, we have implemented a simple hybrid grid and Gaussian likelihood model. By inspecting the shape of the likelihood functions in the grids we will see if smoothness and Gaussian assumptions are appropriate. To enable inference to run in real-time, we restrict the grid to a 'foveal' region centred on the robot's prior mode location, and oriented in egocentric coordinates. We denote likelihoods as $P(s|x, y, \theta)$ where (x, y, θ) each range over 11 discrete values in the grid, and s is the current sensory report. (The grid is 0.25m in diameter, with θ from $\pi/4$ to $+\pi/4$.)

Filtering. A slow method to filter the belief at each step, given sensory reports $s_{1:t}$ over time t would be to work entirely in the grid representation,

$$P(X_t|s_{1:t}) = P(s_t|X_t) \sum_{X_{t-1}} P(X_{t-1}|s_{1:t-1})T(X_t|X_{t-1}, o_t) \qquad (4)$$

where we write $X_t = (x_t, y_t, \theta_t)$, and T is a transition function describing motion probabilities as a function of observed odometry data o_t, and the sums range over all possible states of the grids. This method is slow because it performs 3-dimensional convolutions. We speed up filtering by approximating the posterior and transitions with Gaussians, $N(\mu_t, \Sigma_t)$ and $N(\mu_{o_t}, \Sigma_{o_t})$, at each step as follows. First we compute the grid values,

$$P(X_t|s_{1:t}) = P(s_t|X_t)(N(\mu_{t-1}, \Sigma_{t-1}) * N(\mu_{o,t}, \Sigma_{o,t})) \qquad (5)$$

for each discrete X_t (where $(o, t) = o_t$ to reduce subscripts). Here the computation only ranges over the present 3D grid, rather than the convolution over two such grids. The simpler convolution of Gaussians, $*$, is quickly computed analytically. We then fit a Gaussian to the resulting set of posterior grid points, $\{X_t\}$,

$$P(X_t|s_{1:t}) \approx N(\mu_t = \hat{X}_t, \Sigma_t = cov_{\hat{X}_t}(\{X_t\})) \tag{6}$$

by picking the mode and computing the covariance about the mode with the sum

$$cov_{\hat{X}_t}(\{X_t\}) = \sum_{X_t}(X_t - \hat{X}_t)^T(X_t - \hat{X}_t) \tag{7}$$

This foveal filtering method is illustrated in Fig. 3. Here the agent has moved from a to b and has detected an object with its right whisker. Its previous posterior Gaussian, centred on point c, is used together with the Gaussian transition T to produce a prior Gaussian about point d. This Gaussian is discretised and fused with likelihoods at each grid point to give the posterior shown in the enlargement e. We would then fit a Gaussian to this posterior grid. (This method is similar to the EKF and Unscented Kalman Filters in using Gaussian posteriors, but differs by using a brute-force likelihood. Recall that we wish to examine these likelihoods to see how valid the approximations of EKF and UKF would be.)

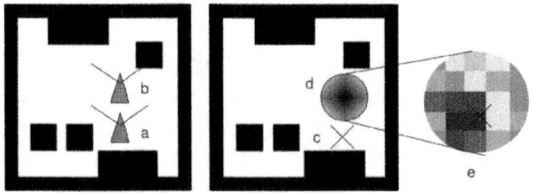

Fig. 3. Foveal grid based navigation

Transition function. Exact transition functions for 2D mobile robots are 'banana shaped' under Gaussian uncertainty about the starting angle, because the direction of travel is uncertain. We use a heuristic Gaussian approximation to this shape, which preserves the mode and tries to approximately minimise the KL divergence from the true distribution. It is also necessary to add a small additional component to each posterior covariance to compensate for quantisation error in the grid and the lack of computation about locations beyond the foveal grid.

Likelihood function. In the likelihood function, $s_t = (\tau_{i=1:2}, r_{i=1:4})_t$ is comprised of information from the two floor sweeping whiskers returning texture classes $\tau_{i=1:2} \in \{VINYL, CARPET, ROUGH\}$ and the four object localisation whiskers returning radial distances to contact, $r_{i=1:4}$ ranging from 1mm to 140mm at contact or null ($\emptyset$) for no contact. We assume these observations are conditionally independent given location and that the radial distance errors are Gaussian. We use the noise levels found from empirical data. We speed up the likelihoods by computing them offline for each grid square in the whole arena and saving them in a hash table. (This requires coordinates in the egocentric grid to be transformed to world-centred coordinates to perform the look-up).

3   Results

Floor texture discrimination. The robot's performance at real-time texture classification was assessed by having the robot follow a course in the arena where it traverses different textures. For an initial investigation, only two surface textures were considered, corresponding to a smooth vinyl and rough carpet. The classification algorithms were trained by presenting the robot with just a single texture, from which its central controller determined the corresponding texture likelihoods for that surface. These were then stored for use in general classification.

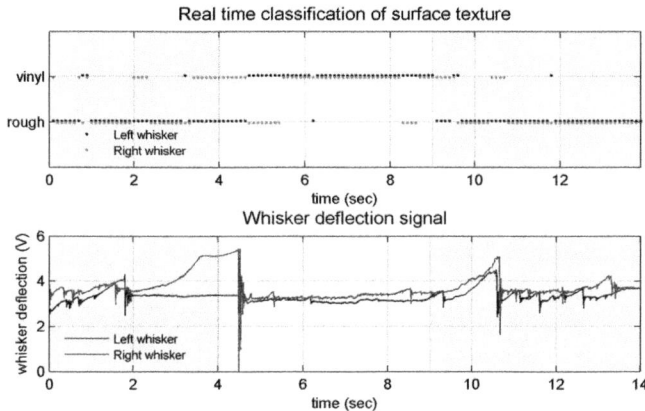

Fig. 4. Real-time classification of surface texture. The top panel shows the classification results for the left (blue) and right (green) whiskers traversing a course over rough carpet (shaded regions), then vinyl (unshaded region) and back to rough carpet. The bottom panel shows the deflections associated with each whisker.

A typical example of the classification performance is shown in Fig. 4. The course consisted of a tile of rough carpet, then a tile of smooth vinyl, and finally a tile of rough carpet. The approximate times when the robot was traversing each texture are marked on the figure, with rough carpet shaded and smooth vinyl unshaded. As is visible from the figure, both whiskers reliably reported the correct texture.

One general feature that we observed is that the texture classifier can produce sporadic results near texture boundaries, with the whiskers disagreeing about which texture is being encountered (Fig. 4 top panel, borders of shaded regions). Examining the whisker deflection traces, reveals that large deflections of the whisker can occur in these regions that last over a second. We diagnosed this issue as due to the whisker catching on the boundary between two tiles of different heights, and so this feature is actually a signature of a change in texture. In general, catching the whisker on the floor surface caused problems for the classification, because these are infrequent events that can last several hundred milliseconds or more. Hence, we sought to position the whisker to angle as far back as possible to minimise such events, which emphasises that the way in which the whisker contacts the floor can be important for how reliable a classifier can perform.

*Object localisation.*Peak deflection magnitude for each contact is shown in Fig. 5. Standard deviation of error for radial distance estimation is shown in the table below.

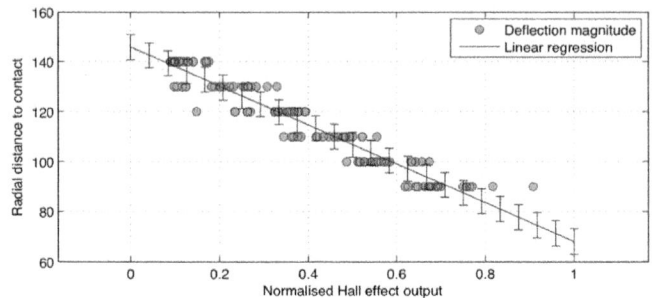

Fig. 5. Peak deflection magnitude for contacts along the shaft of the whisker. Standard error for the regression is 4.98mm.

	Whisker 1	Whisker 2	Whisker 3	Whisker 4	Combined
Std error	5.68mm	2.78mm	1.82mm	4.37mm	4.98mm

Standard classification error is very low, typically less than 5mm over the 60mm range tested. For some whiskers classification error is even lower, below 2mm. These results compare favourably with previous work under highly controlled conditions where speed was variable. This indicates that the noise in the odometry is low enough to ensure a consistent contact force and speed.

Navigation: likelihood function forms. Fig. 6(a) shows typical examples of forms of the likelihood function, from six poses in the arena. (Dimension stacking is used to display each 3D likelihood cube as a row of 2D slices.) It can be seen that these functions are highly non-Gaussian and non-smooth, consisting mostly of multiple straight edges and corners arising from the form of the floor and object layouts in the arena. They include many abrupt discontinuities, because the whiskers only sense a small local area, and neighbouring locations can have completely different textures or object contacts. This is unlike the case of vision or lasers whose likelihoods are locally smooth, and suggests that brute force methods such as the foveal grid (or particle filters) are more appropriate to tactile navigation than methods which assume that the likelihood is Gaussian or otherwise smooth.

Navigation: posterior fusion. Fig. 6(b) shows two typical examples of fusing priors (which are always Gaussian) and likelihood grids, when the prior mode is a short distance away from the ground truth, but still within the fovea radius. In the first example, the likelihood function is extremely focused – because only two points are compatible with the sensors – so it dominates the posterior to

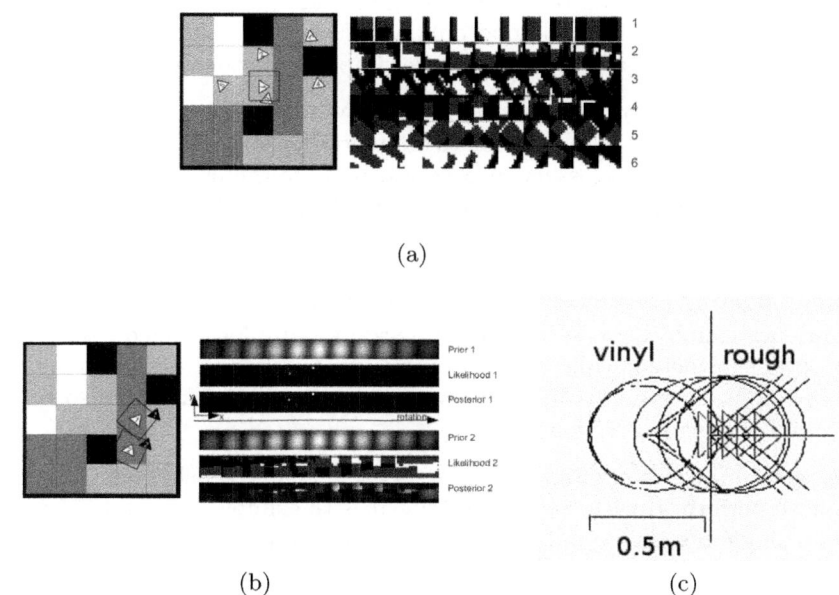

(a)

(b)   (c)

Fig. 6. (a) Example of likelihoods (lighter=stronger) at six locations during an exploration. The white triangles on the map (left) show the fovea locations. At each step we show the likelihood function (right) as a row of 11 × 11 foveal location squares, over a third dimension (along the row) of orientation from $-\pi/4$ to $+\pi/4$. The black squares in the arena are solid objects; white, light and dark grey are VINYL, SMOOTH and ROUGH floor surfaces respectively. The square around pose (1) on the map shows the location and size of the fovea. Foveas are not shown at other locations for simplicity.(b) Two examples of non-smooth likelihoods fusing with Gaussian priors to give non-smooth posteriors. The location of the posterior mode can be highly dependent on the likelihood function, and approximating the likelihood with a Gaussian would change its location. (c) An early test of pose correction in the fully integrated system. (Screenshot from graphical display). The triangles show a sequence of five ground truth locations of the robot moving in a straight line, crossing the boundary from a vinyl tile to a rough tile. The circles show the covariance of the posterior beliefs for the same five steps. The belief is initialised to an incorrect location, 0.2m behing the ground truth. The output shows the beliefs jumping to be closer to the ground truths as the robot crosses the texture boundary.

correct the belief. The second example shows a multi-peaked likelihood function knocking out the region around the prior mode, and producing a multi-peaked posterior. These two typical examples show that using the posterior mode as the next prior mean is an appropriate method, rather than using the posterior mean, or linearising the posterior about the old prior mean as in the EKF. They illustrate the advantage of fitting the Gaussian to the posterior grid, rather than Gaussianising the likelihood only and fusing it analytically with the prior Gaussian (which we attempted in earlier implementations with little success.)

Navigation: odometry. In simulations with low sensor noise, we found that computing the 'banana' approximation to odometry noise under uncertain pose angle noise models was unnecessary in practice, because the use of the foveal grid restricts the posterior Gaussian to the size of the fovea. As long as the likelihood functions are strong enough to move the mode around when fused with the prior, we found little difference in behaviour between using the full approximation and simply inflating every posterior covariance to equal the size of the fovea. The covariance cannot increase beyond this size because locations outside the fovea are never considered; while the full covariance could reduce in size on informative observations, such observations are rare in our tactile task, and also have little effect on the mode positions.

Full system integration. Fig. 6(c) shows a very early screenshot from the first test we have performed with the full system. It is the simplest possible test, moving over a single texture boundary and correcting the posterior belief.

4   Discussion

We have demonstrated an initial implementation of our framework for whiskered perception and navigation, showing how real-time signal processing, texture classification, distance estimation and navigation can be combined on an inexpensive mobile platform.

It is important to consider interactions between components when integrating systems. For example, we found it useful to disable both odometry reports and texture classification while performing mini-whisks. As noted in previous studies, contact location confounds surface property discrimination [8] and we are currently working to implement active sensing behaviours which position the robot so as to obtain standardised contact types to aid perception. Placing action at the heart of the perceptual process in this way is only possible with integrated mobile systems, rather than the statically mounted whiskers that have been used in previous single component laboratory tests [9],[17].

Although introduced purely for computational reasons, both the use of likelihood hash table and the mode-centred Gaussian posterior approximations are similar to, and loosely inspired by, biological models of hippocampal navigation [5]. However use of a single mode-centered Gaussian has lead to problems in pilot runs in which the true posterior becomes strongly multimodal, for example when a focussed likelihood appears far away from the prior mode. Future systems may address this using small particle filters – even just two particles would help. A larger scale solution to navigation is to perform an object recognition step using sensory information, then use the objects as landmarks – such an object recognition system is described in a companion paper [6].

References

1. Diamond, M.E., von Heimendahl, M., Knutsen, P.M., Kleinfeld, D., Ahissar, E.: 'where' and 'what' in the whisker sensorimotor system. Nat. Rev. Neurosci. 9(8), 601–612 (2008)
2. Evans, M., Fox, C.W., Pearson, M.J., Prescott, T.J.: Spectral template based classification of robotic whisker sensor signals. In: Proc. TAROS (2009)
3. Evans, M., Fox, C.W., Pearson, M.J., Prescott, T.J.: Whisker-object contact speed affects radial distance estimation. In: Proc. IEEE ROBIO (2010)
4. Fend, M.: Whisker-based texture discrimination on a mobile robot. Advances in Artificial Life, 302–311 (2005)
5. Fox, C., Prescott, T.: Hippocampus as unitary coherent particle filter. In: Proc. Int. Joint Conference on Neural Networks, IJCNN (2010)
6. Fox, C., Prescott, T.J.: Mapping with sparse local sensors and strong hierarchical priors. In: Groß, R., Alboul, L., Melhuish, C., Witkowski, M., Prescott, T.J. (eds.) TAROS 2011. LNCS (LNAI), vol. 6856, pp. 183–194. Springer, Heidelberg (2011)
7. Fox, C.W., Evans, M., Pearson, M.J., Prescott, T.J.: Towards temporal inference for shape recognition from whiskers. In: Proc. TAROS (2008)
8. Fox, C., Mitchinson, B., Pearson, M., Pipe, A., Prescott, T.: Contact type dependency of texture classification in a whiskered mobile robot. Autonomous Robots 26(4), 223–239 (2009)
9. Gopal, V., Hartmann, M.J.Z.: Using hardware models to quantify sensory data acquisition across the rat vibrissal array. Bioinspir. Biomim. 2(4), S135–S145 (2007)
10. Grant, R.A., Mitchinson, B., Fox, C.W., Prescott, T.J.: Active touch sensing in the rat. Journal of Neurophysiology 101, 862–874 (2009)
11. Hipp, J., Arabzadeh, E., Zorzin, E., Conradt, J., Kayser, C., Diamond, M., Konig, P.: Texture signals in whisker vibrations. J. Neurophysiol. 95(3), 1792 (2006)
12. Kim, D.E., Moller, R.: Biomimetic whiskers for shape recognition. Robotics and Autonomous Systems 55(3), 229–243 (2007)
13. Lepora, N., Evans, M., Fox, C., Diamond, M., Gurney, K., Prescott, T.: Naive Bayes texture classification applied to whisker data from a moving robot. In: Proc. IEEE World Congress on Comp. Int., WCCI 2010 (2010)
14. Melexis, www.melexis.com/assets/mlx90333_datasheet_5276.aspx
15. Mitchinson, B., Martin, C.J., Grant, R.A., Prescott, T.J.: Feedback control in active sensing: rat exploratory whisking is modulated by environmental contact. Proc. Biol. Sci. 274(1613), 1035–1041 (2007)
16. Prescott, T., Pearson, M., Mitchinson, B., Sullivan, J., Pipe, A.: Whisking with robots from rat vibrissae to biomimetic technology for active touch. IEEE Robotics and Automation Magazine 16(3), 42–50 (2009)
17. Russell, R., Wijaya, J.: Object location and recognition using whisker sensors. In: Australasian Conference on Robotics and Automation. Citeseer (2003)

Deformation-Based Tactile Feedback Using a Biologically-Inspired Sensor and a Modified Display

Calum Roke, Chris Melhuish, Tony Pipe, David Drury, and Craig Chorley

Bristol Robotics Laboratory, Bristol, UK
calum.roke@brl.ac.uk

Abstract. Skin deformation has been previously shown as vital for lump detection during the direct manipulation of an object. A deformation-based tactile feedback system is developed and presented that senses and displays this tactile information for palpation in tele-surgical applications. A biologically-inspired tactile sensor modelled from the human fingertip is used to obtain the skin deformation during interaction. It does this by measuring the displacement of the sensor's artificial intermediate epidermal ridges using a simple, computationally-efficient algorithm. The design of a previously-published tactile shape display is then recreated and improved for relaying this sensed information on to a human user's fingertip. This tactile display uses remote actuation to reduce the mass of the display, avoiding the issue of adding a large mass to a tele-operation interface. The tactors within the developed display exhibit 2.5 mm displacement, with a 2.5 mm spacing, 12 Hz bandwidth and a stiffness of 5.0 N/mm. A linear relationship is found between sensor deformation and tactor displacement and the spatial performance of the system is proven by successfully detecting lumps within artificial muscle tissue. This new deformation-based tactile system offers an intuitive sense of touch with minimal processing.

1 Introduction

Laparoscopic tele-surgery, using a machine such as the da Vinci surgical system, has a number of benefits over conventional laparoscopic surgery and open surgery [1]. However, these come at the cost of the complete loss of tactile feedback and extremely limited kinaesthetic feedback [2, 3]. Of these, the loss of tactile feedback in particular limits the ability for the surgeon to identify tissue compliance and lumps, which would be otherwise detected through skin deformation [4]. As a result, the system's suitability for procedures where palpation is required for the identification of such features is reduced. To improve the current surgical systems in this area, the perception of local shape and course textures, both of which rely on spatial information [5], would be of significant benefit.

For a tactile feedback system, a tactile sensor is required for encoding the tissue interactions at the subject and a tactile display is required for rendering the corresponding tactile image, to stimulate the operator's skin. The resulting sensations should be sufficiently correlated to the interaction that the user can either learn to feel through the system or may do so instinctively.

R. Groß et al. (Eds.): TAROS 2011, LNAI 6856, pp. 114–124, 2011.
© Springer-Verlag Berlin Heidelberg 2011

An ideal tactile sensor for compliance and lump perception interactions is a deformable sensor; to measure the temporal and spatial deformation which would have been otherwise applied to a human finger. Such a sensor would offer a more intuitive sense of touch compared to most developed sensors which detect distributed pressure [4]. However, as Ottermo, M. [6] comments, it is expected that such a compliant sensor must have similar properties to that of the human fingertip. Previous examples of compliant sensors applied to tele-tactile systems do not fit this criteria [7][8].

A suitable tactile display for the detected deformation-based interaction is a shape display due to the spatial, low frequency requirements. The criteria for such a display have been found as 2-3 mm tactor (tactile actuator) displacement, 50 N/cm^2 pressure, 30-50 Hz bandwidth for normal finger exploration and an optimal tactor separation of 1 mm [9 - 11]. Additionally for this application, the mass of the display must also be minimized such that the extra weight and inertia on the user interface does not affect the user's integration with the system, or such that it may be reasonably compensated for. A design concept for the tactile display which offers a good basis from which to improve has been produced by Sarakoglou et al. [12]. Their display used remote actuation to offload the mass of the drive motors away from the user's finger, Teflon conduits were used with nylon tendons to transmit the driving forces, and an array of sprung pins were used for tactors. Forces of 1-3 N (25-75 N/cm^2) at full to minimum displacement respectively, were achieved per pin, with a bandwidth of greater than 15Hz and tactor separation of 2 mm. However, due to the mechanics of this design, if the tendons to one of the tactors were to fail then the tactor would be held in the fully extended position pushing against the user's skin. For a high-risk application such as tele-surgery, when the surgeon's finger is be secured against the display, this would be a critical failure. Sarakoglou et al. quote the mean time to failure to be less than 14 hours at a continuous drive of 12 Hz with 2 mm displacement. A tactor that retracts upon failure is a safer alternative, which would result only in the loss of one actuated area, rather than creating significant interference.

The design of a tactile feedback system for palpation during tele-surgery, which furthers the existing display, is presented in this paper. The tactile display is integrated with a novel deformable sensor developed with the intention of it being a reliable platform for future development.

2 System Design

2.1 Sensor

The tactile sensor used for this research is the BRL TacTip sensor [13]. This is a biologically-inspired sensor, based around the deformation of the epidermal layers of the human skin. The deformation from sensor-object interactions is measured optically by tracking the movement of internal papillae (nodule markers) on the inside of the sensor membrane. These papillae are representative of the intermediate epidermal ridges of the skin, whose static and dynamic displacement are normally detected through the skin's mechanoreceptors.

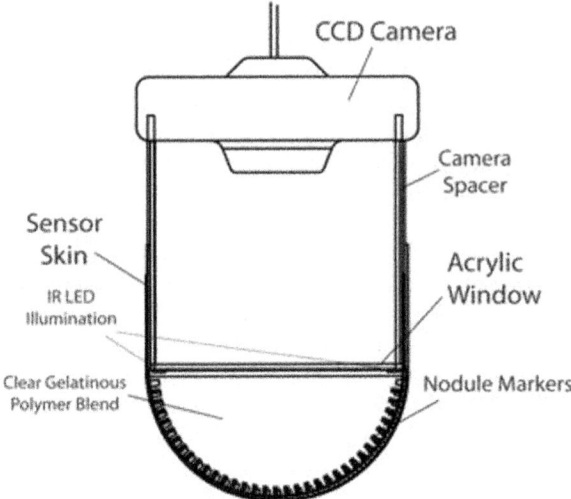

Fig. 1. Cross-section of the BRL TacTip sensor [13] showing the papillae (nodule markers) on the inside of the sensor skin

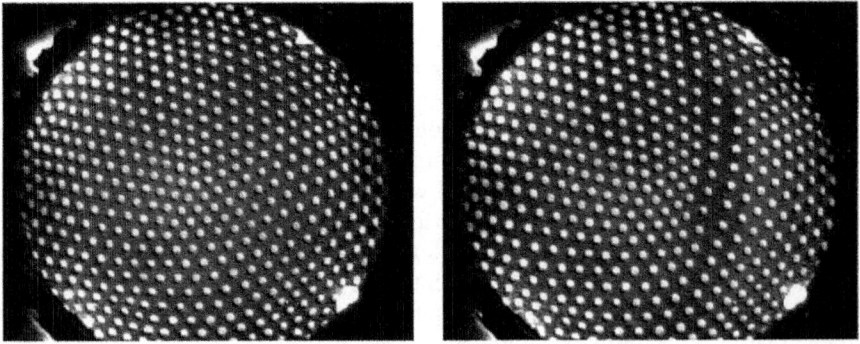

Fig. 2. Image captures showing the illuminated papillae-tips within the sensor. The sensor is shown without any interactions (left) and whilst pressed in to a soft object containing a hard ridge (right).

The sensor has a known compliance and reaction to stimuli similar to that of a human finger. Hence, the sensed deformation on this sensor is directly related to the skin deformation which would have resulted from the direct manipulation of the object with a human finger. This allows the interaction to be efficiently mapped on to the user's finger with minimal calibration and processing. The current size of the sensor area, 40 mm outside diameter, is approximately twice the width of a human fingertip and can discriminate between two points separated by 5 mm, compared to the 2-3 mm of humans [13]. By reducing the distance between the papillae when scaling the sensor down, this two point discrimination threshold can be lowered.

This sensor type is suitable for surgical applications due to its safe, sealed design and cheap durable membrane. As such, the current sensor design already fulfils a number of the criteria outlined by Kattavenos et al. [14] and Schostek et al. [15]. Development would be necessary to scale the sensor down to a useable size and to interface with existing surgical systems. The optoelectronic components within the device could be scaled or relocated away from the sensor tip, potentially using an existing optical-based technique such as those described by Puangmali et al. [16].

The camera used within the sensor for this initial system is a simple off-the-shelf webcam. This camera outputs 29 frames per second, allowing processed deformation data to be sent to the tactile display at 29 Hz.

 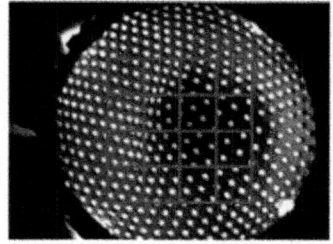 

Fig. 3. Detection of stimuli (left), sensor processing with the boundaries of the papillae regions shown (centre), and graphical representation of the regional membrane deformation corresponding to a 4 x 4 array of output tactors (right)

The deformation of the sensor membrane is determined by grading the regional movement of the internal papillae against calibrated values. The regions of papillae selected are treated as individual sensels (sensor elements) and can be chosen to correspond with the relative locations of the tactor pins on the tactile display.

Within each papillae region, the location and direction of the individual papilla may be used to calculate the applied force upon, and therefore deformation of, the membrane. However, this was found to be unnecessarily computationally heavy. Instead, a simple algorithm is used to average the shade of the pixels within each region of the greyscale image. As an area of the surface is deformed, the underlying papillae deflect away from each other, thereby decreasing the average brightness of the area. This can be seen in figure 2 and 3. Tracking the deformation in this way provides computationally-efficient sensor processing with an accuracy proportional to the number of papilla within each region. Furthermore, due to the ability to select the number and size of the sensels, the sensor can be used with tactor arrangements of differing number and spatial densities.

2.2 Tactile Display

The tactile display system, as described in the introduction section, consists of an actuator housing and tactor assembly mechanically linked using a tendon mechanism.

Fig. 4. Tactile display showing the tactor assembly (left) and actuator housing (right)

The developed tactor assembly utilizes a miniature lever mechanism for the inversion of the mechanical motion. This design adds minimal bulk to the assembly, which weighs 30 g and is 37 x 17 x 40 mm in size. The tactors consist of 1 mm diameter steel rods, with 2.5 mm centre-centre separation. 0.2 N/mm springs are used for tensioning the levers and transmission tendons, to remove any undesired mechanical freedom from within the system.

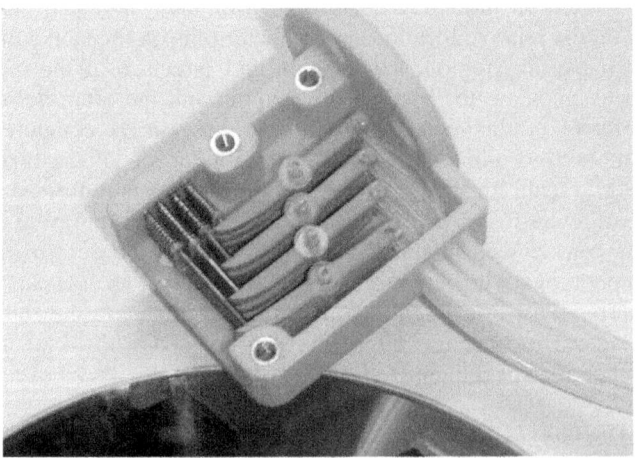

Fig. 5. The tactor assembly showing the internal lever mechanisms

For the actuator assembly, PTFE Teflon conduits are used to house the force-transmitting tendons, similar to that used by Sarakoglou et al.. Nylon tendons were initially also used but these were found to be too elastic to accurately transmit the displacement data. Instead, a single 0.3 mm diameter Dyneema tendon was adopted. The benefit of the Dyneema's larger Young's modulus outweighs the detriment of its higher coefficient of friction.

The actuators for the system consist of 16 RC servomotors similar to those used by Wagner et al. [17]. These servomotors are individually controlled using 50 Hz PWM signals from an Arduino Mega2560 microcontroller board running an Atmel ATmega2560 chip.

Fig. 6. The actuator housing that encloses the actuators and associated hardware. The servomotors can be seen mounted in 1.5 mm vertical steps aligned with the actuator tendons.

2.3 System Integration

The sensor data is passed to the tactile display microcontroller through a serial connection, in a comma separated form. Due to the size difference between the sensor and a human fingerpad, the length of each sensel (5 mm) is double that of the tactor spacing (2.5 mm) for this initial tactile system. This means that through tactile information alone, a detected object feels like it is half its true length and width. Accordingly, the perceived height of the object is also altered and the scaling value between the input deformation and tactor height should be chosen as 2:1 as well. The sensor output is therefore scaled such that a 4 mm displacement of the sensor exerts a 2 mm displacement on the human finger pad.

3 Performance

3.1 Sensor

The linearity of the sensor was tested by deforming the area of the sensor membrane above each sensel and measuring the resultant change in brightness of that region.

A hard flat surface of size equal to an individual sensel region was used to deform the sensor's surface. The position of the flat surface was varied along its normal axis, in the direction parallel with the sensor's body, in 0.5 mm increments. The mean displacement of the sensor membrane was recorded along with the change in brightness of the corresponding sensel region. This process was repeated for each individual sensel and the results averaged across all sensels (Figure 7). A linear relationship was found between the brightness of the sensel region and the surface deformation in that region.

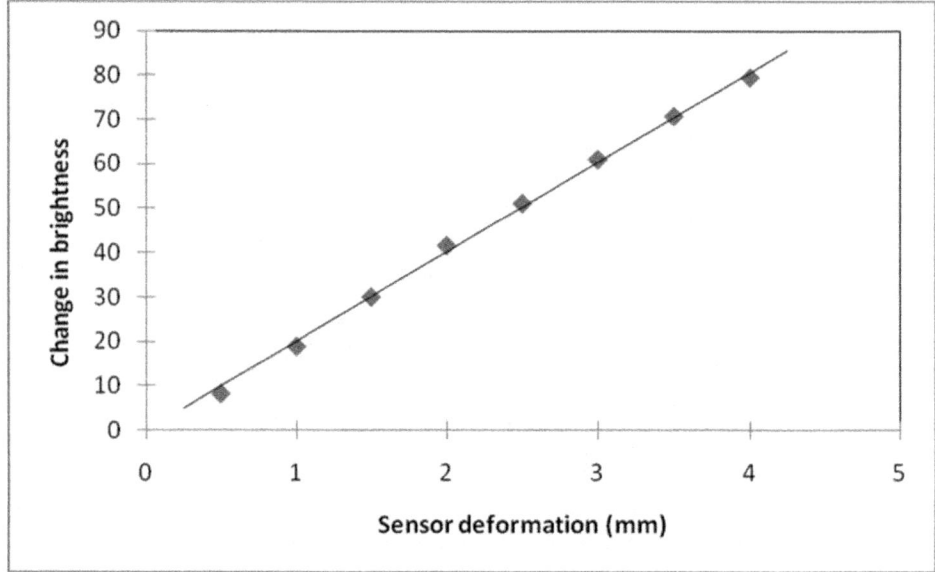

Fig. 7. The change in the sensel region brightness for different amounts of membrane deformation in that area, averaged for all sensels

3.2   Display

The adapted display design produces a maximum tactor height of 2.5 mm and a mean bandwidth of 12 Hz, which is currently limited by the speed of the actuators. The tactor stiffness, found by applying different forces to the tactors whilst at full extension and measuring the resultant displacement (Figure 8), exhibits a non-linear response with a mean tactor stiffness of 5.0 N/mm.

3.3   System

The linearity of the tactile system was tested by displacing the sensor membrane above each of the sensels, using the same method as that in the sensor linearity test. The corresponding tactor displacement was recorded and the results averaged across all of the sensel/tactel pairs.

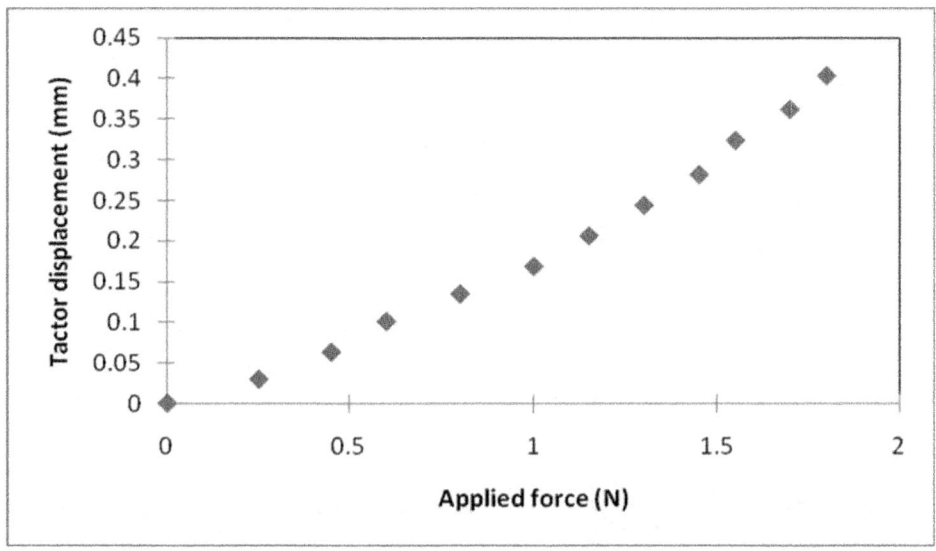

Fig. 8. Tactor displacement due to an applied force, for measuring the tactor stiffness

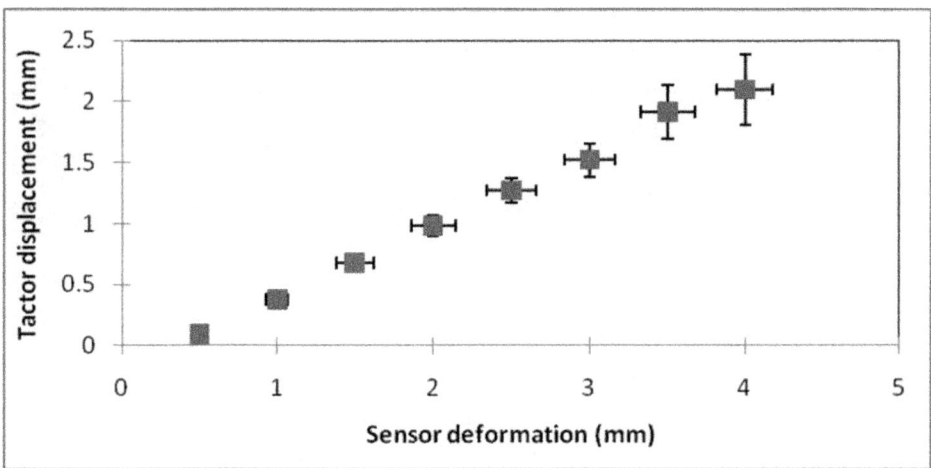

Fig. 9. The tactor displacement against sensor deformation, averaged for all sensel areas and corresponding tactors

The spatial performance of the tactile system was tested for the detection of a hard object within a softer one using artificial tissue containing an embedded object. The artificial tissue is representative of muscular tissue and as such has a similar elastic modulus to that of the sensor. A 15 mm wide wooden cube was used as the embedded object.

The test was carried out by mounting the TacTip sensor in a vertical orientation on a Barrett Technology Inc. robotic arm (WAM) and pushing the sensor down against the tissue to a fixed height, which resulted in an average pressure of 0.8 N/cm^2 across the sensor. The WAM was controlled manually using a simple position program. The test was repeated for different locations along the artificial tissue, with the tactor positions and sensor output being recorded at each. No processing or enhancement was used in addition to the standard displacement calibration within the sensor.

The tactor output for five different positions around the embedded object are shown in Figure 10. The additional deformation of the sensor membrane due to the embedded object is clearly shown by the tactors.

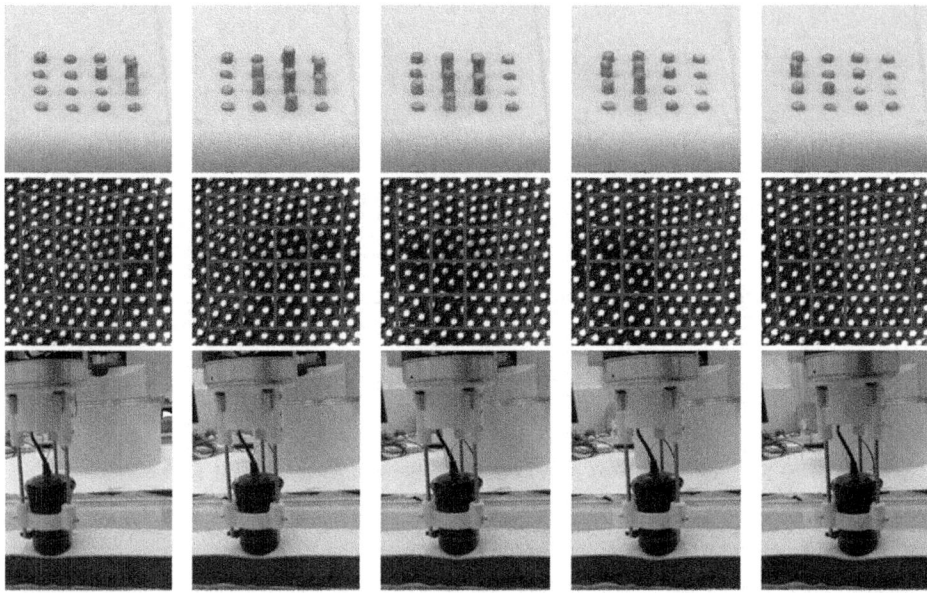

Fig. 10. System testing showing the tactor outputs (top row) and sensor outputs (middle row) for five sensor positions across an area of soft artificial tissue containing a hard embedded object (bottom row)

4   Conclusion

This paper has summarized the initial development of a deformation-based tactile feedback system for tissue palpation during tele-surgery. The ability of the system to relay deformation information has been proven and it has been shown that the system can detect and display deformation-based interactions like those experienced when manipulating hard objects within softer ones, as is often the case during surgical procedures.

Tele-surgery is a novel application for the biologically-inspired sensor and the sensor has been shown to have both theoretical and proven benefits for this application. The optical output of the sensor was split in to sensels (sensor elements) corresponding to the tactor positions in the tactile display. A simple algorithm was then applied to average the brightness of the pixels within each of these areas and use the linear relationship between this quantity and the membrane deformation to calculate the deformation in each of the sensor regions.

The produced tactile display modified an existing design [12] by using tendons with a higher Young's modulus and removing what would be a critical failure point in a high-risk application such as surgery. However, this safer failure mode is at the cost of a higher profile and complexity. RC servomotors were incorporated as the method of actuation following their use in another previous display design [17]. By combining these previous designs, a lightweight and effective tactile display was produced with a centre-centre tactor separation of 2.5 mm, maximum tactor displacement of 2.5 mm, 5.0 N/mm pin stiffness and a mean bandwidth of 12 Hz. This design therefore combines benefits from both of the previous designs apart from the pin spacing, although this is not mechanically limited in this design, so may be reduced. The display offers a cost-effective solution to the tactile problem and the developments have proved effective.

Due to the difference in physical size between the sensor and the tactile display, the tactile image was scaled by 2:1 (the output tactile sensation was half the size of the stimuli). As a result, for this exact system to be later applied to a tele-operated device, the positional and force data would also need to be scaled for the information from the user's kinaesthetic and tactile senses to coincide. The positional ratio would need to be scaled at the same ratio, whereas the ratio of forces would need to take in to account the comparative stiffness of the sensor and the user's fingerpad for the amount of tactile deformation to correlate with the force applied. These factors must be considered further for a different scaling ratio during the future resizing of the sensor for a tele-surgical application.

Future research will also involve investigating this tactile system in a full tele-operated environment, to test the design criteria previously presented. Other variables which have been assumed from previous studies, like the optimal tactor layout and spacing will be investigated further for palpation tasks. Finally, minimisation of the tactile display height must be considered to aid intuitive grasping tasks between actuated fingers.

References

1. Lanfranco, A., Castellanos, A., Desai, J., Meyers, W.: Robotic surgery: a current perspective. Annals of Surgery 239(1), 14–21 (2004)
2. Okamura, A.M.: Haptic Feedback in Robot-Assisted Minimally Invasive Surgery. Curr. Opin. Urol. 19(1), 102–107 (2009)
3. Bholat, O., Haluck, R., Murray, W., Gorman, P., Krummel, T.: Tactile Feedback Is Present During Minimally Invasive Surgery. Journal of the American College of Surgeons 189(4), 349–355 (1999)
4. Peine, W., Howe, R.: Do humans sense finger deformation or distributed pressure to detect lumps in soft tissue? In: Proceedings of the ASME Dynamic Systems and Control Division, Anaheim CA, vol. 64, pp. 273–278 (1998)
5. Bensmaïa, S., Hollins, M.: The vibrations of texture. Somatosensory & Motor Research 20(1), 33–43 (2003)
6. Ottermo, M.: Virtual Palpation Gripper. Ph.D. Thesis of Norwegian University of Science & Technology (2006)

7. Yamauchi, T., Okamoto, S., Konyo, M., Hidaka, Y., Maeno, T., Tadokoro, S.: Real-Time Remote Transmission of Multiple Tactile Properties through Master-Slave Robot System. In: Proceedings of the 2010 IEEE International Conference on Robotics and Automation, pp. 1753–1760 (2010)
8. Sato, K., Minamizawa, K., Kawakami, N., Tachi, S.: Haptic telexistence. In: International Conference on Computer Graphics and Interactive Techniques (2007)
9. Moy, G., Wagner, C., Fearing, R.: A Compliant Tactile Display for Teletaction. In: Proceedings of the 2000 IEEE International Conference on Robotics & Automation, pp. 3409–3415 (2000)
10. Howe, R., Peine, W., Kontarinis, D., Son, J.: Remote Palpation Technology for Surgical Applications. IEEE Engineering in Medicine and Biology Magazine 14(3), 318–323 (1995)
11. Peine, W.J., Wellman, P., Howe, R.D.: Temporal bandwidth requirements for tactile shape displays. In: Proceedings of ASME Dynamic Systems and Control Division, New York, pp. 107–114 (1997)
12. Sarakoglou, I., Tsagarakis, N., Caldwell, D.: A Portable Fingertip Tactile Feedback Array – Transmission System Reliability and Modelling. In: Proceedings of the First Joint Eurohaptics Conference and Symposium on Haptic Interfaces for Virtual Environment and Teleoperator Systems, pp. 547–548 (2005)
13. Chorley, C., Melhuish, C., Pipe, T., Rossiter, J.: Development of a tactile sensor based on biologically inspired edge encoding. In: International Conference on Advanced Robotics, ICAR 2009, Munich, pp. 1–6 (2009)
14. Kattavenos, N., Lawrenson, B., Frank, T., Pridham, M., Keatch, R., Cuschieri, A.: Force-sensitive tactile sensor for minimal access surgery. Min. Invas. Ther. & Allied Technol. 13(1), 42–46 (2004)
15. Schostek, S., Ho, C., Kalanovic, D., Schurr, M.: Artificial tactile sensing in minimally invasive surgery - a new technical approach. Minimally Invasive Therapy 15, 296–304 (2006)
16. Puangmali, P., Althoefer, K., Seneviratne, L., Murphy, D., Dasgupta, P.: State-of-the-art in force and tactile sensing for minimally invasive surgery. IEEE Sensors Journal 4(2), 371–381 (2008)
17. Wagner, C., Lederman, S., Howe, R.: A Tactile Shape Display Using RC Servomotors. In: Proceedings of the 10th Symposium on Haptic Interfaces for Virtual Environment and Teleoperator Systems, vol. 3, pp. 354–356 (2002)

Design and Control of an Upper Limb
Exoskeleton Robot RehabRoby

Fatih Ozkul and Duygun Erol Barkana

Electrical and Electronics Engineering Department,
Yeditepe University, Istanbul, Turkey
fatihozkul85@gmail.com, duygunerol@yeditepe.edu.tr

Abstract. In this work, an exoskeleton type robot-assisted rehabilitation system called RehabRoby is developed for rehabilitation purposes. A control architecture, which contains a high-level controller and a low-level controller, is designed for RehabRoby to complete the rehabilitation task in a desired and safe manner. A hybrid system modeling technique is used for high-level controller. An admittance control with inner robust position control loop has been used for the low-level control of RehabRoby. Real-time experiments are performed to evaluate the control architecture of the robot-assisted rehabilitation system RehabRoby.

Keywords: Robot-assisted rehabilitation system, exoskeleton robot, control architecture.

1 Introduction

There are over 650 million people around the world with disabilities. Although it is accepted as 10% of the whole world population, it is 15.7% in Europe, 12% in USA [1] and 12.29% in Turkey [1]. Physical disability, which occurs by birth or acquired during the life span of the person due to the diseases or a trauma to the central nervous system or musculoskeletal system, affects the functionality of people. The physical therapy and rehabilitation programs are applied to the people with disability to increase their joint range, strength, power, flexibility, coordination and agility of the person, and to improve their functional capacity as well as third level of independence [2], [3]. The availability of such training techniques, however, is limited by a number of factors such as the amount of costly therapist's time they involve and the ability of the therapist to provide controlled, quantifiable, and repeatable assistance to complex movement. Consequently, end-effector and exoskeleton robot-assisted rehabilitation that can quantitatively monitor and adapt to patient progress, and ensure consistency during rehabilitation may provide a solution to these problems and has become an active research area [4]-[8].

Exoskeleton type robots resemble the human arm anatomy and each joint of robot can be controlled separately, which reduces control issue complexity. ARMin [4], T-WREX [5], Pneu-WREX [6], L-Exos [6], and Selford Rehabilitation Exoskeleton [8] are well known exoskeleton type robot-assisted rehabilitation systems. Existing

R. Groß et al. (Eds.): TAROS 2011, LNAI 6856, pp. 125–136, 2011.
© Springer-Verlag Berlin Heidelberg 2011

exoskeleton robot-assisted rehabilitation systems have been developed to provide assistance to patients during the execution of upper-extremity rehabilitation exercises. In this work, an exoskeleton type upper-extremity robot-assisted rehabilitation system, which is called RehabRoby, is developed.

RehabRoby has been designed to implement passive, active-assisted and resistive-assisted therapy modes. RehabRoby has been designed in such a way that it can be easily adjustable for people with different heights and arm lengths. RehabRoby can also be used for both right and left arm rehabilitation. Control of a robot-assisted rehabilitation system in a desired and safe manner is an important issue during the execution of rehabilitation therapies.

Impedance control, position and admittance control have previously been used to control robot-assisted rehabilitation systems. There is a human-robot interaction in the robot-assisted rehabilitation systems, which is an external effect that can cause changes in the dynamics of the robotic systems. The changes in the dynamics of the rehabilitation robotics may result in instability, which indeed may cause unsafe situations for patients during execution of the rehabilitation task. Furthermore, robot-assisted rehabilitation systems, especially exoskeleton types have complex dynamics. Thus, there is a need to design a controller for RehabRoby that compensates changes in the dynamics of the rehabilitation robotics. A controller, which is independent of dynamic model of robot-assisted rehabilitation system, may provide a solution to this problem [9]. Thus, in this work admittance control with inner robust position control loop has been used to control RehabRoby in a desired manner. Note that it is also desirable for a patient to perform the rehabilitation task in a safe manner. A high-level controller, which is a decision making mechanism, has been designed to ensure safety during the execution of the rehabilitation task. The high-level controller presented in this work plays the role of a human supervisor (therapist) who would otherwise monitor the task and assess whether the rehabilitation task needs to be updated.

In this study, control architecture of the robot-assisted rehabilitation system RehabRoby has been evaluated with healthy subjects. Subjects are asked to perform well known rehabilitation tasks elbow flexion with RehabRoby in active-assisted mode and resistive-assisted mode. This assessment may provide us clues on how RehabRoby can be used in upper extremity rehabilitation of stroke patients in different therapy modes.

This paper describes the control architecture of the robot-assisted rehabilitation system RehabRoby in Section 2. An experimental set-up that is used to evaluate RehabRoby system is given in Section 3. Experimental results are presented in Section 4. Discussion of the study and possible directions for future study is given in Section 5.

2   Control Architecture

A control architecture is developed for robot-assisted rehabilitation system RehabRoby to complete the rehabilitation tasks in a desired and safe manner (Fig. 1). Control architecture consists of a robot-assisted rehabilitation system (RehabRoby), low-level and high-level controllers, and a sensory information module. We first describe robot-assisted rehabilitation system RehabRoby, and then present the details of the low-level and high-level controllers that are used to complete the rehabilitation tasks in a desired and safe manner.

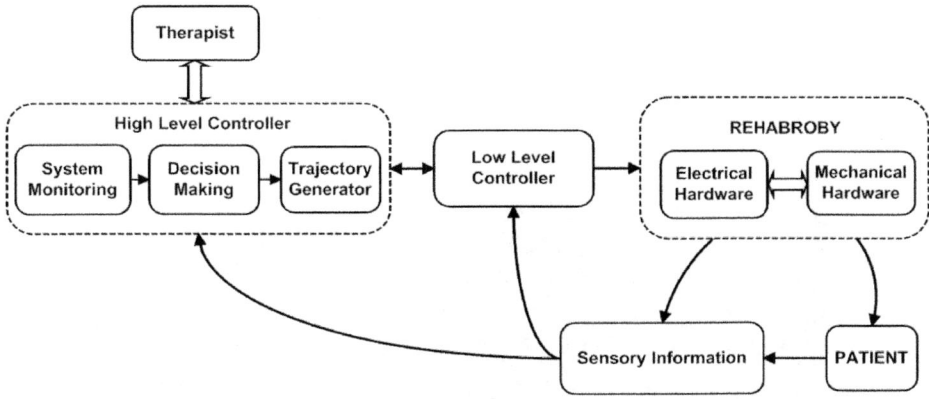

Fig. 1. Control Architecture of the RehabRoby

2.1 RehabRoby

RehabRoby is designed to provide extension, flexion, abduction, adduction, rotation, pronation and supination upper-extremity movements and also combination of these movements that are necessary for activities of daily living (Fig. 2). RehabRoby can provide horizontal abduction/adduction of shoulder rotation (θ_1), shoulder flexion/extension elevation (θ_2), internal and external rotation of shoulder (θ_3), elbow flexion/extension (θ_4), lower arm elbow pronation/supination (θ_5) and wrist flexion/extension (θ_6). RehabRoby has been designed in such a way that it can be easily adjustable for people with different heights and arm lengths. In the design of RehabRoby, anthropometric approaches have been used. The link lengths of RehabRoby are based on the arm lengths of 2100 people in 14 cities in Turkey. RehabRoby can also be used for both right and left arm rehabilitation. Range of motion (ROM), joint torques, velocities and accelerations for RehabRoby have been determined using the measurements of the movements of a healthy subject during two activities of daily living tasks [10], [11].

An arm splint has been designed and attached to RehabRoby (Fig. 2). It has humeral and forearm thermoplastic supports with velcro straps and a single axis free elbow joint. A thermoplastic inner layer covered by soft material (plastazote) is used due to the differences in the size of the subjects' arms. Thus, the total contact between the arm and the splint can be achieved to eliminate loss of movement during the execution of the task. Kistler model press force sensors, which have quite small sizes, are selected to measure contact forces between the subject and RehabRoby. One force sensor has been placed in the inner surface of the thermoplastic molded plate attached dorsally to forearm splint via velcro straps in such a way that the measurement axes of them are perpendicular to each other (Fig. 2). This force sensor is used to measure the applied force during the elbow flexion movement.

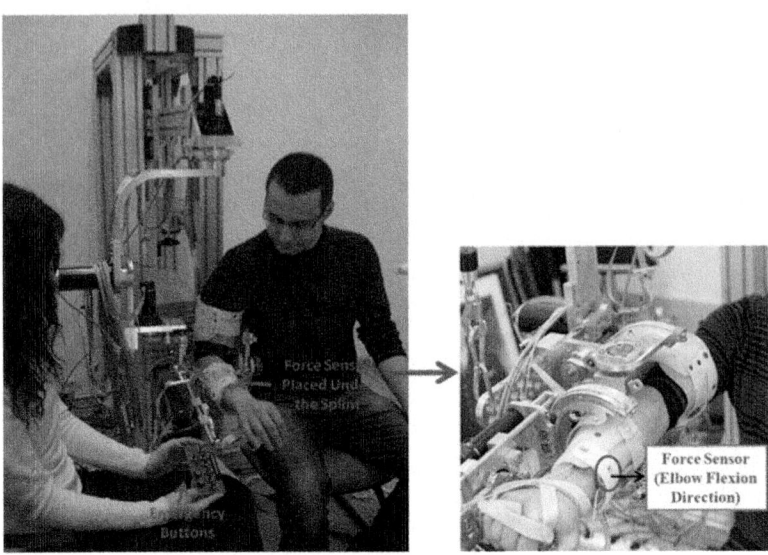

Fig. 2. RehabRoby with Subject

Ensuring safety of the subject is an important issue when designing a robot-assisted rehabilitation system [12]. Thus, in case of emergency situations, the physiotherapist can press an emergency stop button to stop the RehabRoby (Fig. 2). The motor drivers of RehabRoby can be disabled separately or together by pressing the driver enable/disable buttons without disconnecting the energy of the RehabRoby in any case of emergency situations. The power of the system is supported with uninterruptible power supply, thus, there is no power loss and RehabRoby will not collapse at any time. Additionally, rotation angle and angular velocities of each joint of RehabRoby are monitored by the high-level controller which will be described in the next section.

RehabRoby has been interfaced with Matlab Simulink/Realtime Workshop to allow fast and easy system development. Humusoft Mf624 model data acquisition board is selected to provide real time communication between the computer and other electrical hardware. Humusoft Mf624 data acquisition board is compatible with Real Time Windows Target toolbox of MATLAB/Simulink. Digital incremental encoders are coupled with Maxon models of brushed DC motors for joint position measurement. Five of the six encoders have resolutions of 500counts/turn and one of them has a resolution of 1000counts/turn. Encoder data of motors is received through a Humusoft Mf624 with a 500Hz sampling rate. Analog reference current values are converted to digital ones, and then transmitted to the drivers using RS232 serial bus with a baud rate of 115200 using Programmable Interface Controller (PIC) microcontrollers. The current reference values of motors are sent to the microcontroller circuits using the analog outputs of the Humusoft Mf624 card with the same sample rate. Microcontroller circuits are used because four of the six motor drivers of RehabRoby have no analog reference inputs. Analog to digital conversion and serial transmission are completed within 2 milliseconds. A 19'' LCD screen is positioned in front of the subject at a distance of about 1m to display the reference rehabilitation task trajectory and subject's actual movement during the task execution. The force values measured from the force sensors are recorded using the Humusoft Mf624data acquisition card with a sampling rate of 500Hz. The joint torque corresponding to applied forces by the subject is calculated by multiplying the force with the perpendicular distance between the force contact point and the joint axis.

2.2 Low-Level and High-Level Controllers

Low-level controller is responsible to provide necessary motion to RehabRoby so patients can complete the rehabilitation tasks in a desired manner. In this study, admittance control with inner robust position control loop is used as the low-level controller of RehabRoby (Fig. 3).

Admittance control method is a good choice for control applications of the robotic systems which have low back drivability, high inertia and reliable position and force/torque information. Since RehabRoby has complex and uncertain inner dynamics and it is sensitive to external forces during the human-robot interaction, a simple Proportional-Integral-Derivative (PID) or model based position control technique may not be enough to complete the tracking in a desired performance. Thus, a robust position controller has been used in the inner loop of the admittance controller. The effects of the parametric uncertainties in the dynamic model and the external additive disturbances are compensated with an equivalent disturbance estimator in the robust position controller.

Various methods have been previously used to estimate the disturbance in the position control of robotic systems such as adaptive hierarchical fuzzy algorithm [13], model based disturbance attenuation [14]. In this work, we have used discrete Kalman filter based disturbance estimator [15],[16], which is a commonly known and successful technique used to process noisy discrete measurements. Additionally, discrete linear Kalman filter based disturbance estimator estimates the unknown states and parameters in the dynamic model in an accurate manner. To our knowledge admittance control with inner robust position control loop has not been used for control of robot-assisted rehabilitation systems before.

The general structure of the proposed low-level controller for RehabRoby is shown in Fig. 3. The force that is applied by the subject during the execution of the task is measured using the force sensor and this value is then converted to torque using Jacobian matrix. The torque value is then passed through an admittance filter [17], which is used to define characteristics of the motion of the RehabRoby against the applied forces,to generate the reference motion for the robust position controller. The reference motion is then tracked with a robust position control which consists of a linear Kalman filter based disturbance estimator [17].

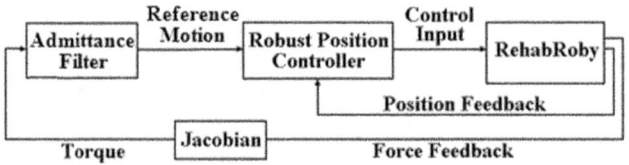

Fig. 3. Block Diagram of Low-level Controller of RehabRoby

High-level controller is the decision making mechanism of RehabRoby. High-level controller decides necessary changes in the low-level controller by analyzing the information that comes from the sensory information module or physiotherapist (system monitoring). The high-level controller presented in this study plays the role of a human supervisor (therapist) who would otherwise monitor the task and assess whether the task needs to be updated. A hybrid system modeling technique is used to design the high-level controller because it is easy to add new rules related to rehabilitation task using the hybrid system modeling technique (Fig. 4).

Initially, states of the high-level controller are defined. When task execution starts, starting and final positions of the joint angles of RehabRoby are initialized in *initialization state*. In this work, three therapy modes, which are passive, active–assisted mode and resistive-assisted mode, have been selected for rehabilitation tasks. *passive state (mode=0)*, *active state (mode=1)* or *admittance controlstate (mode=2)* become active based on the therapist's therapy mode selection. In passive mode, the rehabilitation task is performed only in the passive state in which RehabRoby is responsible to help subject to complete the task while subject is passive. The subject's motion is checked periodically in all therapy modes. If the subject's movement, which is measured as (θ) of RehabRoby, is out of limits ($\theta \geq /\varepsilon/$), then *position control state* becomes active. When *position control state* is active, then RehabRoby provides assistance to the subject's motion until subject's movement is in the desired motion range. When the subject's movement is in the range of limits ($\theta< |\varepsilon/$), then the state, which is active before entering the *position control state*, becomes active again. In any state, the safety conditions of the system are checked periodically and if any unsafe situation is occurred (*e=1*), then *emergency stop state* becomes active and the execution of the rehabilitation task is stopped.

3   Experimental Setup

3.1   Task Design

The objective of this study is to investigate use of a robot-assisted rehabilitation system RehabRoby during the execution of rehabilitation tasks. This assessment may provide us clues on how RehabRoby can be used in upper extremity rehabilitation of stroke patients in different therapy modes. Elbow flexion movement (i.e. reaching towards a glass of water on the table) rehabilitation task has been selected in

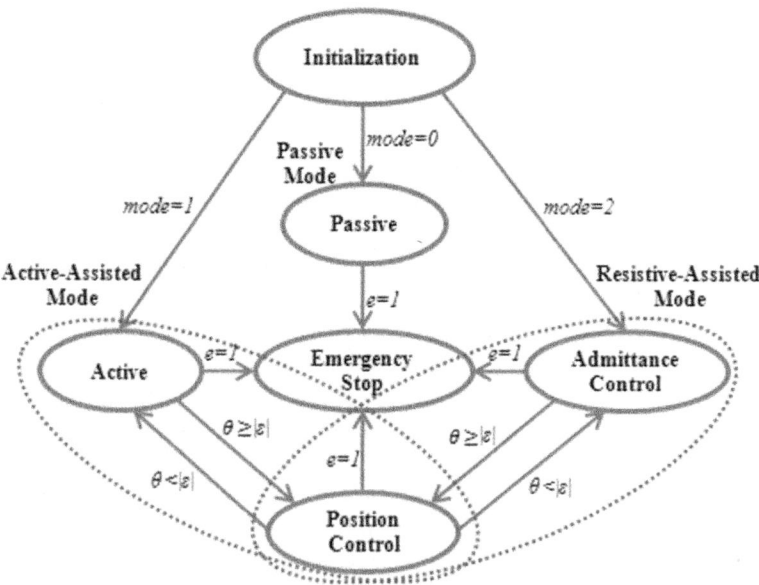

Fig. 4. Control Mechanism for High-Level Controller of RehabRoby

consultation with physiotherapists in Yeditepe University Physical Therapy and Rehabilitation Department. This study has been approved by the Institutional Review Board of Yeditepe University Hospital(IRB #032).

3.2 Experiment Protocol

Subjects were seated in the chair as shown in Fig. 2 and their arms were placed in the splint tightly secured with velcro straps.The height of the RehabRoby was adjusted for each subject to start the task in the same arm configuration. Subject's shoulder was positioned at extension of 90^0, elbow was at neutral position, lower arm was at pronation of 90^0, and the hand and the wrist were free at neutral position as a starting position (Fig. 5a). In elbow flexion movement task (Theta-4 (θ_4)), subjects were asked to flex their elbows to 90^0 in 30 seconds (Fig. 5b and Fig. 5c).

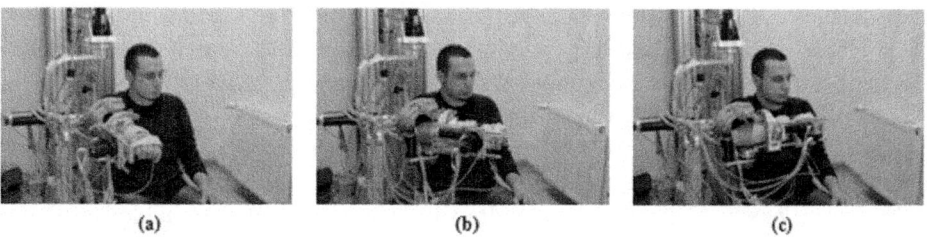

(a)   (b)   (c)

Fig. 5. Subject with RehabRoby. (a) Initial Position of Elbow Flexion, (b) Middle Position of Elbow Flexion Task, (c) Final Position of Elbow Flexion Task.

It is possible for the subjects to perform the rehabilitation in three different therapy modes. However, we only selected active-assisted and resistive-assisted therapy modes in this work. In active-assisted therapy mode, RehabRoby had been kept passive, subjects were asked to perform the rehabilitation tasks by themselves and RehabRoby provided assistance to the subjects when they can not follow the desired movement. No resistance was applied to the subject's movement in active-assisted mode. In resistive-assisted therapy mode, subjects were asked to perform the rehabilitation tasks with a comfortable resistance applied by RehabRoby using admittance control with inner robust position control loop and RehabRoby provided assistance to the subjects when they can not follow the desired movement. The resistance applied in resistive-assisted mode was quite large compared with the resistance that is caused by the inherent dynamics of RehabRoby. Thus, the resistance of the system during performance of the tasks in active-assisted mode is neglected. The parameters in admittance filter, which provided comfortable resistance, had been determined experimentally. The reference trajectories for the rehabilitation tasks were defined using minimum jerk trajectory method. Desired motion range was determined by setting upper and lower limits to the reference trajectories. The upper and lower limits can be adjusted based on the patient's movement capabilities.

Initially, experiment protocol had been explained to each subject, who participated in the study. Then subjects were asked to practice with RehabRoby to become familiar with the rehabilitation task. Subject performed 5 trials to become familiar with the elbow flexion task and RehabRoby before starting real-time experiments.

3.3 Subjects

Totally 9 subjects (4 female and 5 male) whose ages are in the range of 22 to 26 were participated in the study. None of them have any motor impairment in their arms. Two of the subjects were left handed and the others were right handed.

4 Results

We asked subjects to perform elbow flexion rehabilitation task in active-assisted mode and resistive-assisted mode. Subjects were required to move from 0^0 to 90^0 for the elbow flexion task and they were asked to trace a ball (green ball in Fig. 6) that was shown to them on a computer screen. The actual position of the subject (blue ball), the desired position (green ball) and the desired motion (black line) (Fig. 6) were also shown to the subjects.We have only presented a small portion of elbow flexion task (Theta-4 (θ_4)) in here. Perpendicular lines at equal absolute distance from the desired position (green ball) were drawn to represent the upper and lower limits of the desired motion of elbow flexion rehabilitation task (Fig. 6). In this work, visual feedback was selected to keep the concentrations of the subjects at maximum level during the execution of the task.

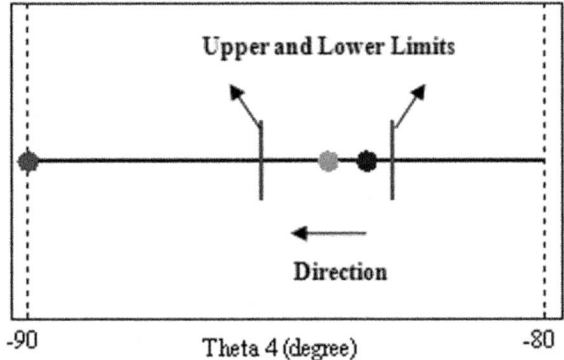

Fig. 6. Visual Feedback Based on Ball Tracking for Elbow Flexion

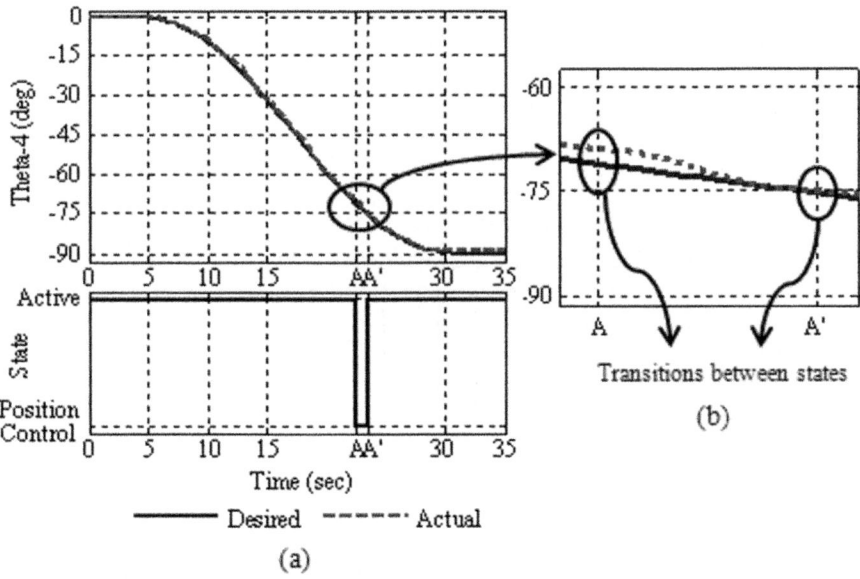

Fig. 7. (a) Motion of Subject 4 During Elbow Flexion Task in Active-Assisted Therapy Mode, (b) A Close Look at the Smoothness of the Transition between Controller States

The subjects were asked to complete the elbow flexion (Theta-4 (θ_4)) in the specified motion ranges using active-assisted and resistive-assisted therapy modes. Allowed maximum deviation from the reference trajectories of each rehabilitation task was selected as 1.5^0. It is possible to increase/decrease the deviation angle depending on the patient's movement capabilities. The subject's movement had been checked in every 2 seconds and if the subject's movement was out of the limits of the desired motion, then RehabRoby became active to provide assistance to the subject to take his/her motion into the desired motion range using the robust position controller with disturbance estimator. Checking time can be changed by the therapist before the therapy starts. The experiments were performed with 9 subjects; however we only presented one of the subjects' data (Fig. 7 and Fig. 8). It could be seen that when the subject was not in the desired motion range, then admittance control with inner robust position control loop became active at A (Fig 7) and the subject came back to the desired motion range at A'. When the subjects were in the desired range, then the RehabRoby became inactive and subject continued execution of the task by his/her effort. Subject needed more number of times of assistance when he/she had performed the elbow flexion task in resistive-assisted therapy mode (Fig. 8). It had also noticed that the transition between the controllers when assistance needed had been smooth (Fig. 7-b). Smooth transitions between the controllers during the execution of the rehabilitation were important to complete the task in a safe manner.

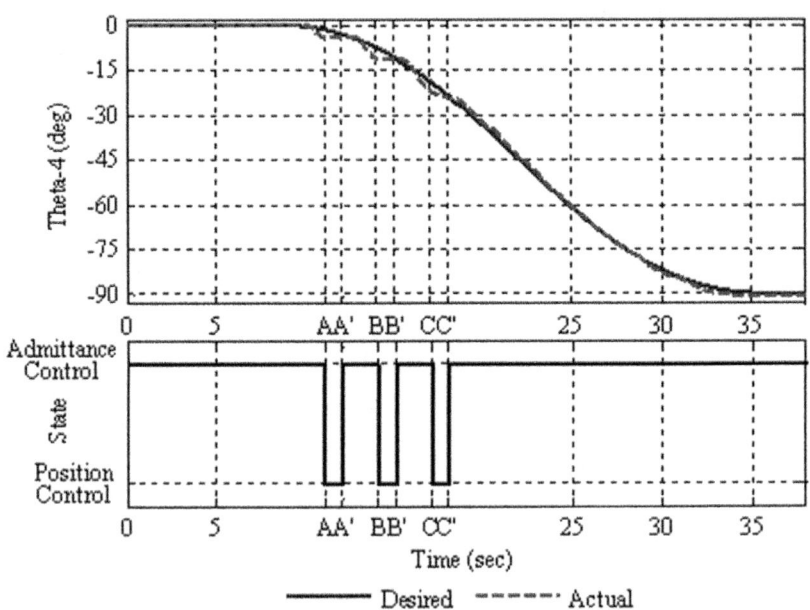

Fig. 8. Motion of Subject 4 During Elbow Flexion Task in Resistive-Assisted Therapy Mode

5   Discussion and Conclusion

We have developed an exoskeleton type upper-extremity robot-assisted rehabilitation system called RehabRoby. RehabRoby can provide passive mode, active-assisted mode and resistive-assisted mode for low-functioning and high-functioning patients. RehabRoby can promote and maintain passive range of motion for low-functioning patient's movements in passive mode. Additionally, RehabRoby can increase strength by providing resistance during the movement of high-functioning patients in resistive-assisted mode. RehabRoby is also adaptable for patients with different gender.

Additionally RehabRoby can be adjusted easily for people with different arm lengths. Furthermore, it can be used for both right and left arm.

A control architecture which consists of a high-level controller and a low-level controller has been developed for RehabRoby. High-level controller is the decision making mechanism that decides necessary changes in the low-level controller according to the sensory information or the therapist's commands. The high-level controller presented in this study plays the role of a human supervisor (therapist) who would otherwise monitor the task and assess whether the task needs to be updated. Admittance control with inner robust position control loop has been used for the low-level controller. Admittance controller has been integrated with a robust position controller which consists of a linear discrete Kalman filter. The evaluation of the proposed robust position controller has shown that discrete linear Kalman filter based disturbance estimator can compensate the effects of the uncertainties in the dynamic model and external disturbances that might happen during human-robot interaction.

As a future work, the robust position controller performance will be improved using adaptive Kalman filter which will adjust the admittance parameters of RehabRoby for each subject. Note also that this is a feasibility study for the proposed robot-assisted rehabilitation system RehabRoby to be used in the future for rehabilitation of stroke patients.

Acknowledgments. We gratefully acknowledge the help of Dr. Serap İnal and Dr. Sule Badilli Demirbas who are in Physiotherapy and Rehabilitation Department in Yeditepe University. The work was supported by the Support Programme for Scientific and Technological Research Projects (TUBITAK-3501) under Grant 108E190.

References

1. Rehabilitation Research and Training Center on Disability Statistics and Demographics, Annual Disability Statistics Compendium (2010),
   http://disabilitycompendium.org/Compendium2010.pdf
2. Turkey Disability Survey, The State Institute of Statistics and The Presidency of Administration (2002)
3. Cromwell, S.A., Owen, P.: Treatment planning in Saunders Manuel of Physical Therapy Practice. In: Myers, R.S. (ed.), pp. 367–374. W.B. Saunders Company, Philadephia (1995)
4. Nef, T., Guidali, M., Riener, R.: ARMin III - arm therapy exoskeleton with an ergonomic shoulder actuation. Applied Bionics and Biomechanics 6(2), 127–142 (2009)
5. Housman, S.J., Le, V., Rahman, T., Sanchez, R.J., Reinkensmeyer, D.J.: Arm-Training with T-WREX After Chronic Stroke: Preliminary Results of a Randomized Controlled Trial. In: Proceedings of the 2007 IEEE 10th International Conference on Rehabilitation Robotics, pp. 562–568. IEEE Press, Noordwijk (2007)
6. Sanchez, R.J., Wolbrecht, E., Smith, R., Liu, J., Rao, S., Cramer, S., Rahman, T., Bobrow, J.E., Reinkensmeyer, D.J.: A Pneumatic Robot for Re-Training Arm Movement after Stroke: Rationale and Mechanical Design. In: Proceedings of the 2005 IEEE 9th International Conference on Rehabilitation Robotics, pp. 500–504. IEEE Press, Chicago (2005)

7. Frisoli, A., Borelli, L., Montagner, A., Marcheschi, S., Procopio, C., Salsedo, F., Bergamasco, M., Carboncinit, M.C., Tolainit, M., Rossit, B.: Arm rehabilitation with a robotic exoskeleleton in Virtual Reality. In: Proceedings of the 2007 IEEE 10th Inter. Conf. on Rehabilitation Robotics, pp. 631–642. IEEE Press, Noordwijk (2007)

8. Kousidou, S., Tsagarakis, N.G., Smith, C., Caldwell, D.G.: Task-Oriented Biofeedback System for the Rehabilitation of the Upper Limb. In: Proc. of the 2007 IEEE 10th Inter. Conf. on Rehabilitation Robotics, pp. 12–15. IEEE Press, Noordwijk (2007)

9. Tsetserukoul, D., Tadakuma, R., Kajimoto, H., Kawakami, N., Tachi, S.: Towards Safe Human-RobotInteraction: Joint Impedance Control Towards Safe Human-Robot Interaction: Joint Impedance Control. In: Proc. of 16th IEEE International Conference on Robot & Human Interactive Communication, pp. 860–865. IEEE Press, Jeju (2007)

10. Fasoli, S.E., Krebs, H.I., Hogan, N.: Robotic technology and stroke rehabilitation: Translating research into practice. Top Stroke Rehabil. 11(4), 11–19 (2004)

11. Oldewurtel, F., Mihelj, M., Nef, T., Riener, R.: Patient-Cooperative Control Strategies for Coordinated Functional Arm Movements. In: Proceedings of the European Control Conference, Kos, Greece, pp. 2527–2534 (2007)

12. Beyl, P., Knaepen, K., Duerinck, S., Van, D.M., Vanderborght, B., Lefeber, D.: Safe and compliant guidance by a powered knee exoskeleton for robot-assisted rehabilitation of gait. Advanced Robotics 25(5), 513–535 (2011)

13. Emara, H., Elshafei, A.L.: Robust robot control enhanced by a hierarchical adaptive fuzzy algorithm. Engineering Applications of Artificial Intelligence 17, 187–198 (2004)

14. Choi, C., Kwak, N.: Robust Control of Robot Manipulator by Model-Based Disturbance Attenuation. IEEE/ASME Transactions on Mechatronics 8(4), 511–513 (2003)

15. Stasi, S., Salvatore, L., Milella, F.: Robust tracking control of robot manipulators via LKF-based estimator. In: Proc. of the IEEE International Symposium, pp. 1117–1124. IEEE Press, Bled (1999)

16. Salvatore, L., Stasi, S.: LKF based robust control of electrical servodrives. IEE Proceedings Electric Power Appl. 3(142), 161–168 (1995)

17. Jung, S., Hsia, T.C.: Neural Network Impedance Force Control of Robot Manipulator. IEEE Transactions on Industrial Electronics 45(3), 451–461 (1998)

Distributed Motion Planning for Ground Objects Using a Network of Robotic Ceiling Cameras⋆

Andreagiovanni Reina, Gianni A. Di Caro,
Frederick Ducatelle, and Luca M. Gambardella

Dalle Molle Institute for Artificial Intelligence (IDSIA), Lugano, Switzerland
{gioreina,gianni,frederick,luca}@idsia.ch

Abstract. We study a distributed approach to path planning. We focus on holonomic kinematic motion in cluttered 2D areas. The problem consists in defining the precise sequence of roto-translations of a rigid object of arbitrary shape that has to be transported from an initial to a final location through a large, cluttered environment. Our planning system is implemented as a swarm of flying robots that are initially deployed in the environment and take static positions at the ceiling. Each robot is equipped with a camera and only sees a portion of the area below. Each robot acts as a local planner: it calculates the part of the path relative to the area it sees, and exchanges information with its neighbors through a wireless connection. This way, the robot swarm realizes a cooperative distributed calculation of the path. The path is communicated to ground robots, which move the object. We introduce a number of strategies to improve the system's performance in terms of scalability, resource efficiency, and robustness to alignment errors in the robot camera network. We report extensive simulation results that show the validity of our approach, considering a variety of object shapes and environments.

1 Introduction

Path planning is a core problem in robotics (see [9] for an overview). In its basic version, the path planning problem consists in the definition of the optimal sequence of rotations and translations needed to move an object of a given geometry from an initial to a target configuration while avoiding collisions with obstacles. If the constraints on the motion only depend on the environment's obstacles and on the relative position of the moving object, the problem is *holonomic*. In *non-holonomic* motion planning also dynamic constraints are considered.

In this work, we focus on holonomic path planning in the following setting. An object of assigned shape has to be moved from an initial to a final location in a large cluttered area. A high movement accuracy is required, up to a few centimeters precision, in order to effectively avoid the existing obstacles. No a

---

⋆ This research was partially supported by the Swiss National Science Foundation through the National Centre of Competence in Research Robotics.

R. Groß et al. (Eds.): TAROS 2011, LNAI 6856, pp. 137–148, 2011.

priori knowledge about the environment is available. In order to acquire and process the information required for planning the motion of the object on the ground, we propose the use of a *network of cooperating cameras* with a *top view* of the area where the object can be moved. We assume the camera network to be able to autonomously take position in the environment. For this purpose, as reference model for the camera nodes, we used a *swarm of flying robots*. The swarm can be deployed in formation such as to cover through the vision system the entire area between the initial and final positions, and to be able to locally communicate. Then, they attach to the ceiling and keep stationary positions for the entire process. More technical details are given in Section 2. This architecture is suitable for path planning problems in *large areas*, where a single camera is not sufficient to effectively cover the entire area, and a ceiling camera network can be effectively deployed. Examples are factories, warehouses, or malls: large indoor areas, characterized by the presence of relatively narrow alleys and turns, and irregularly spread, and often dynamic, obstacles.

In our *distributed architecture*, each camera node plays the role of a *local planner*: it plans the motion relative to the area that it sees, and locally exchanges information with its neighbor nodes through a wireless channel in order to merge and organize the local views into a global feasible plan for the object on the ground. Compared to a single camera solution, this approach determines *sensing errors*, due to intrinsic uncertainties in the calculation of the relative positions of the cameras and the overlapping in their local views, and *efficiency issues*, due to communication and coordination overhead. In this work, implement a fully distributed system for effective path planning building on and extending established approaches for path planning. Then, we consider error and efficiency issues. We cope with the latter by introducing a number of heuristic strategies. On the other hand, since sensing errors are an intrinsic property of the system, we consider them as internal parameters, and we evaluate how they influence the performance. The aim of this work is to show that our distributed planning system can find near optimal paths, makes an efficient utilization of computational and communication resources, is robust to increasing sensing errors, and its performance scales with the size of the environment and the number of cameras. We report extensive simulation results that precisely show these properties considering objects of different shapes and a variety of cluttered environments.

The article is structured as follows. In Section 2 we define the scenario characteristics, the initial assumptions and the camera network model. Section 3 discusses related work. Section 4 explains the planner architecture, describing the different phases of the process and the optimization heuristics we propose. Section 5 shows experimental results, and section 6 concludes the paper.

2   Scenario Characteristics and Camera Network Model

We use a *swarm of flying robots* to implement the camera network used for distributed planning. The robots are modeled after the *eye-bot* [10], an indoor flying robot developed in the EU-funded project *Swarmanoid* (http://www.swarmanoid.org). The eye-bot can passively attach to the ceiling using a magnet (the design assumes the presence of a ferrous ceiling). It has a pan-and-tilt camera, which can be pointed in any direction below it. It can communicate using both wi-fi and a line-of-sight infrared system [10], which also provides it with relative positional information about other eye-bots (distance and angle).

The mobility and autonomy of the robots can be exploited to deploy the camera network over the area where the object moves. The robots can select their stationary positions at the ceiling to let the swarm formation effectively cover the area with the combined camera views, while, at the same time, observing that neighbor robots are in wireless communication range. We do not study how the formation can be obtained; any algorithm that lets robots spread in an area or find a target location from a given source location (e.g., [11]) could be used.

Using the pan-and-tilt unit, the robots orient their camera to look at the area directly below them. Each robot computes a 2D occupancy map for the obstacles below it, and the path of the ground object is computed with respect to this 2D projection.[1] The field of view of each robot must overlap with that of its neighbors. The size of the overlapping area must be greater or equal to the dimension of the moving object. This is required in order to connect locally calculated sub-paths (see Section 4). For the same reason, knowledge of the relative position and orientation between neighbor cameras is also required. The eye-bots can compute this information autonomously, using the range and bearing capabilities of the infrared system.[2] This means that they can also adapt this information when the eye-bot network topology changes. Clearly, the relative distance and angle information acquired through the infrared system are affected by errors (as would be the case with any other relative positioning system). In our simulation model of the eye-bot, we model distance and angle error values using two different zero mean Gaussian distributions with configurable variance values. In this way, we model error estimates in rather general and, at the same time, realistic terms, while avoiding to be too hardware-specific. This approach allowed us to study the impact of these error estimates on distributed path planning performance for a wide range of error values (see Section 5).

Finally, we point out that our focus in this paper is only on the distributed calculation of a path plan. We assume that the start and target locations are given as input, and the object can be moved by a system (e.g., a robot) that can interpret the path planning instructions and can check its correct actuation.

3   Related Work

The path planning problem has been extensively studied considering various formulations and solutions. References and discussions can be found in [8,3,9]. Our distributed planner is based on classical work in path planning [8,1]: it is

---

[1] In this way, we classify the areas covered by table-like obstacles as fully inaccessible, while the object could pass under, depending on its volumetric dimensions.

[2] We assume that all cameras have the same calibration and size of field of view.

derived from the numerical potential field technique for a single camera planner. The potential field is computed using the *wave front expansion with skeleton* on a uniform cell partitioning of the 2D map of the environment. This solution first spreads the potential over a subset of the free space, called *skeleton*, which corresponds to the *Voronoi diagram* [3,5,12]; then, the potential is computed in the rest of the map. The potential descent from the start to the final configuration is performed using A^*[6]. The moving object can have any shape and dimension. We follow the approach proposed in [3,7], modeling the object with a discrete set of *control points*. Various ways to apply A^* to the control points exist; we sum the potential over the different points to compute the distance estimation.

Centralized path calculation is the most common approach to planning, even in the presence of multiple cameras. All partial maps are sent to a central node and are fused to construct a global map. Several methods have been proposed for *map reconstruction* based on feature correlation in partial maps [2]. Compared to these methods, our distributed collaborative approach limits the use of communication and computational resources, and relies on autonomous relative positioning, resulting in more robust, efficient and scalable behavior.

Our system can rapidly build the map of the environment thanks to the camera coverage of the area. Moreover, it has the potential to rapidly react to changes in the environment, and to provide precise localization inside it. In this respect, it is equivalent to a *SLAM* [4] process. In fact, as a possible alternative, a swarm of ground robots could be used for SLAM, and build a free space map of the environment that could be more accurate than the one built by our camera network (see Footnote 1). However, the use of ground robots to build accurate maps of large cluttered environments requires the commitment of extensive resources, for time and computation. Our solution provides better efficiency, but we pay this advantage in terms of a potentially lower map accuracy.

4   Distributed Path Planning

The distributed path planning algorithm that we propose is an extension of the centralized algorithm described in Section 3. The algorithm consists of 3 phases: (i) *neighbor mapping*, (ii) *potential field calculation*, (iii) *path calculation*.

Neighbor mapping. Using the range and bearing system, each node builds a *neighbor table*, in which the relative positions and orientations of neighbor nodes are stored. Assuming that all cameras have a field of view of the same size, each node n uses this information to project the field of view of each neighbor m on its own reference system. This way, n builds an estimate of the *overlapping region* between its own field of view and that of m. In particular, n first segments its local map into a discrete set of cells, and then identifies which cells lay on the edges of the field of view of m, and which cells of its edges lay on m's field of view. Hereafter we refer to the former cells as *shared edges* and to the latter as *open edges*. Figure 1 illustrates this process.

Potential field calculation. The second phase is the calculation of a potential field. It is based on a diffusion process that starts from the destination point

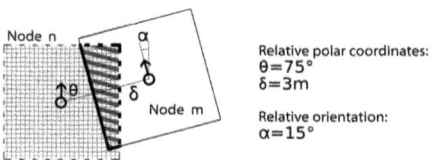

Fig. 1. Estimating θ, δ and α, n builds a projection of m's field of view to define: overlapping (striped region) *shared edges* (solid bold lines), *open edges* (dashed lines)

and iterates between the nodes until the potential is diffused on the entire map. This process is implemented in the following distributed way.

Each node first calculates the local *skeleton*, which corresponds to the *Voronoi diagram* on the 2D occupancy map (i.e., the *maximum-clearance roadmap* for object motion [9]).[3] Once the local skeleton is defined, each node calculates the *local potential field* through a cooperative diffusion process. The potential field defines a force that attracts the object towards the destination, and at the same time repels the object away from obstacles. In our algorithm, the potential is a scalar function that defines the attraction/repulsion intensity for each cell. The function has a global minimum in the destination point, and maximum value on the obstacles. In all other cells the function decreases towards the destination, such that the planner can aim to the goal following the gradient descent direction. The nodes that see the destination begin potential diffusion by assigning a zero value to the destination cell, and then incrementing the potential by one per cell. At first, nodes diffuse the potential only on the set of the skeleton cells; this way, the value assigned to each cell corresponds to the distance from the current cell to the destination over the skeleton. Then, the potential is calculated on the remaining free cells: diffusion starts from the skeleton, and values are increased while moving away from it and getting closer to obstacles.

Once the potential field is completely spread over the local map, each node n sends to its neighbors the values of the shared edges cells that they have in common. To minimize communication overhead, only shared edges cells that belong to the skeleton are transmitted. This way, each neighbor m of n receives a set of n's skeleton cells. Since n has an estimate of the relative position and orientation of m, the coordinates of these cells are expressed relative to m's frame of reference. m replaces the value of the corresponding cells in its own local map, and connects these new skeleton cells to the local existing skeleton. Then, m uses the received skeleton cells as starting diffusion points for spreading and updating the potential field on its local map.

The nodes that can see the destination maintain P potential field maps, one for each of the P control points of the object. This way, these nodes can correctly

---

[3] The environment's skeleton resulting from all local skeletons differs from that which would be calculated in a centralized way using a global map. The differences are due to the fact that, during skeleton calculation, the frontiers of a (local) map need to be considered as obstacles. Therefore, at the corners of each map the local skeleton shows bifurcations that are not present in the centralized skeleton.

position the object to the expected final configuration. The other nodes keep a single potential field, relative to the control point positioned in the object center. This means that these nodes do not take into account the object final orientation. This strategy optimizes the use of computational resources without affecting in practice the quality of the solution.

Path calculation. Once the potential field is spread among all nodes, the node above the start position starts path calculation: it plans the partial path relative to its local map using the values of the potential field. The A^* algorithm is used to identify a gradient descent trajectory in this field. A^* searches the solution by building an *exploration tree* of configurations. Each configuration defines a position and orientation of the object, and has a cost value equal to the sum of the distance from the initial point to the current configuration (measured in number of crossed cells) and the estimated distance to the destination (calculated as the sum of the potential value of the cells occupied by the control points). At each step, the algorithm selects the leaf of the tree that corresponds to the configuration c with the lowest estimated cost. Then, it calculates all the configurations that can be obtained from c by a one-step movement (a rotation or translation)[4] and their estimated costs, and adds them to the tree as leafs of c. The process iterates until the goal configuration is visited (*success*), or the exploration tree is completely expanded (*failure*), or the selected configuration falls outside the area of view laying on an open edge. In the latter case, the node sends the object coordinates to the neighbor m with which it shares the open edge. m continues the process, using as start position the received configuration. Passing object configuration from one node to another is made possible by the overlapping between neighbors' maps. However, this overlapping is subject to errors deriving from camera relative position estimation errors (see Section 2). These errors can affect the system performance, as we study in Section 5.

Since potential diffusion does not explicitly consider the dimensions of the moving object, it is possible to get trapped in local minima during path calculation. This is a well-known problem: following the potential might cause the object to get stuck in an area that is too narrow to let it pass. Our basic algorithm deals with the issue by locally backtracking and trying out alternatives. If a node fails in local planning, it sends to the preceding neighbor in the planning a *local failure* message. At the receiver, this message triggers the resuming of the local tree exploration process to calculate an alternative path. This approach is expensive, both in computation and communications. In order to improve the efficiency of this process and minimize the probability to get trapped in local minima we propose two heuristics (see Section 4.2).

Each node involved in the path calculation process stores: (i) the previous neighbor (from which the node received the entrance path), (ii) the planned local path, and (iii) the next neighbor (to which the node transmitted the object configuration of its local path). The distributed algorithm executes an exhaustive search of the solution: if the process ends with failure it means it has visited all

---

[4] In our system, a one-step translation corresponds to one cell, and a one-step rotation corresponds to an angle α. In the experiments, cell size is 10×10 cm, and $\alpha = 15°$.

possible configurations that are reachable from the starting point but no feasible solution exists given the characteristics of the calculated potential field and the selected search parameters (e.g., cell discretization, minimal rotation angle).

4.1   Adaption to Changes in the Environment

An important advantage of our distributed system is that it can locally and quickly detect and adapt to a change in the environment (e.g., a change in obstacles' position, or the appearing/disappearing of an obstacle). A centralized system would correct the Voronoi skeleton, repeat the potential field diffusion and restart the path planning. In our distributed architecture, the system can reduce the re-initialization costs by replanning only a limited part of the path. A node that detects a change in its local map informs the other nodes only if an alternative local partial path cannot be found.

More specifically, if the change happens when the system has completed the path calculation and the navigation has already started, the desired behavior is a fast response to adapt the path and allow the successful completion of the navigation task. A node n that detects a change in its local map, first tries to locally plan an alternative path by generating a new local potential field, with the constraint of maintaining fixed the original entrance and exit configurations. If a feasible path cannot be found, n notifies the destination node (through multi-hop wireless communication) to trigger a new potential field diffusion process. The destination node decides either to repair the path (from n to the end), or to recalculate the entire path (using the current position of the navigating object as new starting position). The decision for either alternative is taken in relation to the position of n in the sequence of nodes along the path. E.g., a local repair is issued when n is close to the final

4.2   Heuristics to Reduce Local Minima Attraction

The *skeleton pruning* and *local minimum detection* heuristics aim to improve the efficiency of the system minimizing the probability to get stuck in local minima.

Skeleton pruning. We prune the skeleton during potential field diffusion to block passages narrower than a predefined width w_{sp}. We set w_{sp} to the width of the smallest dimension of the moving object. Since size is not the only parameter defining whether an object can cross a passage (e.g., it depends also on the object's morphology) or not, the heuristic does not guarantee to remove all local minima. Setting w_{sp} higher, however, we may remove feasible paths.

Local minimum detection. The heuristic is executed during path calculation, when a node detects a local minimum due to the presence of a too narrow passage. First, the node places a virtual obstacle over the cells of the passage. Then, it triggers a new distributed potential field diffusion step. The assumption behind the heuristic is the following: the final solution obtained by recalculating the whole potential field and restarting the distributed path planning with the new information is expected to be of better quality than that obtained by exhaustively searching a way to escape from the local minimum.

4.3   Heuristics to Optimize the Use of Resources (Efficiency Heuristics)

Even using the above heuristics, dealing with *local minima* causes a high computation and communication load to the system. Moreover, given that the overall path is computed by combining local paths without using any global path information, it can contain *loops*. We propose two *efficiency heuristics* to deal with these issues: *block cells* and *smart loop avoidance*.

Block cells. During path calculation, if a node n has completely expanded its exploration tree in its local area without finding a valid path, it sends a *local failure* message to the preceding neighbor node m in the path. m resumes the path planning process and tries out another exit configuration, which may be in the same direction of attraction toward n. This way, a repeated exchange between n and m can take place until all exit configurations towards n are tried out unsuccessfully. To reduce communication and processing overhead associated to these situations, after a local path failure, m identifies the entry cells that have generated the failure in neighbor n, and avoids to send the object again through the same cells by removing them from the available open edges.

Smart loop avoidance. Given that $(n_1, n_2, \ldots, n_k)$ is the sequence of nodes associated to the computed path, if $n = n_i = n_j$, for any $1 \leq i, j \leq k$, $i \neq j$, then a loop is said to be present in the path if the two configurations to enter n at steps i and j can be connected together within n's local area. Therefore, the sub-path between n_i and n_j can be conveniently removed. This is done in the following way. Controlling its internal path components a node n checks whether a loop is present and can be removed. In this case, the node n sends a *loop* message to its previous node m in the path. After receiving the loop message, m deletes its local path and forwards the loop message to its preceding node. The process is iterated until the loop message reaches again node n and the loop is completely removed from the path. In order to avoid the re-creation of the loop, n perturbs its local potential field, and then resumes the path planning process.

5   Experimental Results

We test our distributed algorithm in simulation considering a large set of sample scenarios with varying area dimensions, position and number of eye-bots, shapes of moving object, and obstacle positions (Figure 2 shows two examples). We studied the performance of the proposed planning solutions in terms of *effectiveness, efficiency, scalability*, and *robustness to alignment errors*. As performance metrics we selected *success ratio*, the percentage of successful runs, and *path quality*, the relative length of the path compared to the path calculated by a *centralized algorithm with complete and perfect knowledge*. We study how performance varies in function of the alignment error between the cameras (angle and distance errors). Errors are modeled using a zero mean Gaussian distribution (in the experiments described in the following, we report the error as the standard deviation of the corresponding distribution).

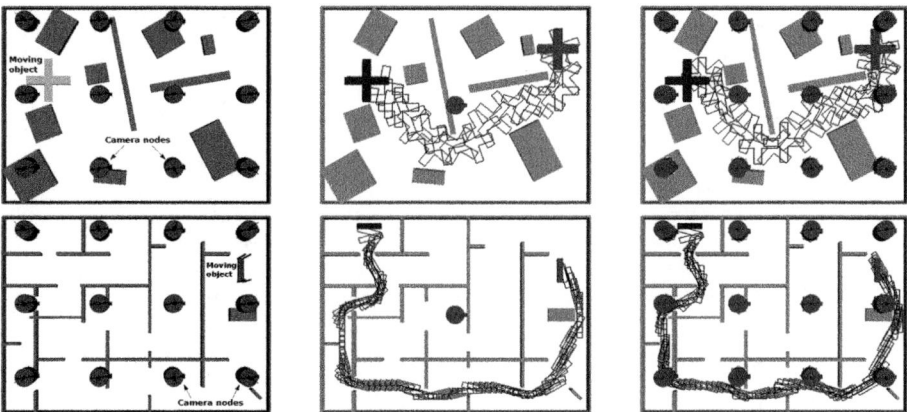

Fig. 2. (left) Examples of the scenarios used for experiments. (center) Reference paths from a global planner with perfect knowledge. (right) Paths of the distributed planner.

In each scenario the robots are deployed in a grid formation that covers the area. The distance between robots is 2 m. The visual field of each robot is the 4×4 m^2 area below their camera. Each point in the data plots is the average of 40 runs over 25 different scenarios including different numbers of robots (ranging from 12 to 36). The spread of the values around the average (not shown) is quite large, due to the strong heterogeneity of the scenarios. However, we assessed the significance of the results and the differences among the different versions of the algorithm through statistical testing. Each scenario is characterized by a different random positioning and structure of the obstacles. The percentage of the area occluded by obstacles vs. the total is approximately 18%. The shape of the moving object is randomly selected among rectangular, cross, and L shapes.

5.1   Robustness to Alignment Errors

The effect of camera alignment errors is shown in Figures 3 (angle error) and 4 (distance error). They report the performance of different versions of the algorithm: without heuristics, only with efficiency heuristics, and with all heuristics. The results show that for relatively low errors the performance is always very close to the reference. For increasing errors, the performance of the algorithm with heuristics degrades rapidly for success rate but slowly for path quality. Without heuristics the behavior is opposite. In both cases, the system is relatively sensitive to errors on the angle, while it is quite robust to distance errors.

In the experiments where the error is zero, the paths of the system without heuristics are on average 25% longer of the reference path. This is partially due to the generation of loops (which can be also caused by the presence of local minima). The use of heuristics mitigates these problems and reduces path length. However, the skeleton difference with respect to a global path planner (see Footnote 3) cannot be avoided, resulting in unnecessary movements for the object when it is moved between nodes, and, therefore, in slightly longer paths.

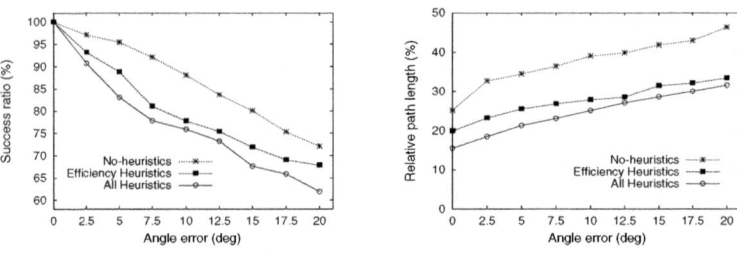

Fig. 3. Performance evaluation vs. increasing error in angle estimation

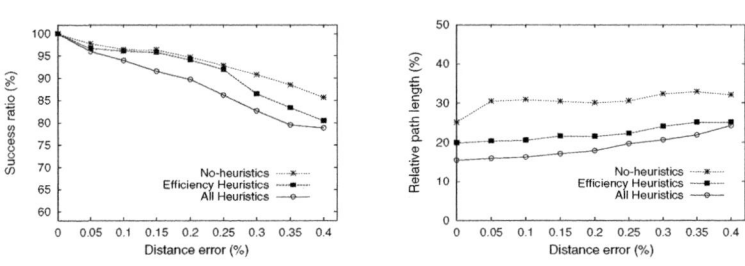

Fig. 4. Performance evaluation vs. increasing percentage error in distance estimation

5.2   Communication Efficiency

Communication efficiency is a fundamental aspect of the system because it impacts scalability. We quantify the use of communication resources through the average number of messages that each robot sends during the path planning process. In these experiments, the robots maintain a 4×3 grid formation.

The plots of Figure 5 show the results for this set of experiments. As expected, the system with *efficiency heuristics* has the best efficiency. It is able to maintain low values of exchanged messages even with the increase of the alignment error. The system with all the heuristics makes more intensive use of communication due to the strategies applied for local minima avoidance.

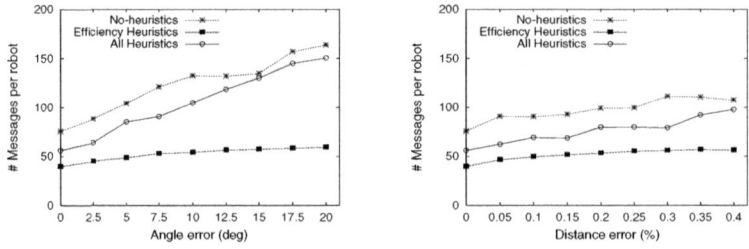

Fig. 5. Use of communication resources vs. error in angle and distance estimation

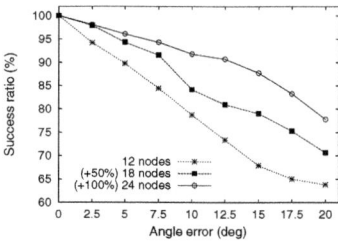 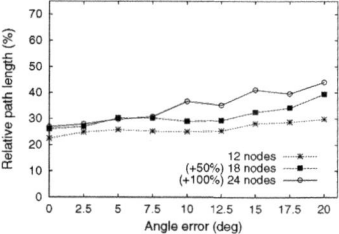

Fig. 6. Planner with heuristics: effect of node density vs. error in angle estimation

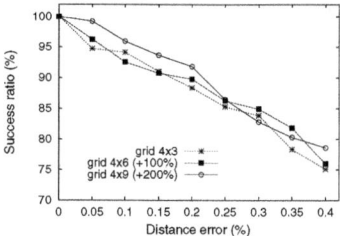 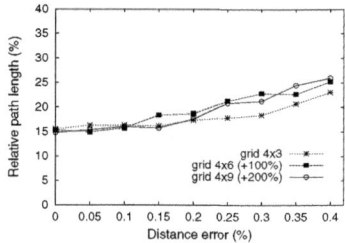

Fig. 7. Planner with heuristics: effect of environment size vs. error in angle estimation

5.3 Scalability: Increasing Robot Density in a Fixed Size Environment

We study the performance of the distributed system with respect to an increase of system's resources: in Figure 6 we show the effect of increasing the density of the nodes over a fixed area while varying angle errors. Increasing node redundancy allows the system to deliver a higher success ratio (Figure 6, left). This effect is more marked in the experiments with larger errors, in which the presence of a larger number of robots can balance the effect of these errors to find alternative valid paths. On the other hand, the increase in the density leads to longer paths, as shown in results of Figure 6 (right).

The results for the performance of the algorithm without the heuristics (not reported) show a similar behavior, with both the graphs being shifted up (higher success ratio and longer paths). The results for the error in distance (not reported) are analogous to those of Figure 6, but show a better robustness of the system to the error, as already observed in the previous experiments.

5.4 Scalability: Increasing Environment Size, Fixed Density of Robots

We studied the scalability of the system increasing environment size and maintaining robot density constant. The area of the basic scenario is 10×8 m^2, with a coverage of 12 robots in a grid formation 4×3. We use this basic scenario and two additional ones in which the area is scaled by a factor 2 and 3 respectively. The plots of Figure 7 show the results: the trends and values in the three scenarios are very similar. This means that the system is able to deal with an increase of the environment size with an increase of resources (number of robots) without degrading performances, showing its scalability.

6   Conclusions and Future Work

We have proposed a novel system for distributed path planning, which calculates with high accuracy the sequence of roto-translations of rigid objects of any shape moving through cluttered areas. The system architecture uses a swarm of flying robots, which deploy in the environment and form a distributed camera network. The robots solve the path planning problem cooperatively, through local calculations and wireless message exchanges. We adapted well-known solutions for path planning to our distributed architecture. Compared to systems with single cameras or centralized computations, the fully distributed approach can be more scalable, flexible, and robust. However it introduces efficiency issues and sensory errors. A number of heuristics were proposed to enhance the system's efficiency and effectiveness. In a wide range of simulation experiments, we show that the system is efficient, scalable, and robust to alignment errors between cameras.

Future work will include in the first place the implementation and testing of the system on real robots. We will also extend it and improve it, and perform tests with dynamic obstacles (e.g., including humans).

References

1. Barraquand, J., Langlois, B., Latombe, J.: Numerical potential field techniques for robot path planning. IEEE Trans. on Sys., Man and Cyb. 22(2), 224–241 (1992)
2. Birk, A., Carpin, S.: Merging occupancy grid maps from multiple robots. Proceedings of the IEEE, Special Issue on Multi-Robot Systems 94(7), 1384–1397 (2006)
3. Choset, H., Lynch, K., Hutchinson, S., Kantor, G., Burgard, W., Kavraki, L., Thrun, S.: Principles of Robot Motion: Theory, Algorithms, and Implementations. MIT Press, Cambridge (2005)
4. Durrant-Whyte, H., Bailey, T.: Simultaneous localisation and mapping (SLAM): The essential algorithms. IEEE Robotics and Automation Magazine (June 2006)
5. Garrido, S., Moreno, L., Blanco, D.: Voronoi diagram and fast marching applied to path planning. In: IEEE Int. Conf. on Robotics and Automation, ICRA (2006)
6. Hart, P., Nilsson, N., Raphael, B.: A formal basis for the heuristic determination of minimum cost paths. IEEE Trans. on Sys., Sc. and Cyb. 4(2), 100–107 (1968)
7. Latombe, J.-C.: A fast path planner for a car-like indoor mobile robot. In: Proceedings of the 9th National Conf. on Artificial Intelligence, pp. 659–665 (1991)
8. Latombe, J.-C.: Robot Motion Planning. Kluwer Academic, Dordrecht (1991)
9. LaValle, S.M.: Planning Algorithms. Cambridge University Press, Cambridge (2006)
10. Roberts, J., Stirling, T., Zufferey, J.-C., Floreano, D.: 3-D range and bearing sensor for collective flying robots. Journal of Field Robotics (submitted, 2011)
11. Stirling, T., Wischmann, S., Floreano, D.: Energy-efficient indoor search by swarms of simulated flying robots without global information. Swarm Intelligence 4(2), 117–143 (2010)
12. Takahashi, O., Shilling, R.J.: Motion planning in a plane using generalized voronoi diagrams. IEEE Trans. on Robotics and Automation 5(2), 143–150 (1989)

Evaluating the Effect of Robot Group Size on Relative Localisation Precision

Frank E. Schneider and Dennis Wildermuth

Fraunhofer Institute for Communication,
Information Processing and Ergonomics (FKIE), Wachtberg, Germany
{frank.schneider,dennis.wildermuth}@fkie.fraunhofer.de

Abstract. Looking on co-operative position estimation in multi-robot systems, the question to what extend the number of robots has an influence on the quality of the resulting localisation is an important and interesting issue. This paper addresses this relation regarding a pure relative localisation approach based only on mutual observations between the robots. The intuitive expectation that more robots should improve the position estimation is motivated and the design of the experiments with special respect to possibly distorting parameters is discussed and reasoned in detail. An in-depth analysis of the collected data explains the only partial conformance of the experimental results with the expected outcome.

1 Introduction

It is an obvious fact that most of the actions performed by mobile robots require some type of localisation. While localisation itself is a field of ongoing research, it is also a vital component for the co-ordination of navigation and movement. This holds especially when dealing with multi-robot systems (MRS).

The problem of localization can be addressed by different approaches. Local localisation evaluates the robot's position and orientation through integration of information provided by miscellaneous encoders and inertial sensors. All these sensors are mounted on the robot itself, and no external information is used [1, 2]. However, due to inherent uncertainty and unbounded error growth this method is usually not simply extendable to MRS.

Global localisation is normally based on some kind of map and uses sensor information to localise the robot with respect to these maps. In recent years, the problem of global localisation as well as typical approaches like Simultaneous Localisation and Mapping (SLAM) was extended to multi-robot localisation [3, 4, 5]. In the related approach of absolute localisation the vehicle determines its position directly through an exterior reference system, usually a satellite-based positioning system, navigation beacons, or passive landmarks. Since at least satellite-based systems do not have the accuracy needed for most robotic tasks, absolute localisation is often combined with other localisation techniques [6].

Relative localisation uses sensor observations to localise the robot with respect to its environment – including other robots – without having an environment model. Most of the authors working on this topic use the mutual observations of the robots as

R. Groß et al. (Eds.): TAROS 2011, LNAI 6856, pp. 149–160, 2011.

means for improving global positioning [7, 3, 4]. A different approach – which is also applied in this paper – is to change the aim of relative localisation and to maintain only a relative positioning between the robots. Hence, the resulting reference co-ordinate system is not global in the sense that it has a fixed reference to world co-ordinates. It is just shared among the members of the MRS and can diverge from world co-ordinates over time [8, 9].

For all multi-robot localisation approaches an interesting topic is the evaluation of the results in terms of precision, stability, scalability, or environment dependency. Thereby, the question whether the number of robots sharing the common co-ordinate system has an influence on the precision of the resulting localisation is one core issue. So far, this question has been studied almost always in terms of global positioning [10, 11], comparing the robots' position estimates with their corresponding world co-ordinates. In [11], for example, an analytical upper bound for the global positioning uncertainty in Cooperative Simultaneous Localization and Mapping (C-SLAM) is obtained. In this paper we are going to examine the relationship between robot group size and localization accuracy with regard to "pure" relative localization.

The remainder of the text consists of four parts. The next chapter shortly introduces the employed relative localisation method. Afterwards, we give a detailed description of the experimental setup. Special emphasis is put on the goal of gathering a good data basis for the evaluation. Design decisions concerning possibly distorting parameters and preconditions are explained and justified. The final chapter presents a detailed analysis of the collected data and tries to explain the only partial conformance of the experimental results with the expected outcome.

2   The Relative Localisation Approach

As introduced in [9] an Extended Kalman Filter (EKF) is used to integrate measurements into a global state estimation. The state vector $x(k) \in \mathbb{R}^{3n}$ describing all important information in a robot group of n robots at time step k is denoted

$$x(k) = [p_1(k) \quad \cdots \quad p_n(k)]^T \tag{1}$$

where $p_i(k) = [p_{i,x}(k) \quad p_{i,y}(k) \quad p_{i,\varphi}(k)]$ means position and orientation of the ith robot. For the initial error covariance P_0 we took $\sigma_t = 3cm$ as translational and $\sigma_o = 3°$ as rotational error.

In order to explain the EKF update step one has to consider the following situation. Each time one of the robots gathers a relative measurement of one of the others this information is deployed into the EKF in order to update the overall system state. For explanation of update step k, without loss of generality we will consider the case that the ith robot has got a new measurement $z = [d \ \alpha]^T$ of the robot with index 1, thereby meaning d the distance between them and α the direction of a vector from the observer to robot no. 1. The information gained from the odometry sensors of the two involved robots is used for the prediction step of the EKF. Let

$$u_i(k+1) = \left[\Delta p_{i,x}(k+1) \quad \Delta p_{i,y}(k+1) \quad \Delta p_{i,\theta}(k+1) \right] \tag{2}$$

the ith robot movement from time step k to $k+1$. Then the state prediction writes to

$$x^-(k+1) = \left[p_1(k) + u_1(k+1) \quad \dots \quad p_i(k) + u_i(k+1) \quad \dots \right]^T. \tag{3}$$

Accordingly, the projection of the error covariance is simply defined as

$$P^-(k+1) = P(k) + Q(k+1). \tag{4}$$

The process noise covariance $Q(k+1)$ in our approach is not constant over time but depends on the distances the two involved robots travelled since their state had been updated last time. The further they moved the larger is the uncertainty on this movement because of possible odometry errors. This consideration led to the following definition:

$$Q(k+1) = \begin{pmatrix} Q_1 & & & 0 \\ & \ddots & & \\ & & Q_i & \\ 0 & & & \ddots \end{pmatrix} \tag{5}$$

a diagonal matrix with

$$Q_i(k+1) = \begin{pmatrix} \left(s_t \Delta p_{i,x}(k+1) + \hat{q}_t\right)^2 & 0 & 0 \\ 0 & \left(s_t \Delta p_{i,y}(k+1) + \hat{q}_t\right)^2 & 0 \\ 0 & 0 & \left(s_o \Delta p_{i,\vartheta}(k+1) + \hat{q}_o\right)^2 \end{pmatrix}. \tag{6}$$

and remaining diagonal elements $\hat{Q} = \begin{pmatrix} \hat{q}_t^{\,2} & 0 & 0 \\ 0 & \hat{q}_t^{\,2} & 0 \\ 0 & 0 & \hat{q}_o^{\,2} \end{pmatrix} \in \mathbb{R}^{3\times3}$, $\hat{q}_t$ and $\hat{q}_t$ very

small positive constants. Thereby, s_t represents the mean error for translational movement of the robots and s_o the same for rotations. As long as all robots show the same error characteristic those values remain constant, otherwise they should depend on the actual robot they belong to. We assume constant values of $s_t = 5cm$ and $s_o = 5°$. The $\hat{q}_t$ and $\hat{q}_o$ represent the inherent uncertainty of the position information for every robot, which grows even if the odometry reports no movement at all.

After this, the EKF prediction step is finished. The new measurement $z(k+1) = [d \; \alpha]^T$ of robot no. 1 obtained by the ith observing robot is now used for the correction step of the EKF:

$$x(k+1) = x^-(k+1) + K(k+1)\left[z(k+1) - h\left(x^-(k+1), 0\right)\right] \tag{7}$$

where the function $h(x, v)$ relates a state vector x, defined as in (1), to a measurement $z = [d \; \alpha]^T$, given the current measurement noise v. For $h(x, v)$ a standard mixed co-ordinate EKF approach [13] is used, which in this case writes to

$$h(x,v) = \begin{pmatrix} h_1(x,v) \\ h_2(x,v) \end{pmatrix} = \begin{pmatrix} \sqrt{(p_{1,x} - p_{i,x})^2 + (p_{1,y} - p_{i,y})^2} + v_d \\ \arctan\left(\dfrac{p_{1,y} - p_{i,y}}{p_{1,x} - p_{i,x}}\right) - p_{i,\vartheta} + v_\vartheta \end{pmatrix}. \tag{8}$$

If we redefine (3) as $\begin{bmatrix} \tilde{p}_{1,x} & \tilde{p}_{1,y} & \tilde{p}_{1,\vartheta} & \cdots & \tilde{p}_{i,x} & \tilde{p}_{i,y} & \tilde{p}_{i,\vartheta} & \cdots \end{bmatrix}^T = x^-(k+1)$ and define

$$P_1 = \sqrt{(\tilde{p}_{1,x} - \tilde{p}_{i,x})^2 + (\tilde{p}_{1,y} - \tilde{p}_{i,y})^2} = \left\| \begin{pmatrix} \tilde{p}_{1,x} \\ \tilde{p}_{1,y} \end{pmatrix} - \begin{pmatrix} \tilde{p}_{i,x} \\ \tilde{p}_{i,y} \end{pmatrix} \right\| \tag{9}$$

then for the Jacobian matrix $H(k+1)$ for the remaining EKF step this leads to

$$H(k+1) = \begin{pmatrix} \dfrac{\tilde{p}_{1,x} - \tilde{p}_{i,x}}{P_1} & \dfrac{\tilde{p}_{1,y} - \tilde{p}_{i,y}}{P_1} & 0 & 0 & \dfrac{\tilde{p}_{i,x} - \tilde{p}_{1,x}}{P_1} & \dfrac{\tilde{p}_{i,y} - \tilde{p}_{1,y}}{P_1} & 0 & 0 \\ \cdots & & & & & & & \cdots \\ \dfrac{\tilde{p}_{i,y} - \tilde{p}_{1,y}}{P_1^2} & \dfrac{\tilde{p}_{1,x} - \tilde{p}_{i,x}}{P_1^2} & 0 & 0 & \dfrac{\tilde{p}_{1,y} - \tilde{p}_{i,y}}{P_1^2} & \dfrac{\tilde{p}_{i,x} - \tilde{p}_{1,x}}{P_1^2} & -1 & 0 \end{pmatrix}. \tag{10}$$

Note that $H_{1,3}(k+1) = 0$ and $H_{2,3}(k+1) = 0$, which means that no update on the orientation of the observed robot takes place during the filter step. This is, of course, due to the fact that in our setup the orientation information cannot be measured by the observer. But as long as all participating robots at least occasionally generate measurements on their own, orientation information for all robots is updated.

The Kalman gain $K(k+1)$ is computed in the usual way with the measurement noise covariance matrix R considered as constant. Due to the characteristics of the laser device and the tracking algorithm for the experiments we used the values $r_t = 0.5cm$ and $r_o = 0.5°$. The filter step ends with the update of the error covariance matrix $P(k+1)$. As a result $x(k+1)$, as computed in (7) contains updates for the position of the first robot, which has been observed, and updates for position and orientation of the ith robot, which in this case played the role of the observer. In other words, starting with initial position estimations the EKF maintains the relative positions for all robots during the ongoing run.

3   Description of the Experiments

Aim of this work is to examine the relationship between robot group size and localization accuracy. On one hand, since each robot measures distance and direction to every other robot in the group, the amount of mutual measurements rapidly increases. On the other hand, apart from the errors caused by the odometry, there are also a possibly growing number of measurement faults due to, for example, temporary occlusions while the robots work together.

The aggregated odometry error for a robot group grows linearly with the number of group members. The frequency of mutual occlusions depends on the kind of task the robots have to fulfil, but normally such occlusions have only a very limited duration, happening for example while two robots are passing each other. Therefore, one can expect that the increasing number of mutual position measurements outweighs the number of possible error sources. Thus, we expected that the overall precision of the relative localisation should increase with the size of the robot group.

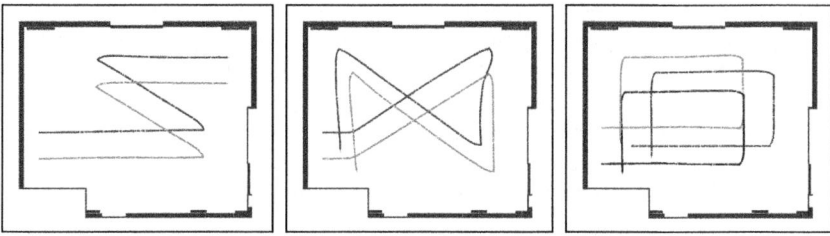

Fig. 1. The three different shapes on which the robots had to move in the large 15 x 18 m experimental hall

3.1 Design Decisions and Necessary Preconditions

The experiments were fully conducted in simulation. Apart from the obvious reason that a simulation with well-known parameters yields reproducible and comparable results, some other important causes lead to this decision. First, the simulation provides an exact ground-truth in form of absolute position information for each robot, far better than any means of positioning for "real" robots.

Additionally, it was planned to compare group sizes of at least five robots, which means a great challenge with experimental robot systems and a lab environment of limited size. And the goal was not to record some impressive demonstration runs but to obtain data from a large number of runs for each scenario in order to establish a meaningful basis for further evaluation.

With different group sizes from two to five robots, at least three different driving scenarios and a minimum of 20 runs for each combination, this already results in a total of 240 experiments, not considering any possible technical troubles or problems in the design of the experiments. Realistically, this easily leads to hundreds of additional trials, consuming weeks of laboratory time. In simulation the original design regarding group size and driving scenarios was retained. Thus, in principle all results of these experiments could be validated using the real platforms and a large experimental hall. In simulation, now each of the scenarios described in the following section was conducted 60 times.

The three different paths on which the robots moved are pictured in figure 1. The left figure shows a group of two robots (red and green line) moving along a Z-like shape. The middle one presents the second scenario, a shape like a horizontal eight with corners. In the right figure a formation of three robots (red, green and blue line) drives on a rectangular path. These scenarios will be mostly referred to as "Z", "eight", and "rectangle" in this paper.

The three paths look similar in terms of driving distance, covered area, or overall turning angle. But there are also relevant characteristics in which the scenarios differ, mainly the kind of turns the vehicles have to follow. For many robot platforms – and for ours as well – any rotation induces a potentially large odometry error. Another frequent observation is the influence of the turning direction on the aggregated odometry error. During longer experiments it sometimes happens that the rotational error increases and decreases in turn. This seems to be because odometry errors caused by clockwise and counter-clockwise turns sometimes eliminate each other.

The multi-robot simulator was configured to use a realistic odometry error model, reflecting the foregoing considerations. Each robot had its own error characteristic with a normally distributed translational error ($\sigma = 3cm$ per 1m translation), and an orientation error spread normally around a small systematic fault ($\mu \in [1.5, 2.5]$, $\sigma = 0.3°$ per 90° turn). Especially the systematic fault causes the typical displaced odometry data. Figure 2 presents some characteristic example runs. In these examples always two robots have been recorded in order to simplify the illustrations.

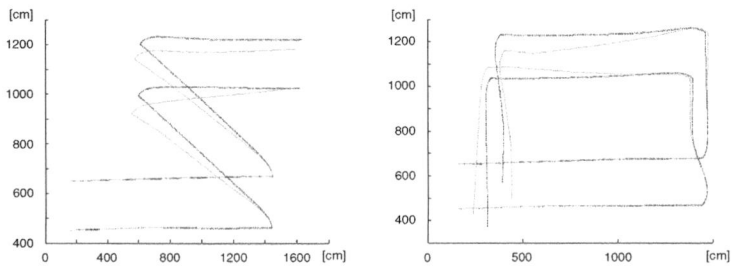

Fig. 2. Comparison of true positions (*blue path*) with information from odometry (*green path*)

The blue lines give the robots' exact positions. The two figures correspond to the "Z" and the "rectangle" scenario described above. In contrast, the green lines show the faulty positions of the robots as they are recorded by the simulated odometry sensors. The left figure gives a good example for the before mentioned decreasing odometry error. The lower of the two robots nearly ends at its exact position, even though with faulty orientation information, and the odometry position of the upper robot also approaches to the real one. This is a quite recurrent behaviour in this first scenario, possibly because in this scenario the robots have to turn exactly the same angle of about 110° in both directions. In the eight and rectangle scenario the overall driving distance for each robot is between 30m and 40m, and the accumulated odometry error at the end is often more then 1m, sometimes even more than 2m. Like the right figure demonstrates, this deviance often points into completely opposite directions, due to the independent error model for each robot. As a result, looking only at the pure odometry data the distance (and orientation) error between any pair of robots easily reaches the magnitude of several metres during and especially at the end of a run. Deviances of this dimension mainly happen in the rectangle scenario. In this scenario the robots have to turn only counter-clockwise, three times about 90°.

Altogether, the average odometry error is smallest in the Z-scenario, largest in the rectangle scenario and somewhere in between for the eight. It is important to mention one more time that, although the presented data is simulation generated, it very well matches our experiences when collecting odometry data from the real robot systems.

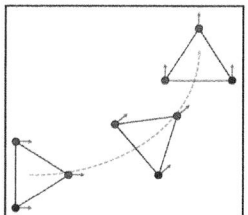 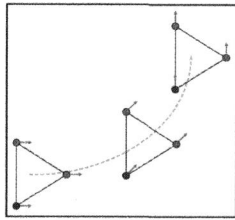

Fig. 3. Different modes of turning a formation: "normal" (*left*) – whole formation turns; "fixed" (*right*) – only the robots turn around

Apart from the impacts of the different paths it is worth discussing the mode in which the robot group had to move during the experiments. Of course, since all robots had to follow the same path, special co-ordination is necessary. Different approaches have been considered. One possibility, for example, is to assign exactly the same path to each robot and then let them start consecutively from the same starting position at a fixed time interval. The disadvantage is the nearly permanent mutual occlusion of robots which are moving along the same line of sight. Apart from the turning points, only direct neighbours can see each other. For relative localisation this is not the best choice for evaluating the overall precision of the approach.

Another idea was to simply give the robots the goal and turning points, leaving the remaining coordination to the collision avoidance. But the lack of a pre-defined and reproducible behaviour for each robot is a problem for the goal of precision measurement. Thus, it was decided to use fixed formations, which the robots had to maintain while following the pre-defined paths. "Fixed" in this context characterises a special mode of movement in formation. Whenever the formation reaches a turning point, instead of completely turning the shape of the formation only the robots themselves turn around (see figure 3). Different approaches to the field of formation navigation are addressed in detail, for example, in [12].

The actual goal formations for the varying group sizes can be seen in figure 4. They are straightforward from the 2-robot line over a triangle and rectangle shape leading to a pentagon for five robots. The minimal distances, as found between neighbouring robots, are 2m for the line shape, 2.25m for the rectangles and about 2.7m for the two other shapes. The largest distance of about 4.25m can be found in the pentagon formation.

As mentioned above the formation algorithm was configured to generate fixed formation shapes in order to obtain regular and straight paths for all robots. The simulated environment was a part of our indoor robotic lab, namely the large experimental hall of about 15 x 18 metres in size. This gives us the possibility to repeat at least some of the trials in a physical environment.

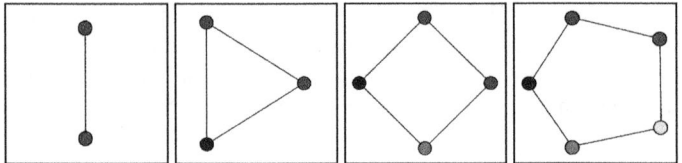

Fig. 4. Formation shapes used for the different robot group sizes

3.2 Collecting Data

Our robots are equipped with standard SICK laser range finders. According to the specification [14], the LMS200 devices deliver distance information with 1cm resolution and an error of less than 1.5cm. In our configuration the angular information has a resolution of 1° and a typical error of about 0.4°. The simulated laser scanners compute their sensor data with the same error characteristic.

In order to derive position information for the other robots from the raw laser readings, a straightforward stateless geometric algorithm was used. We achieve a mean error of about 0.5cm for the relative distance between two robots and less than 0.5° for the measured direction from one robot to another. This is a similar range as it can be achieved for sophisticated probabilistic trackers like Probabilistic Data Association Filtering (PDAF) or Probabilistic Multi Hypothesis Tracking (PMHT), both customized especially for extended targets [15]. Consequently, we omitted the implementation of any other tracking algorithm and continued to use the simple geometric method.

Figure 5 presents exemplary results for the mutual observation and measuring process in terms of quantity of position measurements. We counted the average number of relative position measurements which each robot gathered from all other group members during one experimental run. The y-axis gives this number and the x-axis splits the results for the three different scenarios and the different group sizes. One can expect the double number of measurements with three robots, triple with four and four times more measurements in the five robot group. Actually, in the "Z" scenario for two robots each one measures its counterpart's position at an average of 474 times during one experimental run. In a group of three robots each of them generates an average of 875 measurements, 1340 position values for four robots, and in the five robot group the mutual observation process delivers 1747 measurements per robots. Like figure 5 illustrates the situation is similar for the other scenarios.

In summary, the design of the experiments should be adequate to analyse the influence of the robot group size on the resulting precision of the relative localisation. Due to the way the robots moved, a predictable and comparable behaviour for all scenarios is achieved. The number and duration of mutual occlusions is limited and reproducible. There are linearly growing measurements per robot with growing group size. Although simulation based, the position measurements and the robots' odometry sensors have realistic accurateness and error characteristics. The aggregated odometry error for the robot group differs between the scenarios, being larger for the "rectangle" and smaller for the "Z", but in all cases it grows only linearly with the number of group members. Consequently, we expected the increasing number of mutual position measurements to outweigh the growing odometry error and, thus, the overall precision of the relative localisation to increase with the size of the group.

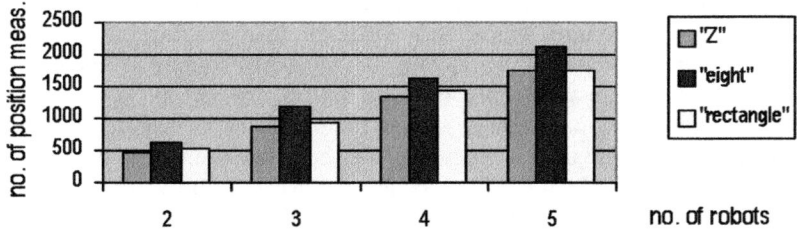

Fig. 5. Amount of position measurements (*y-axis*) for the different number of robots (*x-axis*)

4 Evaluation of the Results

As mentioned earlier a total of 60 runs have been conducted for each group size and each different driving scenario, leading to 720 simulation-based datasets in total. To evaluate the resulting relative localisation we defined a Mean Localisation Error (MLE). For each single localisation step the estimated robot positions were compared to the exact positions as delivered by the simulator. Since we are looking at relative localisation we could not simply compare absolute positions. Instead, for each pair of robots the difference between estimated and true distance was summed up. This deviation was then weighted by the number of robots to get a measure independent of the group size. Finally, an average over the entire dataset was computed, resulting in the MLE per run. One drawback of this approach is that orientation errors are not addressed. Hence, we tried some different approaches, including complex and mathematically demanding methods derived from metrics for robot groups as described in [16] and [17]. But since these approaches produced similar results we kept the computationally simple MLE as described above.

Looking at table 1 one can find the average MLE and the corresponding standard deviation calculated from the 60 runs for each scenario and group size. First of all, it is obvious that the resulting relative localization among the robots is very accurate. The mean error per time step for one robot consistently lies below 1.8cm. Considering only the tracking process with its ±0.5cm distance and ±0.5° angular error, then for typical robot distances between 2m and 4m the angular error delivers displacements

Table 1. Average MLE (in cm) and corresponding standard deviation per 60 runs for different group sizes and the three driving scenarios

	no. of robots	2	3	4	5
"Z"	mean MLE [cm]	1,542	1,552	1,364	1,441
	σ MLE	0,274	0,165	0,106	0,092
"eight"	no. of robots	2	3	4	5
	mean MLE [cm]	1,435	1,496	1,522	1,457
	σ MLE	0,078	0,160	0,132	0,085
"rectangle"	no. of robots	2	3	4	5
	mean MLE [cm]	1,695	1,781	1,501	1,541
	σ MLE	0,294	0,194	0,137	0,095

from 1.74cm in the near case and 3.48cm for the larger distances. Adding the perpendicular distance error this already leads to errors ranging from ±1.81cm up to ±3.52cm, thereby not even taking into account the errors coming from the odometry.

Unfortunately, the results from table 1 do not fully reflect the key expectation that the localisation precision increases with the number of robots. For the "eight" scenario there is no improvement at all, for the other scenarios there is at most a tendency of less than 0.2cm between the average MLE with two robots and with five robots. This might be due to the fact that even with only two robots the accuracy of the laser based mutual tracking outweighs the errors of the odometry sensors. At least the standard deviation of the MLE shows some improvement with larger group sizes. Again, for each combination of group size and scenario the table contains average values per 60 recorded example runs. Except for the "eight" scenario with 2 robots, the standard deviation shows less variation and therefore an improved stability in the relative localisation.

For a detailed analysis of the results we had a closer look at the changes of the Mean Localisation Error (MLE) over runtime. Figure 6 gives two typical examples, plotting the MLE in cm on the y-axis and the number of localisation steps from beginning to the end of the example run on the x-axis. Based on this data it was possible to check whether the error remains constant over time or whether there are peaks corresponding, for example, to turns or temporary occlusions of the robots.

The left part of figure 6 presents one dataset for the "eight" scenario and the two-robot formation, the right chart shows the same for five robots. One can see that, actually, the MLE fluctuates much less if more robots take part in the relative localisation. During a further inspection of the error peaks no general correlation to incidents like turnings or occlusions could be indentified. The variations seem to be arbitrarily spread over the whole runtime.

The only point which occasionally recurs is an increase of the MLE near the end of a run. In the left example in figure 6 the impression of a slightly higher error in the second half starting from time step 700 can be proven by numerical evaluation as well. The MLE is more than 0.5cm higher in the second half of the run. Figure 7 presents more examples for this effect. The left part is taken from a "rectangle" scenario with three robots, in the right part four robots move along the "Z" path. The "rectangle" example is comparable to the before described one. From time step 1700 the standard deviation of the MLE starts growing and from step 2300 the error increases by more then 1cm. The other example gives an even larger growth of the MLE. But in this case this does not go along with larger fluctuations of the localisation error, instead a constant increase of the error can be found. A possible explanation of this occasional effect might be an above-average odometry error aggregation which leads to very bad state predictions in the localisation EKF. But this is subject to ongoing research.

Another important fact currently under investigation is that only very few datasets result in a MLE below 1cm. Even during periods of precise localisation with very small deviation – like in the right example of figure 6 or in the first half of the "Z" run in figure 7 – the error only seldom drops below 1cm. We suspect a systematic fault somewhere in the design of the experiments or the analysis algorithms. But, as the section on the experiments should document, the whole preparation was done with great care. Thus, no such error could be found so far.

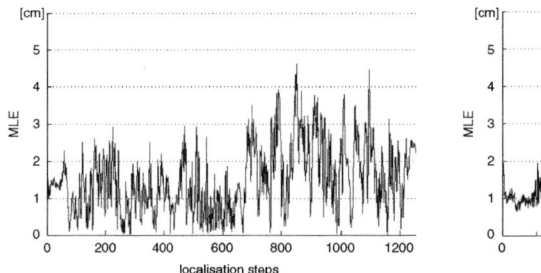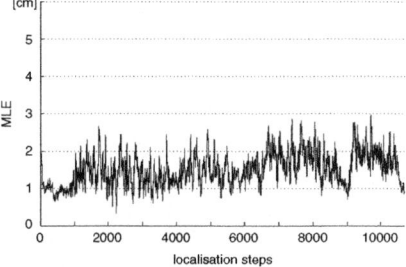

Fig. 6. Example plot of the MLE in cm (*y-axis*) for the "eight" scenario with two (*left*) and five robots (*right*); the *x-axis* runs through the localisation steps resp. the duration of the run. The error for two robots shows more variations, fitting well to the larger standard deviation.

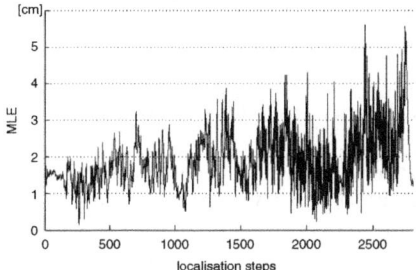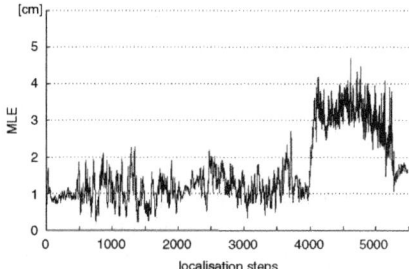

Fig. 7. MLE for the "rectangle" with three (*left*) and the "Z" with four robots (*right*), axes as above, showing larger variations or sections with constantly large errors near the end of a run.

5 Summary

This paper addresses the effect of robot group size on relative localisation precision regarding a pure relative localisation approach based only on mutual observations between the robots. The intuitive expectation that more robots should improve the position estimation is motivated and the design of the experiments with special respect to possibly distorting parameters is discussed and reasoned in detail. A detailed analysis of the collected data shows to what extend the experimental results conform to the expected outcome. Possible explanations for the rather small improvements with larger group sizes are considered.

References

1. Barshan, B., Durrant-Whyte, H.F.: Inertial navigation systems for mobile robots. IEEE Transactions on Robotics and Automation 11(3), 328–342 (1995)
2. Borenstein, J., Feng, L.: Measurement and Correction of Systematic Odometry Errors in Mobile Robots. IEEE Transactions on Robotics and Automation 12(6) (1996)
3. Fox, D., Burgard, W., Kruppa, H., Thrun, S.: A Probabilistic Approach to Collaborative Multi-Robot Localization. Autonomous Robots 8(3), 325–344 (1996)

4. Rekleitis, I.M., Dudek, G., Milios, E.E.: Probabilistic Cooperative Localization and Mapping in Practice. In: Proceedings of the IEEE International Conference on Robotics and Automation (ICRA), Taipei, pp. 1907–1912 (2003)
5. Madhavan, R., Fregene, K., Parker, L.E.: Distributed Cooperative Outdoor Multirobot Localization and Mapping. Autonomous Robots 17(1), 23–39 (2004)
6. Moutinho, A., Azinheira, J.R.: Comparison and fusion of odometry and GPS navigation for an outdoor mobile robot. In: Proceedings of the 11th IEEE International Conference on Advanced Robotics (ICAR), Coimbra (2003)
7. Kurazume, R., Hirose, S.: An Experimental Study of a Cooperative Positioning System. Autonomous Robots 8(1), 43–52 (2000)
8. Howard, A., Matarić, M.J., Sukhatme, G.S.: Cooperative Relative Localization for Mobile Robot Teams: An Egocentric Approach. In: Schultz, A.C., Parker, L.E., Schneider, F.E. (eds.) Multi-Robot Systems: From Swarms to Intelligent Automata, pp. 65–76. Kluwer, Dordrecht (2003)
9. Schneider, F.E., Wildermuth, D.: An Application of Relative Localisation for Multi-Robot Navigation. In: Proceedings of the IASTED International Conference on Automation, Control, and Applications (ACIT-ACA), Novosibirsk (2005)
10. Roumeliotis, S.I., Rekleitis, I.M.: Propagation of Uncertainty in Cooperative Multirobot Localization: Analysis and Experimental Results. Autonomous Robots 17(1), 41–54 (2004)
11. Mourikis, A.I., Roumeliotis, S.I.: Performance Analysis of Multirobot Cooperative Localization. IEEE Transactions on Robotics 22(4), 666–681 (2006)
12. Schneider, F.E., Wildermuth, D.: Directed and Non-Directed Potential Field Approaches to Formation Navigation. In: Proceedings of the 2nd IASTED International Conference on Automation, Control, and Applications (ACIT-ACA 2005), Novosibirsk (2005)
13. Bar-Shalom, Y., Li, X.-R.: Estimation and Tracking: Principles, Techniques and Software. Artech House, Boston (1993)
14. Technische Beschreibung Lasermesssysteme, doc. 8008969/06-2003, SICK AG, Division Auto Ident, Germany (2003)
15. Kräußling, A., Schneider, F.E., Wildermuth, D., Göb, R.: Zur Verfolgung ausgedehnter Ziele – eine Übersicht über ausgewählte Algorithmen und ein Vergleich deren Güte. FKIE-Forschungsbericht, vol. (94). Fraunhofer FKIE, Wachtberg (2005)
16. Schneider, F.E., Wildermuth, D., Kräußling, A.: Discussion of Exemplary Metrics for Multi-Robot Systems for Formation Navigation. International Journal of Advanced Robotic Systems 2(4), 345–353 (2005)
17. Navarro, I., Matía, F.: A Proposal of a Set of Metrics for Collective Movement of Robots. In: Proceedings of the Workshop on Good Experimental Methodology in Robotics, Robotics: Science and Systems Conference (RSS), Seattle (2009)

Instance-Based Reinforcement Learning Technique with a Meta-learning Mechanism for Robust Multi-Robot Systems

Toshiyuki Yasuda, Motohiro Wada, and Kazuhiro Ohkura

Hiroshima University, Higashi-Hiroshima, Japan
{yasuda,wada,ohkura}@ohk.hiroshima-u.ac.jp

Abstract. In recent years, the subject of learning autonomous robots has been widely discussed. Reinforcement learning (RL) is a popular method in this domain. However, its performance is quite sensitive to the discretization of state and action spaces. To overcome this problem, we have developed a new technique called Bayesian-discrimination-function-based RL (BRL). BRL has proven to be more effective than other standard RL algorithms in dealing with multi-robot system (MRS) problems. However, similar to most learning systems, BRL occasionally suffers from overfitting. This paper introduces an extension of BRL for improving the robustness of MRSs. Meta-learning based on the information entropy of firing rules is adopted for adaptively modifying its learning parameters. Physical experiments are conducted to verify the effectiveness of our proposed method.

Keywords: multi-robot system, cooperation, robustness, reinforcement learning, meta-learning.

1 Introduction

Multi robot systems (MRSs) have recently attracted considerable attention from roboticists as these offer the possibility of accomplishing a task that a single robot can not. A robot team may provide redundancy and perform assigned tasks in a more reliable, faster, or cheaper way.

Reinforcement learning (RL) [1] is a frequently used method in the problem domain of learning autonomous robots. Because of the progress in RL research, a mobile robot can acquire appropriate behaviour such as obstacle-avoidance, wall-following or goal-reaching by interaction with an embedded environment. However, the results obtained from single robot systems are not directly applicable to MRSs, mainly because of the following two reasons. First, although RL is quite sensitive to how discretization is performed, its simple form assumes that learning space should be discretized before learning starts. Second, RL assumes a static environment, whereas a robot in an MRS is surrounded by an intrinsically nonstationary environment because of other robots that are learning simultaneously. Nevertheless, RL is often applied to MRS problems successfully because it allows for dynamics up to a certain level in an embedded environment.

R. Groß et al. (Eds.): TAROS 2011, LNAI 6856, pp. 161–172, 2011.
© Springer-Verlag Berlin Heidelberg 2011

Several approaches for solving these problems and for learning in a continuous space have been discussed. A popular method applies function approximation techniques such as artificial neural networks to Q-function. Sutton [2] used Cerebellar Model Articulatory Controller (CMAC) and Morimoto and Doya [3] used Gaussian softmax basis functions for function approximation. Lin represented the Q-function by using multi-layer neural networks called *Q-net* [4]. However, these techniques have an inherent difficulty that a human designer must properly design their neural networks before executing RL. The idea of dimension reduction has been adopted[5,6]. The basic idea of this approach is to explicitly use a simpler representation of data by projecting it to lower dimensional spaces.

Other methods involve the adaptive segmentation of the continuous state space according to the robots' experiences. Asada *et al.* proposed a state clustering method based on a Mahalanobis distance [7]. Takahashi *et al.* used a nearest neighbour method [8]. However, these methods generally require large learning costs for tasks such as the continuous update of data classifications every time new data arrives. Actor-critic algorithms built with function approximators have a continuous learning space and adaptively modify actions [9,10]. These algorithms modify policies based on a temporal difference (TD) error at each time step.

We have previously proposed an instance-based RL method called Bayesian-discrimination-function-based RL (BRL) [11,12,13]. Our preliminary experiments illustrated that BRL exhibits better performance compared with CSCG through the adaptive discretization of state and action spaces. BRL has also proved to be robust against the dynamics in an environment that contains multiple robots. However, similar to other learning systems, we have occasionally observed overfitting problems in BRL. Overfitting is a problem such that a learning robot gradually becomes less robust after stable behaviour is acquired. We consider that this brittleness is critical for MRSs because robots must be essentially situated in a nonstationary environment, which is generally unpredictable. In this paper, an extension of BRL is proposed to overcome the abovementioned problem. The basic idea is that meta-parameters such as learning rate are adaptively coordinated so that each robot modifies its behaviour on the basis of the stability of its own actions.

The rest of this paper is organised as follows. The target problem is introduced in Section 2. The details of BRL and its extensions are explained in Section 3. The results of our experiments are described in Section 4. The conclusions are provided in Section 5.

2   Learning Task

Our target task is a simple MRS consisting of three autonomous mobile robots shown in Fig. 1. This problem is called the *object-orbiting task* and involves requiring the MRS to avoid collision with a wall and to move counterclockwise in a field.

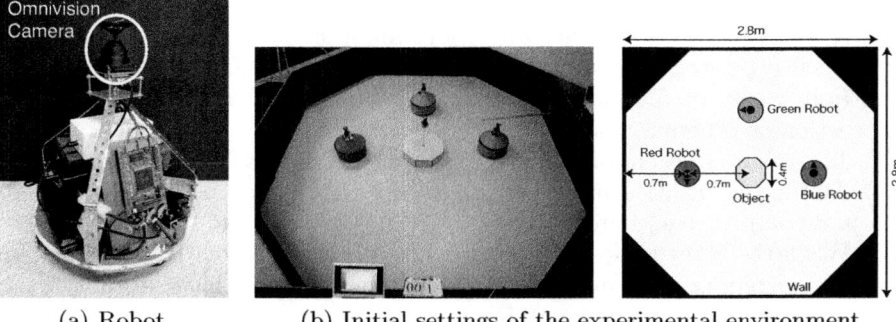

(a) Robot   (b) Initial settings of the experimental environment

Fig. 1. Object-orbiting task

All the robots have the same specifications; each robot is 35 cm in height and 28 cm in length (diameter). They have an omnivision camera at the centre of their body. A robot can detect an object, a wall and the nearest robot. Each robot has three motors for rotating two omnidirectional wheels. A wheel simultaneously provides powered drive in the direction in which it is pointing and passive coasting in an orthogonal direction.

The difficulties in this task can be summarised as follows:

- The robots must cooperate with each other to achieve the given task.
- They begin with no predefined behaviour rule sets or roles.
- They have no explicit communication functions.

3   Extended BRL

3.1   BRL: RL in Continuous Learning Space

Overview. Our approach, called BRL, adaptively updates classifications on the basis of interval estimation, only when such an update is required. In BRL, the state space is covered by multivariate normal distributions, each of which represents a rule cluster C_i. A set of production rules is defined by Bayesian discrimination. This method can assign an input x to the cluster C_i, which has the largest posterior probability $\max \Pr(C_i|x)$. Here, $\Pr(C_i|x)$ indicates the probability (calculated by Bayes' formula) that a cluster C_i holds the observed input x. Therefore, by using this technique, a robot can select a rule that is most similar to the current sensory input. In BRL, production rules are associated with clusters segmented by Bayes boundaries. Each rule contains a state vector v, an action vector a, a utility u and parameters for calculating the posterior probability, $i.e.$ a prior probability f, a covariance matrix Σ and a sample set Φ.

The learning procedure is as follows:

(1) A robot perceives the current sensory input x.
(2) By using Bayesian discrimination, the robot selects the most similar rule from a rule set R. If a rule is selected, the robot executes the corresponding action a, otherwise, it performs a new action.
(3) The robot transfers to the next state and receives a reward r.
(4) All the rule utilities are updated according to r. The rules with utility below a certain threshold are removed.
(5) When the robot performs a new action, it produces a new rule by combining the current sensory input and the executed action. This executed new rule is memorised in the rule set R.
(6) If the robot receives no penalty, an interval estimation technique updates the parameters of all the rules. Otherwise, the robot updates only the parameters of the selected rule.
(7) Go to (1).

Action Selection and Rule Production. In BRL, a rule in the rule set R is selected to minimise a discrimination function g. We obtain g on the basis of the posterior probability $\Pr(C_i|x)$, which is calculated as an indicator of the classification for each cluster by using Bayes' Theorem:

$$\Pr(C_i|x) = \frac{\Pr(C_i)\Pr(x|C_i)}{\Pr(x)}. \tag{1}$$

A rule cluster of the ith rule, C_i, is represented by a v_i-centred Gaussian with covariance Σ_i. Therefore, the probability density function of the ith rule's cluster is represented by:

$$\Pr(x|C_i) = \frac{1}{(2\pi)^{\frac{n_s}{2}}|\Sigma_i|^{\frac{1}{2}}} \cdot \exp\left\{\frac{-1}{2}(x - v_i)^{\mathrm{T}}\Sigma_i^{-1}(x - v_i)\right\}. \tag{2}$$

A robot requires g_i, instead of calculating $\Pr(C_i|x)$[1] by omitting $\Pr(x)$ in Eq.(1) as a common factor for all clusters. A robot must select a rule on the basis of only the numerator. The value of g_i is calculated as follows:

$$\begin{aligned}
g_i &= -\log(f_i \cdot \Pr(x|C_i)) \\
&= \frac{1}{2}(x - v_i)^{\mathrm{T}}\Sigma_i^{-1}(x - v_i) - \log\left\{\frac{1}{(2\pi)^{\frac{n_s}{2}}|\Sigma_i|^{\frac{1}{2}}}\right\} - \log f_i,
\end{aligned} \tag{3}$$

where f_i is synonymous with $\Pr(C_i)$.

After calculating g for all the rules, the winner rl_w with the minimal value of g_i is selected. As mentioned in the learning procedure in Sec. 3.1, the action in rl_w is performed if g_w is lower than a threshold $g_{th} = -\log(f_0 \cdot P_{th})$, where f_0 and P_{th} are predefined positive constants. Otherwise, a new action is produced. This new action is given by comparison with another threshold[2], $g'_{th} = -\log(f_0 \cdot P'_{th})$ in one of the following two ways:

---

[1] The higher the value of $\Pr(C_i|x)$, the lower is the value of g_i.
[2] $P'_{th} < P_{th}$

– $g_{th} \leq g_w < g'_{th}$: The robot executes an action with parameters determined on the basis of rl_w and other rules with g in this range as follows:

$$a' = \sum_{l=1}^{n_r} \left(\frac{u_l}{\sum_{k=1}^{n_r} u_k} \cdot a_l \right) + N(0, \sigma), \tag{4}$$

where n_r denotes the number of referred rules and $N(0, \sigma)$ is a zero-centred Gaussian noise with variance σ. This utility-weighted average action is regarded as an interpolation of previously acquired knowledge.

– $g'_{th} \leq g_w$: The robot generates a random action.

Updating Rule Set. The update phase is performed except when an action by rl_w results in punishment. If a new action is taken (*i.e.* $g_w > g_{th}$), a new rule that is composed of the current sensory input and the executed action is added to R. The parameters for the new rule are defined as follows:

$$v_c = x, \Sigma_c = \sigma_0^2 I, a_c = a_w, u_c = u_0, f_c = f_0. \tag{5}$$

In these equations, σ_0, u_0 and f_0 are constants and I is a unit matrix.

When the action in rl_w is performed as (*i.e.* $g_w \leq g_{th}$), all of its parameters are updated as follows. First, the sample set Φ_w is updated by adding the current sensory input to x. Then, the sample mean $\bar{x} = \{\bar{x}_1, \ldots, \bar{x}_{n_s}\}^T$ and the sample variance $s^2 = \{s_1^2, \ldots, s_{n_s}^2\}^T$ are estimated from the updated set Φ_w. The confidence intervals for $\bar{x}$ and s^2 are also updated. In subsequence, BRL determines whether any component of v and Σ is outside the range of the confidence intervals. If any component is outside that range, the updates are conducted:

$$v_i \leftarrow v_i + \alpha(\bar{x}_i - v_i), \tag{6}$$
$$\sigma_i^2 \leftarrow \sigma_i^2 + \alpha^2[s_i^2 - \sigma_i^2], \tag{7}$$
$$f_w \leftarrow f_w + \beta(1 - f_w), \tag{8}$$

where α and β are constants. For all other rules, the prior probabilities f_i are updated as follows:

$$f_i \leftarrow (1 - \beta)f_i. \tag{9}$$

3.2 BRL with a Meta-learning Mechanism

BRL has a mechanism for adaptively updating rule parameters described in Step (6) in Sec. 3.1. BRL collects input-output data during an experiment. When learning advances, as mentioned in Sec. 3.1, BRL improves the precision of the acquired rules by reducing the value of the Σ component of the rules. However, overfitting occasionally causes the problem of excessively adjusting to the collected input-output data (Fig. 2).

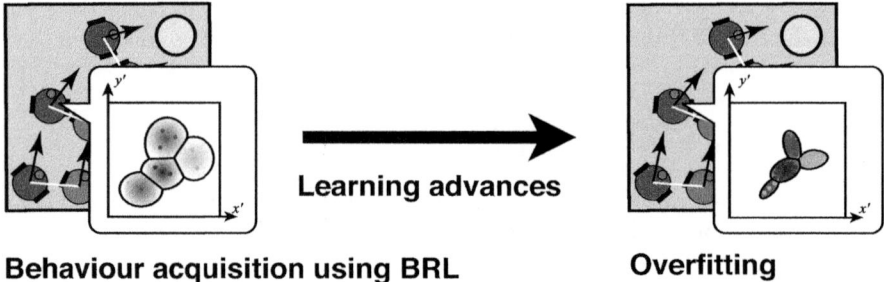

Fig. 2. Overfitting in BRL

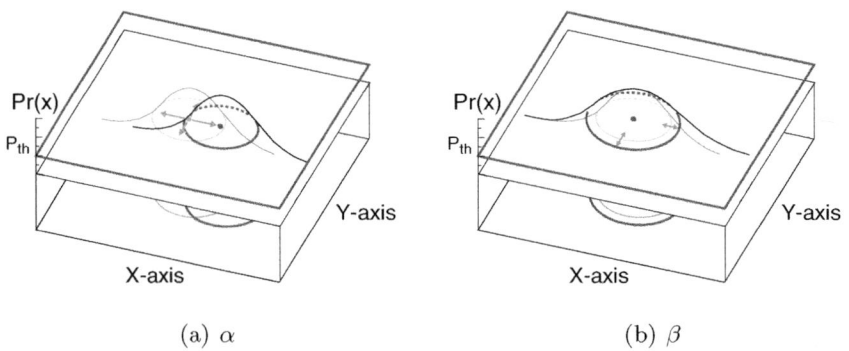

(a) α (b) β

Fig. 3. Learning rates of BRL

To overcome this problem, the present paper proposes another extended BRL that focusses on meta-parameters that modulate the learning and are of crucial importance for RL. In this study, meta-learning [14,15], which is the capability of the learning algorithm to dynamically adjust its meta-parameters, is employed. As for RL robots, Elfwing *et al.* proposed an evolutionary approach to optimise meta-parameters [16].

In this study, the learning rates α and β (Fig. 3), which modulate the centre position and the range of rule clusters, respectively, are coordinated in response to the stability of behaviour S as follows:

$$\alpha \leftarrow (1 - w)\alpha + \frac{w\alpha_{max}}{1 + \exp[-\gamma(S - \delta)]}, \tag{10}$$

$$\beta \leftarrow (1 - w)\beta + \frac{w\beta_{max}}{1 + \exp[-\gamma(S - \delta)]}, \tag{11}$$

where α_{max} and β_{max} denote the maximum values of α and β in the range [0,1], respectively. γ and δ are positive constants and w is an inertia weight in the range [0,1]. Here, S is given on the basis of the transition of the information entropy of the fired rules E:

$$S = |E_t - E_{t-1}|, \tag{12}$$

$$E_t = -\sum Q(i) \log Q(i), \tag{13}$$

where $Q(i)$ is the probability that the ith rule is fired in the tth episode.

4   Real Robot Experiments

4.1   Experimental Settings

The robot makes decisions on the basis of the position (distance r and direction θ) of the object, the nearest robot and the wall by using its omnidirectional camera (Fig. 4). The input to the controller is given by $x = \{r_0,\ \cos\theta_0,\ \sin\theta_0,\ r_1,\ \cos\theta_1,\ \sin\theta_1,\ r_3,\ \cos\theta_2,\ \sin\theta_2\}$, where the suffixes 0, 1 and 2 denote the object, the nearest robot and the wall, respectively. The output to the robot is $a = \{m_{rud},\ m_{th}\}$, where m_{rud} and m_{th} are the motor commands for the rudder and the throttle, respectively.

We represent a unit of time as a *step*. A *step* is a sequence that allows the robot to obtain its own input information, make decisions by itself, and execute its action. An episode of learning continues until two or three robots turn 60° in the counterclockwise direction by maintaining a distance of 1.0 m or less from the nearest robot or until the 100th time-step arrives. The robots are manually transported to their initial position after every 30 episodes. The robots are rewarded when they turn 60° around the object, whereas they are penalised when they collide with the wall or object. If robots complete a lap, the experimental run is regarded as successful. The BRL parameters are the same as the values

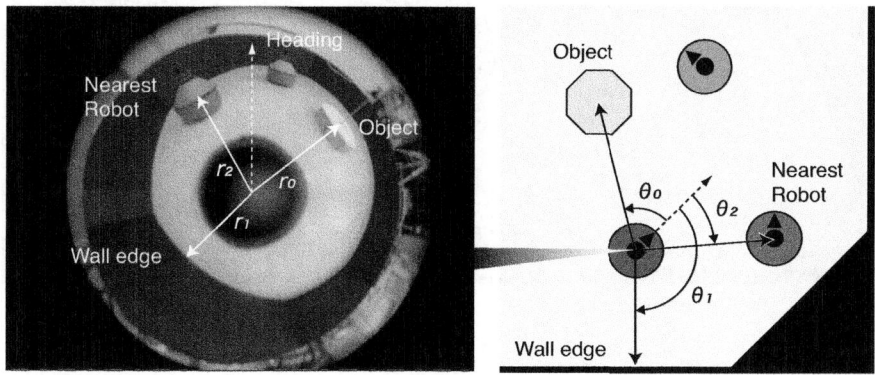

Fig. 4. Input through the omnidirectional camera

proposed in our previous study. The parameters designed for the extended BRL are $\alpha_{max} = 0.001$, $\beta_{max} = 0.01$, $\gamma = 30$, $\delta = 0.7$ and $w = 0.1$.

In addition, Q-learning is adopted for comparison. It has a normalised radial-basis function network to approximate an action value function. The input is the same as BRL, and six primitive actions, such as slowly/quickly moving straight ahead, slowly/quickly moving diagonally forward left and slowly/quickly turning left, are available.

4.2   Results

Five experimental runs were performed for each controller. All the experiments yielded successful results by using Q-learning and BRLs. Figure 5 shows an example of the behaviour during a run for the extended BRL. In this figure, trajectories are depicted from the circles to the triangles. In the early stages, the robots have no knowledge and function by trial and error. During this process, the robots often collide with the wall or object. Then, the robots maintain a regulation distance between each other and completely orbit the object.

The average number of consecutive complete laps and punishments are illustrated in Fig. 6. It is found that a larger number of rewards and a smaller number of punishments are obtained using our extension. These results indicate that for this cooperative task, robots with the extended BRL develop more stable object-orbiting behaviour than those with Q-learning or standard BRL.

Figure 7 illustrates the transition of the learning rate α of the extended BRL in a typical experimental run[3]. learning rate fluctuates and gradually reduces during the experiment. Our proposed meta-learning mechanism enables the learning rates to be modified based on the stability of the behaviour of MRS. This mechanism would provide better, more robust performance in MRSs.

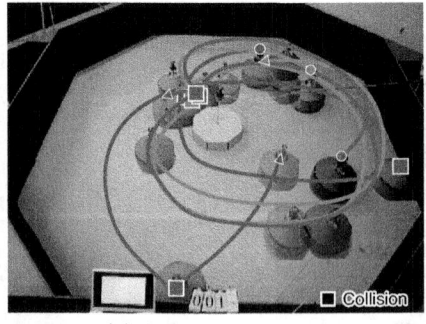

(a) At an early stage   (b) After successful learning

Fig. 5. Examples of behaviour

---

[3] The transition of β is exactly the same as α.

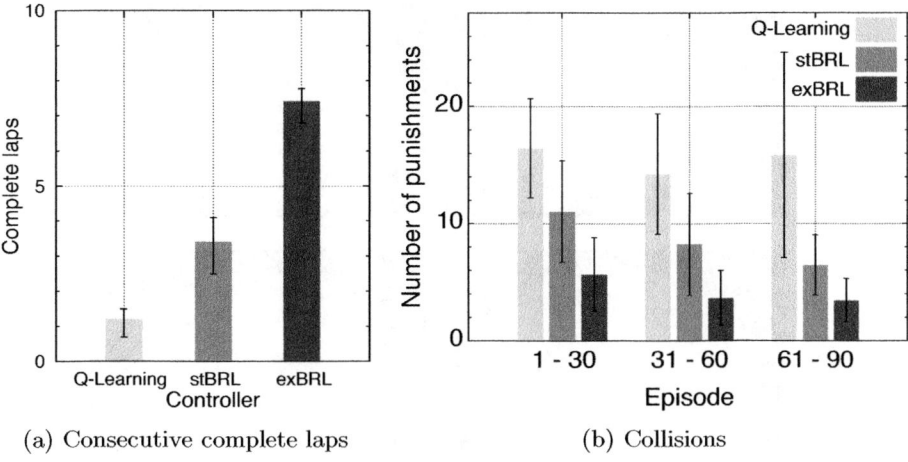

(a) Consecutive complete laps   (b) Collisions

Fig. 6. Performance comparison

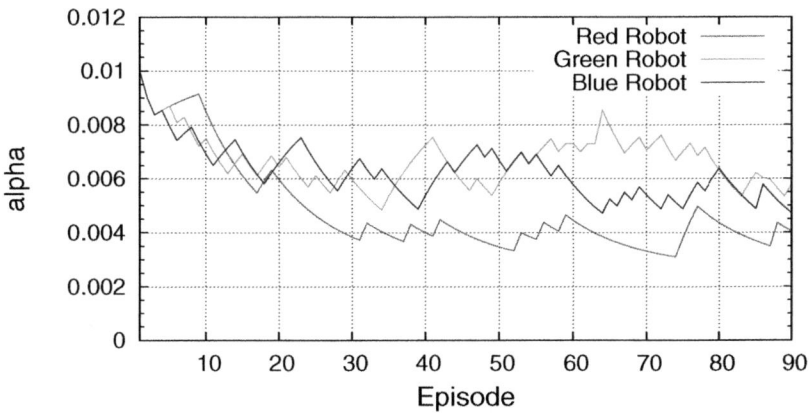

Fig. 7. Transition of learning rate α

4.3   Discussion

To visually grasp how the robots demonstrate their cooperative behaviour, the acquired rules are projected onto a plane using an Isomap [17]. The Isomap can reduce dimensionality on the basis of manifold structures in high-dimensional space and preserve local topological relationships among data.

Figure 8(a) shows an example of successful object-orbiting behaviours. First, the red robot leads the blue robot, which is followed by the green robot. In subsequence, the green robot catches up with the blue robot. Then, the green robot is the red robot's follower and leads the blue robot. Finally, the three

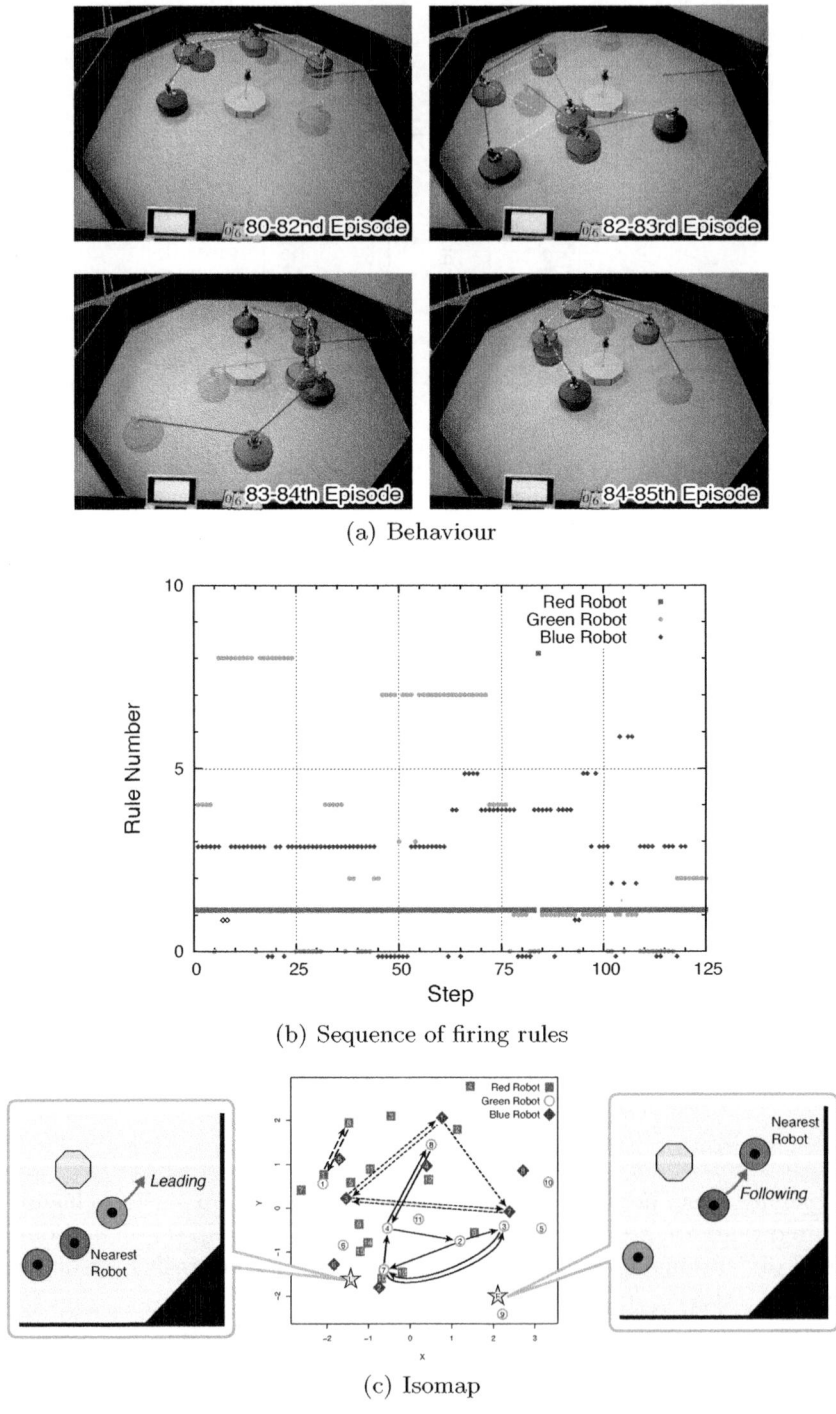

(a) Behaviour

(b) Sequence of firing rules

(c) Isomap

Fig. 8. Acquired rules

robots orbit the object once. The sequence of firing rules in this duration is shown in Fig. 8(b)[4].

Figure 8(c) shows the Isomap based on the input-output data of the three robots and the typical leading and following behaviours. The arrows in this figure indicate the transition of firing rules. The rules for leading and following are located on the left and right, respectively. The rules for the red robot, which always leads, are projected on the left , whereas the rules for the other two are distributed in the map. This indicates that green and blue robots dynamically change their roles of leader and follower.

By observing the acquired behaviour and investigating the rules, it is concluded that the robots developed cooperative behaviour on the basis of autonomous specialisation.

5   Conclusion

We investigated an RL approach for the behaviour acquisition of autonomous MRSs. Our proposed RL technique, BRL, has a mechanism for the adaptive discretization of the continuous learning space and has proven to be effective for an MRS. This paper introduced an extended BRL for improving the robustness of an MRS by providing a meta-learning mechanism that adaptively modifies its learning parameters on the basis of the information entropy of firing rules. The results of the physical experiments demonstrated that our proposed method would improve the robustness of an MRS.

In the future, we plan to investigate the robustness of the extended BRL against environmental change after successful learning. We also plan to conduct experiments employing larger number of robots, especially those with more sensors and actuators.

References

1. Sutton, R.S., Barto, A.G.: Reinforcement Learning: An Introduction. MIT Press, Cambridge (1998)
2. Sutton, R.S.: Generalization in Reinforcement Learning: Successful Examples Using Sparse Coarse Coding. Advances in Neural Information Processing Systems 8, 1038–1044 (1996)
3. Morimoto, J., Doya, K.: Acquisition of Stand-Up Behavior by a Real Robot using Hierarchical Reinforcement Learning for Motion Learning: Learning 'Stand Up' Trajectories. In: Proc. of International Conference on Machine Learning, pp. 623–630 (2000)
4. Lin, L.J.: Scaling Up Reinforcement Learning for Robot Control. In: Proc. of the 10th International Conference on Machine Learning, pp. 182–189 (1993)
5. Kolter, J.Z., Ng, A.Y.: Regularization and Feature Selection in Least-Squares Temporal Difference Learning. In: Proc. of the 26th International Conference on Machine Learning (2009)

---

[4] Rules 0 denote newly produced rules. Although the robots have the same IDs, their components are completely different.

 6. Nouri, A., Littman, M.L.: Dimension Reduction and Its Application to Model-Based Exploration in Continuous Spaces. Machine Learning 81(1), 85–98 (2010)
 7. Asada, M., Noda, S., Hosoda, K.: Action-Based Sensor Space Categorization for Robot Learning. In: Proc. of IEEE/RSJ International Conference on Intelligent Robots and Systems, pp. 1502–1509 (1996)
 8. Takahashi, Y., Asada, M., Hosoda, K.: Reasonable Performance in Less Learning Time by Real Robot Based on Incremental State Space Segmentation. In: Proc. of IEEE/RSJ International Conference on Intelligent Robots and Systems, pp. 1502–1524 (1996)
 9. Doya, K.: Reinforcement Learning in Continuous Time and Space. Neural Computation 12, 219–245 (2000)
10. Peters, J., Schaal, S.: Natural actor critic. Neurocomputing 71(7-9), 1180–1190 (2008)
11. Yasuda, T., Ohkura, K.: Autonomous Role Assignment in Homogeneous Multi-Robot Systems. Journal of Robotics and Mechatronics 17(5), 596–604 (2005)
12. Yasuda, T., Ohkura, K.: Improving Search Efficiency in the Action Space of an Instance-Based Reinforcement Learning. In: Almeida e Costa, F., Rocha, L.M., Costa, E., Harvey, I., Coutinho, A. (eds.) ECAL 2007. LNCS (LNAI), vol. 4648, pp. 325–334. Springer, Heidelberg (2007)
13. Yasuda, T., Ohkura, K.: Reinforcement Learning Technique with an Adaptive Action Generator for a Multi-Robot System. In: Asada, M., Hallam, J.C.T., Meyer, J.-A., Tani, J. (eds.) SAB 2008. LNCS (LNAI), vol. 5040, pp. 250–259. Springer, Heidelberg (2008)
14. Doya, K.: Metalearning and neuromodulation. Neural Networks 15(4-6), 495–506 (2002)
15. Schweighofer, N., Doya, K.: Meta-learning in Reinforcement Learning. Neural Networks 16(1), 5–9 (2003)
16. Elfwing, S., Uchibe, E., Doya, K., Chiristensen, H.I.: Co-evolution of Shaping Rewards and Meta-Parameters in Reinforcement Learning. Adaptive Behavior 16, 400–412 (2008)
17. Tenenbaum, J.B., de Sliva, V., Lagford, J.C.: A Global Geometric Framework for Nonlinear Dimensionality Reduction. Science 290(22), 2319–2323 (2000)

Locomotion Selection and Mechanical Design for a Mobile Intra-abdominal Adhesion-Reliant Robot for Minimally Invasive Surgery

Alfonso Montellano López[1], Mojtaba Khazravi[1], Robert Richardson[1], Abbas Dehghani[1], Rupesh Roshan[2], Tomasz Liskiewicz[2], Ardian Morina[2], David G. Jayne[3], and Anne Neville[2]

[1] Institute of Engineering Systems and Design,
School of Mechanical Engineering, University of Leeds, Leeds, UK
[2] Institute of Engineering, Thermofluids, Surfaces and Interfaces,
School of Mechanical Engineering, University of Leeds, Leeds, UK
[3] Academic Surgical Unit, St. James's University Hospital, Leeds, UK
{mnaml,men3mk,R.C.Richardson,A.A.Dehghani-Sanij,
R.Roshan,A.Morina,T.Liskiewicz,D.G.Jayne,A.Neville}@leeds.ac.uk

Abstract. Miniaturisation of surgical robots combined with bio-inspired adhesive material offer the possibility of a device able to move stably inside the body. In this paper a miniature adhesion-reliant robot is proposed as an alternative to current cumbersome, externally anchored surgical robots. An effective locomotion strategy is selected according to the specific working environment of this application. This environment is the ceiling of the insufflated human abdomen during laparoscopic surgery. Having chosen the most appropriate actuation technology in the market (piezo-electricity), the mechanical design to implement the former locomotion strategy is demonstrated.

Keywords: Locomotion, Mechanical Design, Adhesion, Robot, Intra-abdominal, Minimally Invasive Surgery.

1 Introduction

1.1 Current Developments in Surgical Robotics

Since the first use of a robotic assistant for surgery in the 1980s, the number of robots in the operating theatre has steadily increased over the last decades [1]. Procedures in which robots are currently used include abdominal, colorectal, spinal, cardiac, ear, knee and brain surgery where robots are able to move a camera inside the body, cut, laser-cut, suture, drive needles and drill bones. It is generally accepted that after appropriate training of the surgeon using the device, a robot-assisted intervention is more precise, shorter and reduces the risk of post-operative complications as compared to open surgery and laparoscopy [1, 2]. The focal areas of research and development for surgical robots are: robotic holders of laparoscope and camera,

R. Groß et al. (Eds.): TAROS 2011, LNAI 6856, pp. 173–182, 2011.
© Springer-Verlag Berlin Heidelberg 2011

robotic operators, actuated endoscopes and needle guiding robots. For the most challenging applications, engineers are exploring the potential of new scientific approaches, like bio-inspiration.

1.2 Bio-inspiration for Adhesive Materials

The recent discovery of the use of Van der Waals forces by lizards [3] to climb and attach to virtually any surface motivated the scientific community to artificially replicate a surface with the same properties of dry adhesion. Making use of micro- and nano-fabrication facilities, several surfaces covered with tiny fibres have been developed successfully [4, 5]. Inspiration has also been taken from insects [6] and tree frogs [7], that use capillary forces to enhance adhesion, a more suitable mechanism for a wet environment like inside the body. Drawing inspiration on all these sources, a pad with a micro-structured surface has being developed to provide wet adhesion to the robot.

Interestingly, the unique adhesive ability of the gecko is not only based on the mechanical structure of its feet, but also on the way it is controlled. By applying an appropriate combination of forces (parallel and perpendicular to the surface they are attaching to) geckoes are able to attach, increase adhesion, dramatically decrease it and detach quickly from it [8].

1.3 Intra-corporeal Devices

To overcome the fact that current surgical robots are costly, cumbersome and have limited accessibility and visual feedback, the new trend is to design miniature robots able to intelligently interact with biological tissue. Such capability is expected to solve the previous difficulties. In this way, surgeons will use even less incisions for the operation as the robots will operate from inside the body, with no external fixture. Amongst the projects concerned with the miniature size in robotic surgery, the following are particularly remarkable:

- HeartLander, Carnegie Mellon University [9]. A robot remotely actuated to crawl over the beating heart. Locomotion for this robot is provided by two vacuum suction cups that alternate attachment to the pericardium.
- Nebraska Wheeled Robot, University of Nebraska [10]. Essentially composed of two metallic wheels, the robot moves over the abdominal organs driven by two motors. It is, however, unable to climb or attach to internal walls.
- The ARAKNES project [11], Scuola Superiore Sant'Anna (SSSA) in Pisa with other European partners. The aim of the project is to build an array of micro robots to access the stomach through the mouth, and perform image-guided operations controlled by a console. No locomotion is planned for this robot.

The use of adhesion to facilitate locomotion for this type of robots is a promising solution. But it sets a series of challenges to be addressed by the designer. What is a suitable locomotion strategy for a robot of this kind? How can it be built to the size requires by intra-corporeal operation? This paper shows a simple design based on a miniature motor that is readily available. Before presenting the hardware design and choosing the locomotion mechanism, the system is defined.

2 Mechanical Requirements

2.1 System Definition

The goal of this investigation is to develop a robot able to move inside the abdomen reliant on adhesion to attach to internal tissue. The robot will attach to the peritoneum during a laparoscopic procedure. In this kind of procedure, the abdomen is insufflated with carbon dioxide to create an operating space (see Figure 1). The peritoneum is the ceiling of this space, so the robot has to walk upside down, hence the use of adhesion.

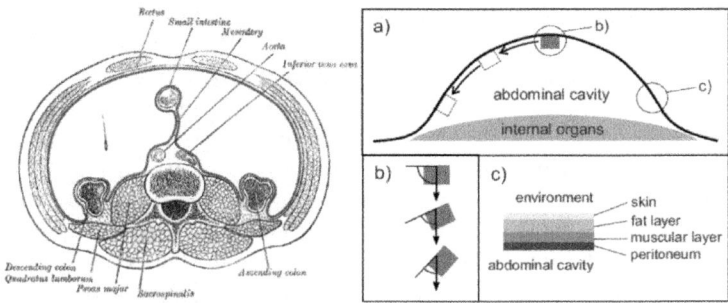

Fig. 1. On the left, transverse section of the abdomen [12]. The peritoneum is represented by a dark red line, enclosing the operating space created by insufflation. On the right, diagram showing (a) the device in the operating environment, (b) working at different angles, (c) the composition of the abdominal wall [7].

Motion above the internal organs allows the robot to reach virtually all points in the cavity. It presents a more flexible system compared to a robot that moves over the organs and can only reach the target from the surrounding ground. Similarly, it offers more manoeuvrability compared to an arm which is externally anchored, where motion is restricted by the insertion port. The actuation mechanism of an adhesion-reliant robot is significantly less complex than other designs. For instance, when using suction cups or magnetic anchoring, a vacuum circuit and a magnetic field generator are respectively required. When interacting with biological tissue, adhesion can be very helpful to avoid causing damage to the surface or deal with an unstructured surface.

2.2 Locomotion

To avoid a fall during operation against gravity, the robot will keep contact with the peritoneum at all times. This means that the robot cannot make its whole body travel at the same time. Part of the robot needs to hold the weight while some other part moves, and then swap roles. An exception to this would be a design where the surface of the robot that detaches is the same as the one that attaches, like in a rolling band. To minimise tissue damage, force has to be applied perpendicularly to the surface as well as in a parallel plane. It has to be applied to each individually moving part of the robot. To position the robot over the peritoneum two degrees of freedom are required.

Because of intra-corporeal mobility, the device must be compact and operate as safely as possible. All reduction in volume will make operation more flexible. Safety measures must be taken into account: temperature within bearable limits, no high voltages, no sharp edges, and sterilisable.

The two main general locomotion structures considered are: rotary motion (wheels, tank) transformed into displacement using friction between the robot and the surface and a mechanism able to coordinate the motion of pads (or feet). These paradigms have been selected from amongst others (serpentine, legged) because of their simplicity and compact development.

- Adhesive wheels or a tank with adhesive tracks (see Figure 2) can be implemented with a continuous adhesive band or dividing this band into individual units or wheels. The main advantage of these designs is the simplicity of their actuation system: one actuator is enough. Nevertheless, they present two major disadvantages. The first one is related to the steering ability desirable for our system. Steering a robot in which its adhesive pads roll over a surface involves rubbing the pads against this surface. For biological tissue, such a method increases the risk of tissue damage as it increases strain on the tissue. Skidding may also occur, making locomotion control more difficult. The second reason concerns the control of the adhesion force. With only one motor, the control actions over the adhesive surface are very limited. This limitation jeopardises stability. This is because the robot will neither be able to precisely manipulate adhesive force nor will it have a flexible reactive behaviour when encountering potential perturbations.

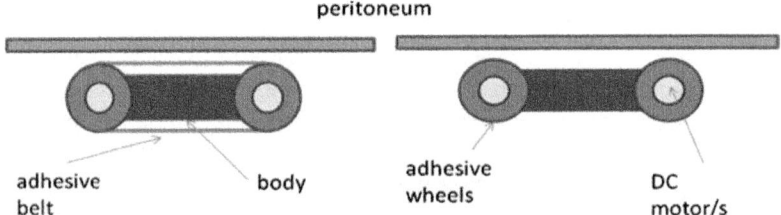

Fig. 2. Tank (left) and wheeled (right) design concept

- The second option is adhesive feet, that is to say, individually actuated units with adhesive properties. These feet will alternate the role of providing stability to the system and advancing part of the system along the surface. The overall result will enable the transportation of the robot to reach any position over the surface. The two main types of locomotion of this kind correspond to the two principal coordinate systems. In a polar walker a foot turns around the body (Θ coordinate) and the foot is linearly actuated to reach the desired distance from the body (r coordinate). The two units then swap body/foot roles, enabling the walker to reach any position in the XY plane. In a Cartesian walker, one linear actuator positions the foot along the Y-axis, another actuator positions the X-axis. Again, the body and foot roles can be swapped between pads to produce displacement of the walker.

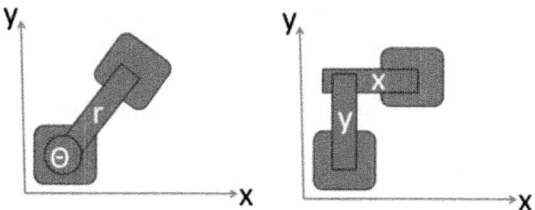

Fig. 3. Polar (left) and Cartesian (right) walkers

This latter strategy is the chosen one.

The weight of the robot is estimated between 20 and 30 gf. Experiments with 113 mm² of the micro-structured polymer developed for this project show adhesion to rat (human-like) peritoneum of 8-9 gf [7]. For the sake of static stability, four or six feet are considered. With the aforementioned area per pad a surface of 452 mm² for four pads and 678 mm² for six is available, both feasible for a surgical port of 20 to 30 mm (Single Incision Laparoscopic Surgery). The more pads, the more complex the design because more actuators and mechanical links are required. On the other hand, the fewer the pads the weaker the available adhesion and therefore the lighter the robot must be. As the weight is expected to be smaller than the minimum force with four pads (32 gf), four is the number of pads chosen. Although four-legged robots have typically a narrower static stability margin, upside-down operation and the use of broad, adhesive feet enables stable robot locomotion.

For the whole Cartesian walker to be positioned anywhere in the XY plane, displacement in both directions is required. In total, two degrees of freedom are needed for each adhesive pad plus the degree of freedom to move towards and away from the peritoneum. The number of degrees of freedom will therefore be: one for each pad on the Z direction plus one for each non-redundant kinematic link in the XY plane.

The final choice is a symmetrical four-padded configuration with eight linear actuators: one between every two pads on the XY plane and one for each pad on the Z direction; and four joints: one for each pad at the union of two actuators.

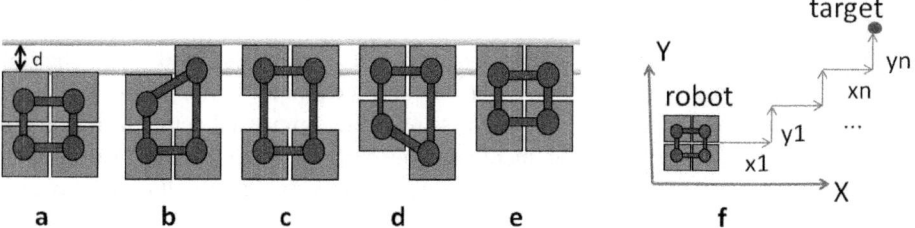

Fig. 4. On the left (a-e), four-pads robot taking one step, travelling a distance d: (a) initial position, (b) top-right pad has travelled a distance d, (c-d) the same for the other three pads until (e) the whole robot reaches the final position. On the right (f), the robot reaches a target position by taking several steps alternating direction X and Y.

Figure 4 illustrates the configuration and design on the XY plane and shows how the robot would travel a certain distance by successively detaching its pads. The symmetry of the structure allow the robot to travel in both directions identically. To reach an (X, Y) position, the trajectory is subdivided into smaller steps as shown.

2.3 Miniature Actuators

To scale down a robot, the first requisite is to have compact enough actuators to fit the dimensional constraints. The trend for miniaturisation has smoothed the progress and development of actuation technologies that were only advantageous for very specific applications on the macro scale. The following technologies were explored because of their performance on the miniature size, their commercial availability and usefulness in bio-medical applications. These are: Ionic Polymeric Metal Composites (IPMC), Shape Memory Alloys (SMA), DC mini-motors and Piezo-electricity. The working principles, construction laws, figures of merit and archetypal applications of these emerging technologies can be found in [13, 14]. Here, all that information is summarised in the next table:

Table 1. Main features of actuation technologies for miniature actuators: compactness, force/ displacement (F/d), speed, voltage, controllability

Actuation technology	Compact.	F / d	Speed	Voltage	Controllability
Electro-magnetic (DC mini-motors)	Bad	Very high / High	Fast	Low	Easy
Piezoelectricity (Piezo-motors)	Good	High / Small	Fast – Very fast	High – Very high	Easy
SMA wire	Good	High – Very high / Medium	Slow	Low	Medium
SMA spring	Good	High / Very large	Slow	Low	Medium
IPMC	Good	Low / Large	Fast	Low	Difficult

Taking into account the scalability, ease of integration and controllability appropriate for the robot, the most mature technology is piezo-electricity. Specifically, ultrasonic motors. The main issue for a commercial actuator of this type is size, as values of force, displacement and speed are sufficiently high. Safety is the most important concern as they usually operate at very high voltages. However, current developments for ultrasonic motors use as low a voltage as for DC motors, for a similar or even higher efficiency.

The smallest size piezoelectric motor found in the market was the linear motor Squiggle RV® by New Scale Technologies Inc. It uses reduced voltage and holds the position when the power is off, avoiding a dangerous fall if power is disrupted. To control it, a very small magnetic linear encoder manufactured by Austria Microsystems is used (Tracker® NS-5310).

3 Mechanical Design

3.1 Motor Integration

The motor is basically composed of four piezoelectric plates bonded to the four faces of a square prism that hosts a threaded shaft. The motor's driver sends a voltage to the plates at their bending resonant frequency causing amplified strain on the material. This brief explanation (more details in [15]) is useful to understand the requirements of the support mechanism that integrates the function of the motor into the whole robot. This support mechanism is composed of:

- a static part attached to the motor's housing: the Mounting Platform (MP) and
- a moving part, travelling with the shaft of the motor: the Ends Supporter (ES).

The function and design specifications of each of the former parts are explained by considering the following features of the motor [16]:

1. Given that the forward and backward motion comes from the rotation caused by the nut on the screw, such rotating movement cannot be altered. No load can be directly coupled to the screw. The tip of the screw has been radiused by the manufacturer to minimise the surface contact between the screw and the load. Thus, the material in contact with the tip of the screw must be very low friction. The angle between the contact surface and the motor shaft should be as perpendicular as possible to prevent the tip from engaging with the material. The load will be attached directly to this ES.

2. Friction between moving parts must be reduced. All friction the motor must overcome reduces the amount of load it can handle. The moving part (ES) is connected to the static one (MP) through two stainless steel rods that slide into the body of the MP. To minimise this effect, the MP will be manufactured from a very low friction material (Delrin Acetal®). Friction inside the two guiding holes in the MP can also be due to misalignment between the axis of the rod and the axis of the hole.

To link the motion we want to acquire on the pad and the motor, we attach the adhesive pad to a supporter. The supporter is directly moved by the motor through two contact points, the two ends of the motor's shaft. The supporter is then connected to the platform that holds the motor's housing, and the platform is connected to the rest of the robot through a joint. To control the motion, the encoder is attached to the platform/housing, measuring the changes in the magnetic field of a magnet fixed to the supporter/adhesive pad. Figure 5 shows this assembly.

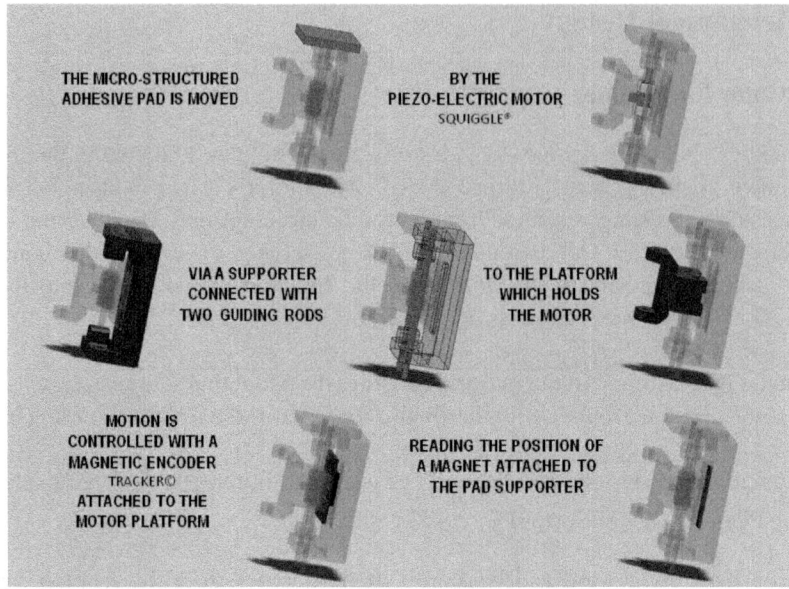

Fig. 5. Step by step assembly of all components in one adhesive unit

Figure 6 shows the assembly of the motor, MP, ES and guiding rods in a vertical, one-axis configuration and in two-axes configuration where horizontal motion can also be obtained.

Fig. 6. 3D CAD design of one vertical axis (left) that moves the pad perpendicularly to the peritoneum and the two-axes configuration (right) to obtain horizontal motion

Two of the horizontal units linked to the vertical one through a joint give the three degrees of freedom required for the pad to be placed anywhere over the peritoneum. Once full control over one adhesive foot is achieved, the same structure is replicated for the other pads until the four of them are inter-connected providing autonomous locomotion to the ensemble.

Figure 7 shows the 3D assembly of the three motor that move the adhesive pad in the three directions and the model of the whole robot, with all pads and actuators inter-connected.

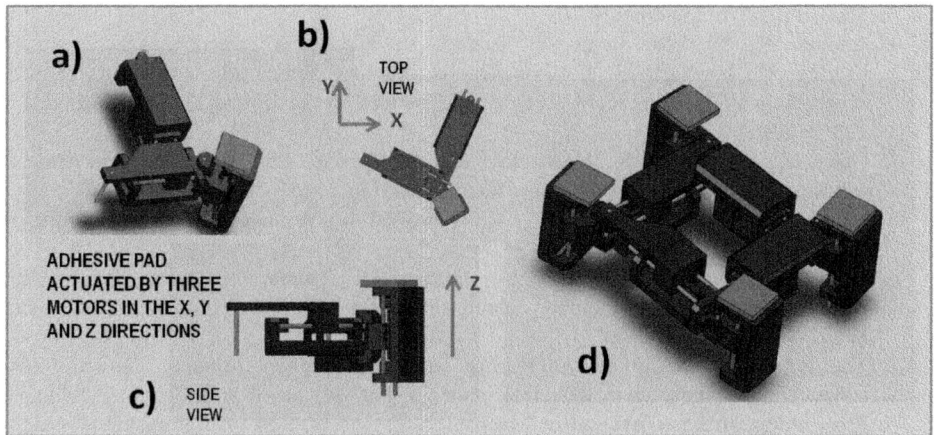

Fig. 7. Adhesive pad actuated by three motors (a), in top (b) and side (c) view. And the whole robot (d).

4   Conclusion

In conclusion, the methodology to build a robot which can go through a surgical incision and operate from inside the abdomen has been shown in terms of the mechanical design. A design that allows the robot to be positioned on the peritoneal surface, independently actuating several adhesive feet has been proposed as a solution.

5   Future Work

The first test to run, once the one-axis configuration is manufactured, is to characterise the force the mechanism can exert and the relation between this force and the adhesion on the pad. The assessment and management of forces and moments on the pad and how they are transmitted to the rest of the mechanical structure are two crucial points to guarantee stability of the robot. Another important factor whose influence must be investigated is inclination of the surface to adhere to as the robot follows the curvature of the abdomen. Along with the hardware implementation an intelligent controller is required to optimise the attachment and detachment cycle on each pad as well as the coordination of the feet.

References

1. Dasgupta, P.: Advanced laparoscopic surgery using tha da Vinci robotic system. In: Engineers, I.o.M. (ed.) Robotics in Surgery. State of the Art., London (2010)
2. Gilbert, J.: Robots in laparoscopic surgery. In: Engineers, I.o.M. (ed.) Robotics in Surgery. State of the Art., London (2010)
3. Autumn, K., Sitti, M., Liang, Y., Peattie, A., Hansen, W., Sponberg, S., Kenny, T., Fearing, R., Israelachvili, J.: Evidence for van der Waals adhesion in gecko setae. Proceedings of the National Academy of Sciences of the United States of America 99, 12252 (2002)
4. Geim, A., Grigorieva, S., Novoselov, K., Zhukov, A., Shapoval, S.: Microfabricated adhesive mimicking gecko foot-hair. Nature Materials 2, 461–463 (2003)
5. Murphy, M., Aksak, B., Sitti, M.: Gecko-inspired directional and controllable adhesion. Small 5, 170–175 (2009)
6. Gorb, S., Sinha, M., Peressadko, A., Daltorio, K., Quinn, R.: Insects did it first: a micropatterned adhesive tape for robotic applications. Bioinspiration & Biomimetics 2, S17 (2007)
7. Taylor, G., Neville, A., Jayne, D., Roshan, R., Liskiewicz, T., Morina, A., Gaskell, P.: Wet adhesion for a miniature mobile intra-abdominal device based on biomimetic principles. Proceedings of the Institution of Mechanical Engineers, Part C: Journal of Mechanical Engineering Science 224, 1473–1485
8. Autumn, K.: Properties, principles, and parameters of the gecko adhesive system. Biological Adhesives, 225–256 (2006)
9. Patronik, N., Ota, T., Zenati, M., Riviere, C.: A miniature mobile robot for navigation and positioning on the beating heart. IEEE Transactions on Robotics 25, 1109–1124 (2009)
10. Rentschler, M., Dumpert, J., Platt, S., Iagnemma, K., Oleynikov, D., Farritor, S.: An in vivo mobile robot for surgical vision and task assistance. Journal of Medical Devices 1, 23 (2007)
11. http://www.araknes.org/home.html (accessed November 2010)
12. http://www.theodora.com/anatomy (accessed November 2010)
13. Dario, P., Valleggi, R., Carrozza, M., Montesi, M., Cocco, M.: Microactuators for microrobots: A critical survey. Journal of Micromechanics and Microengineering 2, 141 (1992)
14. Pons, J.: Emerging actuator technologies: a micromechatronic approach. John Wiley & Sons Inc., Chichester (2005)
15. Henderson, D.: Simple Ceramic Motor... Inspiring Smaller Products. In: ACTUATOR 2006, vol. 50, p. 10 (2006)
16. Design Note: Quick Tips for Integrating SQUIGGLE Motors. New Scale Technologies Inc., http://www.newscaletech.com/app_notes/ DesignNote_QuickTips-for-Integrating-Squiggle-Motors.pdf (accessed November 2010)

Mapping with Sparse Local Sensors and Strong Hierarchical Priors

Charles W. Fox and Tony J. Prescott

Active Touch Laboratory at Sheffield,
University of Sheffield, Sheffield, UK
`charles.fox@sheffield.ac.uk`

Abstract. The paradigm case for robotic mapping assumes large quantities of sensory information which allow the use of relatively weak priors. In contrast, the present study considers the mapping problem in environments where only sparse, local sensory information is available. To compensate for these weak likelihoods, we make use of strong hierarchical object priors. Hierarchical models were popular in classical blackboard systems but are here applied in a Bayesian setting and novelly deployed as a mapping algorithm. We give proof of concept results, intended to demonstrate the algorithm's applicability as a part of a tactile SLAM module for the whiskered SCRATCHbot mobile robot platform.

1   Introduction

The paradigm case for mapping, as in Simultaneous Localisation and Mapping (SLAM) problems [27], considers a mobile robot with noisy odometry and laser scanners. Laser scanners provide large amounts of sensory information, and have effectively unlimited range in indoor environments. Such large quantities of input information allow the use of relatively weak priors, such as independent grid cell occupancy or flat priors over the belief of small feature sets [27].

In contrast, we consider the mapping problem in environments where only sparse, local sensory information is available. For example, a fire-fighting robot building up a map in a smoke-filled house cannot rely on vision or laser scanners functioning at all times, and could instead operate by feeling its way around with touch sensors. Proof that this type of navigation is possible is found in biology: electric fish make use of highly localised electric field sensors [13] and rats navigate through dark underground tunnels using their whiskers [4,2], both having ranges of a few centimetres. In robotics, touch sensors are relatively cheap in both material and computational processing terms, and their use has previously been considered to enhance navigation in cheap household robots [17,6]. (Related work on research robot platforms includes [24,23,15,14,8]).

As an example of this type of mapping, we consider the case of a mobile robot having six whiskers, able to report the (noisy) locations and orientations of contacts with surfaces. Such mechanical sensors and computational classifiers have previously been demonstrated in [17,6,7], and are able to report locations, orientations and textures of contact points (note that textures are

R. Groß et al. (Eds.): TAROS 2011, LNAI 6856, pp. 183–194, 2011.

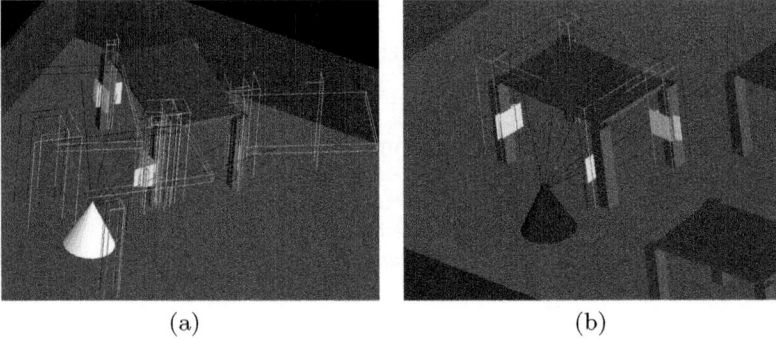

(a)   (b)

Fig. 1. (a) Simulation screen-shot at high annealing temperature. Many hypothesised (wire-frame) tables and legs are on the blackboard, primed by the shapelets (yellow rectangles) contacted by the robot (cone)'s whisker sensors, in an arena containing a physical table (pink). (b) Simulation screen-shot at low annealing temperature. A single table hypothesis remains, aligned correctly with the physical table.

especially difficult to report using other sensor modalities). The present mapping algorithm is intended to form part of a future SLAM navigation module for the whiskered SCRATCHbot hardware platform [21], but here we give a proof of concept mapping-only algorithm in a simulated and simplified microworld. It is the first work to begin fusing whisker contact reports to perform mapping.

To compensate for sparseness of the sensory information available from short-range touch sensors, we make use of strong, hierarchical priors about objects. Hierarchical object recognition models were popular in classical, symbolic AI in the guise of blackboard systems [5,19,3] but have recently been recast in terms of dynamically constructed Bayesian networks [10,18,16,26]. Here we provide a novel application of Bayesian blackboards to the robotic mapping problem.

Object based mapping models have recently appeared [28,11,25,22] which use laser sensors to recognise and learn complex spatial models. However in the sparse local sensor case, this level of detail is unavailable, and only a few contact points may be present. Thus we go beyond the use of individual movable objects, to use strong hierarchical model priors. For example, on recognising a table leg, we may then infer the probable presence the rest of the table, including other leg objects, and edges and corners making up these legs, without ever sensing them directly. To construct hierarchical objects, we use hypothesis priming and pruning heuristics as in blackboard systems. However, following [10], we treat such heuristics as approximations to inference in a dynamically-constructed, Monte Carlo Markov Chain (MCMC) sampling Bayesian network, endowing them with probabilistic semantics.

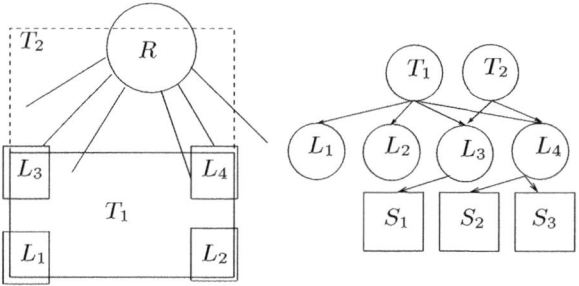

Fig. 2. Hierarchical object recognition. *Left:* Robot R (circle) with six whiskers (lines) makes tactile contact with legs L_j (squares) of a hypothesised table T_1 (rectangle). The two contact points ('shapelets') on the right are sufficient to infer the location of the corner of leg L_4. Coupled with prior knowledge about the shape and size of tables, and the third shapelet, this can be used to infer that there is a table either in the ground truth location or in a second configuration T_2 (dashed rectangle). *Right:* Bayesian network constructed to represent the same scenario. Square nodes are the shapelet observations.

2   Methods

Consider the task of building a map of an arena populated by four-legged table-like objects as in figs. 1(a) and 1(b). (Such objects could include chairs and desks for example). A mobile whiskered agent moves along a predetermined trajectory of location-angle poses, (x^t, y^t, θ^t), around the arena, over discrete time steps t. At each time step, its whiskers, $w \in 1:6$, report egocentric estimates of the radial distance r to, and surface normal ϕ and texture τ of, any contacts made,

$$\hat{r}_w^t = r_w^t + \varepsilon_r, \tag{1}$$

$$\hat{\phi}_w^t = \phi_w^t + \varepsilon_\phi, \tag{2}$$

$$\hat{\tau}_w^t = \tau_w^t + \varepsilon_\tau, \tag{3}$$

where ε are i.i.d. Gaussian noises having zero mean and standard deviations $\sigma_r^w, \sigma_\phi^w, \sigma_\tau^w$ respectively. (Such reports are currently available from the whiskered SCRATCHbot hardware platform, [21,17,6]). Assuming perfect robot localisation in the present study only, these estimates are converted into allocentric Cartesian coordinates to give *shapelet reports*, which are tuples $S(x_S, y_S, \phi_S, \tau_S)$.

2.1   Static Structures: Generative Models

Tables, T, are parametrised by tuples, $T(x_T, y_T, \theta_T, w_T^x, w_T^y, w_T^L, \tau_T)$, where x, y, θ is the pose, w_T^x and w_T^y are width and breadth, w^L is the width of the (square) legs, and $\tau \in (0, 1)$ is a texture parameter describing roughness or smoothness of the material. A generative model of tables is used. We assume a flat prior probability *density* generating tables in the world,

$$p(T(x_T, y_T, \theta_T, w_T^x, w_T^y, w_T^L, \tau_T)|\emptyset) = c_T, \tag{4}$$

where c_T is a (non-normalising) constant.

If a table T exists, it causes (as in [20]) the presence of four legs L,,

$$L(x_L, y_L, \theta_L, w_L, \tau_L, T), \tag{5}$$

where w_L is the width of the square table leg; x_L, y_L, θ_L are its location and rotation, and τ_L is its texture, with probability density

$$p(L(x_L, y_L, \theta_L, w_L, \tau_L, T)|T(x_T, y_T, \theta_T, w_T^L, \tau_T))$$

$$= \alpha_L \exp(-\Delta_{TL}), \tag{6}$$

where α is a (non-normalising) constant, and the distance measure is

$$\Delta_{TL} = \min_i \left(\frac{(x_T^i - x_L)^2 + (y_T^i - y_L)^2}{\sigma_r^2} \right) + \left(\frac{\theta_T - \theta_L}{\sigma_\theta} \right)^2$$

$$+ \left(\frac{w_T^L - w_L}{\sigma_w} \right)^2 + \left(\frac{\tau_T - \tau_L}{\sigma_\tau} \right)^2, \tag{7}$$

where $0 \le i \le 3$, and (x_T^i, y_T^i) are the coordinates of the table's four corners, and σ_w, σ_τ are parameters specifing standard deviations of the leg's w_L, τ_L values conditioned on the table's corresponding w_T^L, τ_T values. The inclusion of T in the parametrisation of L (eqn. 5) means that L is the hypothesis that the leg was caused *only* by table T rather than any other table or cause.

Shapelets are assumed to be generated by nearby legs,

$$p(S(x_S, y_S, \theta_S, \tau_S)|L(x_L, y_L, \theta_L, w_L, \tau_L, T)) = \alpha_S \exp(-\Delta_{LS}) \tag{8}$$

where

$$\Delta_{LS} = \left(\frac{r}{\sigma_r^S} \right)^2 + \left(\frac{f(\theta_L) - \theta_S}{\sigma_\theta^S} \right)^2 + \left(\frac{\tau_L - \tau_S}{\sigma_\tau^S} \right)^2, \tag{9}$$

and r is the shortest radial distance from the perimeter of the leg to (x_S, y_S), computed by basic geometry, $f(\theta_L) = \theta_L + m\pi/2$ picks the angle of the corresponding side m of the leg at this shortest-distance contact point, and $\sigma_r^S, \sigma_\theta^S, \sigma_\tau^S$ model sensor noise.

We also provide small *null priors* to allow legs and shapelets to exist in the absence of any parents. (These are required during construction on the blackboard, so that these objects can survive before their parents are constructed),

$$p(L(x_L, y_L, \theta_L, w_L, \tau_L, \emptyset)|\emptyset) = c_L, \tag{10}$$

$$p(S(x_S, y_S, \theta_S, \tau_S)|\emptyset) = c_S, \tag{11}$$

with constants such that the marginalised densities,

$$p(S(x_L, y_L, \theta_L)) < p(L(x_L, y_L, \theta_L)) < p(T(x_L, y_L, \theta_L)), \tag{12}$$

i.e. larger objects are more probable to exist without high-level causes than smaller objects are.

Unlike the parametrisation of L on T in eqn. 5, shapelets may be caused by *mixtures* of multiple leg hypotheses and by the null prior (eqn. 11). For example if there are two legs very close together then the density for observing shapelets in the area increases. We assume that multiple causal sources combine using noisy-OR semantics,

$$P(x_i|pa(x_i)) = 1 - \prod_{x_j \in pa(x_i)} (1 - P(x_i|x_j)). \tag{13}$$

where $pa(x_i)$ denotes the set of parents of generic node x_i. As we use probability *density* functions we require the continuous version of noisy-OR, proved in the Appendix:

$$p(x_i|pa(x_i)) = \sum_{x_j \in pa(x_i)} p(x_i|x_j). \tag{14}$$

We allow legs to be caused by a mixture of their *single* specified parent (i.e. the T parameter in eqn. 6) and null prior (eqn. 10), using a similar combination rule. Tables are caused by the null prior only (eqn. 4).

Taken together, the equations in this section define a Bayesian network for any given collection of tables, legs and shapelets as shown in fig. 2. However, in addition to the previous causal probabilities, we need to model the following constraints: (a) tables always have four legs; (b) each table leg is at a different corner of the table (we should not see two legs attached to the same corner); (c) two objects of the same type (table or leg) cannot overlap in physical space. Standard Bayesian networks cannot model such relations, as they are limited to joint distributions of the form

$$P(\{x_i\}_i) = \prod_i P(x_i|pa(x_i)), \tag{15}$$

To model these constrains, we extend the Bayesian network to the factor graph,

$$P(\{x_i\}_i) = \frac{1}{Z} \left(\prod_i P(x_i|pa(x_i)) \right) \times \left(\prod_{ij} \phi_c(x_i, x_j)\phi_b(x_i, x_j) \right) \left(\prod_i \phi_a(x_i) \right), \tag{16}$$

where Z is a normalising constant, and ϕ_a, ϕ_b, ϕ_c are unnormalised penalty factors corresponding to the new constraints. Using superscripts for exponentiation, these are

$$\phi_a(x_i) = \epsilon_a^m, \tag{17}$$

$$\phi_b(x_i, x_j) = \epsilon_b^v, \tag{18}$$

$$\phi_c(x_i, x_j) = \epsilon_c^r, \tag{19}$$

where m is the number of missing legs iff x_i is a table, and $m = 0$ otherwise; v is a Boolean (0,1) value, true if hypotheses x_i and x_j are of the same type and overlap in physical space; and r is a Boolean, true if hypotheses x_i and x_j are legs and share the same parent (modelling this parent-sharing is why we parametrise L by T in eqn. 5).

2.2   Inference

For a given set of shapelet observations and candidate hierarchical legs and tables, we may thus construct a factor graph. (We later describe how such a set of is obtained.) Inference becomes highly complicated if the agent has an infinite memory for shapelets, so in the present study we use a working memory (queue) of the seven most recent shapelets, and discard all others.

At each t, new shapelets are read from the sensors, and inference is performed with the aim of obtaining the Maximum A Posterior (MAP) interpretation of their table causes, before the next time step begins,

$$\text{MAP}_t = \arg_{\{T_j\}} \max P(\{T_j\}_j | \{S_k\}_k). \tag{20}$$

Thus we currently – naively – treat each time step as an independent inference problem. Limiting inference to the most recent shapelets also has the effect of working within a local 'fovea' of attention: if no recent shapelets are from distant areas, then only hypotheses around the agent's location will be considered.

There is some subtlety in defining the meaning of MAP states in continuous parameter spaces. In the present study, we assume that discrete hypotheses

---

Algorithm 1. Approximate Metropolis-hasting proposals generation

---
for each time step t **do**
  update shapelet queue S by reading sensors
  for each annealing inverse temperature β **do**
    for each shapelet $S_i \in S$ **do**
      propose and test parent H_i from $Q(pa(S_i))$
      if accepted, add H_i to hypothesis set B
    end for
    for each hypothesis $H_i \in B$ **do**
      $r \leftarrow rand(0,1)$
      if $r < r_1$ **then**
        propose death of H_i. If accepted, remove H_i from B
      else
        if $r < r_2$ **then**
          propose parent change for H_i. If accepted, replace H_i's parent parameter
        else
          if $r < r_3$ **then**
            propose child H_j from $Q(ch(H_i)|H_i)$. If accepted, add H_j to B
          else
            propose parent H_j from $Q(pa(H_i)|H_i)$. If accepted, add H_j to B
          end if
        end if
      end if
    end for
    prune all hypotheses not linked to any shapelet via a common ancestor.
  end for
end for

$H_i(x, y, \theta, \Theta)$ (where $H \in \{S, L, T\}$) represent small but non-infinitesimal collections of possible (x, y, θ) poses, with probability

$$P(H((x - \tfrac{\delta}{2}, x + \tfrac{\delta}{2}), (y - \tfrac{\delta}{2}, y + \tfrac{\delta}{2}), (\theta - \tfrac{\delta}{2}, \theta + \tfrac{\delta}{2}), \Theta))$$

$$= \delta^3 p(H(x, y, \theta, \Theta)), \tag{21}$$

where δ is a small but nonzero constant, Θ are the remaining parameters, and p is the density.

We use the annealed [1] approximate Metropolis-Hastings sampler of algorithm 1 to perform inference. Unlike standard inference problems, object-based mapping is a form of scene analysis task, i.e. the number of objects in the world – and therefore the number and type of nodes in the network – is unknown in advance. Algorithm 1 uses blackboard-like priming and pruning heuristics integrated with the sampling, to control the size of the network. Each hypothesis in the current 'blackboard' set B maintains (amongst other parameters), pose parameters x, y, θ and a current parent. The current parent may be another hypothesis, or may be null. Importantly, hypotheses that are not currently 'true' (according to the sampler) are never stored in B. The set B acts as a factor graph as detailed in the previous section, and may be thought of as the contents of a blackboard [5].

To obtain unbiased samples from the true joint distribution, Metropolis-Hastings sampling requires detailed technical conditions to be met, which are complicated by the jumps between factor graphs of different structures and sizes. Reversible jump methods [12] provide a rigorous theoretical basis from which to define acceptance probabilities based on reweighting proposals. Future work should incorporate such theory, for now we heuristically choose the Q distributions[1] and r_i thresholds; and use the annealed original P distribution from the factor graph as a simple Gibbs [1] acceptance probability,

$$P(\text{accept } H_i) = P^\beta(H_i | mb(H_i)), \tag{22}$$

where $mb(H_i)$ is the Markov blanket of H_i containing its parents, rivals $riv(H_i)$, and children $ch(H_i)$, β is inverse temperature. The Markov blanket conditional is

$$P(H_i | mb(H_i)) = P(H_i | pa(H_i), cop(H_i), ch(H_i), riv(H_i))$$

$$= \frac{1}{Z} \frac{\Phi_a \Phi_b \Phi_c P(H_i | pa) P(ch | H_i)}{P(ch | (H_i)) P(H_i | pa) + P(ch | \neg H_i) P(\neg H_i | pa)}$$

$$= \frac{1}{Z} \frac{\Phi_a \Phi_b \Phi_c p(H_i | pa) p(ch | H_i)}{\delta^3 (p(ch | H_i) p(H_i | pa) + p(ch | \neg H_i) p(\neg H_i | pa))},$$

where Z normalizes the factors contribution $\Phi_a \Phi_b \Phi_c$ only; δ is the constant of eqn. 21; $\Phi_a = \phi_a(H_i) \prod_{j \in pa(i)} \phi_a(H_j)$ includes missing children of H_i and

---

[1] Details can be found in the source code, however note that MH sampling can operate on *any* proposal Q so its precise form is unimportant. Better results are obtained as the approximate Q becomes close to the true P.

also the missing child penalty for each parent of H_i which would have a missing child in the case where H_i is false; $\Phi_b = \prod_{j \in mb(i)} \phi_b(H_i, H_j)$ and $\Phi_c = \prod_{j \in mb(i)} \phi_c(H_i, H_j)$. The update allows computation to proceed using density functions rather than probabilities, but depends on the choice of the small constant, δ.

Newly proposed nodes must be linked to existing ones, so it is necessary to locate all potential parents $pa(H_i)$. A threshold radius in pose space is used, which limits this set to candidates which are close enough to have non-negligible generating probabilities, i.e.

$$pa(H_i) := \{H_j : P(H_i|H_j) \gg 0\}. \tag{23}$$

For computational efficiency it is useful to implement a spatial hash-table to look up these nearby hypotheses. This hash-table may also be reused to look up overlapping hypotheses in the computation of ϕ_c.

3   Results

We have implemented a simple simulation of a whiskered robot in a world populated by six four-legged, table-like objects, in a simple mapping task. The simulation is coded in C++ using the ODE physics engine (www.ode.org) for collision detection. Source code is available on request, and contains handset values for all parameters described in this paper. The sensor noise levels are comparable to those found in strain template shapelet classifiers [9]. The agent follows a fixed sequences of poses around the world and runs algorithm 1 once at each pose. There are $10 \times 10 \times 4$ poses, from 10 discrete x and y positions and four compass θ angles, as shown in fig. 3. To further simplify the present simulation, tables and table hypotheses all have fixed identical w_T^x, w_T^y and τ_T parameters; and physical (but not hypothesis) tables and have fixed identical w_T^L parameters.

Steps in the inference are illustrated in the supplemental video material. The MAP hypothesis sets from all poses are collated and plotted onto a map of the arena in fig. 4. Comparing against the ground truth in fig. 3, the collated plot shows that table hypotheses are usually found in the correct locations, corresponding to the real tables. The average number of whiskers contacting tables at each pose having at least one table contact is 4.2 ± 1.7. As we would expect from such a sparse amount of data, there are thus many incorrect hypotheses found in MAPs of the form shown in fig. 2. These are created from poses which do not provide enough information about the tables to resolve ambiguities, for example when the robot is close enough to touch two legs but no third leg as in fig. 2. Also of interest in the results are the many table hypotheses perceived around the edge of the arena. These are due to the agent observing shapelets from contact with the walls around the arena. The system does not (yet) have perceptual models of walls, so the best available explanations for such shapelets are those which postulate tables with legs at these shapelet locations. (This is

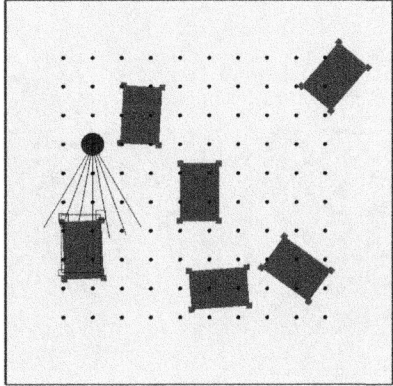

Fig. 3. Overhead view showing ground truth table configuration, and locations (black dots) of the discrete poses occupied by the robot. There are four angle poses at each location, facing in compass directions.

Fig. 4. Montage showing collection of inferred tables from each independent robot pose, for realistic [9] ($\sigma_r = 0.1, \sigma_\theta = \pi/32$) sensors

a form of perceptual relativism: lacking a WALL concept, the system explains the data using its best available TABLE theories.) Similar plots for noiseless and highly noisy sensor cases are shown in figs. 5(a) and 5(b) for comparison. In both cases, the approximate locations of inferred tables are similar, though the accuracy of inferred table poses depends on the noise.

4   Discussion

We have presented a proof-of-concept implementation of a novel framework for hierarchical object-based mapping from sparse local sensors, such as whiskers

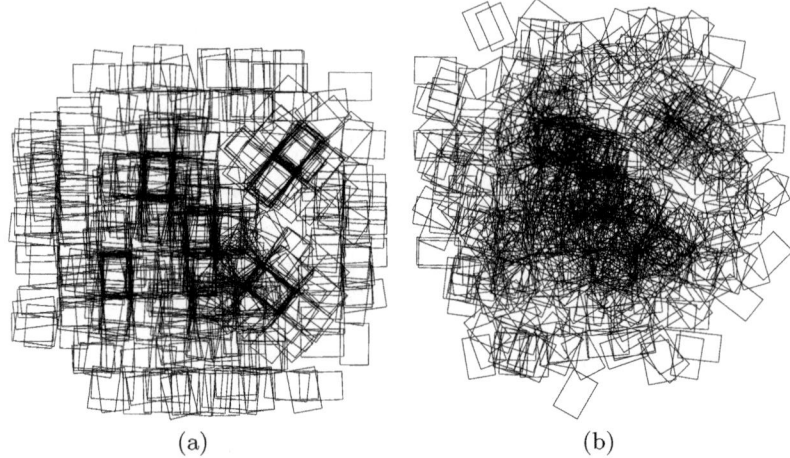

(a)   (b)

Fig. 5. (a) Montage showing the collection of inferred tables from each independent robot pose, for ideal, noiseless sensors. (b) Montage showing the collection of inferred tables from each independent robot pose, for very noisy ($\sigma_r = 0.5, \sigma_\theta = \pi/8$) sensors.

and other types of touch; but also applicable to sensors such as low-power or covert short range scanners; or local field-based sensors as used by electric fish.

Many simplifications were made in this proof-of-concept, which future versions of the system should relax. The results presented here are simply the collation of many independent MAP_t inferences made from the different poses, and no information is shared between poses. Storing longer-term memories of shapelets and fusing them into the inferences would obviously allow a more refined map of the arena to be constructed: at present each table shown in the results has been inferred from typically 4.2 ± 1.7 shapelets only, which is extremely sparse. The present system makes no use of negative evidence, i.e. the observed absence of shapelets on non-contacting whiskers: this could be used to remove some of the ambiguous percepts. The heuristic threshold constants in the proposal distribution should be replaced with Reversible Jump MCMC reweightings to remove bias in the sampling distribution (although in practice the heuristic thresholds can work well, as ultimately only the annealed MAP is sought, rather than an approximation to the whole distribution).

Importantly, the proof-of-concept simulation operates in a world having only one size and texture of table (though tables may have different leg sizes). Enlarging the parameter space to range over tables sizes and textures will allow inference of more realistic four-legged objects such as different kinds of chairs and desks. Other types of objects could also be introduced, such as walls, kitchen units and radiators. The Bayesian blackboard architecture is able to automatically select between rival object models, treating them as rival hypotheses [10]. However, as the number of models and parameters grows, sampling of course becomes less efficient. For example, it becomes less probable that a perfectly-fitting table will ever be proposed. (Even though once proposed, it will tend to remain accepted for having such a good fit.) We plan to investigate the use of 'smart proposals' which are classical heuristic object detectors (e.g. Hough transforms to find edges and corners) but re-purposed as Metropolis-Hastings proposals in the Bayesian Blackboard. When combined with RJ-MCMC acceptance probabilities, this gives a way to speed up the proposals but retain the probabilistic semantics.

We next hope to extend our implementation to recognise several types of object of varying size, and move from simulation to the SCRATCHbot platform [21], which is currently able to report shapelets of the form used in simulation. SCRATCHbot includes noisy odometry, so will require our mapping system to function as part of a SLAM system. New forms of loop-closure in SLAM may become possible by recognising different parts of the same hierarchical object.

Acknowledgements. This work was supported by EU Framework projects BIOTACT (ICT-215910) and ICEA (IST-027819).

References

1. Aarts, E., Korst, J.: Simulated Annealing and Boltzmann Machines. Wiley, Chichester (1988)
2. Ahl, A.: The role of vibrissae in behavior: a status review. Veterinary Research Communications 10(1), 245–268 (1986)
3. Binford, T., Levitt, T.: Evidential reasoning for object recognition. IEEE Tranactions on Pattern Analysis and Machine Intelligence (2003)
4. Carvell, G., Simons, D.: Biometric analyses of vibrissal tactile discrimination in the rat. J. Neurosci. 10(8), 2638 (1990)
5. Erman, L., Hayes-Roth, F., Lesser, V., Reddy, R.: The Hearsay-II system. ACM Computing Surveys 12(2) (1980)
6. Evans, M., Fox, C., Pearson, M., Prescott, T.: Spectral Template Based Classification of Robotic Whisker Sensor Signals. In: Proc. TAROS 2009 (2009)
7. Evans, M., Fox, C., Prescott, T.: Tactile discrimination using template classifiers. In: Doncieux, S., Girard, B., Guillot, A., Hallam, J., Meyer, J.-A., Mouret, J.-B. (eds.) SAB 2010. LNCS, vol. 6226, pp. 178–187. Springer, Heidelberg (2010)
8. Fend, M.: Whisker-based texture discrimination on a mobile robot. Advances in Artificial Life, 302–311 (2005)
9. Fox, C., Pearson, M., Mitchinson, B., Pipe, T., Prescott, T.: Simple features for texture classification. Somatosensory and Motor Research 24(3), 139–162 (2007)
10. Fox, C.: ThomCat: A Bayesian blackboard model of hierarchical temporal perception. In: Proc. FLAIRS (2008)
11. Gallagher, G., Srinivasa, S.S., Bagnell, J.A., Ferguson., D.: Gatmo: A generalized approach to tracking movable objects. In: ICRA (2009)
12. Green, P.: Reversible jump Markov chain Monte Carlo computation. Biometrika 82(4), 711–732 (1995)
13. Heiligenberg, W.: Neural nets in electric fish. MIT, Cambridge (1991)
14. Kaneko, M., Kanayama, N., Tsuji, T.: Active antenna for contact sensing. IEEE Transactions on Robotics and Automation 14(2), 278–291 (1998)
15. Kim, D., Moller, R.: Biomimetic whiskers for shape recognition. Robotics and Autonomous Systems 55(3), 229–243 (2007)

16. Laskey, K.B., da Costa, P.C.: Of starships and klingons: Bayesian inference for the 23rd century. In: Proc. UAI (2005)
17. Lepora, N., Evans, M., Fox, C., Diamond, M., Gurney, K., Prescott, T.: Naive Bayes texture classification applied to whisker data from a moving robot. In: Proc. IEEE WCCI (2010)
18. Milch, B.: Probabilisitic Models with Unknown Objects. Ph.D. thesis, UC Berkeley (2006)
19. Mitchell, M.: Analogy-Making as Perception. MIT, Cambridge (1993)
20. Pearl, J.: Causality. Cambridge University Press, Cambridge (2000)
21. Pearson, M.J., Mitchinson, B., Welsby, J., Pipe, A.G., Prescott, T.J.: Scratchbot: Active tactile sensing in a whiskered mobile robot. In: Doncieux, S., Girard, B., Guillot, A., Hallam, J., Meyer, J.-A., Mouret, J.-B. (eds.) SAB 2010. LNCS, vol. 6226, pp. 93–103. Springer, Heidelberg (2010)
22. Petrovskaya, A., Khatib, O., Thrun, S., Ng, A.Y.: Touch based perception for object manipulation. In: ICRA (2007)
23. Schultz, A., Solomon, J., Peshkin, M., Hartmann, M.: Multifunctional whisker arrays for distance detection, terrain mapping, and object feature extraction. In: Proc. ICRA 2005 (2005)
24. Seth, A., McKinstry, J., Edelman, G., Krichmar, J.: Texture discrimination by an autonomous mobile brain-based device with whiskers. In: Proc. IEEE ICRA (2004)
25. Srinivasa, S., Ferguson, D., Helfrich, C., Berenson, D., Romea, A.C., Diankov, R., Gallagher, G., Hollinger, G., Kuffner, J., Vandeweghe, J.M.: Herb: a home exploring robotic butler. Autonomous Robots 28(1), 5–20 (2010)
26. Sutton, C., Burns, B., Morrison, C., Cohen, P.R.: Guided incremental construction of belief networks. In: Proc. Fifth Int. Symp. Intelligient Data Analysis (2003)
27. Thrun, S., Burgard, W., Fox, D.: Probabilistic Robotics. MIT, Cambridge (2006)
28. Wang, C.-C., Thorpe, C., Thrun, S., Herbert, M., Durrant-Whyte, H.: Simultaneous localization, mapping and moving object tracking. Int. J. Robotics Research 26, 889 (2007)

Appendix: Noisy-OR Density Combination

Let Y_i range over nodes $pa(X)$ in a continuous-valued Bayesian network with noisy-OR parent combinations,

$$P(X|\{Y_i\}_i) = 1 - \prod(1 - P_i). \tag{24}$$

with $P_i = P(X|Y_i)$. Consider the probability of a small range of hypotheses,

$$\delta^3 p(X|\{Y_i\}_i) = 1 - \prod(1 - \delta^3 p_i), \tag{25}$$

where p are probability densities and P are probabilities. Expansion terms with powers of δ that are > 3 vanish, so

$$\delta^3 p(X|\{Y_i\}_i) = \delta^3 \sum p_i. \tag{26}$$

The δ^3 terms cancel to yield

$$p(X|\{Y_i\}_i) = \sum p_i. \tag{27}$$

Multi-rate Visual Servoing Based on Dual-Rate High Order Holds

J. Ernesto Solanes, Josep Tornero, Leopoldo Armesto, and Vicent Girbés

Universitat Politècnica de València, València, Spain
{juasogal,jtornero,leoaran,vigirjua}@upvnet.upv.es

Abstract. This paper describes a multi-rate approach based on the extensive use of Dual-rate High Order Holds for visual servoing systems. Moreover a complete description of a general multi-rate approach, comparing Dual-rate Image-Based Visual Servoing algorithm with Dual-rate PID controller, where the PID is separated into its two different dynamics, using two different sampling periods is presented. In addition, a multi-rate Kalman filter is compared with Dual-rate High Order Holds, as an attempt to extend the use of this kind of interfaces to the estimation process. Results are obtained by simulation and also using a 6-DOF industrial robot (KUKA KR5 sixx R650) for the case of tracking objects with fast movement. This paper has validated the use of Dual-rate High Order Holds in non-linear systems, in general, and in robot visual servoing, in particular.

Keywords: Visual Servoing, Multi-rate control, non-linear systems.

1 Introduction

Visual Servoing uses visual data in order to control the motion of a robot [11], [6]. During the last decade, there has been an increment of visual servoing applications in many industrial sectors. Technological advances have allowed the implementation of visual control systems, since control actions can be generated in real-time based on visual data. Classic techniques follow three approaches: 1) 2D or Image-Based Visual Servoing (IBVS) [14], 2) 3D or Position-Based Visual Servoing (PBVS) [11] and 3) 2 1/2D or Hybrid Visual Servoing (HVS) [5]. This paper is focused on IBVS approach with Eye-in-Hand configuration, although the methodology proposed can be extended to the other approaches. IBVS shows greater precision and robustness, since it is independent of the process calibration, although the convergence is not globally guaranteed [6].

Visual Servoing applications are affected by sensor latency, that is, the time needed for computations, which might be significantly higher than the control period. Moreover there are other aspects on real systems (i.e., data-missing, communication delays, robot control constraints, etc.) that are not usually taken into account. Most of papers try to overcome this problem focusing on modifying control algorithm (i.e. optimization [15], or neural networks, [16]) to improved overall system performance. Our approach focuses on multi-rate control.

R. Groß et al. (Eds.): TAROS 2011, LNAI 6856, pp. 195–206, 2011.
© Springer-Verlag Berlin Heidelberg 2011

Multi-rate systems have been extensively treated in the last four decades and it is possible to find many contributions dealing with modeling and analysis as well as control design of periodic sampled-data systems, see [2] for a survey on modeling approaches. One of the most relevant modeling techniques is the Lifting Technique [12], where an isomorphism between a linear periodic system and an enlarged linear-time invariant (LTI) system is defined via the lifting operator. Another interesting point of view for modeling multi-rate systems is the one provided by [2], where two periodic matrices relate inputs and outputs according to the multi-rate sampling pattern. The fact is that multi-rate control have been heavily studied to linear systems but little to non-linear systems.

Moreover, there are very few papers using multi-rate structures. In [10] and [9], a multi-rate feedback disturbance rejection control is described. The aim of this approach is to use a multi-rate controller based on robust control theory, with multi-rate loop, in order to reduce the sensor latency. The main problem is its internal model of high frequency disturbance damages the closed loop characteristics, which may cause poor stability robustness.

This paper describes how to overcome the problems of delays, data-missing, etc., in visual servoing, using a multi-rate approach based on the extensive use of Dual-rate High Order Holds (DR-HOH) [18], [17]. Given that DR-HOHs do not need the model of the plant, they are good candidates for being applied to non-linear systems, in general, and for robotic systems in particular. Dual-rate Holds are obtained based on Lagrange extrapolation, as well as based on Bezier parametric equations and Taylor series.

The paper provides a complete description of a general multi-rate approach for IBVS-based visual servoing. As an alternative to the IBVS algorithm, the paper also deals with a Dual-Rate PID controller (DR-PID), where the PID is separated into its two different dynamics, using two different sampling periods. It has been compared the performance of multi-rate Kalman filters with DR-HOH, as an attempt to extend the use of this kind of interfaces. In all the cases, the paper shows the benefits of introducing the multi-rate control with respect to the classic single-rate approach.

2   Dual-Rate High Order Holds

A dual-date High-Order-Hold (DR-HOH) is a circuit for generating, from a sequence of inputs sampled at low sampling rates, a continuous signal or a discrete one at higher sampling rate. The mathematical background of DR-HOH is described in [18], [17]. Initially, DR-HOHs were introduced as generalizations of conventional holds such as zero, first, second order hold, etc. Later, in [1] a wide variety of holds were proposed using general primitive functions.

In this paper, DR-Holds circuits are obtained based on Lagrange extrapolator, Bezier parametric equations or Taylor series as primitive functions. The concept of primitive function is used to generate the set of signals at base-period (T) every frame-period ($\bar{T}$) [19] and [1]. The ratio between base and frame periods

Table 1. MR-HOH based on polynomial functions (transfer function representation)

DR-HOLD	TRANSFER FUNCTION ENTRIES $\mathbf{G}_h(z^N)$
DR-ZOH	I
DR-FOH	$\dfrac{\left((N+i)/N\right)z^N - i/N}{z^N}\cdot I$
DR-SOH	$\left(\dfrac{\left(1+(3/2)(i/N)+(1/2)(i/N)^2\right)z^N}{z^N} + \dfrac{2i/N+(i/N)^2}{z^N} + \dfrac{(1/2)\left(i/N+(i/N)^2\right)z^{-N}}{z^N}\right)\cdot I$

Table 2. MR-HOH based on polynomial functions (discrete algorithm representation)

DR-HOLD	DISCRETE ALGORITHM $\mathbf{u}(k,i)$
DR-ZOH	$\mathbf{u}(k,0)$
DR-FOH	$\left(\frac{N+i}{N}\right)\cdot\mathbf{u}(k) - \frac{i}{N}\cdot\mathbf{u}((k-1))$
DR-SOH	$\left(1+\frac{3i}{2N}+\frac{1}{2}\left(\frac{i}{N}\right)^2\right)\cdot\mathbf{u}(k,0) + \left(\frac{2i}{N}+\left(\frac{i}{N}\right)^2\right)\cdot\mathbf{u}(k-1,0) + \frac{1}{2}\left(\frac{i}{N}+\left(\frac{i}{N}\right)^2\right)\cdot\mathbf{u}(k-2,0)$

is N. The primitive function uses the input values inducing an order of function complexity:

$$\mathbf{u}_h(t) = \sum_{l=0}^{n} \mathbf{f}_{n,l}(t,t_k)\cdot\mathbf{u}(t_{k-l}) \tag{1}$$

where $t_k = k\bar{T}$ is the sampling-time for the last available input and $t_{k-l} = t_l - l\bar{T}$ the time previous frame periods. $\mathbf{u}_h(t)$ is the hold output evaluated at time t based on inputs from t_k to t_{k-l}. Therefore, every base period, the output is:

$$\mathbf{u}_h(t_k + i\cdot T) = \sum_{l=0}^{n} \mathbf{f}_{n,l}(i\cdot T)\cdot\mathbf{u}(t_{k-l}) \tag{2}$$

where $i = 0, 1, \ldots, N-1$. The notation for discrete sequence of input/output values uses double index indicating the indexes for frame and base periods, that is $\mathbf{u}(k,i) = \mathbf{u}(k\bar{T} + iT)$. Therefore the hold can be expressed in discrete time as:

$$\mathbf{u}_h(k,i) = \sum_{l=0}^{n} \mathbf{f}_{n,l}^{*}(i)\cdot\mathbf{u}(k-l,0) \tag{3}$$

with $\mathbf{f}_{n,l}^{*}(i) = \mathbf{f}_{n,l}(iT)$. According to [19], the lifted output vector is defined as:

$$\mathbf{F}(k) = \left[\sum_{l=0}^{n} \mathbf{f}_{n,l}^{*}(0)\cdot\mathbf{u}(k-l) \ \sum_{l=0}^{n} \mathbf{f}_{n,l}^{*}(1)\cdot\mathbf{u}(k-l) \ \ldots \ \sum_{l=0}^{n} \mathbf{f}_{n,l}^{*}(N-1)\cdot\mathbf{u}(k-l) \right] \tag{4}$$

and by applying Z-transform, we can obtain the corresponding discrete-time transfer function at frame-period:

$$\mathbf{F}(z,z^N) = \mathbf{G}_h(z,z^N)\cdot\mathbf{U}(z,z^N), \quad with \quad \mathbf{G}_h(z,z^N) = \sum_{i=0}^{N-1} \mathbf{G}_i(z^N)\cdot z^{-i} \tag{5}$$

being $z^N = e^{s\bar{T}}$.

In particular, primitive functions based on Lagrange extrapolator can be obtained from the following general expression:

$$\mathbf{u}_h(t) = \sum_{l=0}^{n} \prod_{\substack{q=0 \\ q \neq l}}^{n} \left[\frac{t - t_{k-q}}{t_{k-l} - t_{k-q}} \right] \cdot \mathbf{u}(t_{k-l}) \tag{6}$$

By using Lagrange extrapolator with different polynomial orders as primitive functions, we can derive classic DR-ZOH ($n = 0$), DR-FOH ($n = 1$) and DR-SOH ($n = 2$), emulating traditional holds such as ZOH, FOH and SOH. Tables 1 and 2 show transfer functions and its equivalent discrete algorithm for implementation of before mentioned DR-HOHs.

3   Multi-rate Visual Servoing

Visual servoing applications have two elements: Vision System relative to image acquisition and processing algorithm, and Control System, relative to the robot control. Fig. 1 shows the block diagram of a multi-rate visual servoing control designed to overcome the problems of delays, data-missing, etc., present in most of all robotics systems.

In this multi-rate scheme, let us denote $N_V T$ as the Vision System sampling period, and $N_C T$ as the Control System sampling period, being T the based period. Given that, industrial robots must be controlled at high frequencies, we will use the base period T. In addition, the visual processing and control algorithms can work at an intermediate frequency given by the sampling period $N_C T$ ($N_V \geq N_C \geq 1$). For simplicity, we will assume a pure dual-rate scheme, where the ratio $N_c/N_v \in Z^+$.

In classic control theory, both systems are sampled at the same frequency ($N_C = N_V$), giving the classic single-rate scheme. In that case, the visual servoing is at the sampling period of the slower element, at $N_V T$.

3.1   Control System

The multi-rate control system is based on DR-Visual Servoing Controller and a single-rate Robot Control Unit, taking input signal at the period of $N_C T$ an output signals at the sampled period T. DR-Visual Servoing Controller can implement any visual servoing: PBVS, IBVS, etc.

Dual-rate Image-Based Visual Servoing Controller. The action control is computed directly from image features [6]. The control law is defined as:

$$\mathbf{u}(k\bar{T} + iT) = -\lambda \cdot \hat{L}_s^* \cdot \left(e(s(k\bar{T} + iT) - s^*(T)) \right) \tag{7}$$

where λ is the controller gain, $\mathbf{e}$ is the features error, s are the current detected features on image plane, s^* is the desired features on image plane, and L_s is

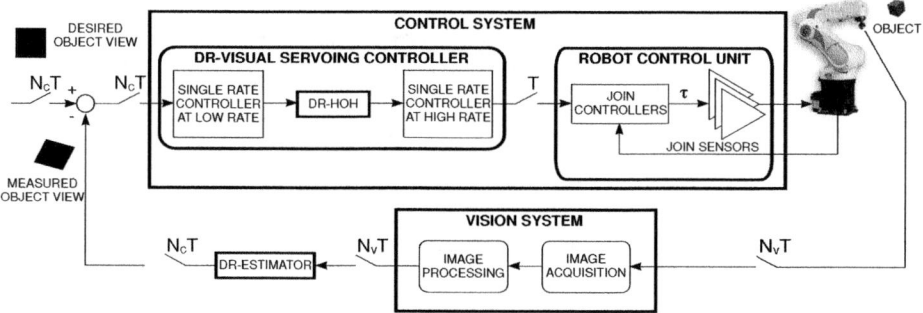

Fig. 1. Visual Servoing System block diagram

the interactin matrix introduced by [11] and $\hat{L}_s^*$ is the Moore-Penrose inverse of L_s. This kind of controllers only have one dynamics so there are two possibilities:1) compute the controller action at low frequency (at $\bar{T} = N \cdot T$), see Fig. 2(b) ; 2) compute the controller action at high frequency (at T), see Fig. 2(c).

Dual-rate PID. As an alternative, it is possible to use a PID controller instead of a visual servoing algorithm. Let a continuous PID controller described in Equation 2(a) in multiplicative form:

$$G_{PID}(s) = K_p \cdot \underbrace{\left(1 + \frac{1}{\tau_i s}\right)}_{PI} \cdot \underbrace{(1 + \tau_d s)}_{PD} \tag{8}$$

where K_P is the proportional gain, and τ_d and τ_i are the derivative and integral time-constants, respectively. The PI can be discretized at a low frequency, the PD can be discretized at a high frequency and a DR-HOH is used in order to change the frequency [20],

$$G_{PI}(z^N) = K_p \cdot \frac{z^N - \left(1 - \frac{N \cdot T}{\tau_i}\right)}{z^N - 1}, \quad \text{(PI low frequency)} \tag{9}$$

$$G_h(z, z^N) = \left(\frac{z^N - 1}{z - 1}\right)^2 \cdot \frac{z^2 - \frac{N-1}{N} \cdot z}{z^{2N}}, \quad \text{(DR-FOH)} \tag{10}$$

$$G_{PD}(z) = \frac{z - \frac{\tau_d/T}{1 + \tau_d/T}}{z}, \quad \text{(PD high frequency)} \tag{11}$$

Robot Unit Controller. It may works in two different ways: if it is possible to act directly to the robot torques/forces, we referred to it as Direct Visual Servoing, on the other hand, in the majority of industrial robots, there are inner loops involved in kinematic and dynamic control. In that case, we have an Indirect Visual Servoing.

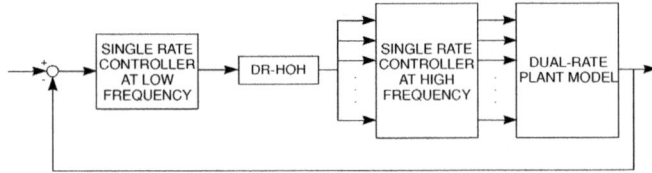

(a) General Dual-rate control structure

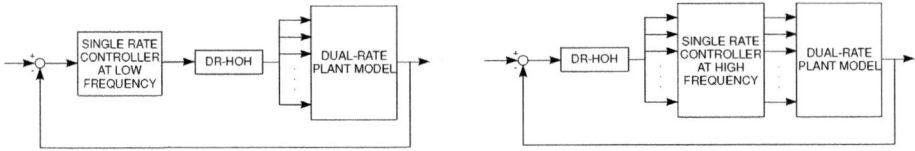

(b) Controller working at low frequency   (c) Controller working at high frequency

Fig. 2. Dual-rate controller possibilities

3.2 Vision System

We have separated the Vision System into two different elements: Image Acquisition, related to cameras, and Image Processing Algorithm, related to features extraction. That is: *Image Acquisition* is the part of vision system responsible to acquire images. Usually, this is only one thread which stores the new acquired image into a buffer in order to accomplish the Vision System real time requirements; *Image Processing* is the part of vision algorithm responsible to the image feature extraction. This part can be implemented on a separate thread which takes images from the buffer.

The single-rate Vision System is working at sampling period $N_V T$, and its output has to be accommodated to the sampling period $N_C T$ in order to close the loop, as can be seen in Fig. 1. A DR-HOH or other multi-rate estimator such as a DR-Kalman Filter [3] can be introduced to work as DR-Estimator. The aim of this element is to extrapolate detected features on images adapting them to the sampling frequency of the reference object view.

4 Analysis

4.1 Simulation Results

In this section, we are going to use Dual Rate High Order Holds (DR-HOHs) for interfacing signals at different frequencies: from $N_C T$ to T inside the control system, and from $N_V T$ to $N_C T$ after the vision system. As we will prove by simulation as well as experimental results, DR-HOHs are playing an important

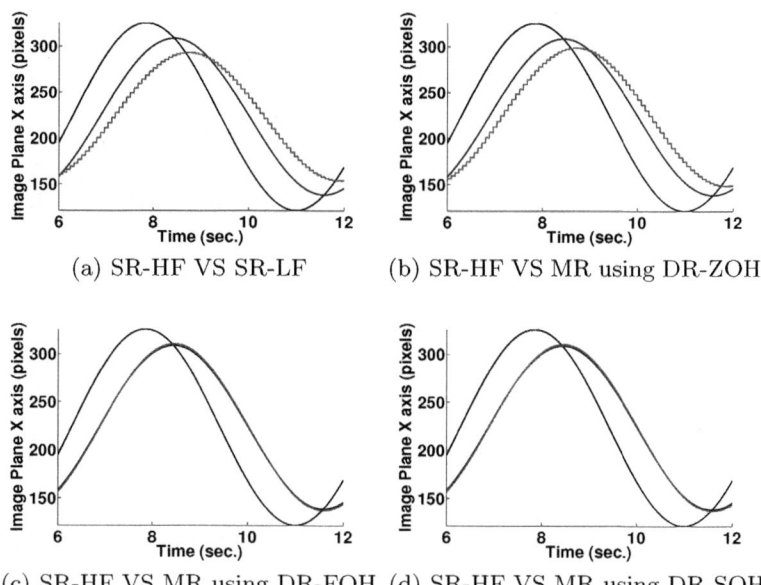

(a) SR-HF VS SR-LF   (b) SR-HF VS MR using DR-ZOH.

(c) SR-HF VS MR using DR-FOH (d) SR-HF VS MR using DR-SOH

Fig. 3. Multi-rate Visual Servoing using DR-HOH versus Single-rate Visual Servoing

role interfaces between linear and non-linear subsystems working at different frequencies.

We have tested several DR-HOH of different orders and types as described in Table 1. Among them, DR-FOH has given better results, being a trade-off between complexity and computational requirements. In particular, for the case of DR-FOH, we have chosen the form of Equation 5, with transfer function,

$$G_h(z, z^{\bar{N}}) = \left(\frac{z^{\bar{N}} - 1}{z - 1} \right)^2 \cdot \frac{z^2 - \frac{\bar{N}-1}{\bar{N}} \cdot z}{z^{2\bar{N}}} \tag{12}$$

Note that, for any DR-HOH corresponding to Visual Servoing Controller ($\bar{N} = N_C$ and $\bar{T} = T$) and for the Dual-Rate Estimator implemented also as DR-HOH ($\bar{N} = \frac{N_V}{N_C}$ and $\bar{T} = N_C T$), we are considering three different sampling rates for each subsystem: Vision System period at $80ms$ ($N_V T$, with $N_V = 8$), a Visual Servoing Control period at $20ms$ ($N_C T$, with $N_C = 2$) and a base period at $10ms$ (T). Therefore, the ratio between the Vision system and Control is $N_V/N_V = 4$. Those rates have been chosen because we are interested in stabilizing IBVS controllers for some extreme situations based on DR-HOH. The simulation is based on an object moving in a sinus trajectory. The features detected are the four vertex of the object. The implementation is based on Robotics [7] and Machine Vision [8] Matlab/ Simulink toolboxes.

Fig. 3 shows the tracking of the object with the robot for four different configurations: dual-rate schemes with DR-ZOH (Fig. 3(b)), DR-FOH (Fig. 3(c))

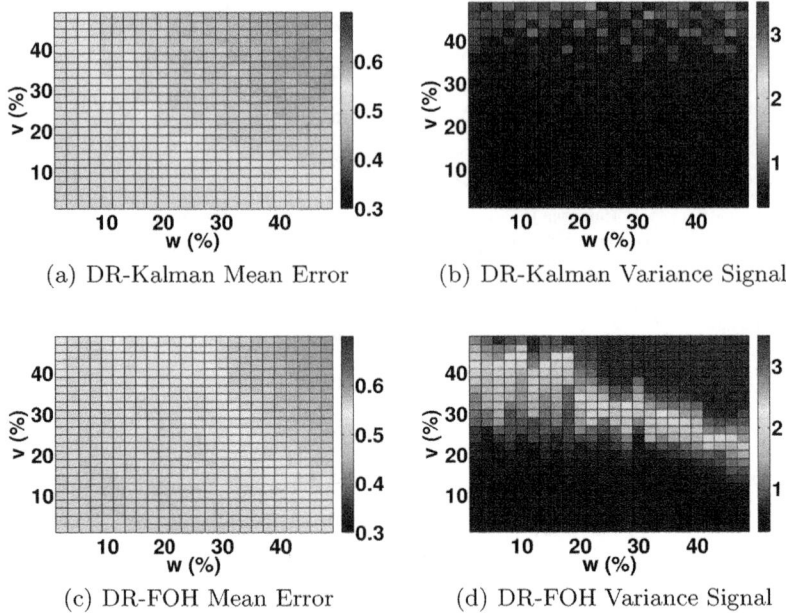

(a) DR-Kalman Mean Error

(b) DR-Kalman Variance Signal

(c) DR-FOH Mean Error

(d) DR-FOH Variance Signal

Fig. 4. DR-HOH versus DR-Kalman Filter

and DR-SOH (Fig. 3(d)), and also single-rate at low and high frequency (Fig. 3(a)). It should be remarked that the low frequency single-rate scheme runs at the frequency of the slowest subsystem, that is the Vision System at $80ms$ sampling period. Also note that the high frequency single-rate scheme is not physically implementable on a real system. In this sense, our aim is to obtain a multi-rate scheme which can produce a similar behavior of the ideal single-rate high-sampling frequency approach.

As expected, DR-HOH approaches provide better responses than low frequency single-rate one and a close response to the high frequency single-rate approach. It is interesting to remark that DR-FOH and DR-SOH has better response than DR-ZOH and Single-rate approach. In fact, these responses look like ideal high frequency response, while DR-ZOH looks like single-rate response at low frequency.

In addition to this, we have also analyzed the image tracking error based on different estimation approaches. In particular, our purpose is to compare multi-rate Kalman Filter and DR-HOH as estimators. In particular, the DR-HOH estimator is implemented as two independent DR-HOHs for each coordinate of detected vertex. It is obvious that, DR-HOHs do not perform filtering and therefore they are not so robust to noises. The aim of the analysis is to determine the maximum level of noise that can be allowed with DR-HOH. Fig. 4 shows the results obtained with both dual-rate estimators in terms of mean estimation error and its covariance, while varying the percentage of contained error on the

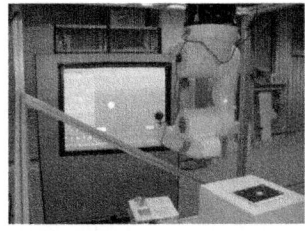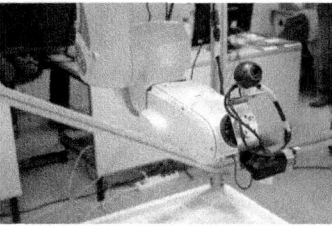

Fig. 5. Visual servoing set-up. Left: Visual Servoing Platform. Right: Kuka industrial robot end-effector.

system (robot pose) and measurement (vision system). For measurement errors less than 15%, the performance of both estimators is very similar. For higher errors, DR-Kalman filter behaves better because its intrinsic filtering procedure, which is not the case of DR-HOH is basically an extrapolator. More conclusions will be drawn from the experimental results.

4.2   Experimental Results

All the algorithms described in simulation have been implemented in a real platform based on an industrial robot (Kuka KR5 sixx R650) using a Logitech C300 webcam in Hand-in-Eye configuration, with a resolution of $640x480$ pixels working at $30fps$, as shown Fig. 5.

Visual Servoing Algorithms are implemented on a WorkStation with Ubuntu 10.04 S.O, using several open libraries (i.e. OpenCV [4] or Visual Servoing Platform [13]). Communication between the Control Unit and the WorkStation is based on Ethernet, with a robot manufacturer's hardware Kuka's Ethernet.RSI card. This card allows data to flow between controllers and external system (in this case the WorkStation) at a maximum rate of $83Hz$.

The industrial robot control unit has a watchdog maximum time of $12ms$. Therefore the Workstation has to send date packages within this time or data will be ignored and the robot will stop. The multi-rate approach tries to avoid this situation generating data packages at high frequency, due to the fact that the rest of the system is working at lower frequencies. For that purpose, we consider the following sampling rates: Vision System running at $60ms$ ($N_V T$, with $N_V = 6$), Visual Servoing Control running at $20ms$ ($N_C T$, with $N_C = 2$) and a base period at 10 milliseconds (T). In spite of the fact that our Vision System algorithm takes $33ms$, we have chosen a period of $60ms$, which may be required for more complex object detection procedures. The moving object is projected on screen keeping an accurate control on its real position and used as ground-truth data. For simplicity, we have used a circular object moving at a velocity of $13cm/s$. The object describes several trajectories: triangle, square, circle and ∞. However, in the paper we only show results for square trajectories.

Table 3 shows quadratic errors ($e^2(s^* - s)$), settling time (t_s) and oversoot (δ) parameters when using Image-Based Visual Servoing controller. In particular, the conventional low frequency single rate approach is compared with the multi-rate one using DR-ZOH, DR-FOH, DR-SOH and DR-Kalman filter. It can be seen that multi-rate approach with DR-FOH produces the best results. In fact, comparing with conventional single-rate IBVS, it has a 50% less quadratic error. In other words, it is possible a tracking at double velocity.

Moreover, Fig. 6, Left shows an object following a rectangular trajectory in order to compare single-rate visual servoing, multi-rate visual servoing with DR-FOH and Multi-rate visual servoing using DR-Kalman filter, in both cases as features estimator. Although in this particular case the DR-HOH's behavior is better than DR-Kalman Filter, it cannot be extended to other cases, since DR-HOHs are much more sensible to noisy data.

As an alternative to conventional visual servoing controllers, a dual-rate PID controller has been considered. According to Fig. 2, the dual-rate PID controller can be implemented in three forms: case 1, formed with the PI working at low frequency and a PD working at high frequency with a DR-HOH in between (Fig. 2(a)); case 2, PID working at low frequency (Fig. 2(b)); case 3,PID working at high frequency (Fig. 2(c)). Fig. 6, Right compares the behavior of the single-rate approach with the three different cases previously described. The better behavior is achieved in case 1, where the dynamic associated to PI and PD are running at their highest frequency. The improvement with respect to the conventional single-rate approach reaches a 75% in terms of quadratic errors.

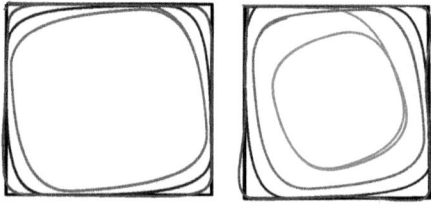

Fig. 6. Single-rate versus Multi-rate approaches. Left: SR-IBVS vs DR-IBVS: black, track reference; magenta, SR-IBVS; red, MR-IBVS using DR-HOH; blue, MR-IBVS with DR-Kalman Filter. Right: SR vs. DR PID using DR-FOH: black, track reference; green, single-rate; magenta, case 3; blue, case 2; red, case 1.

Table 3. Single-rate VS multi-rate IBVS controller: Object moving at 8 cm/s

	$e^2(s^* - s)$ (m)	t_s (ms)	$\delta(\%)$
SINGLE-RATE IBVS	0.02441748	—	$Under-damped$
MULTI-RATE IBVS (DR-ZOH)	0.02131598	—	$Under-damped$
MULTI-RATE IBVS (DR-FOH)	0.01741782	550	12
MULTI-RATE IBVS (DR-SOH)	0.01768391	530	9
MULTI-RATE IBVS (KALMAN)	0.01874451	580	$Under-damped$

5   Conclusions

In this paper, we have proposed a generic multi-rate control approach applicable to non-linear systems. This approach can be applied to robotic systems, in general, and to robot visual servoing, in particular. The approach is based on Dual-Rate High Order Holds, acting as hybrid interfaces, that can generate outputs at higher sampling frequencies than the input signals. Dual-rate HOHs are obtained based on three methods: Lagrange extrapolation, Bezier parametric equations and Taylor series.

First, the paper decomposes a general multi-rate visual servoing overall system into subsystems, where detailed descriptions of possible implementations of each subsystem are provided. It is interesting to remark that we have analyzed several subsystems: Image-based visual servoing (IBVS) and multi-rate PID as Control System, as well as DR-HOH and Dual-rate Kalman filters as Estimator.

The experimentation has been validated on a 6-DOF industrial robot arm with Eye-in-Hand camera, and initially validated on simulations by emulating the hardware setup. The results show that multi-rate approaches get similar performances as a theoretical single-rate system running at the high sampling frequency of the controller, improving significantly the traditional single-rate system at image acquisition rate (low frequency).

Based on our experiments, using DR-FOH, the tracking error can be reduced about 50% for the IBVS and almost 75% for the Dual-rate PID. In particular, the DR-PID has been analyzed in three different configurations, given, as expected, the best performance when the integral action is running at low frequency and the derivative at high.

In the Estimator subsystem, we have compared the performance of multi-rate Kalman filters with DR-HOH, as an attempt to extend the use of this kind of interfaces. For measurement errors less than 15%, the performance of both estimators are very similar. For higher errors Kalman performs better because its inherent filtering effect. Moreover, a comparison between Multi-rate visual servoing using DR-HOHs and using DR-Kalman Filter as Estimator subsystem has been made experimentally, which results shows that, in our particular case, the error is lower using DR-HOHs, although results cannot be extended since DR-Kalman filter is more robust against noises than DR-HOHs.

As future work, asynchronous DR-HOHs applied to Visual Servoing systems we will be considered to deal with non-periodic sampling scheme as well as data-missing problems. In addition, it will be possible to include an explicit the system model into the DR-HOHs, improving significantly the estimation process.

Acknowledgements. This work was supported by VALi+d Program (Generalitat Valenciana), DIVISAMOS Project (Spanish Ministry), PROMETEO Program (Conselleria d'Educació, Generalitat Valenciana) and MAGV Project (PAID-05-10 Program from VIDI UPV).

206   J.E. Solanes et al.

References

1. Armesto, L., Tornero, J.: Dual-rate high order holds based on primitive functions. In: American Control Conf., pp. 1140–1145 (2003)
2. Armesto, L., Tornero, J.: A general formulation for generating multi-rate models. In: American Control Conf., pp. 1146–1151 (2003)
3. Armesto, L., Tornero, J.: Linear quadratic gaussian regulators for multi-rate sampled-data stochastic systems. In: ICINCO, pp. 67–74 (2006)
4. Bradski, G.: The OpenCV Library. Dr. Dobb's Journal of Software Tools (2000)
5. Chaumette, F., Malis, E.: 2 1/2 d visual servoing: a possible solution to improve image-based and position-based visual servoings. In: Proceedings of ICRA 2000 IEEE Int. Conf. on Robotics and Automation, vol. 1, pp. 630–635 (2000)
6. Chaumette, F.: Potential problems of stability and convergence in image-based and position-based visual servoing. In: Kriegman, D., Hager, G., Morse, A. (eds.) The Confluence of Vision and Control. LNCIS, vol. 237, pp. 66–78. Springer, Heidelberg (1998)
7. Corke, P.I.: A robotics toolbox for matlab. IEEE Robotics Automation Magazine 3(1), 24–32 (1996)
8. Corke, P.: The machine vision toolbox: a matlab toolbox for vision and vision-based control. IEEE Robotics Automation Magazine 12(4), 16–25 (2005)
9. Fujimoto, H.: Visual servoing of 6 dof manipulator by multirate control with depth identification. In: Proceedings of 42nd IEEE Conf. on Decision and Control, vol. 5, pp. 5408–5413 (December 2003)
10. Fujimoto, H., Hori, Y.: Visual servoing based on multirate sampling control-application of perfect disturbance rejection control. In: Proceedings 2001 ICRA. IEEE Int. Conf. on Robotics and Automation, vol. 1, pp. 711–716 (2001)
11. Hutchinson, S., Hager, G., Corke, P.: A tutorial on visual servo control. IEEE Transactions on Robotics and Automation 12(5), 651–670 (1996)
12. Khargonekar, P., Poolla, K., Tannenbaum, A.: Robust control of linear time-invariant plants using periodic compensation. IEEE Transactions on Automatic Control AC-30, 1088 (1985)
13. Marchand, E., Spindler, F., Chaumette, F.: Visp for visual servoing: a generic software platform with a wide class of robot control skills. IEEE Robotics and Automation Magazine 12(4), 40–52 (2005)
14. Martinet, P., Gallice, J., Khadraoui, D.: Vision based control law using 3d visual features. In: Committees, Econometrica, pp. 497–502 (1996)
15. Mei, J., HuiGuang, L.: Optimization design for visual servoing in eye-in-hand robotics. In: IEEE Int. Conf. on Automation and Logistics, ICAL 2008, pp. 2064–2068 (September 2008)
16. Siebel, N.T., Kassahun, Y.: Learning neural networks for visual servoing using evolutionary methods. In: Sixth Int. Conf. on Hybrid Intelligent Systems, HIS 2006, p. 6 (December 2006)
17. Tornero, J., Gu, Y., Tomizuka, M.: Analysis of multi-rate discrete equivalent of continuous controller. In: American Control Conf., pp. 2759–2763 (1999)
18. Tornero, J., Tomizuka, M.: Dual-rate high order hold equivalent controllers. In: American Control Conf., pp. 175–179 (2000)
19. Tornero, J., Tomizuka, M.: Modeling, analysis and design tools for dual-rate systems. In: American Control Conf., pp. 4116–4121 (2002)
20. Tornero, J., Tomizuka, M., Camina, C., Ballester, E., Piza, R.: Design of dual-rate pid controllers. In: Proceedings of the IEEE Int. Conf. on Control Applications, pp. 859–865. IFAC Press (2001)

Optimal Path Planning for Nonholonomic Robotic Systems via Parametric Optimisation

James Biggs

Advanced Space Concepts Laboratory, Department of Mechanical Engineering,
University of Strathclyde, Glasgow, UK
james.biggs@strath.ac.uk

Abstract. Motivated by the path planning problem for robotic systems this paper considers nonholonomic path planning on the Euclidean group of motions $SE(n)$ which describes a rigid bodies path in n-dimensional Euclidean space. The problem is formulated as a constrained optimal kinematic control problem where the cost function to be minimised is a quadratic function of translational and angular velocity inputs. An application of the Maximum Principle of optimal control leads to a set of Hamiltonian vector field that define the necessary conditions for optimality and consequently the optimal velocity history of the trajectory. It is illustrated that the systems are always integrable when $n = 2$ and in some cases when $n = 3$. However, if they are not integrable in the most general form of the cost function they can be rendered integrable by considering special cases. If the optimal motions can be expressed analytically in closed form then the path planning problem is reduced to one of parameter optimisation where the parameters are optimised to match prescribed boundary conditions. This reduction procedure is illustrated for a simple wheeled robot with a sliding constraint and a conventional slender underwater vehicle whose velocity in the lateral directions are constrained due to viscous damping.

1 Introduction

In recent years there has been a great deal of research that involves motion planning in the presence of nonholonomic constraints. Several books give a general overview of the nonholonomic motion planning problem (MPP) [22,5] for nonholonomic mechanical systems, and [17,9,14] in the context of robotics. Nonholonomic motion planning is challenging because nonlinear control theory does not provide an explicit procedure for constructing controls. In addition linearisation techniques, effective for nonlinear systems, fail to be useful, as highlighted in [15] that linearisation renders such systems uncontrollable. The computation of feasible trajectories for nonholonomic systems is a complex task and generally treated using numerical methods. However, numerical methods have the drawback that they are inherently local and not guaranteed to find a feasible solution. Additionally, such methods as dynamic programming [16] may provide global optimal solutions, but are often computationally expensive. However, when the configuration space can be represented by a Lie group the motion

R. Groß et al. (Eds.): TAROS 2011, LNAI 6856, pp. 207–218, 2011.

planning algorithms can be designed to exploit the underlying structure of the system. For nonholonomic systems defined on Lie groups, the MPP methodologies are naturally based on Lie-algebraic techniques. The general idea is to use a family of control functions i.e. piecewise constant controls, periodic controls or polynomial controls to generate motions in the directions of iterated Lie brackets, that is, steering the system in directions that are not directly controlled, see for example [13,21,18]. These methodologies provide a constructive procedure for motion planning based on the iterated Lie bracket in that they provide a feasible path between two configurations (at least approximately). However, in general there will be more than one feasible path between two configurations with some more practically desirable than others. As in dynamic programming a specific path can be selected by optimising the manoeuvre with respect to some pre-specified cost function. Despite the numerous optimisation tools available, the use of optimal control theory to tackle the MPP has had little impact on practical applications, presumably because the delicate numerical treatment of optimal control problems is often less suited to practical implementation than other methods. However, since the development of geometric control theory, new approaches have arisen, distinguished from numerical methods, in that they exploit the systems underlying analytic structure.

The use of optimal control theory to tackle the MPP for nonholonomic systems on Lie groups is studied in [1,6,7,11,8,5]. In Brockett's seminal papers [6,7] it is shown that a number of these optimal control problems are completely solvable in closed form where the optimal angular velocities can be expressed as analytic functions such as trigonometric or Jacobi elliptic functions. In this paper the focus is placed on the group $SE(n)$ where $n = 2, 3$ which have direct applications to robotic systems. The cost function used is an integral function of angular velocities. The function yields smooth curves where the velocity along the curve is minimised and therefore the forces required to track are theoretically small. The emphasis in this paper is placed on integrable cases that not only lead to closed form expressions for the optimal velocity inputs but also closed form expression for the corresponding motions. Contrary to integrable systems in mechanics, which are few and far between, this setting allows us to render the closed-loop system integrable by appropriate manipulation of the cost function. This can be achieved by setting weights of the cost function to be equal which essentially introduces enough symmetry into the resulting Hamiltonian system for it to be integrable. However, setting weights to be equal reduces the generality of the cost function and therefore the class of solution. Reducing the MPP to an analytic closed form solution essentially reduces the MPP to a problem of optimising the available parameters of these analytic functions to match prescribed boundary conditions. In addition an iterative approach can be used to select the parameters to avoid stationary and known obstacles.

1.1   Nonholonomic Systems on the Euclidean Groups $SE(n)$

From a practical point of view the kinematics of robotic systems can often be framed as nonholonomic control systems on the Euclidean Lie groups $SE(2)$

(position and orientation in the plane) and $SE(3)$ (position and orientation in space). Examples of these are the "vertical" rolling disk or Unicycle [8], hopping robots [10] on $SE(2)$ and Autonomous Underwater Vehicles [15], Unmanned Air Vehicles [12,4] and robotic manipulators [19] on $SE(3)$. The nonholonomic kinematics of these systems can be expressed as:

$$\frac{dg(t)}{dt} = \sum_{i=1}^{s} u_i(t)X_i \qquad (1)$$

where the curve $g(t) \in SE(n)$ describes the motions of the system in the configuration manifold $SE(n)$. $X_1, ..., X_n$ are arbitrary vector fields in the tangent space $TSE(n)$ at $g(t)$, denoted $T_{g(t)}SE(n)$ and $u_1, ..., u_s$ are the control functions. It follows that $X_1, ..., X_s \in T_{g(t)}SE(n)$ are the controlled vector fields on the manifold $SE(n)$ and $X_0 = \sum_{i=1}^{m} X_i \in T_{g(t)}SE(n)$ is the drift vector on $SE(n)$. For nonholonomic systems $n > s$. The cases considered are when X_i are left (respectively right) invariant vector fields on $SE(n)$, see [11], the vector fields can be expressed as $X_1 = g(t)A_1, ..., X_n = g(t)A_n \in T_{g(t)}SE(n)$, where $A_1, ..., A_n \in T_I SE(n)$ are basis elements of the tangent space at the identity $I \in SE(n)$. The tangent space at the identity is called the Lie algebra denoted $\mathfrak{se}(n)$. It follows that the differential equation (1) can be expressed as:

$$\frac{dg(t)}{dt} = g(t)(\sum_{i=1}^{s} u_i(t)A_i) \qquad (2)$$

where $A_1, ..., A_n \in \mathfrak{se}(n)$ where the Lie algebra $\mathfrak{se}(n)$, is a vector space together with the matrix commutator, the Lie bracket:

$$[X, Y] = XY - YX \qquad (3)$$

where $X, Y \in \mathfrak{se}(n)$ where for $n = 2$ the basis of the Lie algebra is

$$A_1 = \begin{pmatrix} 0 & 0 & 1 \\ 0 & 0 & 0 \\ 0 & 0 & 0 \end{pmatrix}, \ A_2 = \begin{pmatrix} 0 & 0 & 0 \\ 0 & 0 & 1 \\ 0 & 0 & 0 \end{pmatrix}, \ A_3 = \begin{pmatrix} 0 & -1 & 0 \\ 1 & 0 & 0 \\ 0 & 0 & 0 \end{pmatrix} \qquad (4)$$

satisfying the following relations

$$[A_1, A_3] = -A_2, \ [A_2, A_3] = A_1, \ [A_1, A_2] = 0. \qquad (5)$$

and when $n = 3$ the basis for the six-dimensional Lie algebra is:

$$A_1 = \begin{pmatrix} 0 & 0 & 0 & 0 \\ 0 & 0 & 0 & 0 \\ 0 & 0 & 0 & -1 \\ 0 & 0 & 1 & 0 \end{pmatrix}, \ A_2 = \begin{pmatrix} 0 & 0 & 0 & 0 \\ 0 & 0 & 0 & 1 \\ 0 & 0 & 0 & 0 \\ 0 & -1 & 0 & 0 \end{pmatrix}, \ A_3 = \begin{pmatrix} 0 & 0 & 0 & 0 \\ 0 & 0 & -1 & 0 \\ 0 & 1 & 0 & 0 \\ 0 & 0 & 0 & 0 \end{pmatrix}$$

$$A_4 = \begin{pmatrix} 0 & 0 & 0 & 0 \\ 1 & 0 & 0 & 0 \\ 0 & 0 & 0 & 0 \\ 0 & 0 & 0 & 0 \end{pmatrix}, \ A_5 = \begin{pmatrix} 0 & 0 & 0 & 0 \\ 0 & 0 & 0 & 0 \\ 1 & 0 & 0 & 0 \\ 0 & 0 & 0 & 0 \end{pmatrix}, \ A_6 = \begin{pmatrix} 0 & 0 & 0 & 0 \\ 0 & 0 & 0 & 0 \\ 0 & 0 & 0 & 0 \\ 1 & 0 & 0 & 0 \end{pmatrix}$$

and satisfies the commutative table:

[,]	A_1	A_2	A_3	A_4	A_5	A_6
A_1	0	A_3	$-A_2$	0	A_6	$-A_5$
A_2	$-A_3$	0	A_1	$-A_6$	0	A_4
A_3	A_2	$-A_1$	0	A_5	$-A_4$	0
A_4	0	A_6	$-A_5$	0	0	0
A_5	$-A_6$	0	A_4	0	0	0
A_6	A_5	$-A_4$	0	0	0	0

Note that the driftless system in (2) can be augmented to include systems with drift by setting one of the controls u_i to a constant a priori. The method described in this paper is also applicable to these cases. Controllability for systems of the form (2) can be assessed through computations performed at the level of the Lie algebra involving the Lie bracket (3), see [11,5]. For example if $[X, Y] = Z$ for some $Z \in \mathfrak{se}(n)$, then it is possible to move in the direction of the vector field Z by controlling only the vector fields X and Y. From a control theory viewpoint this is particularly useful if $Z \notin span\{X, Y\}$. Indeed if it is possible to directly control the vector fields $X_1, ..., X_s$ then motions can be generated in the direction of its iterated Lie brackets:

$$X_i, [X_i, X_j], [X_i, [X_j, X_k]], ..., \tag{6}$$

$1 \leqslant i, j, k, ... \leqslant s$. In other words, although not all directions are controlled, it may be possible to obtain motions in all of them, by taking sufficiently many Lie brackets. If the system is controllable then it is possible to construct a well defined optimal control problem. This is formalised in the following subsection.

1.2 Problem Statement

Subject to the kinematic nonholonomic constraint given by (2) and given that the system is controllable the problem is then to find a trajectory $g(t) \in SE(n)$ from an initial position and orientation $g(0) \in SE(n)$ to a final position and orientation $g(T) \in SE(n)$ where T is some fixed final time that minimises the functional

$$J = \frac{1}{2} \int_0^T c_i u_i^2 dt \tag{7}$$

where $i = 1, ..., s$ and c_i are constant weights. This cost function is not conventional but it is meaningful as it ensures smooth motions which, given a reasonably long enough fixed final time, will not require large torques and forces to track them. In addition it enables the MPP to be formulated in the context of geometric optimal control and this enables us to ask questions of integrability and in some cases solve the system in closed form. Furthermore, obtaining a closed form solution essentially reduces the MPP to a problem of optimising the available parameters to match the prescribed boundary conditions.

2   Methodology

The methodology for MPP comprises of the following phases:

1. Lifting the optimal control problem on $SE(n)$ to a Hamiltonian setting via the maximum principle of optimal control and Poisson calculus.
2. Solving integrable cases of the Hamiltonian vector fields analytically in the most general form of the cost function (7).
3. Given the optimal velocities derive the corresponding motions in $SE(n)$ analytically reducing the MPP to a parameter optimisation problem.
4. As the boundary conditions are not contained in the cost function it is necessary to numerically optimise the available parameters of the analytic solutions to match the prescribed boundary conditions (This stage is not covered in this paper.)

2.1   General Hamiltonian Lift on SE(n)

The application of the coordinate free Maximum Principle to left-invariant optimal control problems are well known, see [20], [11]. As the Hamiltonian is left-invariant the cotangent bundle $T^*SE(n)$ can be realised as the direct product $SE(3) \times \mathfrak{se}(n)^*$ where $\mathfrak{se}(n)^*$ is the dual of the Lie algebra $\mathfrak{se}(n)$ of $SE(n)$. Therefore, the original Hamiltonian defined on $T^*SE(n)$ can be expressed as a reduced Hamiltonian on the dual of the Lie algebra $\mathfrak{se}(n)^*$ as $T^*SE(n)/SE(n) \cong \mathfrak{se}(n)^*$. The appropriate Hamiltonian for the constraint (2) with respect to minimizing the cost function (7) is given by (see [11] for details):

$$H(p, u, g) = \sum_{i=1}^{s} u_i p(g(t)A_i) - \rho_0 \frac{1}{2} \sum_{i=1}^{s} c_i u_i^2 \qquad (8)$$

where $p \in T^*SE(n)$ (where $n = 1$ or $n = 2$) and $\rho_0 = 1$ for regular extremals and $\rho_0 = 0$ for abnormal extremals. In this paper we consider only the regular extremals, therefore we set $\rho_0 = 1$. The Hamiltonian (8) defined on $T^*SE(n)$ can be expressed as a reduced Hamiltonian on the dual of the Lie algebra $\mathfrak{se}(n)^*$. It follows that $p(g(t)A_i) = \hat{p}(A_i)$ for any $p = (g(t), \hat{p})$ and any $A_i \in \mathfrak{se}(n)$. Defining the extremal (linear) functions explicitly as $\lambda_i = \hat{p}(A_i)$, where $\hat{p} \in \mathfrak{se}^*(n)$ the Hamiltonian (8) can be expressed on $\mathfrak{se}(n)^*$ as

$$H = \sum_{i=1}^{s} u_i \lambda_i - \frac{1}{2} \sum_{i=1}^{s} c_i u_i^2 \qquad (9)$$

Through the Maximum Principle and the fact that the control Hamiltonian (9) is a concave function of the control functions u_i, it follows by calculating $\frac{\partial H}{\partial u_i} = 0$ and $\frac{\partial H}{\partial v_i} = 0$ that the optimal kinematic control inputs are:

$$u_i^* = \frac{1}{c_i} \lambda_i, \qquad (10)$$

where $i = 1, ..., s$. Substituting (10) back into (9) gives the appropriate left-invariant quadratic Hamiltonian:

$$H = \frac{1}{2}(\sum_{i=1}^{s} \frac{\lambda_i^2}{c_i}) \tag{11}$$

where λ_i are the extremal curves. For each quadratic Hamiltonian (11), the corresponding vector fields are calculated using the Poisson bracket $\{\hat{p}(\cdot), \hat{p}(\cdot)\} = -\hat{p}([\cdot, \cdot])$ where $(\cdot) \in \mathfrak{se}(n)$. Then the Hamiltonian vector fields are given by:

$$\frac{d(\cdot)}{dt} = \{\cdot, H\} \tag{12}$$

where $(\cdot) \in \mathfrak{se}(3)^*$. Finally, substituting (10) into (2) yields:

$$\frac{dg(t)}{dt} = g(t)\nabla H \tag{13}$$

where ∇H is the gradient of the Hamiltonian and $g(t) \in SE(n)$ are the corresponding paths. The MPP is thus reduced to solving for $g(t) \in SE(n)$ such that the boundary conditions $g(0) \in SE(n)$ and $g(T) \in SE(n)$ in some final time T are matched.

2.2   A Note on Integrability

The motion planning problem has been reduced to the problem of finding solutions to the equations (13) where $g(t) \in SE(n)$ with $n = 1$ or $n = 2$. The natural question to now ask is if it is possible to reduce the equations to quadratures (as integrals of analytic functions) or better still solve them in closed form. In order for the equations to be at least reducible to quadratures the system has to be integrable. Moreover, a Hamiltonian function on a symplectic manifold N of dimension $2n$ is said to be integrable if there exist constant functions $\varphi_2, ..., \varphi_n$ on N that together with the Hamiltonian $H = \varphi_1$ satisfy the following two properties:

- $\varphi_1, ..., \varphi_n$ are functionally independent i.e the differentials $d\varphi_1, ..., d\varphi_n$ are linearly independent for an open subset of N.
- The functions $\varphi_1, ..., \varphi_n$ Poisson commute with each other.

Thus, in identifying the $(n - 1)$ functions φ_i the Hamiltonian function is completely integrable and in general the system can be reduced to quadratures and in some cases can be solved in closed form. The implications of this are that an approximately optimal analytic solution can be derived (when the system can be reduced to quaratures) and an exactly optimal analytic solution when completely solvable. In terms of the motion planning problem if the system is integrable it means that the problem can be reduced to one of parameter optimisation where the initial values of the extremal curves can be optimised to match prescribed boundary conditions. For all left-invariant Hamiltonian systems defined on $SE(2)$ and $SE(3)$ we can be more specific about integrability:

Lemma 1. *For any left (respectively right) invariant Hamiltonian system defined on $SE(2)$, there exist three integrals of motion, the Hamiltonian H the Casimir function $M = \lambda_1^2 + \lambda_2^2$, and the integral of motion φ_3 corresponding to a right-invariant vector field.*

Lemma 2. *For any left (respectively right) invariant Hamiltonian system defined on $SE(3)$, there exist five constants of motion the Hamiltonian H the Casimir functions $I_2 = \lambda_1^2 + \lambda_2^2 + \lambda_3^2$ and $I_3 = \lambda_1\lambda_4 + \lambda_2\lambda_5 + \lambda_3\lambda_6$ and the right-invariant vector fields φ_4, φ_5.*

details of this can be found in [11]. This implies that left-invariant systems on $SE(2)$ are always integrable and on $SE(3)$ integrable if an additional constant of motion can be found. However, it is easy to induce an additional conserved quantity by introducing a symmetry into the problem by setting two of the weights of the cost function (7) to be equal. Although this restricts the generality of the cost function it ensures that the system is integrable. This procedure of manipulating the cost function is illustrated in example 2 of this paper. For integrable cases on $SE(2)$ and $SE(3)$ it is often possible to solve these systems in closed form as shown in the following example. In addition integrability is an intrinsic property of the system as it implies that all motions will be regular as opposed to chaotic. This ensures that all of the reference motions derived will be regular.

3   Examples

3.1   Example 1-The Wheeled Robot

A wheeled robot's configuration space can be described by a curve $g(t) \in SE(2)$

$$g(t) = \begin{pmatrix} R(t) & x \\ 0 & 1 \end{pmatrix} \tag{14}$$

where $R(t) \in SO(2)$ and $x \in \Re^2$, where

$$R(t) = \begin{pmatrix} \cos\theta & -\sin\theta \\ \sin\theta & \cos\theta \end{pmatrix} \tag{15}$$

we assume that the wheeled robot has a sliding constraint and can move backward or forwards at a velocity s which can be controlled. This velocity constraint can be expressed as:

$$\frac{dx}{dt} = R(t) \begin{bmatrix} s \\ 0 \end{bmatrix} \tag{16}$$

Furthermore, the robot can rotate at an angular velocity $u = \dot{\theta}$. Differentiating equation (14) and taking into the account the constraint (16) it is easily shown that the nonholonomic kinematic constraint can be expressed as a left-invariant differential equation:

$$g(t)^{-1}\frac{dg(t)}{dt} = \begin{pmatrix} 0 & -u & s \\ u & 0 & 0 \\ 0 & 0 & 0 \end{pmatrix} \tag{17}$$

this can be expressed in the form:

$$g(t)^{-1}\frac{dg(t)}{dt} = sA_1 + uA_3 \tag{18}$$

where the Lie algebra is given by (4) and the cost function (7) is given by:

$$J = \int_0^1 s^2 + cu^2 dt \tag{19}$$

where c is a constant weight and the time t is scaled such that in real time τ with final fixed time t_f is $t = \tau/t_f$. In relation to the general form (7) $u_1 = s, c_1 = 1, u_2 = u, c_2 = c$. The Hamiltonian function corresponding to the constraint (17) that minimises the cost function (19) is:

$$H = s\lambda_1 + u\lambda_3 - \frac{1}{2}(s^2 + cu^2) \tag{20}$$

then Pontragin's maximum principle says that if

$$\frac{\partial H}{\partial s} = 0, \frac{\partial H}{\partial v} = 0, \frac{\partial^2 H}{\partial s^2} < 0, \frac{\partial^2 H}{\partial v^2} < 0, \tag{21}$$

then the functions s and u are optimal. These conditions are satisfied if:

$$s = \lambda_1, u = \frac{\lambda_3}{c} \tag{22}$$

substituting these values into (20) yields the optimal Hamiltonian H^*:

$$H^* = \frac{1}{2}\left(\lambda_1^2 + \frac{\lambda_3^2}{c}\right) \tag{23}$$

The corresponding Hamiltonian vector fields which implicitly define the extremal solutions are given by the Poisson bracket $\frac{d\lambda_i}{dt} = \{\lambda_i, H\}$ this yields the differential equations:

$$\begin{aligned} \dot{\lambda}_1 &= \frac{\lambda_2\lambda_3}{c}, \\ \dot{\lambda}_2 &= -\frac{\lambda_1\lambda_3}{c}, \\ \dot{\lambda}_3 &= -\lambda_1\lambda_2. \end{aligned} \tag{24}$$

in addition observe that the Casimir function

$$M = \lambda_1^2 + \lambda_2^2 \tag{25}$$

is constant along the Hamiltonian flow i.e. $\{M, H^*\}$. These extremal curves can be solved analytically and allow us to state the following Lemma:

Lemma 3. *The optimal velocity s in the surge direction and angular velocity u that minimises the cost function (19) subject to the kinematic constraint (17) are Jacobi elliptic functions $sn(\cdot, \cdot), dn(\cdot, \cdot)$ of the form:*

$$\begin{aligned} s &= \sqrt{M} sn\left(\frac{\sqrt{2H^*}}{\sqrt{cH^*}}t, \frac{M}{2H^*}\right) \\ u &= \sqrt{\frac{2H^*}{c}} dn\left(\frac{\sqrt{2H^*}}{\sqrt{cH^*}}t, \frac{M}{2H^*}\right) \end{aligned} \tag{26}$$

where H^* and M are constants defined by (23) and (25) respectively and c is the constant weight in the cost function (19) with the corresponding path:

$$
\begin{aligned}
x_1 &= -\frac{\sqrt{2}\sqrt{cH}}{\sqrt{M}}dn\phi \\
x_2 &= \frac{2Ht}{\sqrt{M}} - \sqrt{M}\left(\frac{\sqrt{cH}E\left(am\phi, \frac{M}{2H}\right)\sqrt{2-\frac{Msn^2\phi}{H}}}{Mdn\phi}\right)
\end{aligned}
\tag{27}
$$

where $E(\cdot,\cdot)$ is the elliptic integral of the second kind and $am(\cdot)$ is the Jacobi amplitude and where the rotation of the body along the path is:

$$
R(t) = \begin{pmatrix} cn\phi & -sn\phi \\ sn\phi & cn\phi \end{pmatrix}
\tag{28}
$$

with $\phi = \left(\frac{\sqrt{2H}}{\sqrt{cH}}t, \frac{M}{2H}\right)$.

Proof.
The conserved quantities (25) can be parameterised by the Jacobi elliptic functions:

$$
\lambda_1 = rsn\left(\alpha t, m\right), \lambda_2 = rcn\left(\alpha t, m\right)
\tag{29}
$$

then substituting (29) into (25) yields $r = \sqrt{M}$. Then (23) can be parameterised by defining:

$$
\lambda_3 = \sqrt{M}pdn\left(\alpha t, m\right)
\tag{30}
$$

substituting (30) and (29) into (23) gives $m = \frac{M}{2H}$ and $p = mc$, then:

$$
\lambda_1 = rsn\left(\alpha t, \frac{M}{2H}\right), \lambda_2 = rcn\left(\alpha t, \frac{M}{2H}\right) \lambda_3 = \sqrt{2H}cdn\left(\alpha t, \frac{M}{2H}\right)
\tag{31}
$$

finally to obtain α substitute (31) into (24) which yields:

$$
\lambda_1 = rsn\left(\frac{\sqrt{2H}}{\sqrt{cH}}t, \frac{M}{2H}\right), \lambda_2 = rcn\left(\frac{\sqrt{2H}}{\sqrt{cH}}t, \frac{M}{2H}\right) \lambda_3 = \sqrt{2H}cdn\left(\frac{\sqrt{2H}}{\sqrt{cH}}t, \frac{M}{2H}\right)
\tag{32}
$$

then from (22) yields (26), as $u = \dot{\theta}$ it follows from (26) that:

$$
\theta = am(\phi) + C_1
\tag{33}
$$

where C_1 is a constant of integration. For simplicity we set $C_1 = 0$ such that the rotation matrix $R(t)$ emanates from the origin. This yields (28) then substituting (28) and (26) into equation (16) yields:

$$
\begin{aligned}
\frac{dx_1}{dt} &= \sqrt{M}sn\phi cn\phi \\
\frac{dx_2}{dt} &= \sqrt{M}sn^2\phi
\end{aligned}
\tag{34}
$$

these can be integrated analytically to yield (27). $\square$ The final step in the procedure would be to optimise the parameters to match prescribed boundary conditions $g(0) = g_0$ and $g(T) = g_T$.

3.2   Example 2-A Conventional Autonomous Underwater Vehicle (AUV)

This example is taken from [2] where a conventional slender AUV travels at arbitrary speed $v = \frac{d\gamma}{dt}$ constrained to travel in the surge direction and where the lateral motions are damped out quickly due to viscous friction. The kinematics of the AUV are then approximately described by

$$\frac{dg(t)}{dt} = g(t)(u_4 A_4 + u_1 A_1 + u_2 A_2 + u_3 A_3) \tag{35}$$

where u_4 is the translational velocity in the surge direction and u_1, u_2, u_3 are the angular velocities in the yaw, pitch and roll directions and $g(t) \in SE(3)$ describes the configuration space of the vehicle (position and orientation). The basis A_1, A_2, A_3, A_4, A_5, A_6 of the lie algebra $\mathfrak{se}(3)$ correspond physically to infinitesimal motion of the AUV in the yaw, pitch, roll, surge, sway and heave directions respectively. An application of the maximum principle described in Section 2 leads to the optimal velocity inputs $u_1 = \lambda_1/c_1, u_2 = \lambda_2/c_2, u_3 = \lambda_3/c_3, u_4 = \lambda_1/c_4$ where λ_i are defined implicitly by the Hamiltonian vector fields:

$$\begin{cases} \dfrac{d\lambda_1}{dt} = \dfrac{-\lambda_2\lambda_3}{c_2} + \dfrac{\lambda_2\lambda_3}{c_3}, & \dfrac{d\lambda_2}{dt} = \dfrac{\lambda_1\lambda_3}{c_1} - \dfrac{\lambda_1\lambda_3}{c_3} + \dfrac{\lambda_4\lambda_6}{c_4} \\[2mm] \dfrac{d\lambda_3}{dt} = \dfrac{-\lambda_1\lambda_2}{c_1} + \dfrac{\lambda_1\lambda_2}{c_2} - \dfrac{\lambda_4\lambda_5}{c_4}, & \dfrac{d\lambda_4}{dt} = \dfrac{-\lambda_2\lambda_6}{c_2} + \dfrac{\lambda_5\lambda_3}{c_3} \\[2mm] \dfrac{d\lambda_5}{dt} = \dfrac{\lambda_1\lambda_6}{c_1} - \dfrac{\lambda_4\lambda_3}{c_3}, & \dfrac{d\lambda_6}{dt} = -\dfrac{\lambda_1\lambda_5}{c_1} + \dfrac{\lambda_4\lambda_2}{c_2} \end{cases} \tag{36}$$

these equations are not integrable as an additional conserved quantity is required for integrability. Therefore, to solve these equations numerical integration methods are required. However, in this case the system cannot be reduced to a problem of parameter optimisation to match the boundary conditions. However, we can manipulate the weights of the cost function to render the system integrable. To demonstrate how this system can be solved by manipulation of the cost function we enforce a constraint on the weights of the cost function $c_1 = c_2 = c_3 = 1$ to yield :

$$\begin{cases} \dfrac{d\lambda_1}{dt} = 0, & \dfrac{d\lambda_2}{dt} = \dfrac{\lambda_4\lambda_6}{c_4} \\[2mm] \dfrac{d\lambda_3}{dt} = -\dfrac{\lambda_4\lambda_5}{c_4}, & \dfrac{d\lambda_4}{dt} = -\lambda_2\lambda_6 + \lambda_5\lambda_3 \\[2mm] \dfrac{d\lambda_5}{dt} = \lambda_1\lambda_6 - \lambda_4\lambda_3, & \dfrac{d\lambda_6}{dt} = -\lambda_1\lambda_5 + \lambda_4\lambda_2 \end{cases} \tag{37}$$

by exploiting the fact that λ_1 is constant call $\lambda_1(0)$ and using the conserved quantity $I_2 = \lambda_1^2 + \lambda_2^2 + \lambda_3^2$ which for this case is $I_2 - \lambda_1^2(0) = \lambda_2^2 + \lambda_3^2$ suggests using polar coordinates $\lambda_2 = r\sin\theta, \lambda_3 = r\cos\theta$. It is then easily shown that the solution to (37) is:

$$\begin{array}{lll} \lambda_1 = \lambda_1(0), & \lambda_2 = r\sin\vartheta t, & \lambda_3 = r\cos\vartheta t \\ \lambda_4 = \lambda_4(0), & \lambda_5 = s\sin\vartheta t, & \lambda_6 = s\cos\vartheta t \end{array} \tag{38}$$

where the constants r, s, ϑ are defined as:

$$r = I_2 - \lambda_1^2(0), \quad s = \frac{rc_4}{\lambda_4(0)}(\lambda_1(0) - \lambda_4(0)r), \quad \vartheta = \lambda_1(0) - \frac{\lambda_4(0)r}{s} \quad (39)$$

to compute the corresponding motions $g(t) \in SE(3)$ analytically can then be obtained by following the procedure in [3]. This essentially reduces the MPP to a problem of parameter optimisation to match the boundary conditions.

4   Conclusion

In this paper a procedure is presented for reducing the complexity of the motion planning problem for some robotic systems (whose kinematics can be represented by left-invariant vector fields). The procedure assigns a meaningful cost function to the systems (nonholonomic) kinematics which ensures feasible and smooth motions. Furthermore, it is shown how it is possible to manipulate the weights of the quadratic cost function to obtain an analytic solution to this optimal control problem. This illustrates that for many robotic motion planning problems it is possible to construct analytically defined optimal motions in closed-form. This essentially reduces the motion planning problem of robotic systems to a problem of optimising the parameters of analytic functions to match prescribed boundary conditions. Future work will develop this method to take into account obstacle avoidance by using an iterative approach to select the parameters of the analytically defined curves.

References

1. Baillieul, J.: Geometric Methods for Nonlinear Optimal control problems. Journal of Optimization Theory and Applications 25(4) (1978)
2. Biggs, J.D., Holderbaum, W.: Optimal kinematic control of an autonomous underwater vehicle. IEEE Transactions on Automatic Control 54(7), 1623–1626 (2009)
3. Biggs, J., Holderbaum, W.: Integrable quadratic Hamiltonians on the Euclidean group of motions. Journal of Dynamical and Control Systems 16(3), 301–317 (2010)
4. Biggs, J.D., Holderbaum, W., Jurdjevic, V.: Singularities of Optimal Control Problems on some 6-D Lie groups. IEEE Transactions on Automatic Control, 1027–1038 (2007)
5. Bloch, A.M.: Nonholonomic Mechanics and Control. Springer, New York (2003)
6. Brockett, R.W.: Lie Theory and Control Systems Defined on Spheres. SIAM Journal on Applied Mathematics 25(2), 213–225 (1973)
7. Brockett, R.W., Dai, L.: Non-holonomic Kinematics and the Role of Elliptic Functions in Constructive Controllability. In: Nonholonomic Motion Planning, Kluwer Academic Publishers, Dordrecht (1993)
8. Bullo, F., Lewis, A.: Geometric Control of Mechanical Systems. Springer, NY (2005)
9. Canny, J.F.: The Complexity of Robot Motion Planning. MIT Press, Cambridge (1998)
10. Carinena, J., Ramos, A.: Lie systems in Control Theory. In: Nonlinear Geometric Control Theory. World Scientific, Singapore (2002)
11. Jurdjevic, V.: Geometric Control Theory. Cambridge University Press, Cambridge (1997)

12. Justh, E.W., Krishnaprasad, P.S.: Natural frames and interacting particles in three dimension. In: Proceedings of 44th IEEE Conf. on Decision and Control and the European Control Conference (2005)
13. Lafferriere, G., Sussmann, H.: Motion Planning for Controllable systems without drift. In: Proc. IEEE Int. Conf. on Robotics and Automation, Sacramento, CA, pp. 1148–1153 (1991)
14. Latombe, J.C.: Robot Motion Planning. Kluwer, Boston (1991)
15. Leonard, N., Krishnaprasad, P.S.: Motion control of drift free, left-invariant systems on Lie groups. IEEE Transactions on Automatic Control 40, 1539–1554 (1995)
16. Luenberger, D.: Linear and Nonlinear Programming. Addison-Wesley Publishing Company, Reading (1984)
17. Murray, R., Li, Z., Sastry, S.: A Mathematical Introduction to Robotic Manipulation. CRC Press, Boca Raton (1994)
18. Murray, R., Sastry, S.: Steering nonholonomic systems using sinusoids. In: Proc. of 29th IEEE Conf. on Decision and Control, Honolulu, Hawaii, pp. 2097–2101 (1990)
19. Selig, J.M.: Geometric Fundamentals of Robotics. Monographs in Computer Science. Springer, Heidelberg (2006)
20. Sussmann, H.J.: An introduction to the coordinate-free maximum principle. In: Jakubczyk, B., Respondek, W. (eds.) Geometry of Feedback and Optimal Control, pp. 463–557. Marcel Dekker, New York (1997)
21. Sussmann, H.J., Liu, W.: Limits of highly oscillatory controls and approximations of general paths by admissible trajectories. In: Proc. of 30th IEEE Conf. on Decision and Control, pp. 437–442 (1991)
22. Zexiang, L., Canny, J.F.: Nonholonomic Motion Planning. Kluwer Academic Publishers, Dordrecht (1993)

Probabilistic Logic Reasoning about Traffic Scenes

Carlos R.C. Souza[1] and Paulo E. Santos[2]

[1] Mercedes-Benz do Brasil, São Paulo, Brazil
carlos.roberto@daimler.com
[2] Centro Universitario da FEI, São Paulo, Brazil
santosp@ieee.org

Abstract. This paper describes a probabilistic logic reasoning system for traffic scenes based on Markov logic network, whose goal is to provide a high-level interpretation of localisation and behaviour of a vehicle on the road. This information can be used by a lane assistant agent within driver assistance systems. This work adopted an egocentric viewpoint for the vision and the reasoning tasks of the vehicle and a qualitative approach to spatial representation. Results with real data indicate good performance compared to the common sense interpretation of traffic situations.

1 Introduction

Lane changing is recognised as a significant cause of automotive accidents [4]. This fact justifies the development of active safety systems by car manufacturers. Active safety systems (e.g., *Lane Assistant System* or LAS) aim to prevent accidents by warning the driver through a visual or audio alarm, or even by controlling the vehicle's actuators. This paper contributes with a part of a LAS that is dedicated to a high-level interpretation scheme.

Commercial lane assistant systems are generally based on a monocular camera relying on quantitative measures of distance between the vehicle and the road divider [11]. In this paper, the perception of the traffic environment is sensed by a webcam attached to a vehicle. The dividers' discrete position (e.g., right, left) and type information (e.g., continuous, dashed) are extracted from an off-the-shelf vision system and used as evidence to infer localisation and behaviour of the vehicle. The uncertainties inherent to the sensors, actuators and to the real-world phenomena are treated with Markov logic networks (MLN) [1], which facilitates a complex representation of high-level knowledge (such as traffic rules) as well as of the domain uncertainties. MLN also permits us to estimate the probability of an event occurrence in the presence of faults or imprecision of the sensors from probabilistic inference.

In [3] Markov logic networks are also used to infer object relations in traffic scenes but, although the agent is immersed in its environment, the reasoning is global, i.e., it is done from a bird's eyes view. In turn, [15] implements an egocentric reasoning using Bayesian networks and description logics to model

R. Groß et al. (Eds.): TAROS 2011, LNAI 6856, pp. 219–230, 2011.
© Springer-Verlag Berlin Heidelberg 2011

the environment. However, the lack of expressiveness of description logics and the acyclicity constraint of Bayesian networks jeopardise scaling the model up to a practical application standard. Some of these issues are overcome in MLN, as we shall see in this paper.

This paper is organised as follows: in the next section we propose a framework for lane divider interpretation. On Section 3 we briefly describe the vision system. On Section 4 we present a brief overview of probabilistic logic and present the Markov logic network approach. Section 5 presents a formalisation of the traffic domain. On Section 6 we show some results with real data applied to our model of traffic environment. Section 7 concludes this work.

2   A Markov Logic Framework for a Lane Assistance System

A reasoning system was developed with Markov logic network (MLN) to infer the egocentric localisation of the vehicle relative to lane dividers and the vehicle's behaviour (e.g., whether it is driving on the wrong way, or crossing the lane). Data input to the reasoning system (identification, type and localisation of dividers) comes from a vision algorithm, which receives information from a webcam attached to a driving car. Both video processing and inference must be real time, with low resolution images from finite (reduced) domains this can be achieved. The proposed framework is designed as shown in Figure 1. This diagram is based on the work described in [11].

We can describe each layer of the diagram as follows:

1. *Perception*: a monocular camera that provides environment information. Here we use a Microsoft VX2000 webcam with 320x240 pixels of video resolution;
2. *Segmentation*: extracts features (edges) from the video obtained on Layer 1;
3. *Line Tracking*: identify lane dividers and track them;
4. *MLN Inference*: comprehends the model of the traffic domain and the inference methods for querying traffic conditions;
5. *Assistant Function*: based on the results of the previous layer, it decides what type of signal or message will be sent to the next layer. The implementation of this layer is outside the scope of this work;
6. *Actuator*: could be an audio alarm, a message or light in the panel or even an output from the main module to actuate on the vehicle itself.

Layers 2 and 3 are processed with Matlab and Layer 4 is an external program invoked by Matlab (as introduced in Section 4.1). A more complete description of each of these layers follows below.

3   Vision System

The video stream used in this work has a resolution of 320x240 pixels and its frame rate is 30 Hz. Assuming the half-half methodology for training and testing

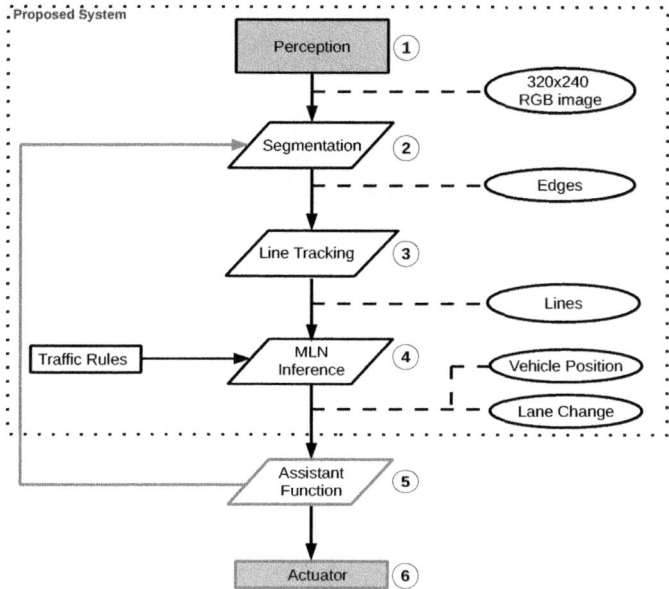

Fig. 1. LAS proposed based on the architecture of [11]

[2], odd frames were used to train the MLN and even frames were used for test and inference. Thus, in practice, we had a frame rate of 15Hz to process.

Basically the vision algorithm includes the following steps:

1. Take the lower half of RGB image (the region of interest (ROI)) convert it to intensity and apply a 2D filter with a vertical mask. For our video stream we changed the original mask to a Sobel vertical mask, which increased intensity of the edges of the image;
2. Binarise the filtered image and apply the Hough transform;
3. Clear the Hough matrix based in a window of ρ and θ and locate two local maxima; where ρ is the perpendicular distance from the lines detected to the image's origin and θ is the angle between ρ and the top of the image;
4. Track a set of lane dividers, matching the current detection with those lines of the previous frame;
5. Detect if there is a lane departure (a divider) under the vehicle. For the reasoning system it was adopted that every time a line passes through the bottom of the image, it is considered to be under the vehicle;
6. Take the RGB image and convert it to YCbCr. Detect the type and the colour of lane dividers by counting the percentage of white or yellow pixels on the lines;
7. Display the type and colour text of the lines and output a lane crossing warning text when it is true.

Fig. 2. Some frames obtained with the vision algorithm

A more detailed description of this vision algorithm can be found at [7]. Figure 2 presents some frames obtained with this algorithm.

The next section makes a brief overview of probabilistic logics and introduces the Markov logic framework.

4   Probabilistic Logics

For a long time in Artificial Intelligence, logic and probabilistic reasoning approaches were treated separately [14]. However, many tasks of the real-world require probabilistic reasoning about relational data representing multiply related objects with large sets of variables. This model of the world needs to be compactly represented in order to minimise representational and inferential complexities. Propositional probabilistic approaches such as Bayesian networks are insufficient to cope with these requirements, because they describe a fixed set of random variables, and specify dependencies and probability distributions for each variable individually. Thus, a significant number of *first-order probabilistic languages* (FOPL) have been proposed over the last decades, since they are able to compactly represent large sets of random variables by abstracting over objects [9].

In [9] these approaches are organised into a taxonomy of the outcome spaces to which the languages assign probabilities. In another work [10], these approaches were divided into two groups: *extensional* and *intensional* systems. The first one propagates truth values generalised from true or false to a scale of varying degrees of certainty. The second group places restrictions on a probability distribution on possible worlds.

There are further divisions within the intensional approaches, but the most predominant subgroup is the family of *Knowledge Base Model Construction* (KBMC), which constructs a propositional graphical model from a first-order language specification that answers a query [14]. Some KBMC approaches are based on Bayesian networks such as *Bayesian Logic* (BLOG) [8] or *Probabilistic Relational Models* (PRM) [6]. Other approaches have Markov networks as probabilistic model, such as *Relational Markov Networks* (RMN) [16] or *Markov Logic Networks* (MLN) [13].

Different from Bayesian networks, Markov networks are undirected models that use weights to define the relative probability of instantiations. In relational domains, some random variables can occasionally depend on one another without a clear causal relation. In this case, Markov network models have an advantage over Bayesian network because they have no acyclicity constraint, which simplifies the modelling of the problem. However, the disadvantage is that learning can be hard in undirected graphical models when using, for example, existentially quantified formulae.

Markov Logic Networks (MLNs) [13] have evolved rapidly in recent years, in general, due to its simple semantics while retaining the expressiveness of first-order logic. Accompanied by a well supported software (*Alchemy* [5]) and its large spectrum of real domain applications, MLNs are in an advanced stage of development in relation to any other FOPL approaches [14].

4.1   Markov Logic and Alchemy

Markov Logic Network (MLN) was developed aiming an unified language to be an interface layer for Artificial Intelligence. This interface should provide the basic infrastructure to the "foundational" areas of AI such as knowledge representation, automated reasoning, probabilistic models and machine learning. [1].

The basic idea of MLNs is to soften the restrictions imposed by a first-order knowledge base (KB). Each first-order formula has an associated weight that reflects how strong is the formula constraint: the higher the weight, the greater the difference in probability between a world that satisfies the formula and one that does not satisfy [1].

A Markov logic network L is a set of formulae F_i in first-order logic (e.g., $divider(YDashed, Right, t) \Rightarrow wrongWay(t))$ with a weight w_i (real number) attached to each formula. This can be viewed as a template for constructing Markov networks $M_{L,C}$, where C is a set of constants (e.g., for position: Left, Right, Centre, Under) and the probability distribution over possible worlds x is given by:

$$P(X=x) = \frac{1}{Z} exp \left(\sum_i w_i f_i(x) \right) = \frac{1}{Z} \prod_i \phi_i(x_{\{i\}})^{n_i(x)},$$

where $n_i(x)$ is the number of true groundings of F_i in x, $x_{\{i\}}$ is the state of the i-th clique which has a corresponding feature $f_i(x) \in \{0.1\}$ and an associated weight $w_i = log\phi_i(x_{\{i\}})$. Z is a normalisation factor also known as the partition function and is given by $Z = \sum_{x \in \mathcal{X}} \prod_i \phi_i(x_{\{i\}})$.

Inference in MLN can be probabilistic or logical, but are #P-Complete and NP-Complete respectively. However, a MLN allows encoding of knowledge, including context-specific independencies which makes inference more efficient. Moreover, it is also possible to use approximate inference methods in Markov networks such as the *Markov Chain Monte Carlo* (MCMC), with Gibbs sampling, to sample each variable in turn given its Markov blanket [13]. These and other algorithms for inference and learning in MLN are implemented in the open-source software *Alchemy* [5] which is used in this work.

Learning in MLN can be discriminative or generative and the algorithm more commonly used for both learning and inference is the MC-SAT. MC-SAT is a slice sampling MCMC which uses a combination of satisfiability testing and simulated annealing to sample from the slice of a distribution [1]. MC-SAT treats deterministic or near-deterministic dependencies by efficiently finding isolated modes in the distribution, resulting in a fast mixing of Markov chain. Alchemy has MC-SAT as default inference, but there are other algorithms available, such as belief propagation, Gibbs sampling and simulated tempering.

In the next section we describe our formalisation of the problem and the contextual information.

5   Contextual Information

Contextual information (in the case of traffic scenes) is the relevant information to understand the environment and it could be related to traffic rules or to attributes of things in the environment. One important characteristic of a Lane Assistance System (LAS) is the fact that the agent of perception is inside the context, e.g., the point of view of the agent is the same of the driver.

In order to model the traffic environment, we first have to answer some questions about normal driving:

- *What is important to know about the environment to keep the car on the lane?*
  Taking into account only the road, without obstacles, we visually follow the lane dividers on both sides of the vehicle, controlling it to stay at the centre of the lane.
- *What sort of identification gives us sense of direction or localisation on the road?*
  Shape and colour of a divider give us the sense of direction and position. We know that because there are traffic rules which give us conditions to reason about the lane dividers we are seeing. The rules concerning our problem domain are described in Table 2 below.

The following variables were defined to identify the environment:

- Type of lane divider: *type={YContinuous, WContinuous, YDashed, WDashed, Merge}*. Where *W* means white and *Y* means yellow. *Merge* is the line representing the access to enter the road and to depart from it, it is a dashed line with shorter intervals than the dashed lane dividers;
- Road direction: *way={One, Two}*, one or two-way road;
- Vehicle position: *lanepos={Left, Centre, Right}*. Vehicle position is relative to the road (e.g., "the vehicle is on the centre lane"). Centre is considered every position that is not in the extreme left or right lanes. Divider position is relative to the vehicle (e.g., "the divider is on the left side of the vehicle");
- Divider status: *status={Ok, Under}*. Indicates if a divider is under the vehicle (e.g., during a crossing) or at the vehicle sides (*Ok*).

Using these variables the following predicates were defined. Notice that they all represent facts about the ego-vehicle, thus the variable for the agent was omitted. In what follows the variable *time* represents the frame number at which the predicates hold.

- *dividerL/R(type,status,time)*: divider identification with type (*Dashed, Continuous*), its status (*ok, under*), its frame number (time) and its relative position to the car (*Left or Right: L/R*). Ground facts of *dividerL*/3 and *dividerR*/3 are output by the vision algorithm;
- *carRelPos(lanepos,time)*: car's relative position wrt the road (*lanepos*);
- *crossingLeft(time)*: vehicle is crossing to the left lane;
- *crossingRight(time)*: vehicle is crossing to the right lane;
- *emergencyLane(time)*: vehicle is on the emergency stop lane;
- *prohibitedManoeuvre(time)*: vehicle is doing a prohibited manoeuvre (e.g., crossing a yellow continuous divider);
- *wrongWay(time)*: vehicle is on the wrong way of the road.

With these predicates, MLN formulae were constructed to encode traffic rules (about right-handed traffic) and the knowledge about the environment as shown on Table 2. Note that initially these formulae are only first-order logical sentences without weight, the weights will be learnt from the data using MC-SAT on the related Markov network. Due lack of space we abbreviate the predicates in the formulae as shown in Table 1.

Table 1. Abbreviation of predicates and constants

dL	*dividerL*	pM	*prohibitedManoeuvre*	Le	*Left*
dR	*dividerR*	emg	*emergencyLane*	Ri	*Right*
cRP	*carRelPos*	t	*time*	WC	*WContinuous*
xL	*crossingLeft*	ty	*type*	YC	*YContinuous*
xR	*crossingRight*	s	*status*	WD	*WDashed*
rW	*roadWay*	Un	*Under*	YD	*YDashed*
wW	*wrongWay*	Ce	*Centre*	M	*Merge*

Predicates without specific constants (e.g., s, ty, t) will be grounded for all constants. The negation of some predicates on each formula was necessary to specify unchanged characteristics of the domain given the evidences.

In the next section we show some results of training and inference applying this model on real data.

6   Results

The weight learning for each formula was executed with MC-SAT algorithm [1] for the 1800 odd frames. The weight w_i learnt means that a world where n clauses of a formula are true is $e^{w_i n}$ less probable than a world where all clauses

Table 2. MLN formulae for LAS

---

1. If vehicle is on a yellow continuous divider or there is one at the right side, it is doing a prohibited manoeuvre.
$dL(YC, Un, t) \lor dR(YC, Ok, t) \Rightarrow pM(t) \land wW(t) \land rW(Two, t) \land cRP(Le, t) \land \neg cRP(Ce, t) \land \neg cRP(Ri, t) \land \neg emg(t)$

2. If there isn't evidence of a two way road, consider it to be a one way road.
$\neg dL(YC, s, t) \lor \neg dL(YD, s, t) \lor \neg dR(YC, s, t) \lor \neg dR(YD, s, t) \Rightarrow rW(One, t) \land \neg rW(Two, t)$

3. If there is a yellow continuous or yellow dashed divider at any position, the road has two ways.
$dL(YC, s, t) \lor dL(YD, s, t) \lor dR(YC, s, t) \lor dR(YD, s, t) \Rightarrow rW(Two, t) \land \neg rW(One, t) \land cRP(Le, t) \land \neg cRP(Ce, t) \land \neg cRP(Ri, t) \land \neg emg(t)$

4. If the left divider is white continuous and the right divider is white dashed, the road has one way and the vehicle is on the left lane.
$dL(WC, Ok, t) \land (dR(WD, Ok, t) \Rightarrow rW(One, t) \land \neg rW(Two, t) \land cRP(Le, t) \land \neg cRP(Ce, t) \land \neg cRP(Ri, t) \land \neg pM(t) \land \neg wW(t)$

5. If the divider is dashed on both sides of the view, the vehicle is on the central lane. Centre is considered any position that isn't the extreme right or left lanes.
$dL(WD, Ok, t) \land dR(WD, Ok, t) \Rightarrow cRP(Ce, t) \land \neg cRP(Le, t) \land \neg cRP(Ri, t) \land \neg wW(t) \land \neg xL(t) \land \neg xR(t) \land \neg pM(t) \land \neg emg(t)$

6. If the left divider is not white dashed and the right divider is white dashed, the vehicle is on the left lane.
$\neg dL(WD, Ok, t) \land dR(WD, Ok, t) \Rightarrow cRP(Le, t) \land \neg cRP(Ce, t) \land \neg cRP(Ri, t) \land \neg xL(t) \land \neg xR(t) \land \neg pM(t) \land \neg emg(t)$

7. If the divider is yellow dashed at the right side, the car is on the wrong way. It is not a prohibited manoeuvre because this condition is permitted for overtaking.
$dR(YD, Ok, t) \land dL(ty, Ok, t) \Rightarrow wW(t) \land cRP(Le, t) \land \neg cRP(Ce, t) \land \neg cRP(Ri, t) \land \neg xL(t) \land \neg xR(t) \land \neg pM(t) \land \neg emg(t)$

8. If the left divider is white dashed and the right divider is white continuous or merge, the vehicle is on the right lane.
$(dR(WC, Ok, t) \lor dR(M, Ok, t)) \land dL(WD, Ok, t) \Rightarrow cRP(Ri, t) \land \neg cRP(Ce, t) \land \neg cRP(Le, t) \land \neg wW(t) \land \neg xL(t) \land \neg xR(t) \land \neg emg(t)$

9. If a divider is white continuous and another is not white dashed, the vehicle can be on the emergency lane.
$dL(WC, Ok, t) \land \neg dR(WD, s, t) \Rightarrow emg(t) \land cRP(Ri, t) \land \neg cRP(Le, t) \land \neg cRP(Ce, t) \land \neg pM(t) \land \neg wW(t)$
$dR(WC, Ok, t) \land \neg dL(WD, s, t) \Rightarrow emg(t) \land cRP(Le, t) \land \neg cRP(Ri, t) \land \neg cRP(Ce, t) \land \neg pM(t) \land \neg wW(t)$

10. If the vehicle is over a right divider or over a left divider with a right divider Ok, the car is crossing to the right (and analogously for crossing to the left). This prevents one predicate overlapping with another when, as given by the vision system, the divider changes from left to right and vice-versa at the middle of the screen.
$dR(ty1, Un, t) \lor (dL(ty2, Un, t) \land dR(ty2, Ok, t)) \Rightarrow xR(t) \land \neg xL(t)$
$dL(ty1, Un, t) \lor (dR(ty2, Un, t) \land dL(ty2, Ok, t)) \Rightarrow xL(t) \land \neg xR(t)$

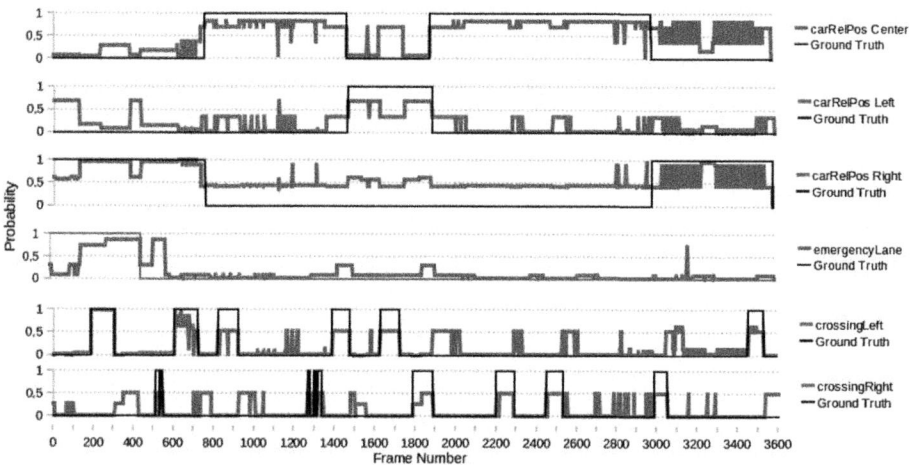

Fig. 3. Position and crossing probabilities querying the system at the even frames and the relative ground truth

are true. This is specially important to a lane assistance system, since it allows the system to provide plausible answers even with misclassified evidences.

During inference MC-SAT is also used with evidences generated by the vision algorithm on the 1800 even frames as we previously described on Section 2. The query was made on the predicates: carRelPos/2, crossingLeft/1, crossingRight/1, emergencyLane/1, prohibitedManoeuvre/1, wrongWay/1 and roadWay/2. The inference results were plotted on Figure 3 along with their relative ground truth.

We have used as decision threshold the probability greater than 50%. As we can note on Figure 3, this threshold would be higher for position predicates (*CarRelPos*), but the behaviour predicates (*crossingLeft/Right*) had lower probabilities than the others. However, the distinction between them were clearly defined (as shown on the two bottom graphs of Figure 3). The low probabilities of *CrossingLeft/Right* predicates is due to the fact that the vision system only sees the nearest lane dividers, thus in the middle of a *CrossingLeft* event (for instance) there is also evidence that this might be a *CrossingRight* event.

In the framework presented on Figure 1, we could treat this specific problem on layer five, analysing which predicate started the lane change and ignoring the other predicates right after its next occurrence. This, however, is an issue for further research.

The ground truth was manually labelled, considering a lane departure every time the inferior extremity of a divider appears at the bottom line of the frame. Crossing a divider was adopted until the divider changes the side wrt the vehicle (i.e., when it crosses the centre of the frame), analogously for lane position changes. We verified (as expected) that a specification of position during a lane change in the model degraded the prediction of *crossing* predicates. This is due to the nature of the Markov network, where the position predicates become part of the same clique of crossing predicates, thus contributing to the same probability distribution.

6.1   Evaluation

In order to evaluate our model we use a confusion matrix as defined in [2]: $\left(\begin{smallmatrix} tp & fn \\ fp & tn \end{smallmatrix}\right)$, where tp stands for true positive, fp is false positive, tn is true negative and fn is false negative. With a confusion matrix, for each of our queries, we measured the accuracy, sensitivity, precision and specificity of prediction.

Accuracy is given as: $accuracy = \frac{tp+tn}{(tp+tn+fp+fn)}$. The accuracy is the measure of reliability of the detection, i.e., from the total number of predictions how much were correct.

The sensitivity (or recall) is the fraction of the existing true predicates that were detected as such and is given by: $sensitivity = \frac{tp}{(tp+fn)}$.

The precision is a measure of accuracy of a specific class, in our case we are interested in the detection of true positive predicates, i.e., the algorithm reports position and behaviour of the vehicle without false detection. The precision is given as: $precision = \frac{tp}{(tp+fp)}$.

The specificity is a measure of how frequently the algorithm reports false predicates as true negative and is given by: $specificity = \frac{tn}{(tn+fp)}$.

On Table 3 we present the results obtained for the inferences using the 1800 even frames.

Table 3. Results obtained for the inference. The symbol * means that there was no true grounded predicates on the data and # means the same for false grounded predicates.

Ground Predicate	Accuracy (%)	Sensitivity (%)	Precision (%)	Specificity (%)
carRelPos(Centre)	75.5	77.8	74.9	73.1
carRelPos(Left)	91.0	56.2	61.0	95.4
carRelPos(Right)	84.2	55.1	96.2	98.9
crossingLeft	90.5	97.3	64.2	89.2
crossingRight	89.8	86.4	53.5	90.3
emergencyLane	96.0	66.7	82.9	98.7
prohibitedManoeuvre	100	*	*	100
wrongWay	100	*	*	100
roadWay(One)	100	100	100	#
roadWay(Two)	100	*	*	100
Mean	89.6	77.1	76.1	90.9

Although our video stream does not present situations of prohibited manoeuvre, wrong way driving or two way road, their predicates were 100% true negative. It is worth pointing out that there were misclassified evidences of yellow dividers in a few frames (in 10 consecutive frames of the dataset), and that even with this misclassified data the true negative rate was 100%.

For the roadWay(One) ground predicate, its precise detection can be attributed, in general, to the assertion: "if there isn't evidence for a two-way road, consider it as a one-way road" (Statement 2 in Table 2). It is indeed reasonable to say that if we are not seeing a yellow divider, we do not need to worry if the road has two ways.

In the next section, we discuss the results obtained and conclude this paper.

7   Concluding Remarks

In this paper, we proposed a formalisation in probabilistic logic of a lane assistance system. The results showed that this model presents coherent values given the evidences and that the distinction between cases were clearly defined. Low precision values for crossing predicates were due the change of divider's side during the crossing action. Low sensitivity for lateral lane positions (Left / Right) and emergency lane, can be associated to the misclassification of dividers by the vision algorithm.

The inference in our model considers only the current evidences, for that reason, some influence of misclassified dividers is observable primarily on lateral lanes where continuous dividers are sometimes detected as dashed.

Future work shall refine the model by using the event calculus [12] to reason about actions, in this work we focused on spatial reasoning (mainly on reasoning about the divider's positions). Further study can be done with Markov conditions on inference, i.e, with current and past evidences. In parallel, the vision algorithm has space for improvement. For example, type detection and tracking could have as parameters the vehicle's speed in order to automatically adjust values of filters or of decision thresholds.

Inference on Alchemy with this model was efficient but not real-time with MC-SAT algorithm (approximately 0.15 seconds per frame). We achieved real time processing changing the inference algorithm to belief propagation (0.02 seconds per frame) whereas analogous true (false) positive rates were obtained with both algorithms on this domain.

Acknowledgements. Paulo Santos acknowledges support from FAPESP project LogProb, 2008/03995-5, São Paulo, and also travel support from CNPq, Brazil.

References

1. Domingos, P., Lowd, D.: Markov Logic: an interface layer for artificial intelligence. Morgan & Claypool (2009)
2. Fitzpatrick, J.M., Sonka, M.: Handbook of Medical Imaging, 1st edn. Medical Image Processing and Analysis, vol. 2, ch. 10, pp. 567–605. SPIE Press, San Jose (2000)
3. Hensel, I., Bachmann, A., Hummel, B., Tran, Q.: Understanding object relations in traffic scenes. VISAPP (2010)
4. Holzmann, F.: Needs of improved assistant systems. In: Adaptive Cooperation between Driver and Assistant System, pp. 3–10. Springer, Heidelberg (2008)

5. Kok, S., Sumner, M., Richardson, M., Singla, P., Poon, H., Lowd, D., Domingos, P.: The alchemy system for statistical relational ai. Technical report, Department of computer science and engineering, university of Washington (2007), http://alchemy.cs.washington.edu
6. Koller, D., Pfeffer, A.: Probabilistic frame-based systems. In: Proceedings of AAAI, pp. 580–587. AAAI Press, Menlo Park (1998)
7. MathWorks: Lane departure warning system (May 2010), http://www.mathworks.com/products/viprocessing/demos.html?file=/ products/demos/shipping/vipblks/vipldws.html#5, revision: 1.1.6.3
8. Milch, B., Marthi, B., Russell, S., Sontag, D., Ong, D.L., Kolobov, A.: Blog: probabilistic models with unknown objects. In: IJCAI, pp. 1352–1359 (2005)
9. Milch, B., Russell, S.J.: First-order probabilistic languages: into the unknown. In: Muggleton, S.H., Otero, R., Tamaddoni-Nezhad, A. (eds.) ILP 2006. LNCS (LNAI), vol. 4455, pp. 10–24. Springer, Heidelberg (2007)
10. Pearl, J.: Probabilistic reasoning in intelligent systems: networks of plausible inference. Morgan Kaufmann, San Francisco (1988)
11. Reif, K. (ed.): Automobilelektronik, 3rd edn. Vieweg Teubner, Wiesbaden (2009)
12. Reiter, R.: Knowledge in action: logical foundations for specifying and implementing dynamical systems. The MIT Press, Massachusetts (2001); illustrated edition edn.
13. Richardson, M., Domingos, P.: Markov logic networks. Machine Learning 62(1-2), 107–136 (2006)
14. de Salvo Braz, R., Amir, E., Roth, D.: A survey of first-order probabilistic models. In: Holmes, D.E., Jain, L.C. (eds.) Innovations in Bayesian networks. SCI, vol. 156, pp. 289–317. Springer, Heidelberg (2008)
15. Santos, P., Cozman, F., Pereira, V.F., Hummel, B.: Probabilistic logic encoding of spatial domains. UniDL (2010)
16. Taskar, B., Abbeel, P., Koller, D.: Discriminative probabilistic models for relational data. In: Darwiche, A., Friedman, N. (eds.) UAI, pp. 485–492. Morgan Kaufmann, San Francisco (2002)

Real-World Reinforcement Learning for Autonomous Humanoid Robot Charging in a Home Environment

Nicolás Navarro, Cornelius Weber, and Stefan Wermter

Knowledge Technology, Department of Computer Science,
University of Hamburg, Hamburg, Germany
{navarro,weber,wermter}@informatik.uni-hamburg.de

Abstract. In this paper we investigate and develop a real-world reinforcement learning approach to autonomously recharge a humanoid Nao robot [1]. Using a supervised reinforcement learning approach, combined with a Gaussian distributed states activation, we are able to teach the robot to navigate towards a docking station, and thus extend the duration of autonomy of the Nao by recharging. The control concept is based on visual information provided by naomarks and six basic actions. It was developed and tested using a real Nao robot within a home environment scenario. No simulation was involved. This approach promises to be a robust way of implementing real-world reinforcement learning, has only few model assumptions and offers faster learning than conventional Q-learning or SARSA.

Keywords: Reinforcement Learning, SARSA, Humanoid Robots, Nao, Autonomous Docking, Real World.

1 Introduction

Reinforcement learning (RL) is a biologically supported learning paradigm [13,14,3], which allows an agent to learn through experience acquired by interaction with its environment. Reinforcement learning neural network architectures have an input layer, which represents the agent's current state, and an output layer, which represents the chosen action given a certain input.

Reinforcement learning algorithms usually begin with the agent's random initialization followed by many randomly executed actions until the agent eventually reaches the goal. Following a few successful trials the agent starts to learn action-state pairs based on its acquired knowledge. The learning is carried out using positive and negative feedback during the interaction with the environment in a trial and error fashion. In contrast with supervised and unsupervised learning, reinforcement learning does not use feedback for intermediate steps, but rather a reward (or punishment) is given only after a learning trial has been finished. The reward is a scalar and indicates whether the result was right or wrong (binary) or how right or wrong it was (real value). The limited feedback characteristics of this learning approach make it a relatively slow learning mechanism, but attractive due to its potential to learn action sequences.

R. Groß et al. (Eds.): TAROS 2011, LNAI 6856, pp. 231–240, 2011.
© Springer-Verlag Berlin Heidelberg 2011

In the literature, reinforcement learning is usually used within simulated environments or abstract problems [15,5,11]. Those kinds of problems require a model of the agent-environment dynamics, which it is not always available or easy to infer. Moreover, a number of assumptions, which are not always realistic, have to be made, e.g. action-state transition model, design of reward criterion, magnitude and kind of noise if any, etc.

On the other hand, real-world reinforcement learning approaches are scarce [2,6,7], mostly, because RL is expensive in data or learning steps and the state space tends to be large. Moreover, real-world problems present additional challenges, such as safety considerations, real time action execution, changing sensors, actuators and environmental conditions, among many others.

Several techniques to improve real-world learning capabilities of RL algorithms exist. *Dense reward functions* [2] provide performance information in intermediate steps to the agent. Another frequently used technique is manual *state space reduction* [2,7], which is a very time consuming task. Other approaches propose modification and exploitation of the agent's properties [6], which is not always possible. *Batch reinforcement learning* algorithms [7] use information from past state transitions, instead of only the last transition, to calculate the prediction error function; this is a powerful approach but a computationally demanding technique. A final example of these techniques is supervised reinforcement learning algorithms [2]. The supervision consists of human-guided action sequences during initial learning stages.

The proven value of RL techniques for navigation and localization tasks [10,2] motivates us to develop a RL approach to navigate autonomously into a docking station used for recharging. This approach makes use of a supervised RL algorithm and a Gaussian distributed state activation that allows real-world RL. Our approach proves to work with a reduced number of training examples, and is robust and easy to incorporate into conventional RL techniques such as SARSA.

2   Problem Overview

There are a number of research approaches studying domestic applications of humanoid robots, in particular using the Nao robot [9,8]. One of the Nao's limitations for this kind of environment is due to its energetic autonomy, which typically does not surpass 45 min. This motivates the development of strategies to increase the robot's operational time minimizing human intervention. In this work we develop a real-world reinforcement learning based on SARSA learning, see section 3, applied to an autonomous recharging behavior. This work is validated using a real Nao robot inside a home-like environment.

Several docking station designs and/or recharging poses are possible. The proposed solution is intended to increase the energetic capabilities of the Nao without major interventions on the robot's hardware or affecting its mobility or sensory capabilities. Despite the challenge to maneuver the robot backwards, we chose a partial backward docking. This offers advantages such as easy mounting on the Nao, it does not limit the robot mobility, nor obstructs any sensor, nor

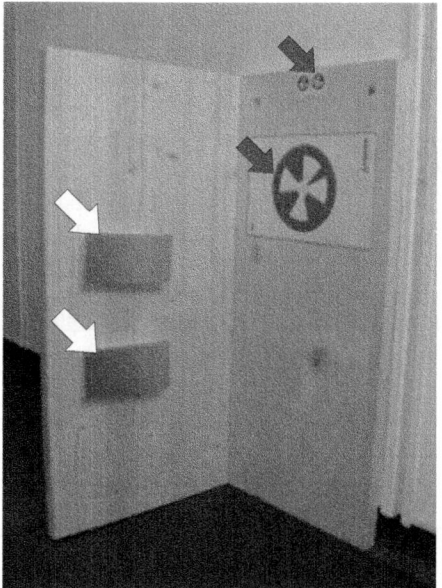 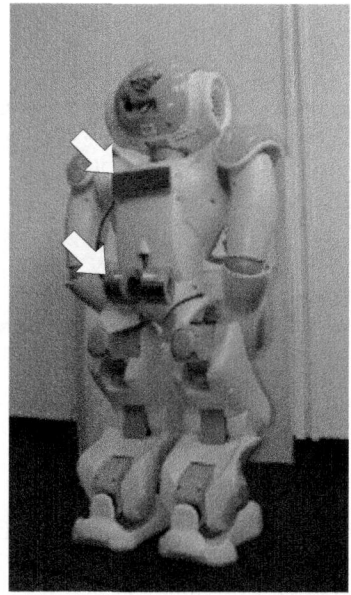

(a) Charging station   (b) Nao robot with electrical
                                        contacts

Fig. 1. (a) White big arrows indicate the electrical contacts placed on the docking station and gray arrows indicate the landmarks position. (b) Robot's electrical
connections.

requires long cables going to the robot extremities and allows a quick deployment
after the recharging has finished or if the robot is asked to do some urgent task.

The prototype built to develop the proposed autonomous recharging is shown
in figure 1(a). White arrows indicate two metallic contacts for the recharging,
and gray arrows indicate three landmarks (naomarks)[1] used for navigation. The
big landmark is used when the robot is more than 40 cm away from the charging station, while the two smaller landmarks are used for an accurate docking
behavior.

The autonomous recharging was split into four phases. During the first phase
a search and approach hard-coded algorithm searches for the charging station
via a head scan followed by a robot rotation. The robot estimates the charging station's relative position based on geometrical properties of landmarks and
moves towards the charging station. This approach places the robot approximately 40 cm away from the landmarks, see figure 2(a). In the second phase
the robot re-estimates its position and places itself approximately parallel to the
wall as shown in figure 2(b).

The third phase uses the reinforcement learning (SARSA) algorithm to navigate the robot backwards very close to the electric contacts as presented in

---

[1] 2-dimensional landmark provided by Aldebaran-Robotics.

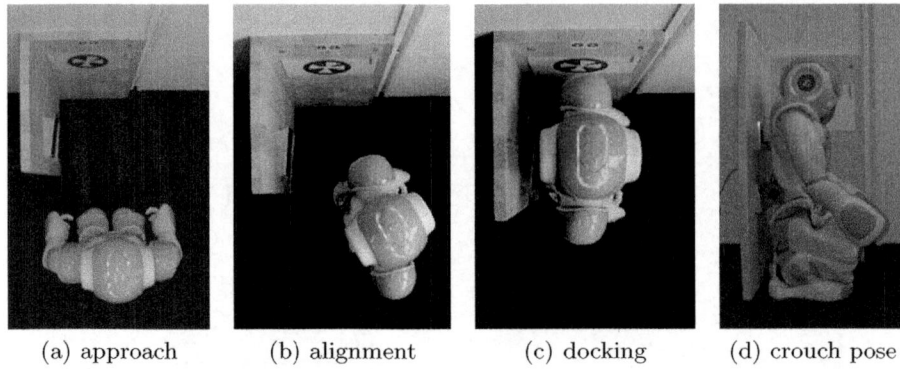

(a) approach   (b) alignment   (c) docking   (d) crouch pose

Fig. 2. Top view of the autonomous robot behavior in its four different phases (approaching, alignment, docking and recharging)

figure 2(c).[2] After reaching the final rewarded position, the fourth and final phase starts. A hard-coded algorithm moves the robot to a crouch pose, see figure 2(d), in which the motors are deactivated and the recharging starts.

3   Network Architecture and Learning

We use a fully connected two layer neural network, see figure 3. The input layer (1815 neurons) represents the robot's relative distance and orientation to the landmarks. The output layer (6 neurons) represents the actions that can be performed: move forward and move backward 2.5 cm, turn left or right 9° and move sideward to the left or right 2.5 cm. These values were adjusted empirically as a trade-off between speed and accuracy.

As mentioned in section 2, the robot starts to execute the SARSA algorithm approx. parallel to the wall and 40 cm away from the landmark.[3] During docking the minimal measured distance of the robot's camera to the landmark is approx. 13 cm, which corresponds to the robot's shoulder size plus a small safety distance. The state space is formed by the combination of three variables. These are the angular sizes of the two small naomarks and the yaw (pan) head angle. They represent the robot's relative distance and orientation, respectively.

Those three values are discretized as follows: The angular size of each landmark within the visual field is discretized into 10 values for each landmark. These values represent distances from [13, 40] cm in intervals of 2.7 cm. We add 2 values to indicate the absence of the corresponding landmark. This leads to a total of 11 values per landmark. The third variable is the head's pan angle. An internal routine permanently turns the robot's head to keep the interesting landmark centered in the visual field. The head movements are limited to [70°,

---

[2] In this docking phase, Nao's gaze direction is oriented towards the landmarks.

[3] Distance measured from the landmark to the robot's camera.

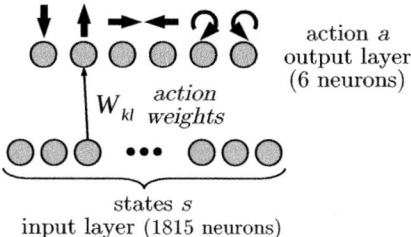

action a
output layer
(6 neurons)

W_{kl} action weights

states s
input layer (1815 neurons)

Fig. 3. Neural network schematic overview. An example of connections in the used neural network.

120°[and the values are discretized with intervals of 3.3° yielding 15 new values. Hence, the total number of used states is obtained by the combination of all the values, i.e. $11 * 11 * 15 = 1815$.

The learning algorithm is based on SARSA [13,14] and summarized as follows.

For each trial the robot is placed at an initial random position within the detection area. It permanently turns its head towards the landmarks. The landmarks sizes and the head pan angle are used to compute the robot internal state. Here, instead of using a single state activation of SARSA, where only a single input neuron has maximal activation ($S_i = 1$) at the time, we use a Gaussian distributed activation of states [4], which is centered in the current robot internal state ("*SARSA active state*"). The Gaussian is normalized, i.e. the sum over the state space activations is 1.

$$S_j = \frac{1}{\sigma^3 (2\pi)^{2/3}} \cdot e^{-\frac{(x_j - \mu_x)^2 + (y_j - \mu_y)^2 + (z_j - \mu_z)^2}{2\sigma^2}} \qquad (1)$$

We use $\sigma = 0.85$, which effectively "blurs" the activation around the "*SARSA active state*". In this way generalization to states that have not been visited directly is possible.

μ_x represents the current size value for "*landmark 1*", μ_y represents the current size value for "*landmark 2*" and μ_z represents the current value for the head yaw angle. The variables x_j, y_j and z_j take all the possible values of the respective dimension, i.e. size "*landmark 1*", size "*landmark 2*" and head's pan angle, respectively. In this way a normalized state activation is computed centered around (μ_x, μ_y, μ_z) and extend to the entire state space.

One of the motivations of having a Gaussian state activation is that states closer to the current internal state are more likely to generate the same action than farther states. Using this idea, we can extend and spread what we know about a state over larger regions of the state space.

Poor sampling from the state space during training will lead to poor generalization. This can be compensated using a representative training example set generated by tele-operation, what is termed supervised reinforcement learning [2]. A representative training example set should consist not only of the most frequent trajectories but it should cover also less frequently visited regions of the state space. A practical way to build a representative training example set is in an incremental fashion, i.e. generate a training set, train the network and test the output placing the robot in a random position (ideally not contained in the training examples). If the result is unsatisfactory then generate a few additional training examples containing this troubled case and re-train the network. These steps should be repeated until the results are satisfactory.

With the input from equation (1), the net activation of action unit i is computed as:

$$h_i = \sum_l W_{il} S_l \qquad (2)$$

W_{ij} is the connection weight between action i and state l. Connection weights are initially set to zero. Next, we used a softmax-based stochastic action selection:

$$P_{a_i=1} = \frac{e^{\beta h_i}}{\sum_k e^{\beta h_k}} \qquad (3)$$

β controls how deterministic the action selection is, in other words the degree of exploration of new solutions. Large β implies a more deterministic action selection or a greedy policy. Small β encourages the exploration of new solutions. We use $\beta = 70$ to prefer new routes. Based on the activation state vector (S_l) and on the current selected action (a_k), the value $Q_{(s,a)}$ is computed:

$$Q_{(s,a)} = \sum_{k,l} W_{kl} a_k s_l \qquad (4)$$

A binary reward value r is used. If the robot reaches the desired position it is given $r = 1$, zero if it does not. The prediction error based on the current and previous $Q_{(s,a)}$ value is given by:

$$\delta = (1 - r)\gamma Q_{(s',a')} + r - Q_{(s,a)} \qquad (5)$$

The time-discount factor γ controls the importance of proximal rewards against distal rewards. Small values are used to prioritize proximal rewards. On the contrary, values close to one are used to consider equally all rewards. We use $\gamma = 0.65$. The weights are updated using a δ-modulated Hebbian rule with learning rate $\epsilon = 0.5$:

$$\Delta W_{ij} = \epsilon \delta a_i S_j \qquad (6)$$

4   Supervised Reinforcement Learning and Experimental Results

In real-world scenarios, the random exploration of the state space, common in reinforcement learning, is prohibitive for several reasons such as real time action execution, safety conditions, changing sensors, actuators and environmental conditions, among many others.

In order to make the docking task feasible in a real-world RL approach, we skip the initial trial and error learning as presented in [2]. We tele-operate the robot from several random positions to the goal position saving the action state vectors and reward value. This training set with non-optimal routes is used for *offline* learning. Specifically 50 training examples with an average of 20 action steps were recorded. Then, using this training set, 300 trials were computed. Within each trial, the SARSA learning algorithm was performed as described in equations (1)-(6), however in equation (3) the selected action was given by the tele-operation data. We refer to this procedure as supervised RL. The state activation was tested for three cases: using conventional single state activation, using Gaussian distributed state activation and using a truncated Gaussian state activation. The truncated Gaussian state activation is obtained by limiting the non-zero values of x, y and z to a neighborhood of 1 state radius around μ_x, μ_y, and μ_z respectively and then normalized, in other words apply the Gaussian distribution to a neighborhood of one state radius around the "*SARSA active state*" instead of applying the activation over the entire state space.

We compare results obtained with the weight for each case after 300 trials. After the training phase using single states activation, the robot is able to reach the goal imitating the tele-operated routes. However, the robot's actions turn random in the states that have not yet being visited. In contrast, after training with a Gaussian distributed state activation the robot is able to dock successfully from almost every starting point, even in those cases where the landmarks are not detected in one step. This provides the Gaussian state activation with a clear advantage in terms of generalization. Thus faster learning than in case of SARSA or Q-learning is obtained. For the truncated Gaussian activation we observe slightly better results than using single state activation. A partial Gaussian activation may be useful for instance when the states are very different to each other and thus different actions are required.

In table 1, we compare the performance of the three methods starting from ten different positions. We present the number the steps executed in each trial and the corresponding stopping condition. *Single* stands for SARSA state activation. *Truncated* stands for Gaussian state activation limited to a neighborhood of one state radius around SARSA active state. *Gaussian* stands for Gaussian distributed state activation. We consider as *Success*, when the robot reaches successfully the desired goal position; as *False Pos.* (false positive) when the robot's measurement indicates that it is in the goal position but is not touching the metallic contacts. *Blind* indicates when the robot hadn't seen the landmarks for three or more consecutive actions. *Collision* indicates that the robot was crashing against the docking station. Under a detected *Blind* or *collision* event the respective trial was aborted.

Table 2 summarizes the obtained results. We present the average number of steps needed to reach the goal after training. The learned action-state pairs indicate the percentage of network weights that differ from their initialization values.

Table 1. Results of the three tested methods for ten different trajectories

Starting position	Single		Truncated		Gaussian	
	Stop condition	Step Nr.	Stop condition	Step Nr.	Stop condition	Step Nr.
1	Success	15	False Pos.	7	Success	9
2	False Pos.	5	Success	9	Success	17
3	Success	28	Success	29	Success	10
4	Success	20	Success	33	Success	13
5	Blind	17	Blind	7	Success	22
6	Blind	17	Blind	17	Success	32
7	Success	32	Success	48	Collision	49
8	Success	33	Success	23	False Pos.	37
9	Collision	154	Success	9	Success	24
10	Success	15	Success	14	Success	27

Table 2. Summary of ten trials for the three tested methods

State activation	Action-state pairs learned (%)	Nr. of success	Nr. false positive	Nr. aborted	Avg. nr. steps on success	Std. deviation
Single	4	6	1	3	23.8	8.23
Truncated	34	5	3	2	23.6	14.3
Gaussian	100	8	1	1	19.3	8.35

(a) Sample of receptive fields of *"Move to the Left"* after 300 trials with single state activation

(b) Sample of receptive fields of *"Move to the Left"* after 300 trials with Gaussian states activation restricted to one states radius

(c) Sample of receptive fields of *"Move to the Left"* after 300 trials with Gaussian states activation, $\sigma = 0.85$ without cutoff

Fig. 4. Receptive fields (RFs) of one action unit (*Move to the Left*) after 300 trials. Dark color represents the weight strength. From left to right the RFs for 8 of the 15 possible head rotations are presented.

Examples of the obtained receptive fields (RFs) after 300 trials are presented in figure 4. The goal position is shown in the upper left corner of each picture. White pixels represent unlearned action-state pairs. Darker gray represent a stronger action-state binding and thus the action is more likely to be selected when the robot is in this state. The eight different pictures for each case correspond to the different action-state pairs for particular head angles.

5   Conclusions

Motivated by the limited energetic capabilities of the Nao robot and our need for studying humanoid robots within home environments, we developed an autonomous navigation procedure for recharging the Nao, which does not require human assistance. Autonomous docking for a Nao robot was achieved for a real home like environment. Initial training examples, together with a Gaussian distributed states activation made real-world learning successful.

The use of appropriate training examples proved to be a key factor for real-world learning scenarios, reducing considerably the required learning steps from several thousand to a few hundred. Additionally, Gaussian distributed states activation demonstrated to be useful for generalization and eliciting a state space reduction effect. The use of these techniques is straightforward to SARSA learning. Promising results were presented, which suggest further opportunities in real-world or simulated scenarios.

We see at least two possible extensions of Gaussian distributed states activation. We believe that a conservative version, extending only to a small neighborhood called *truncated* in section 4, could help to increase learning speed even without tele-operated examples. Alternatively, the use of a memory of successful action sequences may be of great utility in other applications. This memory could be generated independently by tele-operation or fully automatic. Then these examples can be used for automatic offline training, while the robot is executing less demanding tasks.

During the experimental phase, we noticed that 2-dimensional landmarks can be detected only from within a small angle range, i.e. when the robot sees them without much distortion, and detection is very noise susceptible. For future work a docking procedure using a 3-dimensional landmark is under development. Additionally, forward, backward and turn movements will be preferred, because of the limited performance of sideward movements due to slippage of the Nao.

To obtain a more robust solution using this approach, we suggest adding a final module after the reinforcement learning module. The objective of this module will be to check sensor values, including sensors not considered in the current implemented, to determine whether the robot is in a false positive position. In this case, corrective actions could be learnt.

Acknowledgments. This research has been partly supported by the EU project RobotDoc [12] under 235065 ROBOT-DOC from the 7th Framework Programme, Marie Curie Action ITN and by the KSERA project funded by the European

Commission under the 7th Framework Programme (FP7) for Research and Technological Development under grant agreement n° 2010-248085.

References

1. Nao academics edition: medium-sized humanoid robot developed by Aldebaran Robotics, http://www.aldebaran-robotics.com/
2. Conn, K., Peters, R.A.: Reinforcement learning with a supervisor for a mobile robot in a real-world environment. In: International Symposium on Computational Intelligence in Robotics and Automation, CIRA, pp. 73–78. IEEE, Los Alamitos (2007)
3. Dorigo, M., Colombetti, M.: Robot shaping: An experiment in behavior engineering (intelligent robotics and autonomous agents). The MIT Press, Cambridge (1997)
4. Foster, D., Morris, R., Dayan, P.: A model of hippocampally dependent navigation, using the temporal difference learning rule. Hippocampus 10(1), 1–16 (2000)
5. Ghory, I.: Reinforcement learning in board games. Tech. rep., Department of Computer Science, University of Bristol (2004)
6. Ito, K., Fukumori, Y., Takayama, A.: Autonomous control of real snake-like robot using reinforcement learning; abstraction of state-action space using properties of real world. In: Palaniswami, M., Marusic, S., Law, Y.W. (eds.) Proceedings of the 3rd International Conference on Intelligent Sensors, Sensor Networks and Information, ISSNIP, pp. 389–394. IEEE, Los Alamitos (2007)
7. Kietzmann, T.C., Riedmiller, M.: The neuro slot car racer: Reinforcement learning in a real world setting. In: International Conference on Machine Learning and Applications, ICMLA, pp. 311–316. IEEE, Los Alamitos (2009)
8. The KSERA project (Knowledgeable SErvice Robots for Aging), http://ksera.ieis.tue.nl/
9. Louloudi, A., Mosallam, A., Marturi, N., Janse, P., Hernandez, V.: Integration of the humanoid robot Nao inside a smart home: A case study. In: Proceedings of the Swedish AI Society Workshop (SAIS). Linköping Electronic Conference Proceedings, vol. 48, pp. 35–44. Uppsala University, Linköping University Electronic Press (2010)
10. Muse, D., Wermter, S.: Actor-Critic learning for Platform-Independent robot navigation. Cognitive Computation 1(3), 203–220 (2009)
11. Provost, J., Kuipers, B.J., Miikkulainen, R.: Self-organizing perceptual and temporal abstraction for robot reinforcement learning. In: AAAI Workshop on Learning and Planning in Markov Processes (2004)
12. The RobotDoC collegium: The Marie Curie doctoral training network in developmental robotics, http://robotdoc.org/
13. Sutton, R.S., Barto, A.G.: Reinforcement learning: An introduction (adaptive computation and machine learning). The MIT Press, Cambridge (1998)
14. Weber, C., Elshaw, M., Wermter, S., Triesch, J., Willmot, C.: Reinforcement learning embedded in brains and robots. In: Reinforcement learning: Theory and applications, pp. 119–142. InTech Education and Publishing (2008)
15. Weber, C., Triesch, J.: Goal-directed feature learning. In: Proceedings of the International Joint Conference on Neural Networks, IJCNN, pp. 3355–3362. IEEE Press, Piscataway (2009)

Robot Routing Approaches for Convoy Merging Maneuvers

Fernando Valdes[1], Roberto Iglesias[2], Felipe Espinosa[1],
Miguel A. Rodríguez[2], Pablo Quintia[2], and Carlos Santos[1]

[1] Electronics Department, University of Alcala, Alcalá de Henares, Spain
{fernando.valdes,espinosa,carlos.santos}@depeca.uah.es
[2] Department of Electronics and Computer Science, USC, Santiago de Compostela, Spain
roberto.iglesias.rodriguez@usc.es,
mrodri@dec.usc.es, pablo.quintia@usc.es

Abstract. Autonomous and cooperative guidance strategies for a convoy of electric vehicles in an urban context are a challenging research topic in robotics and intelligent transportation systems. The vehicles that form the convoy eventually will have to leave it to perform a mission and return to the convoy formation once the mission has been accomplished. Nevertheless, the merging maneuvers amongst the convoy and the units returning to it (pursuing unit) is a complex task that involves the determination of the best merging point and the route across the city to reach it. This paper tackles with this routing problem of a robot located in a map and that is trying to join a convoy in constant movement along a peripheral trajectory. We have developed two search strategies able to determine the optimal merging point and the best route to reach it: on one hand we describe a *basic solution* able to solve the problem when the time spent by the robot travelling along every street of the map is considered to be known and constant. On the other hand, we extended this *basic approach* to provide a new search strategy that considers uncertainty in travelling times. This increases considerably the complexity of the problem and makes necessary the inclusion of a risk factor that must be considered when determining the best route and the merging point for the maneuver. We also put our search strategies into practice, simulating the behaviour of both, the convoy and the robot trying to joining it, using Player&Stage. In this article we show the results we achieved and that validate both of the aforementioned solutions.

Keywords: Robotic route planner, optimal meeting point, merging convoy maneuver, dynamic programming, uncertainty travelling time.

1 Introduction

The development of a convoy of autonomous transport units is an excellent proposal to reduce traffic jams and air pollution, especially in urban areas. This is one of the reasons why the development of these convoys has become an important challenge in the context of intelligent transportation systems (ITS) [1-2]. Basically, we can picture a group of electric vehicles moving in a coordinated way along a specific route (convoy formation), so that when an individual unit is required for hire, it would leave

R. Groß et al. (Eds.): TAROS 2011, LNAI 6856, pp. 241–252, 2011.
© Springer-Verlag Berlin Heidelberg 2011

the convoy to collect passengers, would drive them to a specific destination and, once the mission is finished, would return to the convoy as quickly as possible. There is a central fleet management system (remote centre) to collect passengers' requests, manage the fleet of vehicles, and dynamically assign routes and missions to them.

Nevertheless, an important challenge in this scenario is related with the return to the convoy of a single unit that has ended its mission. Since the convoy is moving continuously, the meeting point will have to be selected so that the link-up takes place in the shortest possible time, and the convoy always arrives at the meeting point after the pursuer. Therefore the destination will be the first point in the trajectory of the convoy that is reachable for the pursuer before the convoy. In this article we propose a solution for this challenge. Like many other research groups have done before, we will use a robotic demonstrator to evaluate our proposal.

Classic route search algorithms are based on graphs for finding the optimal route between and initial position and a destination, both of which are static and known. These proposals are extremely precise, but they entail excessive convergence times when conducting the search, hampering their real-time application. Worthy of note among the classic routing algorithms are Dijkstra [3, 4], A* [5-7], RTA* [8], LRTA* [9], LPA* [10] and D* Lite [11]. Only a small number of proposals, such as MTS [12], tackle the case of routing towards a known but non-static destination. In this case the pursuing vehicle will be able to reach a moving destination provided that its speed is greater than that of the destination change.

Problem Setup

To address the routing problem of a pursuer trying to reach a convoy of robots we shall assume the following conditions:

- Availability of a metric map that shows the roads in the setting where the convoy moves, their length, and the crossroads. Starting from this map we can easily draw a topological map. This map lacks scale but the relationship between nodes (crossroads) is maintained. The crossroads of the metric map are represented as nodes in the topological map, and the streets are represented as arcs. Each arc is weighted with the time required by a robot to travel along it.
- There must be at least one pursuing robot, located in a node on the topological map, and that is attempting to join in the convoy.
- There will be a convoy of robots travelling continuously along a road around the transport scenario.
- When the pursuer decides to join the convoy we shall assume that both, the pursuer's position and the convoy's position are known by the pursuer, and that the pursuer also knows the speed of the convoy.

In this article we describe a novel search algorithm specifically designed to operate in real time and to find the optimal meeting point for the pursuer and the convoy. Optimality here means that the time elapsed since the pursuer starts moving until it reaches the convoy must be minimum.

To illustrate some of the most significant variables involved in our algorithms, we consider a map of a simplified transport setting, such as that in Figure 1. The transport setting is delimited by a curved route along which the convoy travels continuously in an anti-clockwise direction. Two types of nodes can be distinguished: inner (n_i) and peripheral (s_i). All the peripheral nodes that can be reached by the pursuing unit and that are traversed by the convoy are candidate for executing the merging maneuver.

In the example shown in Figure 1 the initial position of the pursuing robot is represented as n^0. The set of nodes $\{s_1..s_{15}\}$ delimit the convoy's peripheral route. Eventually, it is important to be aware of the following notation:

P : is the total number of different peripheral nodes which comprise the convoy's route; in the example appearing in the figure P=15.

$s_i \in S = \{s_1..s_P\}$: is a peripheral node. The convoy traverses all s_i nodes consecutively. Since the route is circular it is true to say that $s_0 = s_P, s_1 = s_{P+1}.....$.

M : is the total number of inner nodes in the map.

$n_i \in N = \{n_1..n_M\}$: is a node from the interior of the transport setting. In general we denote parent node as the node where the pursuer is or the node that is being analysed, while successor nodes n^h are those to which the pursuing robot can travel from the parent node.

s^0: represents the initial position of the convoy, i.e. the position of the convoy when the pursuing unit is at n^0.

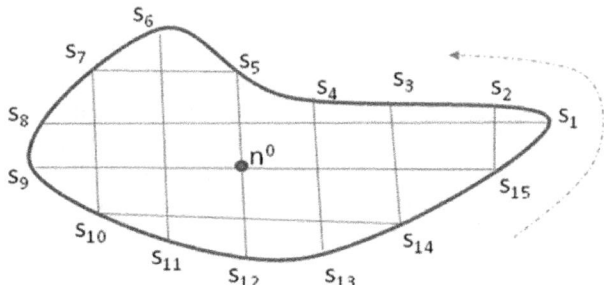

Fig. 1. Simplified map of the environment showing the initial position of the pursuing robot and the convoy's peripheral route

In general, the time required to travel along any street of the city (arc in the topological map) is not always the same. We will model this with Gaussian distributions, although other alternatives will be considered in our future work. Then, the following terms should be clarified:

$t(n_i,n^h)$: travelling time experienced by any transport unit when travelling between two consecutive inner nodes of the map, n^h is one of the successors of n_i. Whenever one of the robots travels from n_i to n^h, it records the time taken $t(n_i,n^h)$ and remits it to the RC (remote control in charge of the robots' fleet management).

$T(n_i,n^h)$: average travelling time between two consecutive internal nodes. This value is obtained from the statistical analysis of all the recordings: $t(n_i,n^h)$.

$Var(n_i,n^h)$: variance associated to the average value $T(n_i,n^h)$.

$T(n^h, s_i)$: estimated time for the pursuing unit to reach node s_i from n^h, these two nodes need not be consecutive justifying the estimation concept.

$Var(n^h, s_i)$: estimated variance associated to the time $T(n^h, s_i)$.

$t(s_i, s_{i+1})$: travelling time experience by the convoy when travelling between two consecutive peripheral nodes.

$T(s_i, s_j)$: estimated time it will take the convoy to travel from s_i to s_j, these being non-consecutive nodes. If s_i and s_j are consecutive ($s_j=s_{i+1}$) this value will be the average of the travelling times experienced by the convoy when moving amongst them.

$Var(s_i, s_j)$: estimated variance associated to the time $T(s_i, s_j)$.

The rest of the paper is structured in the following sections: Section 2 describes the solution we propose to solve the routing problem for convoy merging maneuvers when the travelling times are considered deterministic. Section 3 shows an extension of the previous algorithms when the travelling times among different nodes are stochastic (which is the normal situation in a real transport scenario). Finally, Section 4 summarizes and concludes the key aspects of the work described in this article.

2 Simplified Approach

Due to the complexity of the problem we proceed in two stages. In the first one (described in this section) we develop and evaluate an algorithm able to work on idealistic conditions, i.e. the time required by any robot to travel among two nodes of the map is always the same. This means that all variances are null.

Like most standard route planning algorithms, the search algorithm we have developed generates a minimum cost solution based on a predetermined cost function. This cost function uses two terms to assess whether a node is part of the optimal route (Eq. 1). The first term determines the time needed to travel from the initial node n^0 to the node under consideration, n^h. The second term estimates the travelling time from this node n^h to the destination s_e.

$$f(n^h) = T(n^0, n^h) + T(n^h, s_e) \qquad (1)$$

Regarding the meeting node s_e in Eq.1, to define it we must consider that the pursuer vehicle needs to reach the meeting node before the convoy and, on the other hand, the pursuer's joining up with the convoy must take place in the shortest possible time. For this reason the meeting node s_e must be the one which fulfills Eq.2:

$$s_e = \arg \min_i \{T(s^0, s_i)/T(s^0, s_i) > T(n^0, s_i)\}, i = \{1,..,P\} \qquad (2)$$

Therefore, considering Eq.1 and Eq.2, all problems now reside in determining two sets of values:

$$
\begin{aligned}
a)\, & T(n_i, s_j), \; \forall n_i \in N, \forall s_j \in S \\
b)\, & T(s_i, s_j), \; \forall s_i, s_j \in S
\end{aligned}
\qquad (3)
$$

To calculate these values we resort to Dynamic Programming (DP) [13-15] Dynamic Programming is a well-known methodology for representing optimization problems in terms of functional equations and for solving them. Generally, in DP formulations there is a discrete-time system whose state evolves according to given transition probabilities that depend on a decision/control action.

Basically, if we consider any inner node n_i, and any peripheral node s_j, it is true to say that:

$$T(n_i, s_j) = \min\{T(n_i, n^h) + T(n^h, s_j)\}, \forall n_i \in N, s_j \in S, n^h \in successors(n_i) \qquad (4)$$

Obviously, the aforementioned relations are also valid when the peripheral node s_j is also the meeting node s_e:

$$T(n_i, s_e) = \min\{T(n_i, n^h) + T(n^h, s_e)\}, \forall n_i \in N, n^h \in successors(n_i) \qquad (5)$$

Thus, we reach a relationship between the travelling times that satisfy a Bellman's equation [16]. The recursive application of the previous equations will enable us to estimate the travelling time and its variance from any n_i to any s_j (Figure 2.left), or between any two non-consecutive nodes on the periphery (Figure 2.right). It is also important to notice that the calculation of times is performed rapidly from a computational point of view.

Once we get these T-values we can apply Eq. 2 to get the meeting node s_e, and the search process described in Figure 3 to reach the optimal route from the initial pursuer's position to the meeting node s_e. Our solution fulfills the conditions that guarantee the optimality of the achieved solution [17]:

- All arcs in the graph must have costs greater than some positive number, ε. It is important to remember that an arc cost is the time spent to travel along the street represented by the arc.

- The heuristic term needs to be admissible, i.e., the estimated cost of reaching the target from the node being evaluated should never overestimate the real cost.

```
for i = 1 to P
  for j = 1 to P
    if j ≠ i+1
      T(s_i,s_j) = 0 ;
while ( changes ≠ 0 )
  changes = 0 ;
  for i = 1 to P
    for j = 1 to P
      if i ≠ j
        Δ = T(s_i,s_j) ;
        T(s_i,s_j) = T(s_i,s_{i+1}) + T(s_{i+1},s_j) ;
        if Δ - T(s_i,s_j) ≠ 0 ;
          changes ← changes+1 ;
      otherwise
        T(s_i,s_j) = 0 ;
```

```
for i = 1 to M
  for j = 1 to P
    T(n_i,s_j) = 0 ;
while ( changes ≠ 0 )
  changes = 0 ;
  for i = 1 to M
    for j = 1 to P
      Δ = T(n_i,s_j) ;
      T(n_i,s_j) = min_{successors(n_i)} [ T(n_i,n^h) + T(n^h,s_j) ] ;
      if Δ ≠ T(n_i,s_j) ≠ 0 ;
        changes ← changes+1 ;
```

Fig. 2. Algorithm description for T-values calculation of peripheral nodes (left) and of inner nodes (right)

```
Include the start node n⁰ in the achieved solution so far:
    k = 0 ;
    sol(k) = n⁰ ;
while ( sol(k) ≠ sₑ )
    Evaluate next candidates for the solution:
        f(nʰ), ∀nʰ ∈ successors{sol(k)} ;
    Select the next node as part of the solution:
        sol(k) = arg minₕ[f(nʰ)] ;
    k ← k + 1 ;
```

Fig. 3. Pseudocode of the proposed search algorithm

Example of Application

This example describes how a pursuer located in the inner area of the map shown in Figure 4.left will join a convoy that moves along the peripheral path. To simulate the movement of the robot and the convoy we have used Player&Stage [18-20].

As we already pointed out in the introduction, most classic routing strategies are designed to find the optimal path amongst an origin and a destination that are, both, static and well-known. Nevertheless, in our case the destination is part of the problem itself: determining the meeting node amongst a pursuer and a convoy that is moving continuously is an important challenge that brings in time restrictions, and makes the problem clearly harder. It is important to remember that the running stage starts once the T-values have been estimated and stored in a lookup table; it is then when any pursuer can use these T-values to determine the best meeting node with the convoy and the route to it. We aim to prove that the determination of the route to the meeting node is done very quickly and hence our solution is suitable for real-time computation.

As can be seen in Table 1, we have applied our search algorithm and the A* routing strategy to find the best routes amongst different initial positions n^0 and destinations s_e. We must be aware of the fact that the A* is not valid for problems where minimum searching time is required. In this case we run A* when the meeting node s_e has been already determined using dynamic programming. What we want to see is how our algorithm, described in Figure 3, is able to take advantage of the T-values and the particular characteristics of the problem, to get a computationally efficient search process. Once our search strategy and the A* have been implemented using the C language, we used Player&Stage to simulate the movement of the mobile units and compare them.

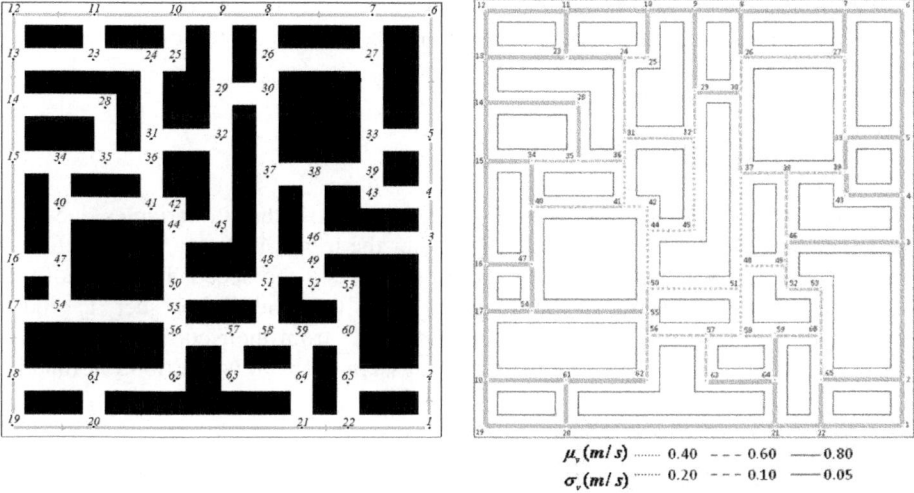

$\mu_v\,(m/s)$ .... 0.40 - - - 0.60 ——0.80
$\sigma_v\,(m/s)$ .... 0.20 - - - 0.10 ——0.05

Fig. 4. Map used in our experiments (left). The same map, but with areas of different traffic densities and hence different robot speeds, has been used to validate the extended algorithm described in section 3 (right).

Table 1. Results of the comparison of A* and proposed algorithm for different pairs (n^0, s_e), according to the simplified approach

Test	n^0	s^0	s_e	Expanded nodes on A*	Expanded nodes on Proposed Algorithm	Search execution time on A* (*ms*)	Search execution time on Proposed Algorithm (*ms*)
1	24	1	8	10	7	0,0165	0,0103
2	29	1	7	6	5	0,0138	0,0080
3	35	1	8	14	8	0,0229	0,0127
4	37	1	4	5	5	0,0131	0,0078
5	44	1	8	16	7	0. 0247	0,0099
6	50	1	9	21	6	0,0406	0,0095
7	54	1	9	20	10	0,0339	0,0120
8	61	1	13	27	13	0,0470	0,0180
9	63	1	13	35	13	0,0594	0,0199
10	65	1	9	19	11	0,0383	0,0147

Table 1 shows 10 different results achieved for different combinations (n^0, s_e). Focusing on this table, we must notice several relevant aspects, such as the number of nodes that were evaluated and included as part of the solution somewhere along the search processes (5th and 6th columns). In the case of the A* we can see that there were nodes which were initially considered as part of the solution, but which were later discarded because of the discovery of better alternatives; this is the reason why despite of the achievement of the same solution the number of expanded nodes is usually lower with our search strategy than with A*. The last two columns of the table allow a comparison of the computing times that A* and our proposal required to find the solutions. The running time using Player&Stage is significantly reduced when our search strategy is used. This shows the advantage of our solution for real-time searches, especially when real maps are used (often more complex and larger than the one in this example).

3 Extended Approach

This new approach considers the travelling times as stochastic variables due to the different sources of uncertainty that might alter the traffic intensity and therefore the travelling times (weather conditions, type of road and the speed limit thereof, etc). Given the stochastic nature of the travelling times that both the pursuer t_p and the convoy t_c will take to reach the meeting node s_e from their current positions, we shall assume that both times are given by Gaussian probability distribution functions [21].

$$f_p(t_p) = \frac{1}{\sqrt{Var(n_i,s_e)}\sqrt{2\pi}} e^{-\frac{1}{2}\left(\frac{t_p - T(n_i,s_e)}{\sqrt{Var(n_i,s_e)}}\right)^2} \quad f_c(t_c) = \frac{1}{\sqrt{Var(s_i,s_e)}\sqrt{2\pi}} e^{-\frac{1}{2}\left(\frac{t_c - T(s_i,s_e)}{\sqrt{Var(s_i,s_e)}}\right)^2} \quad (6)$$

Where $f_p(t_p)$ and $f_c(t_c)$ represent the probability distribution of each time value for both, the pursuer and the convoy. Every time a robot travels along a street, it sends the time it took to do it to the remote control centre. This information will be used to update the variance and the average travelling time associated to the street travelled by the robot. On the other hand, we will use the iterative algorithms shown in Figure 5 to estimate the travelling times and their variances among not consecutive nodes. These algorithms are a extended version of the ones shown in Figure 2. From Eq.6 we can get the cumulative distributions:

$$F_p(t_p) = \int_{-\infty}^{t_p} f_p(t)dt \qquad\qquad F_c(t_c) = \int_{-\infty}^{t_c} f_c(t)dt \qquad (7)$$

$F_p(t_p)$ describes the probability of obtaining a travelling time for the pursuing unit less than or equal to t_p, while $F_c(t_c)$ gives the same information but for the convoy. The probability of the pursuing unit arriving in a time shorter than or equal to $t_p^* = T(n_i,s_e) + 2.33\sqrt{Var(n_i,s_e)}$, is 99.01% (this can be deduced considering the properties of the cumulative distribution functions). As it has already been pointed out, the maneuver is successful when the pursuing unit arrives first to the meeting node s_e. This leads us to define the risk factor δ (maximum risk), as the probability of the maneuver being unsuccessful, or in other words, that the convoy reaches the meeting node in a time equal to or shorter than the pursuing unit:

$$F_c(t_c \le t_p^*) = \int_{-\infty}^{t_p^*} f_c(t)dt = \delta \qquad (8)$$

For example if the risk factor is 4.95%, it is true to say that:

$$T(s_i,s_e) - 1.65\sqrt{var(s_i,s_e)} > T(n_j,s_e) + 2.33\sqrt{var(n_j,s_e)} \qquad ,i = \{1,...,P\}, j = \{1,...,M\} \qquad (9)$$

```
for i = 1 to P
  for j = 1 to P
    if j ≠ i+1
      T(sᵢ,sⱼ) = 0 ;  Var(sᵢ,sⱼ) = 0 ;
while ( changes ≠ 0 )
  changes = 0 ;
  for i = 1 to P
    for j = 1 to P
      If j > i+1
        Δ = T(sᵢ,sⱼ) ;
        T(sᵢ,sⱼ) = T(sᵢ,sᵢ₊₁) + T(sᵢ₊₁,sⱼ) ;
        Var(sᵢ,sⱼ) = Var(sᵢ,sᵢ₊₁) + Var(sᵢ₊₁,sⱼ) ;
        if Δ − T(sᵢ,sⱼ) ≠ 0 ;
          changes ← changes+1 ;
```

```
for i = 1 to M
  for j = 1 to P
    T(nᵢ,sⱼ) = 0 ;  Var(nᵢ,sⱼ) = 0 ;
while ( changes ≠ 0 )
  changes = 0 ;
  for i = 1 to M
    for j = 1 to P
      Δ = T(nᵢ,sⱼ) ;
      T(nᵢ,sⱼ) = min_{successors(nᵢ)} [T(nᵢ,nᵏ) + T(nᵏ,sⱼ)] ;
      Var(nᵢ,sⱼ) = Var(nᵢ,nᴺ) + Var(nᴺ,sⱼ)
      where nᴺ: successor(nᵢ)|T(nᵢ,nᴺ)+T(nᴺ,sᵢ)=T(nᵢ,sⱼ)
      if Δ ≠ T(nᵢ,sⱼ) ≠ 0 ;
        changes ← changes+1 ;
```

Fig. 5. Algorithm description for: T and Var updating related with peripheral nodes (left side); T and Var initial updating related with inner nodes (right side)

The Eq.9 means that in 99% of cases, which is the probability of the pursuer arriving at s_e earlier than $T(n_j, s_e) + 2.33\sqrt{\mathrm{var}(n_j, s_e)}$, the convoy reaches the meeting node with a risk factor lower than 4.95% . That is the probability of having a convoy that arrives to s_e before $T(s_i, s_e) - 1.65\sqrt{\mathrm{var}(s_i, s_e)}$.

According to this, we re-define the meeting node as the one that verifies Eq.10, and which guarantees the solution in 99 % of all possible cases:

$$s_e = \arg\min_{s_e} \left\{ (T(s_i, s_e)) / \left(T(s_i, s_e) + z_c(\delta)\sqrt{Var(s_i, s_e)} \right) > \left(T(n_j, s_e) + 2.33\sqrt{Var(n_j, s_e)} \right) \right\}, \quad (10)$$

$$i = \{1,..,P\}, j = \{1,..,M\}$$

The value of the risk factor determines the value of $z_c(\delta)$. Once s_e has been established, the next challenge for the pursuing unit is to verify the most efficient route to this peripheral node. This is a cyclical process; starting with the current location of the pursuing unit n_i, the proposed search algorithm evaluates each node n^h, which is a successor of n_i, and selects the one with the lowest cost function $f(n^h)$, see Eq.1. At this moment, the successor node becomes a parent node, and the process is repeated as shown in Figure 3.

Example of Application

As in the previous example, the Player&Stage tool has been used to evaluate the extended search proposal described in the previous section. During the simulation the velocities of the robots will be randomly selected from a normal probabilistic density function $f(V)$, with μ_v as mean value, and σ_v as standard deviation, thus emulating the stochastic character inherent to a real transport setting.

In the same way as the travelling times are different for different streets of a city, the value of (μ_v, σ_v) varies depending on the geographical location of each street in the map (Figure 4, right side):

- Sections in the central area of the map, are slower than in the outskirts, since more irregular traffic is expected: $\mu_v = 0.4m/s, \sigma_v = 0.2m/s$.
- Sections in the outer, peripheral area, where faster moving traffic with fewer variations is expected: $\mu_v = 0.8m/s, \sigma_v = 0.05m/s$.
- Sections in the intermediate area: $\mu_v = 0.6m/s, \sigma_v = 0.1m/s$.

In this example of application, the pursuing unit starts from node $n^0 = 55$ and the convoy from node $s^0 = 1$. Before starting a new stretch, the pursuing unit checks whether the estimated time of arrival for the pursuer is shorter than that estimated for the convoy; if not, it recalculates a new merging node and the route to it. The convoy follows an anti-clockwise peripheral route. The dynamic evaluation of objectives is implemented considering a risk factor $\delta = 0.05$ (that means 5%). Forty simulations with Player&Stage were conducted, and for each one, the following information is shown in Table 2:

- Meeting node s_e. We can realize the dynamic evaluation of objectives performed by the pursuing unit. It can be observed how the meeting node s_e changes as the experiment proceeds. Whenever the probability for the convoy arriving earlier than the pursuing unit is above the risk factor (5%), a new meeting node and the route to it are estimated. As it can be appreciated, in this example the maximum number of re-planning is 3 (simulations 2, 11, 20, 27, 28, 38 and 40).
- Indication of whether merging is achieved (ok) or not (not). In failed attempts, the table shows the final value for the estimated meeting node.

At the beginning of this simulation process, the only information available on the estimated times is the statistical distribution of velocities shown in Figure 4 (right side). This is the reason why the meeting node envisaged in the first simulation is $s_e = 9$. In the subsequent simulations the travelling times for each section of the map have been adapted due to the experiences recorded by the pursuing unit.

From the results shown in Table 2, we can draw the next considerations:

- The meeting node initially envisaged becomes more conservative as the number of experiments increases, although this tends to stabilise.
- In 82.5% of the experiments the node where the pursuing unit and the convoy finally meet is closer to the initial position of the convoy than the meeting node considered at the beginning of the simulation.

Only in 2 of the 40 simulations (5%) was the objective not attained. In these cases, the pursuing unit did not arrive before the convoy despite the dynamic re-planning of s_e, (without retracing the route it already followed).

Table 2. Summary of the simulations related to the extended approach

N° Sim	1	2	3	4	5	6	7	8	9	10	11	12	13	14	15	16	17	18	19	20
Node Se	9	8	10	10	11	14	14	14	13	14	13	13	13	13	13	13	13	14	13	13
	13	20			13	13	13	13	14	13	14	11	10	11	11	11	11	13	11	14
	9	18				10	10	10		10	13				10	10		10	10	13
		20									11									10
Merge	ok	ok	ok	ok	ok	ok	not	ok	ok	ok	ok	ok	ok	ok	ok	ok	ok	ok	ok	ok

N° Sim	21	22	23	24	25	26	27	28	29	30	31	32	33	34	35	36	37	38	39	40
Node Se	13	13	13	13	13	13	13	13	13	13	13	13	13	13	13	13	13	13	13	13
	11	11	14	11	10	11	11	14	11	10	14	11	10	10	11	10	14	14	11	14
				10		10	13	13	10			10							13	13
						25	10												10	11
Merge	ok	ok	ok	ok	ok	ok	ok	ok	ok	ok	ok	ok	ok	ok	not	ok	ok	ok	ok	ok

4 Conclusions

In this paper we have investigated the use of dynamic programming to obtain an efficient and optimal routing proposal constrained to merging maneuvers between a moving convoy and a transport unit. The algorithm we propose guarantees the optimality of the route to a destination which depends on both, the pursuing unit and the convoy. The algorithm itself is able to find the appropriate destination that minimizes the time elapsed from when the pursuing unit starts moving towards the convoy until it finally reaches it.

Another important contribution of the work described in this paper consists on the analysis of the effect of the risk factor used to limit the probability of a convoy reaching the merging node before the pursuer (failed merging maneuver). This risk factor takes into account the uncertainty inherent to the travelling times along the different streets of the map. The higher the values of the risk factor the faster the merging maneuver. Nevertheless the assumption of a high risk also entails an increase in the number of re-plannings and the number of failed attempts.

The algorithms described in this paper have been implemented and validated on a demonstrator using Player&Stage.

Acknowledgments. The realization of this research has been made possible thanks to funding from the VISNU (Ref. TIN2009-08984) project. The participation of researchers R. Iglesias and M. A. Rodriguez has been possible thanks to the research grants: TIN2009-07737 and INCITE08PXIB262202PR.

References

1. Europoean transport policy for 2010: time to decide. White paper,
   http://www.central2013.eu/fileadmin/user_upload/Downloads/
   Document_Centre/OP_Resources/EU-transportpolicy2010_en.pdf
   (accessed on March 11, 2011)

2. Thinkev Press. Silence on the streets of Oslo as Think and the BBC lead a convoy of electric cars through the city in celebration of Norways's EV culture leadership. Oslo. Norway, June 11 (2010),
   `http://www.thinkev.com/Press/Press-releases/Silence-on-the-streets-of-Oslo-as-THINK-and-the-BBC-lead-a-convoy-of-electric-cars-through-the-city-in-celebration-of-Norway-s-EV-culture-leadership` (accessed on March 11, 2011)
3. Dijkstra, E.W.: A note on two problems in connection with graphs. Numerische Mathematik 1, 269–271 (1959)
4. Chiong, R., Sutanto, J.H., Japutra, W.: A comparative study on informed and uninformed search for intelligent travel planning in Borneo Island. In: Int. Symp. on Information Technology, Malaysia (2008)
5. Nilsson, N.: Problem-solving methods in artificial intelligence. McGraw Hill, New York (1971)
6. Yue, H., Shao, C.: Study on the application of A* shortest path search algorithm in dynamic urban traffic. In: IEEE Int. Conference on Natural Computation (2007)
7. Goto, T., Kosaka, T., Noborio, H.: On the heuristics of A* algorithm in ITS and robot path-planning. In: Proceed. of the IEEE/RSJ Int. Conf. on Intelligent Robots and Systems, Las Vegas (2003)
8. Korf, R.E.: Real - Time Heuristic Search. Artificial Intelligence (1990)
9. Koenig, S.: A comparison of fast search methods for real-time situated agents. In: International Conference on Autonomous Agents and Multi-Agent Systems, pp. 864–871 (2004)
10. Koenig, S., Likhachev, M., Furcy, D.: Lifelong Planning A*. Artificial Intelligence Journal, 93–146 (2004)
11. Koenig, S., Likhachev, M.: D* Lite. In: National Conference on Artificial Intelligence, pp. 476–483 (2002)
12. Ishida, T., Korf, R.E.: Moving target search: A real-time search for changing goals. IEEE Trans. Pattern Analysis and Machine Intelligence, 09–97 (1995)
13. Yu, S., Ye, F., Wang, H., Mabu, S., Shimada, K., Yu, S., Hirasawa, K.: A global routing strategy in dynamic traffic environments with a combination of Q value-based dynamic programming and Boltzmann distribution. In: SICE Annual Conference, Japan (2008)
14. Santos, L., Coutinho-Rodrigues, J., Current, J.R.: An improved solution algorithm for the constrained shortest path problem. Transportation Resesarch, Part B 41 (2007)
15. Nilsson, N.: Artificial Inteligence. A New Synthesis. Morgan Kaufmann Publishers, San Francisco (1998)
16. Bellman, R.: Dynamic Programming. Dover Publications, New York (2003)
17. Mainali, M.K., Shimada, K., Mabu, S., Hirasawa, K.: Optimal route of road networks by dynamic programming. In: IEEE International Joint Conference on Neural Networks, Hong Kong (2008)
18. Mainali, M.K., Shimada, K., Mabu, S., Hirasawa, K.: Multi-objective optimal route search for road networks by dynamic programming. In: SICE Annual Conference, Japan (2008)
19. Player&Stage Project, http://playerstage.sourceforge.net
20. Espinosa, F., Salazar, M., Valdes, F., Bocos, A.: Communication architecture based on Player&Stage and sockets for cooperative guidance of robotic units. In: 16th Mediterranean Conference on Control and Automation, pp. 1423–1428 (2008)
21. Nehmzow, U.: Scientific methods in mobile robotics. In: Quantitative Analysis on Agent Behaviour. Springer, Heidelberg (2006) ISBN: 10-1-84628-019-2

Sensing with Artificial Tactile Sensors: An Investigation of Spatio-temporal Inference

Asma Motiwala, Charles W. Fox, Nathan F. Lepora, and Tony J. Prescott

Adaptive Behaviour Research Group, Department of Psychology,
University of Sheffield, Sheffield, UK
{a.motiwala,c.fox,n.lepora,t.j.prescott}@sheffield.ac.uk

Abstract. The ease and efficiency with which biological systems deal with several real world problems, that have been persistently challenging to implement in artificial systems, is a key motivation in biomimetic robotics. In interacting with its environment, the first challenge any agent faces is to extract meaningful patterns in the inputs from its sensors. This problem of pattern recognition has been characterized as an inference problem in cortical computation. The work presented here implements the hierarchical temporal memory (HTM) model of cortical computation using inputs from an array of artificial tactile sensors to recognize simple Braille patterns. Although the current work has been implemented using a small array of robot whiskers, the architecture can be extended to larger arrays of sensors of any arbitrary modality.

Keywords: Pattern recognition, Cortical computation, Hierarchical Temporal Memory, Bayesian inference, Tactile perception.

1 Introduction

Interacting with any environment, real or artificial, involves extracting meaningful information about features in the environment, often from noisy and ambiguous sensory inputs, to guide behavior and appropriately respond to and/or control the environment. Implementing these operations in artificial systems has been a persistent challenge. However, biological organisms implement these with relatively much greater ease, in much less controlled natural environments. Given how difficult it is to design artificial systems that can perform with similar adaptability and precision with real world problems as biological systems, there is increasing motivation to address these problems using strategies that are algorithmically similar to those used by biological systems to solve similar problems in their natural environments.

Although, generating appropriate behavior and motor outputs have been argued to be the main function of brains in biological systems, they can only follow from an accurate understanding of the *state* of the agent's internal and external environment. Hence, the first step in appropriately interacting with any environment is to use sensory information available through different sensors to extract properties of how things change in the world.

R. Groß et al. (Eds.): TAROS 2011, LNAI 6856, pp. 253–264, 2011.

Several algorithms of pattern recognition and feature extraction have been suggested to extract meaningful information from sensory data in different contexts and for different types of problems. It has been previously suggested that no algorithm is inherently superior to other algorithms at solving all learning problems, and that the superiority of any algorithm depends largely on the degree to which the assumptions in the algorithm match the properties of the problem space [11]. Hence, given the efficiency with which biological organisms interact with their environment, it can be assumed that their brains have evolved to implement algorithms that best exploit the properties of real world objects and events.

2   Theoretical Background: Hierarchical Temporal Memory and Cortical Computation

The key problem an organism faces in learning about its environment is to extract consistent and stable features in the environment rather than features in the sensory data itself, *i.e.* the organism needs to learn an internal model of *hidden causes* in the world based on information available from its noisy percepts.

The hierarchical temporal memory (HTM) model is a theory of cortical computation. It is a hierarchical probabilistic model, that uses assumptions based on the properties of objects and events in the world to suggest how the cortex learns internal representations of causes from its inputs, and how it uses these for inference with ambiguous sensory information [10]. Based on the *common cortical algorithm assumption* [16], the HTM model attempts to characterize a generic algorithm with which any arbitrary region of the cortex learns and infers, irrespective of the modality or the level of processing. In other words, it aims to characterize a cortical algorithm that best exploits the statistical properties of causes in the world, using a tight set of constraints regarding the known anatomy and physiological behavior of the cortex.

The Problem of Invariance in Learning Representations: Why the Hierarchy Needs to be Spatio-temporal

The HTM model argues that representing a spatio-temporal hierarchy of causes is critical for the brain's ability to infer with ambiguous sensory information. It has been previously shown in the visual cortex that detecting spatial coincidences in its inputs allows a cortical region to represent more complex spatial patterns than those in lower levels in the cortical hierarchy [14]. However, the problem in learning consistent and stable representations of causes is that a single cause can give rise to several spatially dissimilar percepts. For example: people are exceptionally good at recognizing faces. Face recognition is invariant to the size, position and view of the face, despite the fact that different combinations of these generate widely different visual patterns.

The *temporal slowness principle* argues that causes in the world change slower than noisy percepts [13]. Based on this property of temporal slowness, it has been suggested that the cortex learns to represent invariant and complex representations not just by matching spatial similarity of its inputs but by generalizing over successive spatial patterns in time [8,3]. This *temporal proximity constraint*, that spatial patterns that consistently occur close in time are likely to be associated with the same cause in the world, is a key principle in the HTM model [10].

Spatio-temporal Representations

The cortical hierarchy has been modeled as a hierarchy of HTM nodes where every node implements the same algorithm, and at every level in the hierarchy each node pools information over a sub-set of nodes in lower levels. To maintain a spatio-temporal hierarchy of representations, each HTM node represents recurrent spatial patterns in its inputs, as well as sequences of these spatial patterns. These spatial patterns are patterns of activity in the lower level nodes, that any node pools over (*i.e.* its *children*), that are coincident in time and which recurrently occur in the nodes bottom-up inputs.

According to the temporal proximity constraint, sequences of spatial patterns that consistently occur in sensory inputs are likely to represent a single/coherent cause in the world. Thus by encoding sequences of patterns in its inputs, every node maintains representations of causes that get progressively more complex and invariant as we move up the hierarchy. This is similar to the spatio-temporal properties of cortical representations and mirrors the statistical properties of causes in the world.

Inference

Inference in HTM is probabilistic and implemented using *Bayesian belief propagation* [10]. A key feature of Bayesian probabilistic inference is that it is not only based on estimates of the system's current inputs but uses *prior* information about the statistical properties of patterns in its inputs to compute *posterior probabilities* or *beliefs* over states and causes of the inputs. Probabilistic inference is a powerful way of dealing with the inherent ambiguity in sensory information. Moreover, work on perceptual inference and sensory-motor integration is increasingly suggesting that inference in the cortex is probabilistic [15].

The HTM Model as a Biologically Inspired Technology

Not only is the HTM model a neurobiological theory, that attempts to characterize a cortical algorithm which is hypothesized to best exploit the statistical properties of causes in the world, but it has been developed as a platform to be suitable for several real world applications [18]. The only assumptions in an HTM model are those regarding the nature of the spatio-temporal structure of causes, which are assumed to be invariant features of patterns in the real world, and hence can be applied to a wide range of problems.

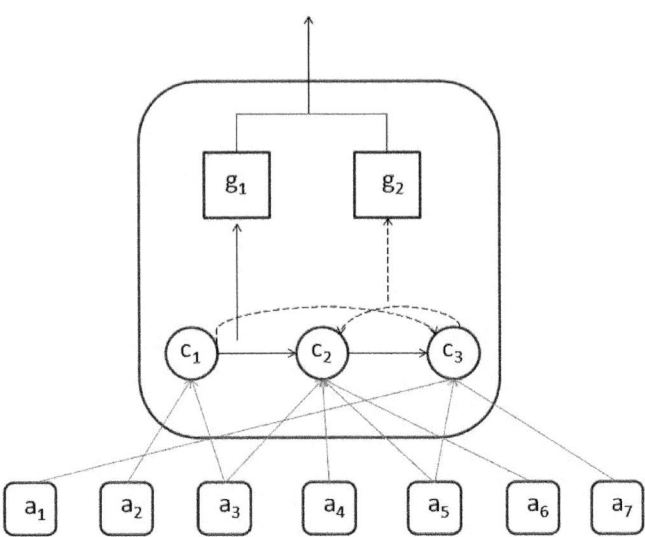

Fig. 1. The figure shows the internal structure of an arbitrary HTM node. The spatial patterns in the node are shown as circles and the sequences as square boxes. Each spatial pattern in the node represents coincident activity in a subset of the node's children and each of the sequences in the node represent sequences of these coincident patterns.

HTM has been previously tested on visual object recognition [10], the model has been extended and tested on sign language recognition [20], as well as for automated design evaluation [12]. HTM has also been tested against standard neural network architectures and has been shown to "produce recognition rates that exceed a standard neural network approach. Such results create a strong case for the further investigation and development of Hawkins' neocortically-inspired approach to building intelligent systems" [21]. In addition, HTM is already being commercially used by Vitamin D, EDSA and Forbes.com, and several other applications are being developed [2].

3   Methods

The current work aimed to investigate how inference using the HTM architecture could be implemented to recognize simple braille patterns using sensory inputs from an array of artificial whiskers.

Probabilistic inference and hierarchical representations are not new concepts, in either biology or pattern recognition. Cortical regions have been shown to be connected in a hierarchical manner [7] and several neurobiological models as well as statistical models of pattern recognition use hierarchical organization [17]. In addition, the idea that inference in the cortex is probabilistic has been popularized as the *Bayesian brain* hypothesis and belief propagation has generated great interest as a biologically plausible inference strategy [5]. However, HTM is unique in that it uses spatio-temporal hierarchical representations for probabilistic inference, in contrast to most neurobiological and statistical models which either have spatial or temporal structure [17].

This spatio-temporal architecture was particularly well suited for the investigation of tactile perception since neither the spatial nor the temporal components in the sensory data can be ignored. The amount of information available from any single tactile sensor at any given instant is extremely sparse. Hence it is critical to pool over multiple sensors as well as consider how the inputs evolve over time to learn and infer about the properties of different stimuli.

Data Acquisition

The sensory input to the model was obtained from an array of three artificial whisker sensors that have been developed on the BIOTACT project [1]. The stimuli used were braille-like patterns (shown in Figure 2) and the size of the braille patterns was chosen to match the spatial resolution of the sensor array. The stimulus was swept over the whisker sensors using the XY positioning robot in a plane perpendicular to the shaft of the whiskers. The arrangement of the whiskers and the stimulus being swept over it are shown in Figure 2.

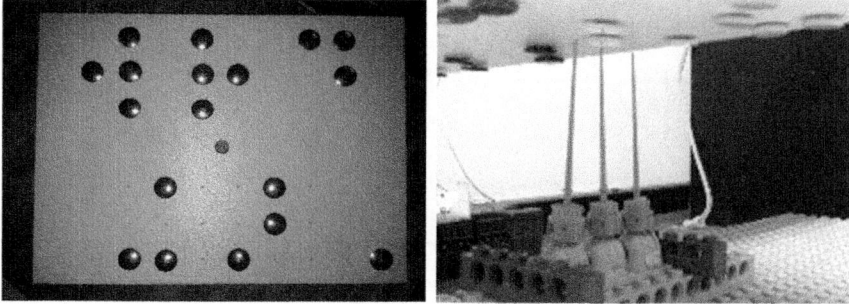

Fig. 2. The left panel shows an example of a braille pattern used as a stimulus. The right panel shows how the robot whiskers were set up and how the braille stimulus was swept over them for the current implementation.

The sensor, which has been previously described in [6], detects deflection in the whisker from its resting position on two orthogonal axes x and y and converts it into two voltage signals which are proportional to the whisker displacement on the two axes. For the current implementation, the displacement amplitude in the whiskers, when the stimulus was moved over them, was used. The signal from the sensors was segmented and an *activation* $a_m(t)$, for each sensor m at timestep t, was defined as a function of the displacement amplitude $d_m(t)$ over the tth segment,

$$a_m(t) = \frac{1}{1 + e^{-d_m(t)}} \tag{1}$$

where $d_m(t)$ is the deflection in the whisker sensor m at time t and $a_m(t)$ is the corresponding activation of the sensor. The whisker deflection signal was segmented and converted into an activation for each sensor to simplify the implementation of the model. The duration of the time window for each segment was defined to match the relative frequency with which non-zero patterns tend to occur in the inputs. Defining the activation function is not critical to the working of HTM, but since braille patterns are essentially binary, this transformation was used to make the *activation patterns* more representative of the characteristics of the stimuli (as shown in the lower panel in Figure 3).

The arrangement of the whisker sensors used in the current implementation is similar to the way in which the microvibrissae of the rat are used to explore patterns for shape recognition (see Figure 4). It has been shown in behavioral studies that the microvibrissae are critically involved in object/pattern recognition, as

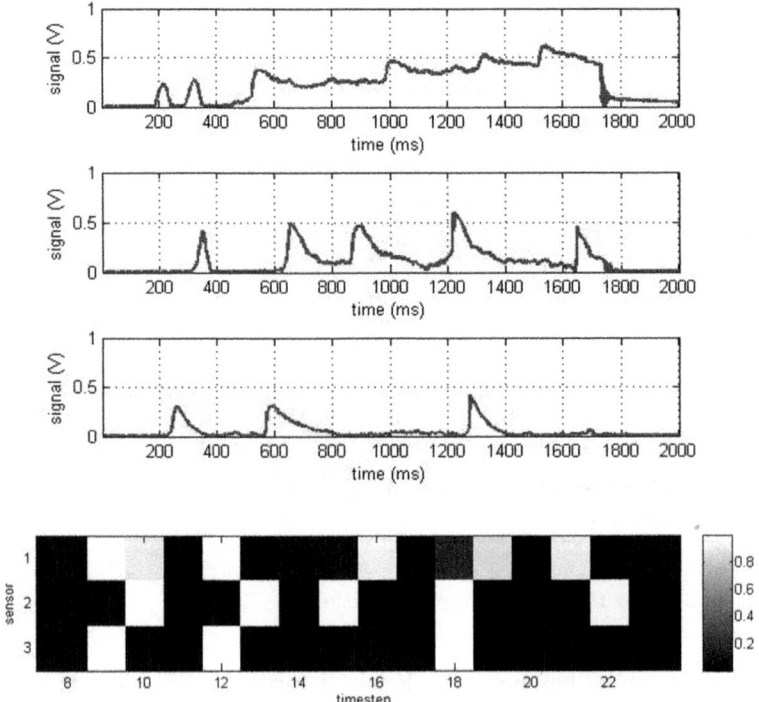

Fig. 3. The upper panels in the figure shows an example of a set of unprocessed signals from the whisker sensors. The top panel is an example of noisy sensory data, corrupt with a significantly large low frequency component and the second panel is an example of data which clearly reflects the features in the stimulus. These whisker deflection timeseries were filtered to remove low frequency components, segmented and the corresponding activation patterns for each sensor at timesteps corresponding to the segments of the timeseries are shown in the lower panel.

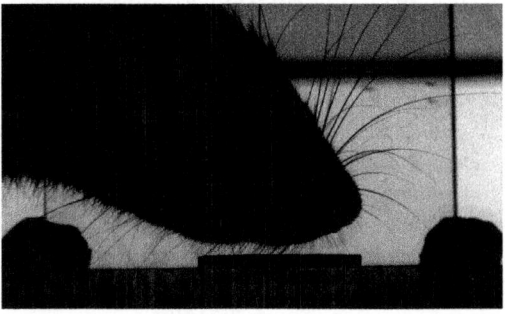

Fig. 4. The figure shows an image of a rat using its microvibrissae to explore a coin

opposed to the role of the macro-vibrissae in object localization [4]. And hence, the current implementation of HTM is also being suggested as a model of processing in the primary sensory cortex associated with the microvibrissae.

HTM Implementation

Inference in HTM nodes is with probabilistic message passing based on the Bayesian Belief Propagation algorithm [19] and the complete mathematical description of inference in HTM has been previously presented [10].

A key feature in Bayesian probabilistic inference is that it is based not only on current evidence, but also past evidence to resolve ambiguity. To this end, current evidence, of the form of a *likelihood* distribution, is combined, using the

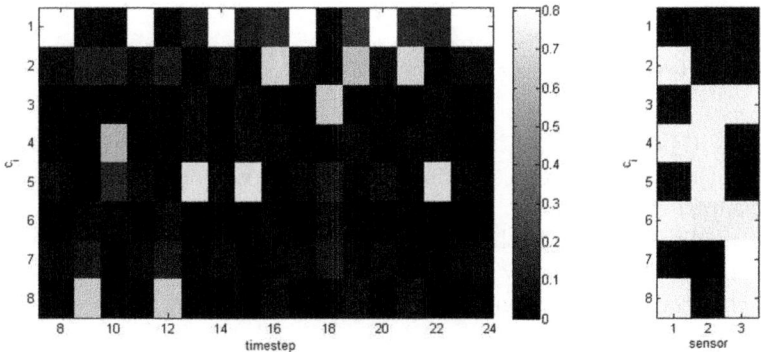

Fig. 5. The left panel in the figure shows the likelihood $y(i)$ which the node receives over its coincidence patterns. Each column in the image represents a probability distribution over spatial patterns c_i for each time step. The right panel shows the activation expected in each sensor corresponding to each of the spatial patterns c_i represented in the node.

Bayes Theorem, with a *prior* distribution, to obtain a *belief* distribution over states or events.

The feedforward input to the HTM node at each time step is the likelihood of the spatial pattern of activation in the whisker sensors that the node pools over. The activation patterns can be represented in m-dimensional space, where m is the number of sensors/children the spatial pattern is detected over. The likelihood y can then be defined by soft classification of data points in this space into clusters associated with the node's spatial patterns using any measure of degree belonging of data points to each of the clusters,

$$y_t(i) = P(A(t)|c_i) \qquad (2)$$

where c_i represents coincident pattern i and $A(t) = a_1(t), a_2(t)...a_m(t)$. The coincident patterns represented in the node correspond to sub-characters in the braille pattern.

The likelihood y is combined with a prior distribution $\pi(c)$ to obtain a belief $Bel(c)$ over the node's spatial patterns.

$$Bel_t(c_i) = \frac{1}{Z} y_t(i) \pi_t(c_i) \qquad (3)$$

where Z is the normalization constant.

Sequences represented in the node are assumed to be Markov chains, *i.e.* they are sequences of states in which the state of the system at any given time t is dependent on the previous states of the system, hence, the prior $\pi(c)$ and the likelihood $\lambda(g)$ are computed from the prior $\pi(g)$ and the likelihood y, respectively, using transition probabilities between the node's spatial patterns.

$$\pi_t(c_i) = \sum_{c_j} \sum_{g_r} P(c_i(t)|c_j(t-1), g_r) \pi(g_r), \qquad (4)$$

and

$$\lambda_t(g_r) = P(A(0 \to t)|g_r) = \sum_{c_i} y(i) \sum_{c_j} P(c_i(t)|c_j(t-1), g_r) \qquad (5)$$

where $P(c_i(t)|c_j(t-1), g_r)$ is the transition probability represented within the node. All the internal representations of the node were predefined in the current implementation.

This likelihood $\lambda_t(g)$ is the feedforward message that the node sends to the node above it in a hierarchy (*i.e* its *parents*) and the prior, which is sent to the node's children, is a function of the belief $Bel_t(c)$.

4   Results

For the current implementation the working of a single HTM node was instantiated. The coincidence patterns in the node represent single columns of Braille alphabets (see Figure 2), and the sequences represent entire Braille alphabets.

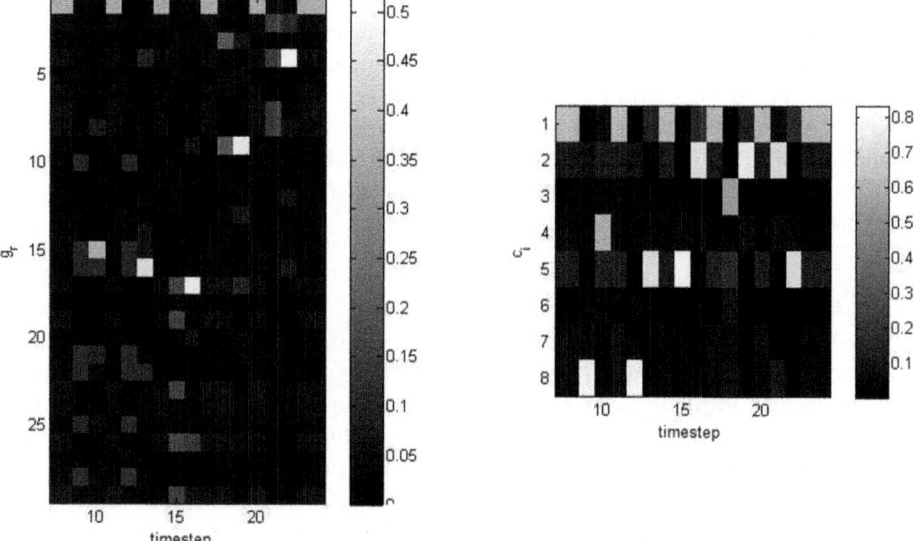

Fig. 6. The right panel in figure shows the node's belief over its coincidence patterns $Bel(c)$ and the left panel shows the feedforward likelihood over its sequences $\lambda(g)$. Each column in both images represent a probability distribution at that time step.

The results from the current implementation show how the probability distributions over the node's spatial patterns and sequences, *i.e.* the distributions over the nodes internal representations, change over time in response to the inputs from the whisker sensors (as shown in Figure 6).

Since the spatial patterns and sequences were predefined in the node, the accuracy with which the node inferred could be easily verified by matching the patterns in the node, over which the probability distributions peaked, with the patterns in the stimulus. For example, the first spatial pattern as well as the first sequence represented in the node correspond to blank patterns in the stimulus. The time steps at which the distributions peak on the first pattern indicate that the the node has inferred blank patterns in the stimulus at those time steps (see Figure 5). It is evident from number of time steps between blank patterns that sequences of non-zero patterns in the data are only two patterns long. This is a true of all braille characters which are formed by precisely arranged raised dots which occur only in six positions (two columns of three rows each) for each character (as shown in Figure 2). In other words, the patterns associated with the peaks in the belief distribution over the node's spatial patterns and the likelihood over its sequences match those in the stimulus, *i.e.* the patterns over which the node infers the highest degree of certainty match the actual patterns in the stimuli.

The results also show that the likelihood distribution over the sequences has a relatively high level of ambiguity, *i.e.* it is relatively broad, on the first time step when any non-zero sequence begins in the data. However, the distribution over the sequences sharpens over the next time step. Even in the general case, with sequences of any arbitrary length, it would be expected that the distribution over the node's sequences would become progressively sharper as more evidence is accumulated. In other words, since it is assumed that causes can share common sub-components, the degree of certainty in the state of a cause increases as more evidence becomes available.

5   Discussion

The sensor array, stimulus as well as the model architecture were all kept minimal in the current implementation to investigate thoroughly the inner workings of the HTM model. Although the data set of stimulus patterns used here was very simple, the spatio-temporal structure of braille patterns as a tactile stimuli is sufficiently rich to use them to test pattern recognition in HTM. The results show that the discussed architecture can successfully use spatio-temporal patterns in inputs to infer complex causes. This architecture can further be extended to receive sensory inputs from a larger array of sensors and by implementing a more elaborate hierarchy of nodes, to represent and infer with more complex spatio-temporal patterns. It is anticipated that insights from the current work are likely to be instrumental, within the general framework of developing robotic multiwhisker touch, for recognizing complex behaviorally relevant objects and patterns.

Key Challenges in Using HTM

For the current implementation, since the structure of the stimulus used was extremely simple, it was relatively straightforward to predefine the internal representations in the HTM node with which inference was implemented. However, a key challenge with more extended HTM hierarchies is that the nature of the node's internal representations become less intuitive as we go further up in the hierarchy. Hence, it becomes crucial that adequate learning mechanisms are defined for the model to extract these representations directly from its inputs.

The temporal proximity constraint is a powerful idea in learning invariant representations of stable causes in the world based on rapidly changing, ambiguous sensory inputs. Pooling over local spatial coincidences in a hierarchical arrangement allows representing progressively more complex spatial patterns from simpler components/features. By analogy, patterns of transitions between spatial patterns can be used to learn hierarchical representations of larger temporal patterns in inputs to represent stable/invariant features of causes in the world. However, learning complex sequences from transition probabilities between spatial patterns is not the same as learning complex spatial patterns from local coincidences. Information about inputs at previous time steps is not available to the system from successions in the same sort of way as is the information about larger spatial patterns from local spatial coincidences. Hence, learning associations between complex sequences of patterns and higher level causes is still a key challenge.

Finally, it has been pointed out that, "if it were not for the fact that we can see, we might reasonably think it is impossible" [9]. This is true not only for vision but for several problems that nature seems to have found elegant solutions for and this becomes strikingly evident in trying to replicate these functions in artificial systems. Biomimetic robotics is a powerful tool to reverse engineer solutions available in biological systems, that have evolved to efficiently handle several hard problems, and to optimize these strategies for specific problems of interest.

Acknowledgments. The authors would like to thank Alex Cope for comments on the manuscript and for help on several occasions in the implementation of the presented work. They would also like to thank Stuart Wilson and Mat Evans of ATLAS (the Active Touch Laboratory at Sheffield) for their help with data collection, as well as, Javier Caballero and Darren Lincoln, at the university of Sheffield, for several helpful discussions. This work was supported by EU Framework project BIOTACT (ICT-215910).

References

1. Biotact consortium, http://www.biotact.org
2. Numenta customers, http://www.numenta.com/about-numenta/customers.php
3. Berkes, P., Wiskott, L.: Slow feature analysis yields a rich repertoire of complex cell properties. Journal of Vision 5(6) (2005)
4. Brecht, M., Preilowski, B., Merzenich, M.: Functional architecture of the mystacial vibrissae. Behavioural Brain Research 84(1-2), 81–97 (1997)
5. Doya, K.: Bayesian brain: Probabilistic approaches to neural coding. The MIT Press, Cambridge (2007)
6. Evans, M., Fox, C.W., Pearson, M.J., Lepora, N.F., Prescott, T.J.: Whisker-object contact speedaects radial distance estimation. In: IEEE International Conference on Robotics and Biomimetics (2010)
7. Felleman, D.J., Van Essen, D.C.: Distributed hierarchical processing in the primate cerebral cortex. Cerebral Cortex 1(1), 1 (1991)
8. Földiák, P.: Learning invariance from transformation sequences. Neural Computation 3(2), 194–200 (1991)
9. Frisby, J., Stone, J.: Seeing: the computational approach to biological vision. The MIT Press, Cambridge (2009)
10. George, D., Hawkins, J.: Towards a mathematical theory of cortical micro-circuits. PLoS Comput. Biol. 5(10), e1000532 (2009)
11. George, D.: How to make computers that work like the brain. In: Proceedings of the 46th Annual Design Automation Conference, DAC 2009, pp. 420–423. ACM, New York (2009)

12. Hartung, J., McCormack, J., Jacobus, F.: Support for the use of hierarchical temporal memory systems in automated design evaluation: A first experiment. In: Proceedings of the ASME 2009 International Design Engineering Technical Conferences & Computers and Information in Engineering Conference, San Diego, CA, USA (2009)

13. Hinton, G.: Connectionist learning procedures. Artificial Intelligence 40(1-3), 185–234 (1989)

14. Hubel, D., Wiesel, T.: Receptive fields, binocular interaction and functional architecture in the cat's visual cortex. The Journal of Physiology 160(1), 106 (1962)

15. Knill, D., Pouget, A.: The Bayesian brain: the role of uncertainty in neural coding and computation. Trends in Neurosciences 27(12), 712–719 (2004)

16. Mountcastle, V.: An organizing principle for cerebral function: The unit model and the distributed system. In: Edelman, G., Mountcastle, V. (eds.) The Mindful Brain. MIT Press, Cambridge (1978)

17. Numenta: Hierarchical temporal memory: Comparison with existing models. Tech. rep., Numenta (2007)

18. Numenta: Problems that fit htm. Tech. rep., Numenta (2007)

19. Pearl, J.: Probabilistic reasoning in intelligent systems: networks of plausible inference. Morgan Kaufmann, San Francisco (1988)

20. Rozado, D., Rodriguez, F., Varona, P.: Optimizing hierarchical temporal memory for multivariable time series. In: Diamantaras, K., Duch, W., Iliadis, L.S. (eds.) ICANN 2010. LNCS, vol. 6353, pp. 506–518. Springer, Heidelberg (2010)

21. Thornton, J., Gustafsson, T., Blumenstein, M., Hine, T.: Robust character recognition using a hierarchical Bayesian network. In: Sattar, A., Kang, B.-h. (eds.) AI 2006. LNCS (LNAI), vol. 4304, pp. 1259–1264. Springer, Heidelberg (2006)

Short-Range Radar Perception
in Outdoor Environments

Giulio Reina[1], James Underwood[2], and Graham Brooker[2]

[1] Department of Engineering for Innovation, University of Salento, Lecce, Italy
giulio.reina@unisalento.it
[2] Australian Centre for Field Robotics, University of Sydney, Sydney, Australia
{j.underwood,gbrooker}@acfr.usyd.edu.au

Abstract. For mobile robots operating in outdoor environments, perception is a critical task. Construction, mining, agriculture, and planetary exploration are common examples where the presence of dust, smoke, and rain, and the change in lighting conditions can dramatically degrade conventional vision and laser sensing. Nonetheless, environment perception can still succeed under compromised visibility through the use of a millimeter-wave radar. This paper presents a novel method for scene segmentation using a short-range radar mounted on a ground vehicle. Issues relevant to radar perception in an outdoor environment are described along with field experiments and a quantitative comparison to laser data. The ability to classify the scene and significant improvement in range accuracy are demonstrated showing the utility of millimeter-wave radar as a robotic sensor for persistent and accurate perception in natural scenarios.

1   Introduction

Accurate and robust perception is critical for an autonomous robot to successfully accomplish its tasks in challenging environments. Imaging sensors can provide obstacle avoidance, task-specific target detection and generation of terrain maps for navigation. However, visibility conditions are often poor in field scenarios. Day/night cycles change illumination conditions. Weather phenomena such as fog, rain, snow and hail impede visual perception. Dust clouds rise in excavation sites, agricultural fields, and they are expected during planetary exploration. Smoke also compromises visibility in fire emergencies and disaster sites. While laser scanners and (stereo) cameras are common imaging sensors affected by these conditions [11], radar operates at a wavelength that penetrates dust and other visual obscurants and it can be successfully used as a complementary sensor to conventional range devices. In addition, radar can provide information of distributed and multiple targets that appear in a single observation. However, radar has shortcomings as well, such as large footprint, specularity effects, and limited range resolution, all of which may result in poor environment survey or difficulty in interpretation. Typically, radar outputs power-downrange arrays, i.e. a single sensor sweep contains n samples at discrete range increments dR along

R. Groß et al. (Eds.): TAROS 2011, LNAI 6856, pp. 265–276, 2011.

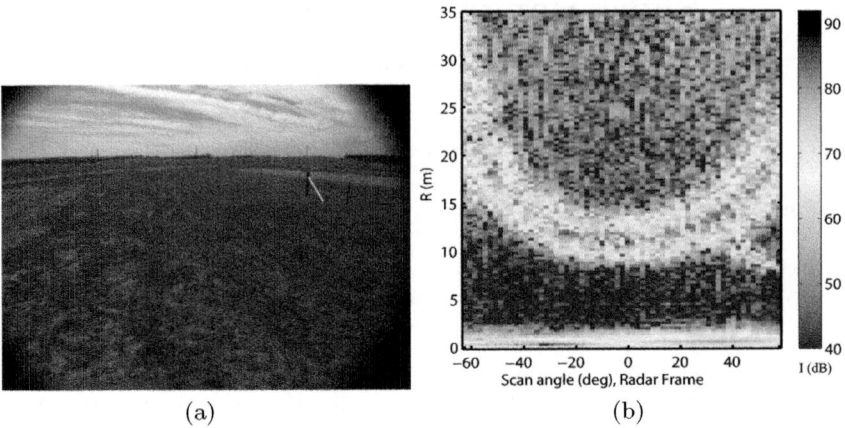

Fig. 1. A sample radar image acquired from a large flat area: camera image approximately colocated with the radar (a), azimuth angle-range image (b)

each azimuth or scan angle. As an example, Fig. 1(b) shows a bidimensional intensity graph of the radar data (radar image) acquired from a large, relatively flat area (Fig. 1(a)). The abscissas in Fig. 1(b) represent the horizontal scanning angle. The ordinates represent the range measured by the sensor. Amplitude values above the noise level suggest the presence of objects with significant reflectivity. Amplitude close or below the noise level generally corresponds to the absence of objects but exceptions exist. These include specular reflecting surface aligned to reflect the signal away, a highly absorbing material, or a total occlusion of radiation. One interesting feature of the radar image is the ground echo, i.e. the intensity return scattered back from the portion of terrain that is illuminated by the sensor beam. In the presence of relatively flat terrain, the ground echo appears as a high-intensity parabolic sector in the radar image (see Fig. 1(b)). This sector is referred to as the radar image background throughout the paper. The ability to automatically identify and extract radar data pertaining to the ground and project them onto the vehicle body frame or navigation frame would result in an enabling technology for all visibility-condition navigation systems. In previous research by the authors, a theoretical model describing the geometric and intensity properties of the ground echo in radar images was described [10]. Here, this model serves as a basis for the development of a novel method for Radar Ground Segmentation (RGS), which allows classification of observed ground returns in three broad categories, namely ground, non-ground, and unknown. The RGS system also improves the accuracy in range estimation of the detected ground for enhanced environment mapping. In addition, non-ground (i.e. obstacles) present in the foreground can also be detected and ranged independently of the ground as high-intensity peaks. Detection and segmentation of ground in a sensor-generated image is a challenging problem with many

Fig. 2. The CORD UGV employed in this research (a), and its sensor suite (b)

applications in perception. This is a key requirement for scene interpretation, segmentation and classification, and it is important for autonomous navigation [9]. Obstacle detection and avoidance has been commonly performed using ranging sensors such as stereo vision, laser or radar to survey the 3-D shape of the terrain. Some features of the terrain such as slope, roughness, or discontinuities have been then analyzed to segment the traversable regions from the obstacles [7]. In addition, some visual cues such as color, shape and height above the ground have also been employed for segmentation in [3], [6]. Relatively little research has been devoted to investigate millimeter-wave radar for short-range perception and three-dimensional terrain mapping. For example, a millimeter-wave radar-based navigation system detected and matched artificial beacons for localization in a two-dimensional scan [2]. Radar capability was demonstrated in a polar environment [5] and for mining applications [1].

In this investigation, a mechanically scanned millimeter-wave radar is employed. The sensor, custom built at the Australian Center for Field Robotics (ACFR), is a 95-GHz Frequency Modulated Continuous Wave (FMCW) millimeter-wave radar that reports the amplitude of echoes at ranges between 1 and 120 m. The wavelength is $\lambda=3mm$, and the 3 dB beamwidth is about 3.0 deg in elevation and azimuth. The horizontal field of view is 360 deg with an angle scan rate of about 3 rps. The range resolution is about 0.32 m at 20 m [1]. For the extensive testing of the system during its development, we employed the CAS Outdoor Research Demonstrator (CORD) that is an 8 wheel skid-steering all-terrain unmanned ground vehicle (UGV) (see Fig. 2(a)). In Fig. 2(b), the radar is visible, mounted to a frame attached to the vehicle's body and tilted forward so that the center of the beam intersects the ground at a look-ahead distance of about 11.4 m. The remainder of this paper is organized as follows. The model of ground echo is recalled in Section 2 and the RGS method is described in detail in Section 3. In Section 4, the RGS system is validated in field tests performed with the CORD UGV. Relevant conclusions are drawn in Section 5.

2   Ground Echo Modeling

For robot perception, an accurate range map of the environment can be obtained from a radar through the scanning of a narrow beam, which is usually referred to as a pencil beam. In the proposed system, the radar is directed at the front of the vehicle with a constant forward pitch to produce a grazing angle β of about 11 deg, as shown in Fig. 3(a). The origin of the beam at the center of the antenna is O. The proximal and distal borders of the footprint area illuminated by the divergence beam are denoted with A and B, respectively. The height of the beam origin with respect to the ground plane is h. The slant range of the radar bore sight is R_0, the range to the proximal and distal borders is denoted with R_1 and R_2, respectively. The theoretical model of the ground echo in the radar image was previously developed by the authors. It provides prediction of the range spread of the ground return along with its expected power spectrum. We recall the important properties of the model and refer the reader to [10] for more details. The expected slant distance R_0 can be expressed as a function of the azimuth angle α and the tilt of the robot

$$R_0 = \frac{h}{\cos \theta \cdot \sin \alpha \cdot \sin \phi - \sin \theta \cdot \cos \alpha} \tag{1}$$

where h is the height of the radar from the ground (see Fig. 3(a)), and ϕ and θ are the roll and pitch of the robot, respectively. Similarly, the range of the proximal and distal borders R_1 and R_2 (points A and B in Fig. 3(a)), can be obtained

$$R_1 = \frac{h}{\cos \theta_{el}(\cos \theta \cdot \sin \alpha \cdot \sin \phi - \sin \theta \cdot \cos \alpha) - \cos \theta \cdot \cos \phi \cdot \sin \theta_{el}} \tag{2}$$

$$R_2 = \frac{h}{\cos \theta_{el}(\cos \theta \cdot \sin \alpha \cdot \sin \phi - \sin \theta \cdot \cos \alpha) + \cos \theta \cdot \cos \phi \cdot \sin \theta_{el}} \tag{3}$$

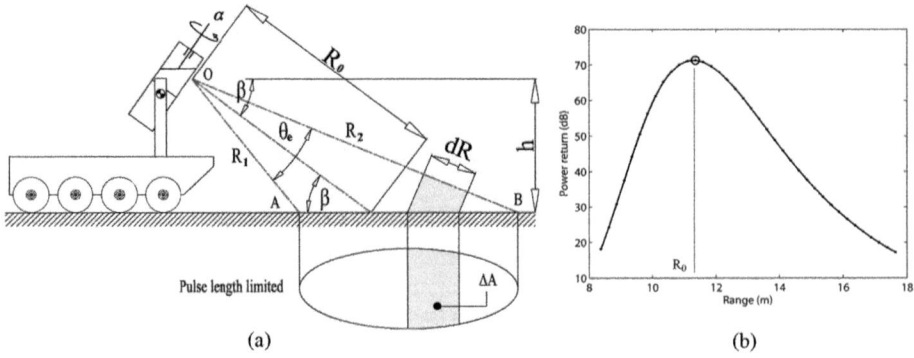

(a)   (b)

Fig. 3. Scheme of a pencil beam (of beamwidth θ_e) sensing terrain at grazing angle β (a). Simulated power return of the ground echo (b); the following parameters were adopted in the simulation: k=70dB, R_0=11.3m, h=2.2m

Under the assumption that the radar-illuminated ground falls within the so-defined near-field region (i.e. approximately $R < 16m$, [10]), the power return of the ground echo can be expressed as a function of the range R

$$P_r(R, R_0, k) = k \frac{G(R, R_0)^2}{\cos \beta} \qquad (4)$$

where k is a constant quantity, G is the antenna gain (usually modeled as gaussian) and β is the grazing angle, as explained in Fig. 3(a). Figure 3(b) shows a simulated wide pulse of the ground return using (4). The model is defined by the two parameters k and R_0 that can be determined in practice by fitting the model to experimental data, as explained later in the paper. In summary, (1) and (4) represent two pieces of information defining the theoretical ground echo in the radar image. Any deviation in the range or intensity shape suggests low likelihood of ground return in a given radar observation.

3 The Radar Ground Segmentation System

A radar image can be thought of as composed of a foreground and a background. The background is referred to as the part of the image containing ground echoes that appear as wide pulses due to the high-incident angle surface. Conversely, obstacles present in the foreground can be detected and ranged as high-intensity narrow pulses. The RGS system aims to assess ground by looking at the image background obtained by a millimeter-wave radar mounted on a mobile robot. It performs two main tasks:

- Background extraction from the radar image.
- Analysis of the power spectrum across the background to perform ground segmentation.

3.1 Background Extraction

Prediction of the range spread of the ground echo as a function of the azimuth angle and the tilt of the vehicle can be obtained using the geometric model presented in Section 2. It should be recalled that the model is based on the assumption of globally flat ground. Therefore, discrepancies in the radar observations may be produced by the presence of local irregularities or obstacles in the radar-illuminated area. In order to relax the assumption of global planarity and compensate for these effects, a change detection algorithm is applied in the vicinity of the model prediction to refine the search. For the mathematical details, we refer the reader to [10], rather than repeating this material here. A typical result is shown in Fig. 4. The radar signal obtained from a single azimuth observation ($\alpha=32$deg) is denoted by a solid gray line. The theoretical prediction of the range spread of the ground return is shown by black points in the bottom of Fig. 4 representing the range of the central beam, the proximal and distal borders, points R_0, R_1 and R_2, respectively. When a positive change in the

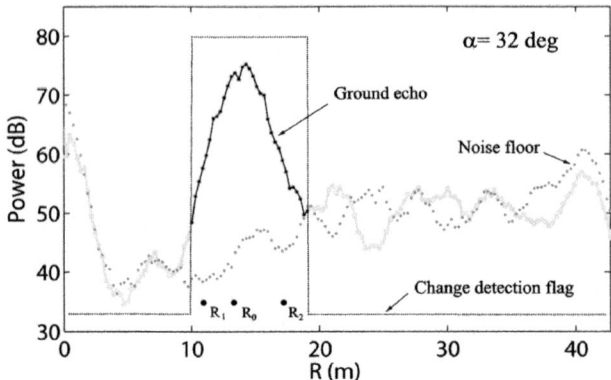

Fig. 4. Ground echo extraction in the radar signal through a change detection approach (a): radar signal at scan angle α =32 deg (gray solid line), extracted ground echo (solid black line), change detection flag (dotted black line). Note that the opposite (i.e. 180-deg scan angle difference) radar signal is also plotted (gray dotted line), it points skyward and no obstacle is detected in it, thus showing the typical noise floor in the radar measurement.

radar signal is found in the vicinity of the proximal border (in practise within a 1-m window centered on R_1), a flag is raised (dotted black line). The alarm is lowered when a negative change is detected in the vicinity of the distal border. The ground echo can, then, be extracted from the given observation (portion of the signal denoted in black). The process can be repeated for the whole scanning range, and the background of the radar image can be successfully extracted.

3.2   Ground Segmentation

The image background contains ground candidates. In order to define a degree of confidence in actual ground echo, the power return model, presented in Section 2, can be fitted to a given radar observation. The hypothesis is that a good match between the parametric model and the data attests to a high likelihood of ground. Conversely, a poor goodness of fit suggests low likelihood due, for example, to the presence of an obstacle or to irregular terrain. We recall that $P_r(R)$, is a function defined by the parameters R_0 and k. k can be interpreted as the power return at the slant range R_0 and both parameters can be estimated by data fitting for the given azimuth angle. By continuously updating the parameters across the image background, the model can be adjusted to local ground roughness and produce a more accurate estimation of R_0, as shown later in the paper. A non-linear least squares approach using the Gauss-Newton-Marquardt method is adopted for data fitting. The initial parameter estimates are chosen as the maximum measured power value and the predicted range of the central beam as expressed by (1), respectively, limiting the problems of ill conditioning and divergence. Output from the fitting process are the updated parameters $\bar{R}_0$ and $\bar{k}$ as well as an estimate of the goodness of fit. The coefficient of efficiency was found to be well suited for this application

$$E = 1 - \frac{\sum(t - y)^2}{\sum(t - \bar{t})^2} \tag{5}$$

t being the data point, $\bar{t}$ the mean of the observations, and y is the output from the regression model. E ranges from $-\infty$ to 1, as the best possible value. By evaluating the coefficient of efficiency and the model parameters, ground segmentation can be effectively performed and radar observations can be labeled as ground, unknown, and non-ground object (i.e. obstacle). Typical results are shown in Fig. 5. Specifically, in Fig. 5(a), the model matches very well the experimental data with a high coefficient of efficiency $E{=}0.96$, thus attesting to the presence of ground. Conversely, Fig. 5(b) shows an example where the goodness of fit is poor ($E < 0$); in this case a low confidence in ground echo is associated with the given observation. In practice, a threshold Th_E is experimentally determined and the observation i is labeled as ground if E_i exceeds Th_E. However, relying on the coefficient of efficiency, may be misleading in some cases. Figure 5(c) shows an example where a ground patch would be seemingly detected according to the high coefficient of efficiency ($E = 0.91$), when there is actually no ground return. In order to solve this issue, a physical consistency check can be performed by looking at the updated value of the proximal and central range as estimated by the fitting process. For this case, they result almost coincident ($\bar{R}_0{=}10.82\ m$ and $\bar{R}_1{=}10.42\ m$, respectively), and certainly not physically consistent with the model described in Section 2. Therefore, the radar observation is labeled as uncertain ground, if the difference between the central and proximal range is lower than an experimentally-defined threshold Th_R. An analogue comparison is done between the distal and central border, as well. In case of uncertain terrain, an additional check is performed to detect possible obstacles present in the region of interest, which would appear as narrow pulses of high intensity. In this respect, it should be noted that, during operation, the RGS system records the value of $\bar{k}$, defining a variation range for the ground-labeled observation. Typically, $\bar{k}$ was found to range from 73 to 76 dB. If a percentage relative change in the maximum intensity value between the uncertain-labeled observation t_{max}, and the model y_{max} is defined, $\Delta P = \frac{t_{max} - y_{max}}{t_{max}}$, then an obstacle is detected when ΔP exceeds an experimentally defined threshold Th_p and, at the same time, t_{max} is greater than the maximum value of $\bar{k}$. An example of obstacle (labeled as non-ground) detection is shown in Fig. 5(d). In summary, the classification approach shown in Table 1 can be defined.

4   Experimental Results

In this section, field results are presented to validate our approach for ground segmentation using a millimeter-wave radar. The test field was located in a rural

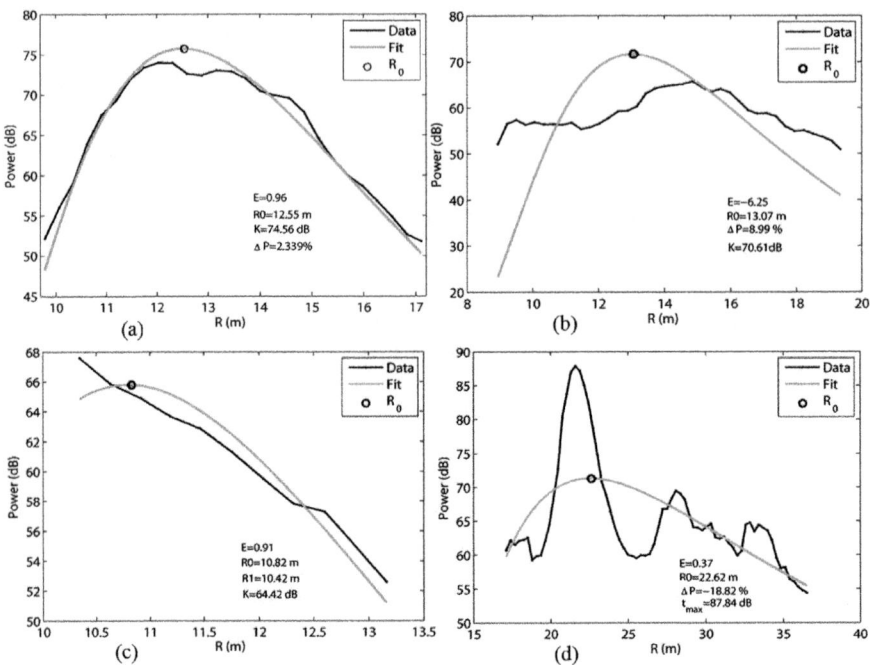

Fig. 5. Ground segmentation by model fitting: good fit labeled as ground (a), poor fit labeled as uncertain ground (b), seemingly high fit labeled as uncertain ground due to physics inconsistency with the model (c), narrow pulse labeled as obstacle (d)

Table 1. Set of rules used by the RGS system for background classification

Class	Goodness of fit $(Th_E = 0.8)$	Parameters of the regression model $(Th_R = 1.5m, Th_P = 10\%)$			
	E	$(\bar{R}_0 - \bar{R}_1)$	$(\bar{R}_2 - \bar{R}_0)$	ΔP	$\bar{k}$ (dB)
Ground	$\geq Th_E$	$> Th_R$	$> Th_R$	$< Th_P$	73–76
Unknown	$< Th_E$	-	-	$< Th_P$	-
Non-ground	$< Th_E$			$\geq Th_P$	$> \bar{k}_{max}$

environment at the University of Sydney's test facility near Marulan, NSW, Australia. It was mainly composed of a relatively flat ground with sparse low grass delimited by fences, a static car, a trailer and a metallic shed, as shown in Fig. 6. During the experiment, the CORD vehicle was remotely controlled to follow an approximately closed-loop path of 210 m with an average travel speed of about 0.5 m/s. In this experiment, the RTK DGPS/INS unit and a high-precision 2D SICK laser range scanner provided the ground truth. The full data set is public and available online [8].

Fig. 6. The Marulan test field is generally flat with a notable exception due to a road running through the side including a significant berm on each edge

4.1   Ground Segmentation and Range Accuracy

Figure 7 shows a typical result obtained during the experiment with a stationary car to the right of the robot. Figure 7(a) shows the radar intensity image overlaid with the results obtained from the RGS system. Ground labels are denoted by black dots, a black cross marks uncertain terrain, and non-ground belonging to the foreground (high-intensity narrow pulses) is denoted by a black triangle. No obstacle belonging to the background is detected in this scan. In Figure 7(b), the results are projected over the image plane of the camera for visualization purposes only. Finally, a comparison with the laser-based ground truth is provided in Figure 7(c), which demonstrates the effectiveness of the proposed approach for ground segmentation. As it can be seen from this figure, the RGS system correctly detected the flat ground area in front of the robot and the obstacle to the right. Uncertain terrain was flagged along the portion of the background occluded by the car and to the far left due to the presence of highly irregular terrain. Overall, the RGS system was tested over 1,100 radar images each containing 63 azimuth observations for a total of 69,300 classifications. As a measure of the segmentation performance, the false positive and false negative rate incurred by the system during classification of ground and non-ground was evaluated by comparison with the ground truth laser data. To this aim, a previously proposed method for segmentation of laser data (GP-INSAC, [4]) was applied to the ground-truth map to extract the true ground and true obstacles. As described in Section 3.2, whenever the RGS system labels data along a particular scan azimuth as ground or non-ground, a range reading is returned. When combined with the localisation estimation of the vehicle, this provides a 3-D georeferenced position for the labeled point. A ground-labeled observation is counted as a false positive if a closest neighbor can not be found in the true ground data within a minimum distance (less than 0.5 m in our case). Similarly, a non-ground-labeled point is counted as a false positive if it is not sufficiently close to the nearest true obstacle datum. The rate of false negatives is evaluated by manual inspection in each radar image, instead. The results are collected in Table 2. The percentage of false positives in ground-labeled observations was 2.1%, likely due to seeming matches produced by obstacles present

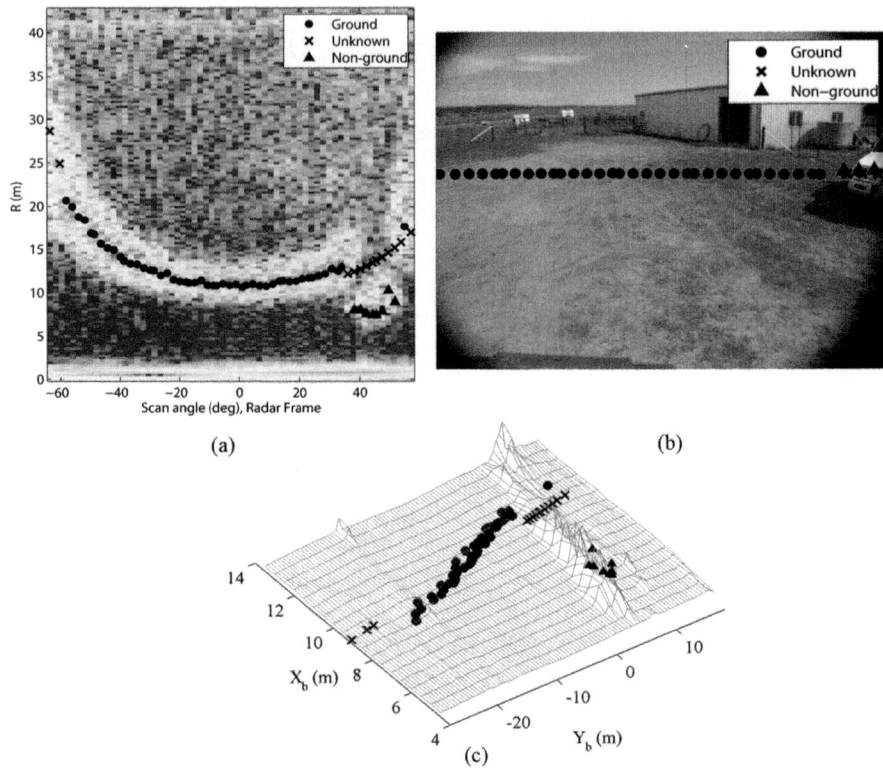

Fig. 7. Output of the RGS system (a), results overlaid over the camera image (b), and over the laser-generated ground-truth map (c)

in the illuminated area. No false positives were detected in the non-ground la-beled observations. The false negative rate for the ground-labeled observations was 3.5%. For non-ground-labeled observations, the false negative rate was 5.6%. The average percentage of unknown labels in a single radar image was 19.8%, including occluded areas and false negatives due to radar misreading or low res-olution (i.e., footprint overlapping part of an object and ground). It should be noted that false negatives mostly appear in the radar image as spurious observa-tions that do not affect the general understanding of the scene. The accuracy of the RGS system in mapping the ground was assessed through comparison with the true ground map. For the ground-labeled observation i, the RGS system outputs the relative slant range $\bar{R}_{0,i}$. Through geometric transformation, it is possible to estimate the corresponding 3-D point in the world reference frame P_i and to compare it to the closest neighbor of the ground truth map P_i^{gt}. Since the laser-generated map is available as a regularly-sampled grid with square cells of 0.3 m where the center of the grid represents the average height of the cell points, we refer to a mean square error in the elevation E_z. In this experiment, the RGS system detected ground returns in $n = 40150$ observations with an

Table 2. Segmentation results obtained from the RGS system

Class	Observations	False Positives (%)	False Negatives (%)	Accuracy
Ground	40150	2.1	3.5	$E_z=0.051m$
Non-ground	657	0.0	5.6	$E_{xy}=0.065m$

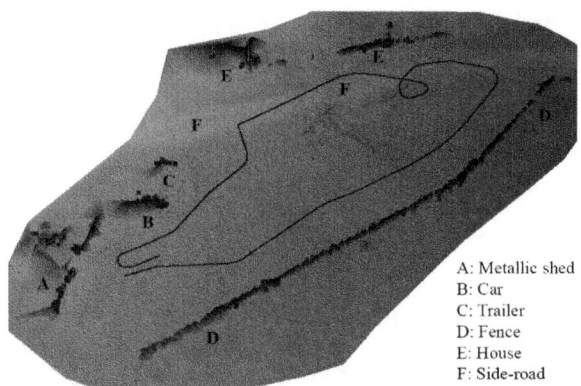

A: Metallic shed
B: Car
C: Trailer
D: Fence
E: House
F: Side-road

Fig. 8. Radar-generated map after Delaunay triangulation

error of $E_z = 0.051m$ and an associated variance of $\sigma_z = 0.002m^2$. If the value of R_0 is measured conventionally taking the intensity peak of the ground return, the error grows to $E_z = 0.251m$ and $\sigma_z = 0.181m^2$. Similarly, the accuracy of the system in measuring the position of detected obstacles can be evaluated by comparison with the nearest datum in the true obstacle map. A mean square error E_{xy} can be defined this time in terms of $x - y$ coordinates. The RGS system measured non-ground returns in $n_1 = 657$ observations with an error of $E_{xy} = 0.065m$ and a variance of $\sigma_{xy} = 0.0015m^2$, thus proving the effectiveness of the proposed approach. For a complete overview of the system performance, the results obtained from the RGS module along the entire experiment are used to build a map of the environment, as shown in Fig. 8 after applying a Delaunay triangulation. The ground labeled-observations are denoted by gray scale dots colored according to the elevation, whereas the obstacle-labeled points are shown by black points for higher contrast. The path followed by the robot is also shown by a solid black line. This figure demonstrates that the RGS system is capable of providing a clear understanding of the environment.

5   Conclusions

In this paper, a novel method for performing ground segmentation was presented using a MMW radar mounted on a off-road vehicle. It was based on the development of a physical model of the ground echo that was compared against a given radar observation to assess the membership confidence to one of the three broad categories of ground, non-ground, and unknown. In addition, the RGS system provided improved range estimation of the ground-labeled data for more accurate environment mapping, when compared to the standard highest intensity-based approach.

Acknowledgements. The Australian Department of Education, Employment and Workplace Relations is thanked for supporting the project through the 2010 Endeavour Research Fellowship 1745_2010. This research was undertaken through the Centre for Intelligent Mobile Systems (CIMS), and was funded by BAE Systems as part of an ongoing partnership with the University of Sydney. The financial support of the ERA-NET ICT-AGRI is also gratefully acknowledged.

References

1. Brooker, G., Hennessey, R., Bishop, M., Lobsey, C., Durrant-Whyte, H., Birch, D.: High-resolution millimeter-wave radar systems for visualization of unstructured outdoor environments. Journal of Field Robotics 23(10), 891–912 (2006)
2. Clark, S., Durrant-Whyte, H.F.: The design of a high performance mmw radar system for autonomous land vehicle navigation. In: Int. Conf. Field and Service Robotics, Sydney, Australia, pp. 292–299 (1997)
3. DeSouza, G., Kak, A.: Vision for mobile robot navigation: A survey. IEEE Transactions on Pattern Analysis and Machine Intelligence 24(2), 237–267 (2002)
4. Douillard, B., Underwood, J., Kuntz, K., Vlaskine, V., Quadros, A., Morton, P., Frenkel, A.: On the segmentation of 3-d lidar point clouds. In: IEEE Intl. Conf. on Robotics and Automation, Shanghai, China (2011)
5. Foessel-Bunting, A., Chheda, S., Apostolopoulos, D.: Short-range millimeter-wave radar perception in a polar environment. In: International Conference on Field and Service Robotics, Leuven, Belgium, pp. 133–138 (1999)
6. Jocherm, T., Pomerleau, D., Thorpe, C.: Vision-based neural network road and intersection detection and traversal. In: IEEE/RSJ Intl. Conf. on Intelligent Robots and Systems, Osaka, Japan (1995)
7. Pagnot, R., Grandjea, P.: Fast cross-country navigation on fair terrains. In: IEEE Intl. Conf. on Robotics and Automation, Hiroshima, Japan, pp. 2593–2598 (1995)
8. Peynot, T., Scheding, S., Terho, S.: The marulan data sets: Multi-sensor perception in a natural environment with challenging conditions. The International Journal of Robotics Research 29(13), 1602–1607 (2010)
9. Reina, G., Ishigami, G., Nagatani, K., Yoshida, K.: Odometry correction using visual slip-angle estimation for planetary exploration rovers. Advanced Robotics 24(3), 359–385 (2010)
10. Reina, G., Underwood, J., Brooker, G., Durrant-Whyte, H.: Radar-based perception for autonomous outdoor vehicles. Journal of Field Robotics 28 (2011)
11. Vandapel, N., Moorehead, S., Whittaker, W., Chatila, R., Murrieta-Cid, R.: Preliminary results on the use of stereo, color cameras and laser sensors in antarctica. In: Int. Sym. on Experimental Robotics, Sydney, Australia, pp. 1–6 (1999)

Smooth Kinematic Controller vs. Pure-Pursuit for Non-holonomic Vehicles

Vicent Girbés, Leopoldo Armesto, Josep Tornero, and J. Ernesto Solanes

Universitat Politècnica de València, València, Spain
{vigirjua,leoaran,jtornero,juasogal}@upvnet.upv.es

Abstract. The paper introduces a method for generating trajectories with curve sharpness and curvature constraints. Trajectories are based on continuous-curvature paths using Single & Double Continuous-Curvature paths. The paper also proposes a multi-rate control scheme by considering different dynamics with different sampling frequencies: global path planner at low frequency; target update based on Look-Ahead distance and path computation at medium frequency; curvature control at high frequency. An exhaustive analysis to evaluate the performance of the new method with respect to Pure-Pursuit method has been carried out, showing that the proposed method has better performance in terms of settling time, overshoot and robustness against design parameters. As a conclusion, we can say that our method introduces improvements in comfort and safety because the sharpness, normal jerk and mean abruptness are lower than with the Pure-Pursuit control.

Keywords: Path Generation, Path Following, Kinematic Control.

1 Introduction

In path following problems the goal is to develop a kinematic control law to guide a vehicle to converge to a path. This problem has been applied in common situations such as parking [8,5], overtaking and lane changing [12,10,16] and vision-based line following [9,2]. One well-known approach in order to solve such problem is based on the Pure-Pursuit method [15,11] which determines an appropriate curvature so that the vehicle is able to reach the path. Other approaches include linear or non-linear kinematic control law based on robot kinematic model to guarantee convergence.

It is well known that wheeled mobile robots following a path with continuous-curvature can get benefits in comfort and safety and may also reduce wheels slippage and odometry errors, since transitions are softer with constant curvature rates. In order to generate continuous-curvature paths some researchers used clothoids in navigation problems [1,17,7]. In [13], Elementary and Bi-Elementary paths were introduced, a combination of two/four symmetrical clothoids with the same homothety factor. In Bi-Elementary paths the initial and final configurations are not necessarily symmetric, but the loci of the intermediate configuration is restricted to a circle with specific orientations to ensure that each Elementary

R. Groß et al. (Eds.): TAROS 2011, LNAI 6856, pp. 277–288, 2011.

path contains symmetrical clothoids. Obviously, the solution space is significantly limited in those cases and Elementary and Bi-Elementary paths might not be appropriate to solve specific problems, specially the obstacle avoidance problem or the line following problem with bounded sharpness and curvature. Dubin's curves were the inspiration in [14] to create the SCC-paths (simple continuous-curvature paths).

Although comfort and safety increase when continuous curvature paths and trajectories are generated. These aspects are crucial in transporting people or dangerous goods. However, all these aspects have shown little attention and most of well known kinematic controllers do not consider continuous curvature profiles in path generation neither limit the curvature derivative bounds or sharpness.

In [4,3] we introduced the Double Continuous-Curvature path (DCC) and it was shown how to use it in path following problems within pure-pursuit framework. The contribution of the paper with respect to our previous work [4,3] is to provide an exhaustive benchmark to evaluate the performance of the new method with respect to Pure-Pursuit method. The main advantage is that the proposed curvature profile is always continuous, taking into account actual robot curvature as well as curvature and sharpness constraints. This aspect is exploited by providing a multi-rate control scheme by considering different dynamics involved in path following problems. In this sense, a high level path planner requires the slowest frequency update, a kinematic controller runs an intermediate frequency and low-level motor's curvature controller establishes the continuous curvature profile proposed by the kinematic controller. The paper also analyses the multi-rate ratio among the different inner loops in the overall system.

2   Double Continuous-Curvature Generation

Let $\mathcal{R}$ be a non-holonomic wheeled robot moving on a 2D plane with state space $\mathbf{q} = (x, y, \theta, \kappa)^T \in \Re^2 \times \mathcal{S} \times \Re$ containing the robot Cartesian positions x and y, the robot orientation θ and the curvature κ. The kinematic model for $\mathcal{R}$ is:

$$\dot{\mathbf{q}}(t) = \left(\dot{x}(t), \dot{y}(t), \dot{\theta}(t), \dot{\kappa}(t) \right)^T = \left(v(t)\cos(\theta(t)), v(t)\sin(\theta(t)), v(t)\kappa(t), v(t)\sigma(t) \right)^T \quad (1)$$

being $v(t)$, $\sigma(t)$ the velocity and sharpness to follow a path, respectively.

Let $\mathcal{R}$ be a non-holonomic wheeled robot moving on a 2D plane with bounded curvature $\kappa(t) \in [-\kappa_{max}, \kappa_{max}]$ due to mechanical constraints and with bounded sharpness $\sigma(t) \in [-\sigma_{max}, -\sigma_{min}] \cup [\sigma_{min}, \sigma_{max}]$ introduced to increase safety by satisfying comfort limits.

The goal is to generate a continuous-curvature path $\mathcal{P}$ connecting the robot pose $\mathbf{q}_R = (x_R, y_R, \theta_R, \kappa_R)^T$ to a target configuration $\mathbf{q}_T = (x_T, y_T, \theta_T, \kappa_T)^T$. Such target configuration can be dynamically recomputed from a set of way-points, $w_p = \{w_{p_0}, w_{p_1}, \ldots, w_{p_N}\}$ based on a global planner method such as wavefront planner as depicted in Figure 1, Left. In that case, the path formed

Fig. 1. Waypoints generation either a global planner or a vision system. Left: Wavefront planner. Right: Vision system.

by the poly-line as a consequence of joining way-points might be considered as the reference path to be followed, while the target configuration can be chosen to be any point of the reconstructed poly-line, based on the current robot pose and a given Look-Ahead (LA) distance, a common parameter used in pure-pursuit methods that forces the robot to move ahead. In addition to this, the target configuration can be chosen from detected road profile as shown in Figure 1, Right.

Double Continuous-Curvature paths (DCC) use two concatenated Single Continuous Curvature paths (SCC) [4,3] which are composed of a line segment, a first clothoid, a circle segment (arc) and a second clothoid. Additionally, a final line segment is added at the end of the second SCC-path to provide a set of general solutions. In a DCC-path, the first SCC-path is noted with subscript A and starts at configuration $\mathbf{q}_A = (x_A, y_A, \theta_A, \kappa_A)^T$ with $\kappa_A = 0$, while the second SCC-path is noted with subscript B finishing at configuration $\mathbf{q}_B = (x_B, y_B, \theta_B, \kappa_B)^T$, with $\kappa_B = 0$. The configuration joining both SCC-paths is $\mathbf{q}_C = (x_C, y_C, \theta_C, \kappa_C)^T$, with $\kappa_C = 0$. Figure 2(a) shows an example of a DCC-path together with its curvature profile. It can be appreciated that in that case a general DCC-path is composed of four clothoids named as **A1**, **A2**, **B1** and **B2**, two circular segments $\mathbf{\Omega_A}$ and $\mathbf{\Omega_B}$ and three line segments with length $\mathbf{l_A}$, $\mathbf{l_B}$ and $\mathbf{l_C}$ to properly guarantee the appropriate changes on the curvature.

Let us assume for simplicity that the sharpness of each clothoids pair $<$ **A1,A2** $>$ and $<$**B1,B2**$>$ is the same and given by:

$$\sigma_A = \alpha_A(\sigma_{max} - \sigma_{min}) + \sigma_{min}, \text{with } \alpha_A \in [0, 1] \qquad (2)$$

$$\sigma_B = \alpha_B(\sigma_{max} - \sigma_{min}) + \sigma_{min}, \text{with } \alpha_B \in [0, 1] \qquad (3)$$

being α_A, and α_B design parameters.

A standard DCC-path considers the case where $\kappa_R = 0$ and $\kappa_T = 0$. In that case, the length l_A of the first line (**l_A**) can take any arbitrary positive value or zero, i.e. $l_A \geq 0$. The same can be said for the length of the last line segment $l_B \geq 0$. In those cases, the $\mathbf{q}_A = \mathbf{q}_R$ and/or $\mathbf{q}_B = \mathbf{q}_T$. Another different situation is given when $\kappa_R \neq 0$ or $\kappa_T \neq 0$ since then $l_A = 0$ or $l_B = 0$. In that case, our assumption is to consider that the vehicle starts on a point of the first clothoid

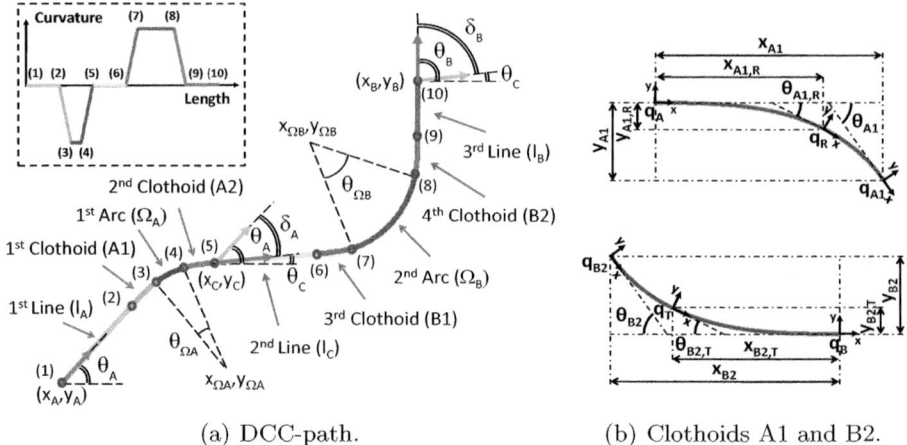

(a) DCC-path.   (b) Clothoids A1 and B2.

Fig. 2. Double Continuous-Curvature path definition and its curvature profile (a). Robot and Target configurations inside clothoids A1 and B2, respectively (b).

or the target point is contained on the last clothoid, see Figure 2(b). In those cases, the computation of $\mathbf{q}_A$ and $\mathbf{q}_B$ is:

$$\mathbf{q}_A = \mathbf{q}_R + \left(-\mathbf{R}(\theta_A)\left[x_{A1,R},\, s_A \cdot y_{A1,R}\right]^T,\, -s_A \cdot \theta_{A1,R},\, -\kappa_{A1,R}\right)^T \tag{4}$$

$$\mathbf{q}_B = \mathbf{q}_T + \left(\mathbf{R}(\theta_B)\left[x_{B2,T},\, -s_B \cdot y_{B2,T}\right]^T,\, s_B \cdot \theta_{B2,T},\, -\kappa_{B2,T}\right)^T \tag{5}$$

where $\{x_{A1,R}, y_{A1,R}\}$ is the clothoid points of $\mathbf{q}_R$ with respect to the clothoid frame as shown in Figure 2(b) and $\{x_{B2,T}, y_{B2,T}\}$ is the corresponding point for $\mathbf{q}_T$. In addition to this, $\theta_A = \theta_R - s_A \cdot \theta_{A1,R}$ and $\theta_B = \theta_T + s_B \cdot \theta_{B2,T}$, where $\theta_{A1,R}$ and $\theta_{B2,T}$ are the clothoid tangent angles and $s_A = \operatorname{sign}(\sigma_A)$ and $s_B = \operatorname{sign}(\sigma_B)$. For instance, clothoid points of $\mathbf{q}_R$ for a given curvature and sharpness can be computed as follows:

$$x_{A1,R} = s_A\sqrt{\frac{\pi}{\sigma_A}}\int_0^{\gamma_{A1,R}} \cos\frac{\pi}{2}\xi^2 d\xi \quad y_{A1,R} = s_A\sqrt{\frac{\pi}{\sigma_A}}\int_0^{\gamma_{A1,R}} \sin\frac{\pi}{2}\xi^2 d\xi \tag{6}$$

with $\gamma_{A1,R} = \sqrt{\frac{\kappa_R^2}{\pi\sigma_A}}$ and $\theta_{A1,R} = s_A\frac{\kappa_R^2}{2\sigma_A}$.

To obtain a standard DCC-path we have an additional degree of freedom which implies the computation of the angle for configuration $\mathbf{q}_C$. Once the angle θ_C is set, the remainder of variables can be computed based on geometric properties. For instance, deflection angles are obtained from configuration pairs $\delta_A = |\theta_C - \theta_A|$ and $\delta_B = |\theta_B - \theta_C|$. If $\delta_A > |\frac{\kappa_{max}^2}{\sigma_A}|$ or $\delta_B > |\frac{\kappa_{max}^2}{\sigma_B}|$ then the corresponding arc segments will cover the remainder angle. Otherwise, the DCC-path will not include arc segments and the clothoid-pairs segments will compensate such deflection angles. The remainder parameters to be computed are lengths l_A, l_B and l_C of line segments.

Let us assume that $\{x(\theta_C), y(\theta_C)\}$ is the point of a DCC-path without line segments. Hence, in order to reach the final configuration $\mathbf{q}_B$, the following equation must be satisfied:

$$x_B = x(\theta_C) + l_A \cos\theta_A + l_B \cos\theta_B + l_C \cos\theta_C \qquad (7)$$

$$y_B = y(\theta_C) + l_A \sin\theta_A + l_B \sin\theta_B + l_C \sin\theta_C \qquad (8)$$

Therefore, we can select θ_C to minimize the overall DCC-path length:

$$\theta_C^* = \arg\min_{\theta_C} L = l_A + l_{A1} + l_{\Omega_A} + l_{A2} + l_C + l_{B1} + l_{\Omega_B} + l_{B2} + l_B \qquad (9)$$

subject to Equations (7) and (8). Clothoid lengths and arc segments can be obtained from clothoid properties [3]:

$$l_{A1} = l_{A2} = \kappa_A(\theta_C)\sigma_A^{-1}, \; l_{B1} = l_{B2} = \kappa_B(\theta_C)\sigma_B^{-1} \qquad (10)$$

$$l_{\Omega_A} = \Omega_A(\theta_C)\kappa_A^{-1}(\theta_C), \; l_{\Omega_B} = \Omega_B(\theta_C)\kappa_B^{-1}(\theta_C) \qquad (11)$$

where $\kappa_A = \min\{\sqrt{\sigma_A \delta_A}, \kappa_{max}\}$ and $\kappa_B = \min\{\sqrt{\sigma_B \delta_B}, \kappa_{max}\}$ are the maximum clothoid's curvatures.

In order to solve this minimization problem, we propose an heuristic criteria by forcing one of the line segment's length to be zero:

$$\text{if } \theta_C \neq \theta_A \text{ and } \kappa_R = 0 \Rightarrow \begin{cases} l_A = (X \cdot \sin(\theta_C) - Y \cdot \cos(\theta_C)) \sin^{-1}(\theta_C - \theta_A) \\ l_B = 0 \\ l_C = (Y \cdot \cos(\theta_A) - X \cdot \sin(\theta_A)) \sin^{-1}(\theta_C - \theta_A) \end{cases},$$

$$\text{if } \theta_A \neq \theta_B \text{ and } \kappa_R = 0 \Rightarrow \begin{cases} l_A = (X \cdot \sin(\theta_B) - Y \cdot \cos(\theta_B)) \sin^{-1}(\theta_B - \theta_A) \\ l_B = (Y \cdot \cos(\theta_A) - X \cdot \sin(\theta_A)) \sin^{-1}(\theta_B - \theta_A) \\ l_C = 0 \end{cases},$$

$$\text{if } \theta_C \neq \theta_B \text{ and } \kappa_R \neq 0 \Rightarrow \begin{cases} l_A = 0 \\ l_B = (X \cdot \sin(\theta_C) - Y \cdot \cos(\theta_C)) \sin^{-1}(\theta_C - \theta_B) \\ l_C = (Y \cdot \cos(\theta_B) - X \cdot \sin(\theta_B)) \sin^{-1}(\theta_C - \theta_B) \end{cases} \qquad (12)$$

Figure 3 shows some representative examples of DCC-paths, in function of the maximum allowable curvature and sharpness. It is well known that settings for curvature and sharpness bounds can seriously affect to wheel slippage and stability, specially for vehicles carrying heavy loads such as industrial forklifts. Therefore, it is crucial to provide appropriate bounds in order to guarantee stability conditions. Although all these aspects are out of the scope of the paper, it should be remarked that such considerations can be directly taken into account in the proposed method. As a consequence, depending on such bounds we can obtain a set of trajectories (for the same start and target configurations) which will imply different curvature profiles. On the other hand, maintaining normal accelerations constant may require modify curvature and sharpness bounds along the overall trajectory.

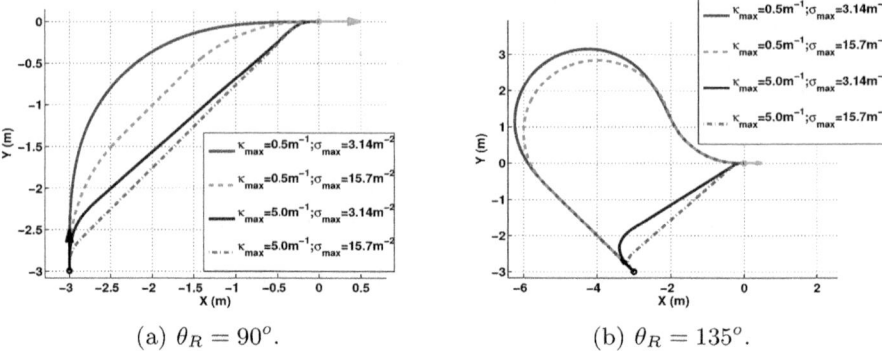

(a) $\theta_R = 90^\circ$.   (b) $\theta_R = 135^\circ$.

Fig. 3. XY traces from different θ_R and with κ_{max} and σ_{max} variation

3   Multi-rate Kinematic Control Based on DCC-Paths

This section describes the application of DCC-paths to the kinematic control problem on static environments. The contribution is to consider a multi-rate sampling scheme for kinematic control, analysing the influence of multi-rate sampling together with other control parameters. Our postulate assumes that the robot performs a low level control loop every T_R seconds, which is in general high frequency sampling period. In addition to this, every T_T seconds we receive a target and pose updates, which are in general at a intermediate frequency. For simplicity, we assume that $T_R = T_T/N \in \mathcal{Z}^+$, which is known as dual-rate ratio between the low-level control loop and the intermediate-level control loop. In addition to this, a higher control loop will consider the robot pose and map updates and will re-plan way-points if needed for a specific goal. The influence of the frequency of this higher control loop is out of the scope of the paper and for simplicity will be considered static (no replanning). Figure 4 shows the multi-rate hierarchical control architecture for mobile robot planning, where for simplicity obstacle avoidance has not been taken into account (pure kinematic control with static environment) and multiple arrows between blocks represent the frequency of the signal.

As mentioned before, $\mathbf{q}_T$ is obtained from a given sequence of way-points to be reached based on a Look-Ahead (LA) distance. This can be seen as the configuration within a given way-points pair w_{P_i} and $w_{P_{i+1}}$, such as:

$$\mathbf{q}_T = [\lambda^*(w_{P_{i+1}} - w_{P_i}) + w_{P_i}, \theta_{w_{P_i}}, 0]^T \tag{13}$$

$$\lambda^* = \arg \min_{\lambda \geq \lambda_\perp} \left| \|\lambda(w_{P_{i+1}} - w_{P_i}) + w_{P_i} - [x_R, y_R]^T\|_2 - LA \right| \tag{14}$$

$$\lambda_\perp = \arg \min_{0 \geq \lambda \geq 1} \|\lambda(w_{P_{i+1}} - w_{P_i}) + w_{P_i} - [x_R, y_R]^T\|_2 \tag{15}$$

where $\theta_{w_{P_i}} = \arctan((y_{w_{P_{i+1}}} - y_{w_{P_i}})/(x_{w_{P_{i+1}}} - x_{w_{P_i}}))$.

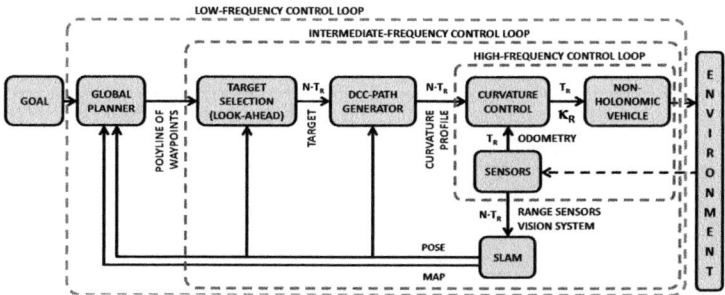

Fig. 4. Multi-rate Continuous-Curvature scheme based on DCC-paths

When a new DCC-path is generated we can consider that the vehicle is always located on the first clothoid **A1** and it has already travelled l_{A0} from DCC-path's origin q_A. From that moment the distance along the DCC-path is computed as follows: $l_R(t) = \int_0^t v_R(t)dt + l_{A0}$. The curvature controller uses the odometry to compute the real travelled distance and it is reset every time the DCC-path is recomputed. Therefore, once we know the current robot configuration along the generated DCC-path, from Figure 2(a) we can get that the curvature control low as a piecewise-function:

$$
\kappa_R(t) = \begin{cases}
0 & \text{if } 0 \leq l_R(t) \leq l_1 \\
(l_R(t) - l_1)\sigma_A & \text{if } l_1 < l_R(t) \leq l_2 \\
l_{A1}\sigma_A & \text{if } l_2 < l_R(t) \leq l_3 \\
l_{A1}\sigma_A - (l_R(t) - l_3)\sigma_A & \text{if } l_3 < l_R(t) \leq l_4 \\
0 & \text{if } l_4 < l_R(t) \leq l_5 \\
(l_R(t) - l_5)\sigma_B & \text{if } l_5 < l_R(t) \leq l_6 \\
l_{B1}\sigma_B & \text{if } l_6 < l_R(t) \leq l_7 \\
l_{B1}\sigma_B - (l_R(t) - l_6)\sigma_B & \text{if } l_7 < l_R(t) \leq l_8 \\
0 & \text{if } l_8 < l_R(t)
\end{cases}
\tag{16}
$$

with $l_1 = l_A$, $l_2 = l_1 + l_{A1}$, $l_3 = l_2 + l_{\Omega_A}$, $l_4 = l_3 + l_{A2}$, $l_5 = l_4 + l_C$, $l_6 = l_5 + l_{B1}$, $l_7 = l_6 + l_{\Omega_B}$, $l_8 = l_7 + l_{B2}$ and $l_9 = l_8 + l_B$.

The discrete implementation of Equation (16) requires the evaluation of $\kappa_R(t)$ at discrete steps "k", such $\kappa_{R,k} = \kappa_R(t + k \cdot T_R)$. In the limit, discrete-time control law converges to the continuous-time one when $N \to \infty$. The cost for implementing Equation (16) is cheap and therefore we can get better performance when applying a profile as shown in Section 4.

The implementation of Equation (12) can generate a singular solution if $\theta_A \approx \theta_B \approx \theta_C$. To avoid this, when the robot is aligned with the target a simpler solution based on classic Pure-Pursuit method is used, where:

$$
\kappa_R(t) = 2 \cdot \frac{(x_T - x_R)\cos\theta_R - (y_T - x_R)\sin\theta_R}{(x_R - x_T)^2 + (y_R - y_T)^2}
\tag{17}
$$

4   Benchmark

Now, we perform an exhaustive analysis of the proposed kinematic controller based on DCC-paths. In particular, the effects of the tracking velocity v_R, the

Table 1. Control variables *vs* control parameters, with Pure-Pursuit (PP) and Double Continuous-Curvature (DCC) Control Methods

		t_s[s]		δ[%]		NBE[m^{-1}]		NA[m^{-4}]		$J_{N_{max}}$[m/s^3]	
		PP	DCC	PP	DCC	PP	DCC	PP	DCC	PP	DCC
$v_R\left[\frac{m}{s}\right]$	0.2	18.90	14.26	10.06	0.56	0.0007	0.0009	0.04	0.01	1.60	0.07
	0.5	7.49	5.58	10.37	0.60	0.0010	0.0011	0.61	0.04	51.20	2.41
	1.0	3.67	2.70	10.51	0.61	0.0021	0.0018	2.18	0.26	418.4	31.88
	2.0	1.79	1.24	10.97	0.67	0.0044	0.0030	7.86	1.57	3467	480.8
LA[m]	0.5	2.45	2.14	4.15	4.87	0.0055	0.0015	10.86	0.26	1000	95.93
	1.0	3.73	2.47	9.72	0.32	0.0013	0.0015	0.38	0.37	200.0	76.06
	2.0	9.71	5.78	46.28	0.31	0.0007	0.0017	0.05	0.49	81.03	50.60
	4.0	15.0	12.01	143.4	0.19	0.0004	0.0017	0.02	0.41	47.28	44.55
N	1	3.79	2.75	10.20	0.57	0.0024	0.0042	0.68	1.72	200.0	76.22
	20	3.51	2.40	8.25	0.78	0.0027	0.0045	0.85	1.06	200.0	60.59
	50	3.25	1.59	5.83	1.36	0.0038	0.0058	1.38	0.47	200.0	33.68

Look-Ahead distance LA and the periodicity ratio N of the multi-rate control structure. In addition to this, we compare the benefits introduced with the proposed method against the classic Pure-Pursuit (PP) algorithm stated in Equation (17).

It is interesting to remark that, without loss of generality, following experiments have been developed for the case in which the start robot configuration is $q_R = (0, 0, \pi/2, 0)$ and the path to follow is a straight line $\rho_l = x \cos \phi_l + y \sin \phi_l$ with $\rho_l = 1$m and $\phi_l = \pi/2$rad (line $y = 1$m), which represents one of the most difficult cases.

To evaluate the performance of the methods several metrics have been used to characterize the resulting path $\{[x_{R,0}, y_{R,0}]^T, [x_{R,1}, y_{R,1}]^T, \ldots\}$. In particular, we consider the settling time $t_s = k_s T_R$ such as $\left|\frac{\rho_l - y_{R,i}}{\rho_l}\right| < 0.02 \; \forall i \geq k_s$, the overshoot $\delta = \max\{y_{R,i}\}$, the mean error $\bar{e} = \sum_i |\rho_l - (x_{R,i} \cos \phi_l - y_{R,i} \sin \phi_l)|$, but also the normalized bending energy $NBE = \frac{1}{n}\sum_{i=0}^{n}\kappa_i^2 dl$, the normalized abruptness $NA = \frac{1}{n}\sum_{i=0}^{n}\sigma_i^2 dl$, where $dl = v \cdot dt$, and the maximum normal jerk $J_{N_{max}} = v^3 \sigma_{max}$, as defined in [18].

It can be appreciated from the results of Table 1 that DCC-path method shows much better by obtaining a lower settling time and therefore reaches the path earlier. DCC-path generally shows an under-damped trajectory, therefore it rarely overpass the line, while the PP method clearly overpasses the path as shown in Figure 5. The velocity has in general little influence on the metric, but obviously it affects the settling time. The NBE is slightly higher in the DCC which implies that the method applies, in general, higher curvatures, which can be seen as a trade-off between overshoot-settling time and bending energy. A peculiar result is the inverse relation of the abruptness metric with respect to the LA distance. In DCC method the abruptness slightly increases with LA, while the PP decreases exponentially. This result seems to be related high overshoot δ performance for higher LA values. Finally, the normal jerk is clearly lower in

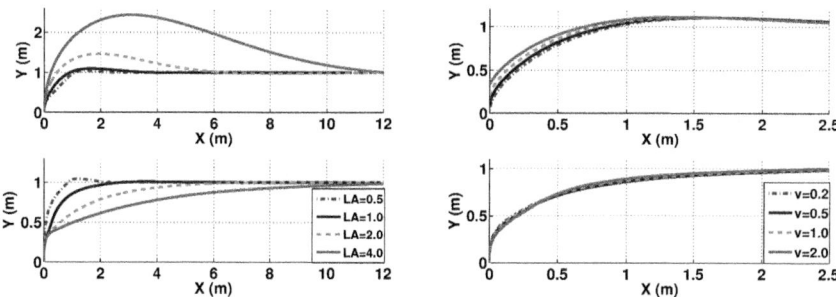

Fig. 5. DCC (below) against PP (above), for different values of LA and v_R. Left: XY vs. $LA(m)$. Right: XY vs. $v_R(m/s)$.

DCC method which supports the idea that continuous-curvature methods can produce smoother and safer trajectories.

We can see from Figure 6, Left that control parameter α_A has direct influence on the abruptness. The higher the clothoid sharpness the higher the abruptness. Moreover, in Figure 6, Right it is shown that the higher the clothoid sharpness the "noisier" the curvature profile. Despite of this, the curvature is still continuous, while the PP method clearly shows discontinuities (discretization effect). Indeed, if we analyse the effect of such ratio, we can see in Figure 7, Right that PP method would obtain a quasi-continuous curvature profile for $N = 1$, but a clearly discretized profile for higher values. On the other hand, the DCC method always has a quasi-continuous curvature profile. Results are specially relevant with high ratio values which is in concordance with the idea that in most real implementations the low-level controller is several times faster than the intermediate-level controller.

Now, we will analyze a square-path following problem for both methods. It can be appreciated from Figure 8(a) that DCC does not have overshoot and the

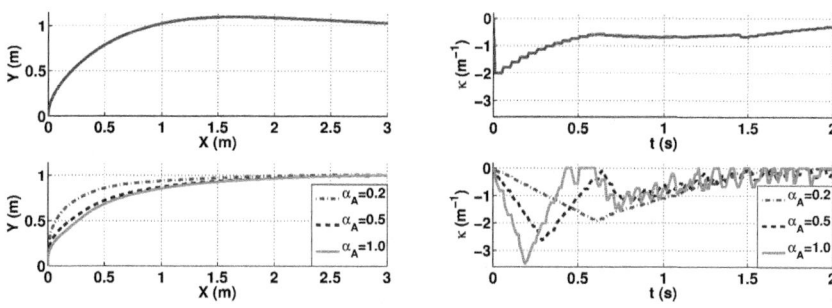

Fig. 6. DCC (below) against PP (above), for different values of α_A (with $\alpha_B = 1$). Left: XY vs. α_A. Right: κ profile vs. α_A.

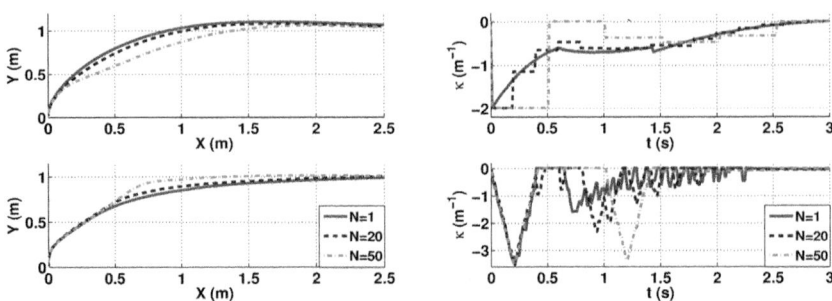

Fig. 7. DCC (below) against PP (above), for different values of N. Left: XY vs. N. Right: κ profile vs. N.

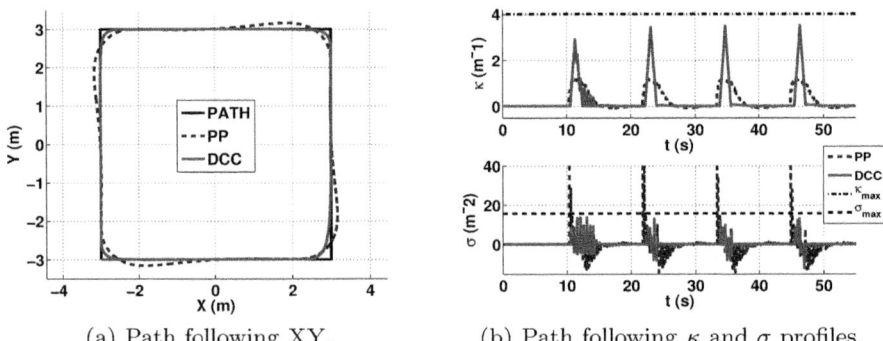

(a) Path following XY.   (b) Path following κ and σ profiles.

Fig. 8. DCC vs. PP, following a square path with $v_R = 0.5m/s$ and $LA = 1m$, considering that kinematic contraints are $\kappa_{max} = 5m^{-1}$ and $\sigma_{max} = 15.7m^{-2}$

Table 2. Control variables vs σ_{max} and κ_{max}

		t_s[s]	δ[%]	NBE[m^{-1}]	NA[m^{-4}]	J_N[m/s^3]
$\sigma_{A_{max}}$[m^{-2}]	3.14	2.24	0.29	0.0012	0.02	12.65
	7.85	2.45	0.36	0.0014	0.12	29.57
	15.7	2.43	0.27	0.0018	0.48	75.87
κ_{max}[m^{-1}]	2.0	7.67	0.20	0.0029	0.036	15.18
	3.0	8.19	0.21	0.0031	0.084	18.17
	5.0	8.54	0.25	0.0039	0.210	33.68

settling time is faster. The trade-off we need to pay is higher curvature values but always satisfying maximum curvature constraints, as shown in Figure 8(b).

Finally, we have extended the result to analyse the performance when varying the maximum allowable sharpness and curvature in computing the DCC-path.

The results obtained in Table 2 lead to the conclusion that the kinematic constraints affect directly to the robot behaviour. In fact, NBE, NA and J_N are proportional to the maximum curvature and curve sharpness, as expected.

5   Conclusions

In this paper, we have introduced a method for generating smooth trajectories considering constraints on maximum curve sharpness and maximum curvature. Double Continuous-Curvature paths (DCC) use two concatenated Single Continuous Curvature paths (SCC) [4,3] which are composed of a line segment, a first clothoid, a circle segment (arc) and a second clothoid. Additionally, a final line segment is added at the end of the second SCC-path to provide a set of general solutions. As a main advantage, the trajectory takes into account lower and upper bounds of curvature and its derivative, the sharpness, which implies smoother and safer trajectories.

The proposed method generates an appropriate curvature profile so that a continuous-curvature path joins a start configuration and a target configuration. We can handle cases where initial and final curvature can take non zero values, which is not the case of other continuous curvature paths such as Elementary or Bi-Elementary paths. The method computes the shortest path to join both configurations and uses few design parameters which are related with the sharpness of clothoids. An heuristic criteria establishes the length of the line segments, clothoids and arc segments as well as an intermediate configuration based on a minimization procedure. In this sense, we have shown a set of representative cases to qualitatively validate the type of solutions provided by the method.

In addition, the paper proposes a multi-rate control scheme by considering different dynamics, with high, intermediate and low level control loops. Different sampling rates are considered: global path planner (high level); target update based on Look-Ahead distance (intermediate level); curvature control (low level). The influence of the ratio between the two lower levels is analysed and brought to the limit. Global path planner may be combined or substituted by vision-based road following application, where a vision system detects the road used as path to follow.

The paper also provides an exhaustive analysis to evaluate the performance of the new method with respect to Pure-Pursuit method. It is shown that the proposed method has better performance with respect to settling time and overshoot, while it has little influence on design parameters. As a conclusion, we can say that the new method introduces improvements in comfort and safety because sharpness and curvature are always within bounds, according to worldwide horizontal alignment criteria [6].

Acknowledgements. This work was supported by VALi+d Program (Generalitat Valenciana), DIVISAMOS Project (Spanish Ministry), PROMETEO Program (Conselleria d'Educació, Generalitat Valenciana) and MAGV Project (PAID-05-10 Program from VIDI UPV).

References

1. Fraichard, T., Ahuactzin, J.-M.: Smooth path planning for cars. In: IEEE Int. Conf. on Robotics and Automation, vol. 4, pp. 3722–3727 (2001)
2. Girbés, V., Armesto, L., Tornero, J.: Pisala project: Intelligent sensorization for line tracking with artificial vision. In: International Symposium on Robotics, pp. 558–563 (2010)
3. Girbés, V., Armesto, L., Tornero, J.: Continuous curvature control of mobile robots with constrained kinematics. In: IFAC World Congress (accepted, 2011)
4. Girbés, V., Armesto, L., Tornero, J.: On generating continuous-curvature paths for line following problem with curvature and sharpness constraints. In: IEEE Int. Conf. on Robotics and Automation, pp. 6156–6161 (May 2011)
5. Jiang, K., Zhang, D.Z., Seneviratne, L.D.: A parallel parking system for a car-like robot with sensor guidance. Journal of Mechanical Engineering Science 213(6), 591–600 (1999)
6. Krammes, R.A., Garnham, M.A.: Worldwide review of alignment design policies. In: Int. Symp. on Highway Geometric Design Practices, pp. 1–17 (1998)
7. Labakhua, L., Nunes, U., Rodrigues, R., Leite, F.S.: Smooth trajectory planning for fully automated passengers vehicles. spline and clothoid based methods and its simulation. In: Informatics in Control Automation and Robotics, Springer, Heidelberg (2008)
8. Laumond, J.P., Jacobs, P., Taix, M., Murray, R.: A motion planner for nonholonomic mobile robots. IEEE Trans. on Robotics and Automation 10(5), 577–593 (1994)
9. Manz, M., von Hundelshausen, F., Wuensche, H.-J.: A hybrid estimation approach for autonomous dirt road following using multiple clothoid segments. In: IEEE Int. Conf. on Robotics and Automation, pp. 2410–2415 (2010)
10. Montés, N., Mora, M.C., Tornero, J.: Trajectory generation based on rational bezier curves as clothoids. In: IEEE Intelligent Vehicles Symposium, pp. 505–510 (2007)
11. Ollero, A., Heredia, G.: Stability analysis of mobile robot path tracking. In: Int. Conf. on Intelligent Robots and Systems, vol. 3, pp. 461–466 (1995)
12. Papadimitriou, I., Tomizuka, M.: Fast lane changing computations using polynomials. In: American Control Conf., vol. 1, pp. 48–53 (2003)
13. Scheuer, A., Fraichard, T.: Collision-free and continuous-curvature path planning for car-like robots. In: IEEE Int. Conf. on Robotics and Automation, vol. 1, pp. 867–873 (1997)
14. Scheuer, A., Fraichard, T.: Continuous-curvature path planning for car-like vehicles. In: IEEE Int. Conf. on Intelligent Robots and Systems, vol. 2, pp. 997–1003 (1997)
15. Wallace, R., Stentz, A., Thorpe, C.E., Moravec, H., Whittaker, W., Kanade, T.: First results in robot road-following. In: Int. Conf. on Artificial Intelligence (1985)
16. Wilde, D.K.: Computing clothoid segments for trajectory generation. In: IEEE Int. Conf. on Intelligent Robots and Systems, pp. 2440–2445 (2009)
17. Yang, K., Sukkarieh, S.: Real-time continuous curvature path planning of uavs in cluttered environments. In: Int. Symp. on Mechatronics and its Applications, pp. 1–6 (2008)
18. Yuste, H., Armesto, L., Tornero, J.: Benchmark tools for evaluating agvs at industrial environments. In: IEEE Int. Conf. on Intelligent Robots and Systems, pp. 2657–2662 (2010)

Supervised Traversability Learning for Robot Navigation

Ioannis Kostavelis, Lazaros Nalpantidis, and Antonios Gasteratos

Laboratory of Robotics and Automation,
Department of Production and Management Engineering,
Democritus University of Thrace, Xanthi, Greece
{gkostave,lanalpa,agaster}@pme.duth.gr

Abstract. This work presents a machine learning method for terrain's traversability classification. Stereo vision is used to provide the depth map of the scene. Then, a v-disparity image calculation and processing step extracts suitable features about the scene's characteristics. The resulting data are used as input for the training of a support vector machine (SVM). The evaluation of the traversability classification is performed with a leave-one-out cross validation procedure applied on a test image data set. This data set includes manually labeled traversable and non-traversable scenes. The proposed method is able to classify the scene of further stereo image pairs as traversable or non-traversable, which is often the first step towards more advanced autonomous robot navigation behaviours.

Keywords: traversability classification, robot navigation, SVM, machine learning, stereo vision, v-disparity image

1 Introduction

The development of an efficient method for inspecting the traversability of a scene is an active research topic [5,2]. Obstacle detection and traversability evaluation are important, as they both provide crucial information for the navigation of mobile robots. The goal of this work is the development of a trained system capable of detecting non-traversable scenes using solely stereo vision input. This task demands reliable and robust machine vision algorithms. Towards this direction, a stereo vision algorithm is used, which retrieves information about the environment from a stereo camera and produces the disparity map of the scene. The v-disparity image is then calculated based on the disparity map [17] and a feature extraction procedure extracts useful information from the v-disparity image. This procedure is repeated for all the available v-disparity images in order to form a data set. The data set is then utilised for the training and the testing phase of a support vector machine (SVM) classifier. Fig. 1 summarises the steps of the proposed methodology.

R. Groß et al. (Eds.): TAROS 2011, LNAI 6856, pp. 289–298, 2011.

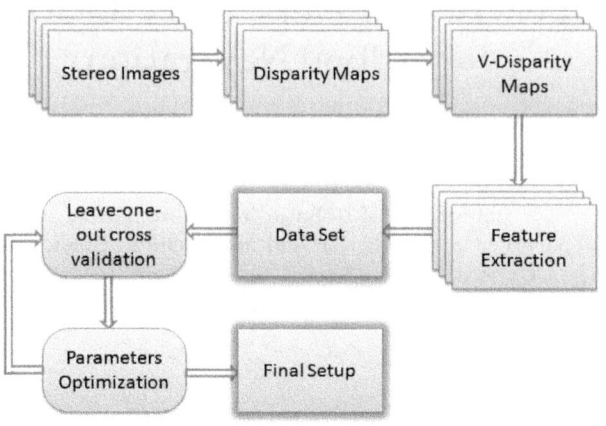

Fig. 1. Flow chart of the proposed methodology

1.1   Related Work

Stereo vision is often used in vision-based robotics, instead of monocular sensors, due to the simpler calculations involved in the depth estimation. A correspondence search between the two stereo images can provide dense information about the depth of the depicted scene. Recently, efficient stereo algorithms have been presented that can provide accurate and reliable depth estimations in frame rates suitable for autonomous robotic applications [11].

Estimation of terrain traversability has been of interest for the mobile robotics community during the last decades and in 1994 a statistics-based method for classifying field regions as traversable or not was proposed in [7]. More recently, the development of autonomous planetary rovers and the DARPA Grand Challenge have triggered rapid advancements in the field [5,13,15]. Machine learning methodologies have often been employed [12] and stereo vision has been widely used as input for such systems [4,6]. One of the most popular methods for terrain traversability analysis is the initial estimation of the v-disparity image [2]. This method is able to confront the noise in low quality disparity images [17,14] and model the terrain as well as any existing obstacles. Several researchers have proposed robot navigation methods based on terrain classification. These methods use features derived from remote sensor data such as colour, image texture and surface geometry. Initially, colour-based methods have been proposed for the classification of outdoor scenes using a mixture of Gaussians models [8]. Additionally, in [3] a terrain classification method based on texture features of the images was introduced. A more sophisticated and computationally demanding method based on 3D point clouds that uses the statistical distribution in space, has been proposed in [9].

Those navigation methods are applied both in traversable and non-traversable scenes consuming valuable resources and time. For efficient robot navigation non-traversable scenes should not be examined in detail. Therefore, scenes should firstly be inspected and classified according to their overall traversability. Supervised machine learning techniques have the advantage that while the training phase is slow, the classification of new unseen instances is performed very fast. The main contribution of the present work is that the training phase can be significantly accelerated by using features of the v-disparity image, rather than features of the input images or the depth map. This is a more abstract description of the scene and reduces the general problem into a simpler one. Once the scene is classified as traversable then higher level navigation algorithms can be applied.

2 Algorithm Description

2.1 Stereo Vision and V-Disparity Image Computation

The disparity maps are computed using a local stereo correspondence algorithm [10]. The utilised stereo algorithm combines low computational complexity with appropriate data processing. Consequently, it is able to produce dense disparity maps of good quality in frame rates suitable for robotic applications. The main attributes that differentiate this algorithm from the majority of the other ones is that the input images are enhanced by superimposing the outcome of a Laplacian of Gaussian (LoG) edge detector and that the matching cost aggregation step consists of a sophisticated gaussian-weighted sum of absolute differences (SAD) rather than a simple constant-weighted one. Furthermore, the disparity selection step is a simple winner-takes-all choice, as the absence of any iteratively updated selection process significantly reduces the computational payload of the overall algorithm.

The results of the per pixel optimum disparity values are filtered at two consequent steps. Firstly, the reliability of the selected disparity value is validated. That is, for every pixel of the disparity map a certainty measure is calculated indicating the likelihood of the pixel's selected disparity value to be the right one. The certainty measure $cert$ is calculated for each pixel (x, y) as in Eq. 1

$$cert(x, y) = \left| SAD(x, y, disp(x, y)) - \frac{\sum_{z=0}^{d-1} SAD(x, y, z)}{d} \right| \qquad (1)$$

According to this, the certainty $cert$ for a pixel (x, y) that the computed disparity value $disp(x, y)$ is actually right is equal to the absolute value of the difference between the minimum matching cost value $SAD(x, y, disp(x, y))$ and the average matching cost value for that pixel when considering all the d candidate disparity levels for that pixel. What the aforementioned measure evaluates is the amount of differentiation of the selected disparity value with regard to the rest candidate ones. The more the disparity value is differentiated, the most possible it is that the selected minimum is actually a real one and not owed to noise or other effects.

A threshold is applied to this metric and only the pixels whose certainty to value ratio $\frac{cert(x,y)}{SAD(x,y,disp(x,y))}$ is equal to or more than 30% are counted as valid ones. The value of this threshold has been chosen after exhaustive experimentation so as to reject as many false matches as possible while retaining the majority of the correct ones. Moreover, a bidirectional consistency check is applied. The selected disparity values are approved only if they are consistent, irrespectively to which image is the reference and which one is the target. Thus, even more false matches are disregarded.

The outcome of the presented stereo algorithm is a sparse disparity map, as shown in Fig. 2(c), containing disparity values only for the most reliable pixels. The rest pixels, shown in black in Fig. 2(c) are not considered at all.

    (a) Left image   (b) Right image   (c) Disparity map

Fig. 2. A stereo image pair and the resulting disparity map

Using the sparse disparity map obtained from the stereo correspondence algorithm a reliable v-disparity image can be computed, as shown in Fig. 3(a). In a v-disparity image each pixel has a positive integer value that denotes the number of pixels in the input image that lie on the same image line (ordinate) and have disparity value equal to its abscissa. The terrain in the v-disparity image is modelled by a linear equation. The parameters of this linear equation can be found using Hough transform [2], if the majority of the input images' pixels belong to the terrain and not to obstacles. A tolerance region on both sides of the terrain's linear segment is considered and any point outside this region can be safely considered as originating from a barrier. The linear segments denoting the terrain and the tolerance region overlaid on the v-disparity image are shown in Fig. 3(b).

2.2   Feature Extraction and SVM-Based Learning

A feature extraction procedure is then applied to the v-disparity image. For each scanline of a v-disparity image, e.g. that of Fig. 3(b), the values of the pixels lying outside the tolerance region are aggregated. The outcome of this procedure is a feature vector for each v-disparity image (or equivalently, for each stereo image

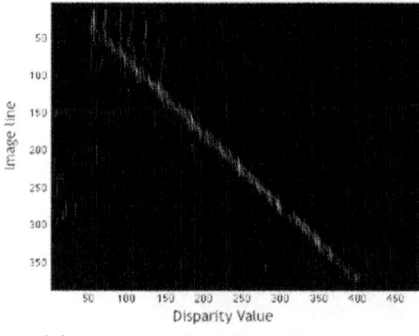
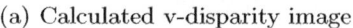
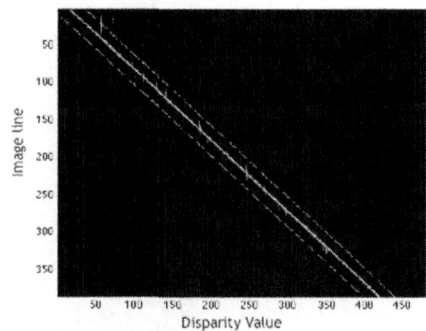

(a) Calculated v-disparity image

(b) V-disparity image with terrain modelled by the continuous line and the tolerance region shown between the two dashed lines

Fig. 3. V-disparity images for a stereo image pair

pair) that has as many dimensions as the number of the image's scanlines and each value stem from the aforementioned values' aggregation. More specifically, for a given v-disparity image of $M \times N$ dimensions, the output of the feature extraction method is a vector $\mathbf{x} = [x_1, x_2, ..., x_M]$ where, x_i denotes the sum of the pixels lying outside the tolerance region for the row i and $i = 1, 2, ..M$ indicates the number of rows of the disparity map. As an example, the pixels of the v-disparity image whose values are aggregated so as to obtain the x_{50} component of the feature vector are the ones lying within the red rectangular regions of Fig. 4. In this figure the red rectangular regions are exaggerated for the purpose of readability.

The feature vectors for all the the stereo pairs of the used data set comprise a data matrix $\mathbf{D} = [\mathbf{x_1}, \mathbf{x_2}, ..., \mathbf{x_L}]^T$, where each vector $\mathbf{x}_j$ corresponds to a stereo pair and $j = 1, 2, ..L$ denotes the number of samples. For the evaluation of our method a diminished data set constituted of 23 traversable and 10 non-traversable indoor scenes was used. One traversable and one non-traversable sample reference image are given in Fig. 5(a) and Fig. 6(a) respectively. The traversability of those scenes is deduced by the distance of the closest object to the camera, as it will be discussed in the next section.

The next step of the proposed methodology deals with the training procedure for the aforementioned data set. This is an off-line context and, therefore, non time-critical part. Thus we chose to use a SVM classifier [16]. For the classification method, the LIBSVM library was used. More detailed information about the selected library and definitions concerning the selection of the optimal parameters can be found in [1]. The SVM approach constructs a binary classifier for each pair of classes by building a function that will be positive for one class (i.e. traversable) and negative for the other one (i.e. non-traversable). Linear,

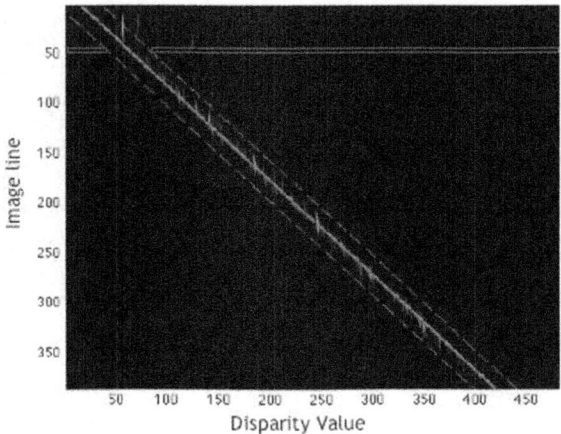

Fig. 4. Features extraction for the 50th image line

polynomial and Gaussian kernels have been tested. The model regularisation parameter C, which penalises large errors, is chosen equal to 100 in order to optimise data separation. The optimal parameter γ used to control the width of the Gaussian distribution was set to the value of 0.1. As for the polynomial kernel, the second order polynomial function was chosen for testing, whereas the third order polynomial didn't offer any additional classification gain.

In order to validate the results, the proposed method has also been tested using alternatively a k-nearest neighbour (k-nn) classifier. The limited size of the data set enforces this additional examination of the method's efficiency. Towards this direction, overtraining results and polarised classification, which may stem from SVM classification, were examined. The parameter of the k-nearest neighbour classifier, which gave the greatest separability between the two classes, was also selected using a leave-one-out cross validation procedure and set to $k = 5$ neighbours.

3   Experimental Validation

The methodology described in this paper does no require a balanced data set for the SVM training phase. In the used input image pairs the traversable scenes were more numerous than the non-traversable ones (i.e. 23 traversable scenes and 10 non-traversable scenes). The used images had 512×384 pixels resolution. Each scene was manually labeled according to the distance of the nearest depicted obstacle. If that distance was less than a threshold value, set for our experiments to 50 cm, the scene was labeled as non-traversable. Otherwise, the scene was considered to be a traversable one. Fig. 5 and Fig. 6 present the

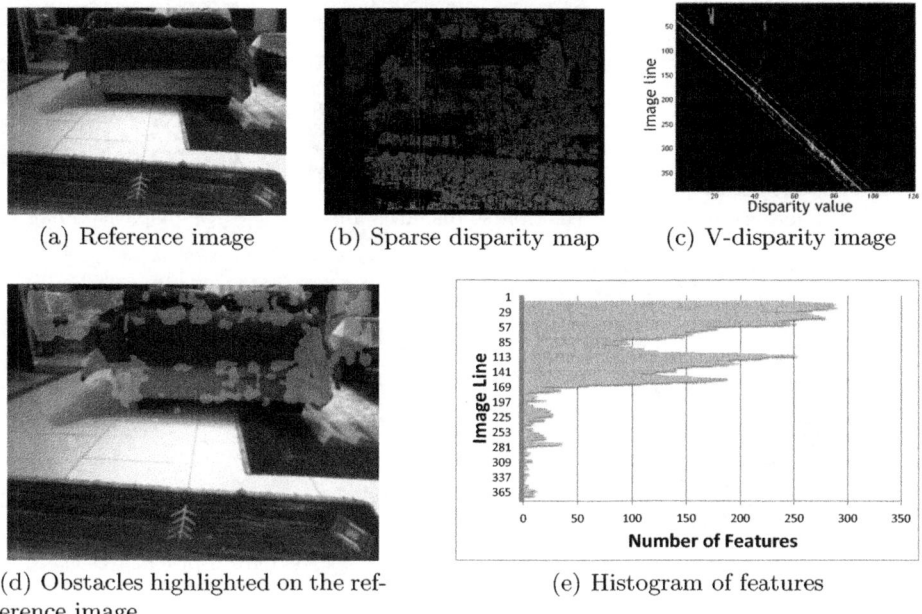

(a) Reference image   (b) Sparse disparity map   (c) V-disparity image

(d) Obstacles highlighted on the reference image   (e) Histogram of features

Fig. 5. Reference image and experimental results for a traversable scene tested

experimental result for an indicative traversable and non-traversable scene, respectively. The reference (left) images of each stereo image pair is given in Fig. 5(a) and Fig. 6(a). The corresponding sparse disparity maps are given in Fig. 5(b) and Fig. 6(b). These disparity maps are used for the computation of the v-disparity images, given in Fig. 5(c) and Fig. 6(c). The obstacles indicated by these v-disparity images are highlighted in red colour in Fig. 5(d) and Fig. 6(d). Finally, Fig. 5(e) and Fig. 6(e) show the results of the v-disparity image's feature extraction process as a histogram of the detected features in each scanline.

The k-nearest neighbour classifier manages to achieve a $74,3\%$ classification rate in leave-one-out cross validation with $k = 5$ nearest neighbours. Considering that this classifier is inherently prone to errors, it can be deduced that the preprocessing procedures are efficient and the feature extraction method indeed creates features which contain crucial information about the traversability of the scenes. Additionally, the SVM classifier succeeded a $91,2\%$ classification rate using the second order polynomial kernel. This proves great separability between the two classes, taking into consideration that this classification rate corresponds to 30 correct out of 33 classified samples. The SVM classifier using a linear kernel as well as a Gaussian one achieved success rates somewhere between the other two classifiers. Table 1 presents the classification rates for the different classifiers that have been tested.

(a) Reference image   (b) Sparse disparity map   (c) V-disparity image

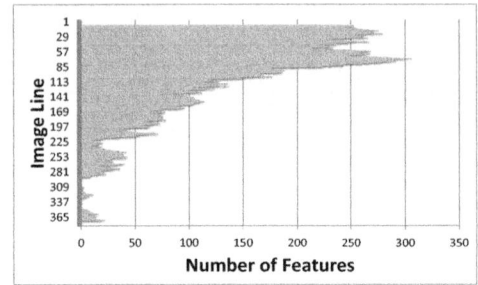

(d) Obstacles highlighted on the reference image   (e) Histogram of features

Fig. 6. Reference image and experimental results for a non-traversable scene tested

Table 1. Classification rate for different classifiers

Classifier Type	Classification Rate
k-nn	74,30%
Linear SVM	87,88%
Gaussian SVM	81,83%
Polynomial SVM	**91,20%**

4   Conclusion

A traversability classification system for autonomous robot navigation has been proposed. The system consists of an optimised stereo algorithm that ultimately produces v-disparity images. The v-disparities are then coded, in a novel and simple way, so as to form the feature vectors of a two-class data set. The primitive v-disparity maps are manually labeled as traversable and non-traversable and a SVM classifier is trained to separate the two different classes. The efficiency of our method was first tested using a k-nearest neighbour classifier, which succeeded 74.3% classification rate and finally using a SVM classifier that employed a second order polynomial kernel. This classifier achieved a classification rate of 91,2%. The significantly high classification ability achieved stems from the production of noise-free disparity maps, which result in reliable v-disparity images.

This initial step is very important for the success of the proposed method as the input feature vectors of the SVM classifier contain crucial and concise information for the traversability of the scene. The high efficiency rate encourages the use of the traversability inspection system for the primary exploratory analysis of a scene. The trained system can be used for autonomous robot navigation diminishing the computational cost and minimising the demanded on-line execution time. It should be noted that the aforementioned classification rates were obtained using a very limited set of input images, i.e. only 33 stereo image pairs. Moreover, the set of input images was not balanced, but the traversable scenes were more than twice the number of the non-traversable ones.

To conclude, the experimental results of the proposed methodology are encouraging. The training of the system can be performed off-line and the separation ability is high. The separation ability is expected to improve even more if a larger and balanced set of stereo input images is applied. Such a trained system is expected to be able to perform terrain traversability classification with limited on-line effort and with high success rates.

References

1. Chang, C., Lin, C.: LIBSVM: a library for support vector machines (2001), software available at http://www.csie.ntu.edu.tw/~cjlin/libsvm
2. De Cubber, G., Doroftei, D., Nalpantidis, L., Sirakoulis, G.C., Gasteratos, A.: Stereo-based terrain traversability analysis for robot navigation. In: IARP/EURON Workshop on Robotics for Risky Interventions and Environmental Surveillance, Brussels, Belgium (2009)
3. Dima, C.S., Vandapel, N., Hebert, M.: Classifier fusion for outdoor obstacle detection. In: IEEE International Conference on Robotics and Automation, vol. (1), pp. 665–671 (1994)
4. Happold, M., Ollis, M., Johnson, N.: Enhancing supervised terrain classification with predictive unsupervised learning. In: Robotics: Science and Systems, Philadelphia, USA (August 2006)
5. Howard, A., Turmon, M., Matthies, L., Tang, B., Angelova, A., Mjolsness, E.: Towards learned traversability for robot navigation: From underfoot to the far field. Journal of Field Robotics 23(11-12), 1005–1017 (2006)
6. Kim, D., Sun, J., Min, S., James, O., Rehg, M., Bobick, A.F.: Traversability classification using unsupervised on-line visual learning for outdoor robot navigation. In: IEEE International Conference on Robotics and Automation (2006)
7. Langer, D.: A behavior-based system for off-road navigation. IEEE Transactions on Robotics and Automation 10(6), 776–783 (1994)
8. Manduchi, R.: Learning outdoor color classification from just one training image. In: European Conference on Computer Vision, vol. 4, pp. 402–413 (2004)
9. Vandapel, N., Huber, D., Kapuria, A., Hebert, M.: Natural terrain classification using 3-D ladar data. In: IEEE International Conference on Robotics and Automation, vol. 5, pp. 5117–5122 (2004)
10. Nalpantidis, L., Sirakoulis, G.C., Carbone, A., Gasteratos, A.: Computationally effective stereovision SLAM. In: IEEE International Conference on Imaging Systems and Techniques, Thessaloniki, Greece, pp. 453–458 (July 2010)

11. Nalpantidis, L., Sirakoulis, G.C., Gasteratos, A.: Review of stereo vision algorithms: from software to hardware. International Journal of Optomechatronics 2(4), 435–462 (2008)
12. Shneier, M.O., Shackleford, W.P., Hong, T.H., Chang, T.Y.: Performance evaluation of a terrain traversability learning algorithm in the DARPA LAGR program. In: Performance Metrics for Intelligent Systems Workshop, Gaithersburg, MD, USA, pp. 103–110 (2006)
13. Singh, S., Simmons, R., Smith, T., Stentz, A., Verma, I., Yahja, A., Schwehr, K.: Recent progress in local and global traversability for planetary rovers. In: IEEE International Conference on Robotics and Automation, vol. 2, pp. 1194–1200 (2000)
14. Soquet, N., Aubert, D., Hautiere, N.: Road segmentation supervised by an extended V-disparity algorithm for autonomous navigation. In: IEEE Intelligent Vehicles Symposium, Istanbul, Turkey, pp. 160–165 (2007)
15. Thrun, S., Montemerlo, M., Dahlkamp, H., Stavens, D., Aron, A., Diebel, J., Fong, P., Gale, J., Halpenny, M., Hoffmann, G., Lau, K., Oakley, C., Palatucci, M., Pratt, V., Stang, P., Strohband, S., Dupont, C., Jendrossek, L.E., Koelen, C., Markey, C., Rummel, C., van Niekerk, J., Jensen, E., Alessandrini, P., Bradski, G., Davies, B., Ettinger, S., Kaehler, A., Nefian, A., Mahoney, P.: Stanley: The robot that won the DARPA grand challenge: Research articles. Journal of Robotic Systems 23(9), 661–692 (2006)
16. Vapnik, V.N.: The Nature of Statistical Learning Theory. Springer, New York (1995)
17. Zhao, J., Katupitiya, J., Ward, J.: Global correlation based ground plane estimation using V-disparity image. In: IEEE International Conference on Robotics and Automation, Rome, Italy, pp. 529–534 (2007)

Task Space Integral Sliding Mode Controller Implementation for 4DOF of a Humanoid BERT II Arm with Posture Control

Said Ghani Khan[1], Jamaludin Jalani[2], Guido Herrmann[2],
Tony Pipe[1], and Chris Melhuish[3]

[1] Bristol Robotics Laboratory, University of the West of England, Bristol, UK
{said.khan,tony.pipe}@brl.ac.uk
[2] Department of Mechanical Engineering and
Bristol Robotics Laboratory, University of Bristol, Bristol, UK
{meyjj,g.herrmann}@bristol.ac.uk
[3] Bristol Robotics Laboratory, University of Bristol and
University of the West of England, Bristol, UK
chris.melhuish@brl.ac.uk

Abstract. This paper presents the implementation (real time and simulation) of an integral sliding mode controller (ISMC) for the four degrees of freedom (DOF) of the humanoid BERT II[1] robot arm, in order to deal with the inaccuracies and unmodelled nonlinearities in the dynamic model of the robot arm. This is a task space controller, tracking Cartesian coordinates x and y. The controller has been implemented using shoulder flexion, shoulder abduction, humeral rotation and elbow flexion joints of the BERT II right arm. The main controller is the combination of a feedback linearization (FL) scheme and an ISMC. The redundant DOF are controlled by a bio-mechanically inspired posture controller, to generate human like motion pattern based on recent work. Good real-time tracking results demonstrates effectiveness of the scheme.

1 Introduction

Dynamic modeling inaccuracies and uncertainties are almost unavoidable for complex systems such as a multi-link humanoid robot arm. On the other hand, sliding mode controllers [12,15,3] are known for robustness, hence, they are a good choice to overcome deficiencies in the dynamic models. An enhanced version of the sliding mode controller has an integral part to the sliding mode variable. This is known as integral sliding mode controller (ISMC). ISMC can be advantageous over a traditional sliding mode controller (SMC), because it guarantees robustness from the start, by eliminating the sliding mode reaching stage [11,16], in contrast to the traditional SMC.

Note that, although it is a powerful technique, there are very few published ISMC applications. The latest research has been presented successfully in [11]

---

[1] The mechanical design and manufacturing for the BERT II torso including hand and arm has been conducted by Elumotion (www.elumotion.com).

R. Groß et al. (Eds.): TAROS 2011, LNAI 6856, pp. 299–310, 2011.
© Springer-Verlag Berlin Heidelberg 2011

by comparing SMC and ISMC through simulation for a two-link rigid manipulator while [9] have applied the ISMC to a power-assisted manipulator with a single degree of freedom through simulation and experiment. Eker and Akmal [2] have suggested a similar scheme using an integral sliding surface in combination with the conventional sliding mode control. They have tested the scheme on a rather simple electro-mechanical system. Recently, in [5,4], we have simulated and experimentally tested ISMC for under-actuated robotic fingers. Thus, this paper presents practical implementations of the Cartesian (x & y) ISMC control in combination with a feedback linearization (FL) and a posture controller for the 4DOF of the humanoid BERT II right arm (see Figure 1). This control scheme has been used because of its robustness to uncertainties and unmodelled nonlinearities. The dynamic model of the BERT II robot arm for the FL is obtained via MapleSim (a package of Maple). The posture controller [13,14,7,6,1] controls the redundant DOF to generate human like motion by minimizing a gravity dependent cost function based on the 4DOF dynamic model of the BERT II arm.

2   Integral Sliding Mode Controller

Motivated by the recent development of the ISMC on a two-link rigid robotic manipulator through simulation, the ISMC control law by [11] is implemented here experimentally for the 4DOF of the BERT II arm which has been also successfully tested for under-actuated robotic fingers [4,5]. The integral sliding mode variable s is defined as follows:

$$s = \dot{e}_x + K_s e_x + K_i \int_0^t e_x(\xi)d\xi - \dot{e}_x(0) - K_s e_x(0). \tag{1}$$

The Cartesian position error vector is given by $e_x = X_d - X$, where $X_d = [x_d, y_d]^T$ the Cartesian demand position (as we are controlling x & y only). The vectors $\dot{e}_x(0)$ and $e_x(0)$ are the initial values of the Cartesian velocity and position error respectively. It should be noted that the values of K_s and K_i define the desired behaviour of the control scheme once sliding motion $s = 0$, is achieved. Hence, it acts like a reference model for the control scheme. The matrices K_s and K_i are:

$$K_s = \begin{pmatrix} K_{s1} & 0 \\ 0 & K_{s2} \end{pmatrix} \tag{2}$$

and

$$K_i = \begin{pmatrix} K_{i1} & 0 \\ 0 & K_{i2} \end{pmatrix} \tag{3}$$

where K_{s1}, K_{s2}, K_{i1} and K_{i2} are positive scalars. As mentioned earlier, ISMC approach has an advantage in relation to other SMC methods, see [11,16]. From equation (1), it is easily seen that if the correct initial values, i.e. $\dot{e}_x(0)$ and $e_x(0)$,

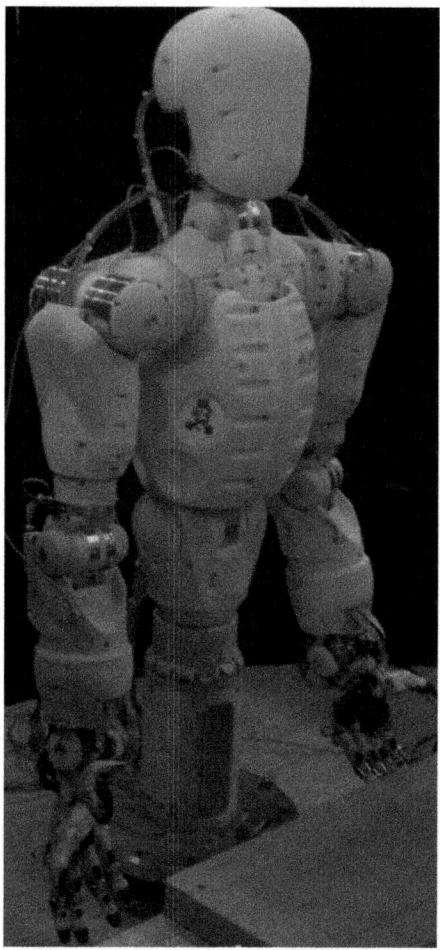

Fig. 1. BRL BERT II torso with two arms and hands

are used then $s(t = 0) = 0$. This is important, as sliding motion is secured from the start and aggressive controller action is kept minimal.

We assume the general structure of the robot dynamics is given by:

$$M(q)\ddot{q} + V(q, \dot{q}) + G(q) = T \tag{4}$$

where $M \in \Re^{n \times n}$ is the inertia matrix, a function of the n joint angles q. $V \in \Re^{n \times 1}$ is the coriolis/centripetal vector. $G \in \Re^{n \times 1}$ is the gravity vector. T is the input torque. The dynamic model given by equation (4) for 4DOF of the BERT II arm has been obtained with the help of MapleSim. The Cartesian space dynamics are now given as follows: Instead of joint torques, the dynamics consider the forces, acting on the end effector:

$$A(q)\ddot{X} + \mu_{cc}(q,\dot{q}) + f(q) = F \tag{5}$$

where $A = (JM^{-1}J^T)^{-1}$, $\mu_{cc} = \bar{J}^T V - A\dot{J}\dot{q}$, $f = \bar{J}^T G$, $F = \bar{J}^T T$ and X is the robot end-effector Cartesian position, i.e. $X = [x, y]^T$ in our case, and J is the Jacobian of the kinematics $X = h(q)$, i.e. $J = \frac{\delta X}{\delta q}$. Hence, the Cartesian velocities are defined as $\dot{X} = J\dot{q}$. The matrix $\bar{J}$ is the inertia weighted pseudo Jacobian inverse adopted from [8], (see also [10]) :

$$\bar{J} = M^{-1}J^T(JM^{-1}J^T)^{-1} \tag{6}$$

The Cartesian position (x and y) of the end effector is used here to define the motion task of the robot. The combine feedback linearization and ISMC Cartesian/task control law is:

$$F = A\ddot{X}_d + AK_P e_x + AK_D \dot{e}_x + f(q) + \mu_{cc}(q,\dot{q}) + \gamma \frac{s}{(||s|| + \delta)} \tag{7}$$

The scalars K_P and K_D are the proportional and the derivative gains respectively. The scalar $\gamma > 0$ is usually chosen large enough to counteract uncertainty and achieve robustness. The scalar δ, is introduced to reduce chattering [5]. The applied torque for the feedback linearization and ISMC control is $T = J^T F$, which is sufficient to control the Cartesian dynamics (5).

The robot arm Cartesian dynamics (x & y) given by equation (5) represent only 2DOF and other control terms have to be provided to ensure stability of the other two degrees of freedom, representing the posture or null-space dynamics in relation to the Cartesian dynamics.

3   Posture Torque Controller

As the main control approach is applied to a multi-redundant system (i.e. 4DOF, shoulder flexion, shoulder abduction, humeral rotation and elbow flexion are used, while, only 2DOF (x & y) of the end-effector are controlled), the motion is under-constrained, and some links may follow bounded but apparently random trajectories for a Cartesian position demand. Therefore, a posture torque controller is used which deals with the redundant motion, to generate a human like movement pattern. This human like motion is achieved by minimizing the effort (a function of gravity) during reaching to a particular point in the work space of the robot arm. The method here is adopted from the previous work by [13,14] (see also [1]). The full description of the 'posture' controller scheme with practical and simulation results are given in [13,14]. The scheme is similar to our recent work [7,6]. The 'posture' controller T_p is in the null space of the main controller given by (7) (Cartesian controller i.e. FL+ISMC); hence, it does not affect the main controller:

$$T = J^T F + N^T T_p, \ N^T = (I - J^T \bar{J}^T) \tag{8}$$

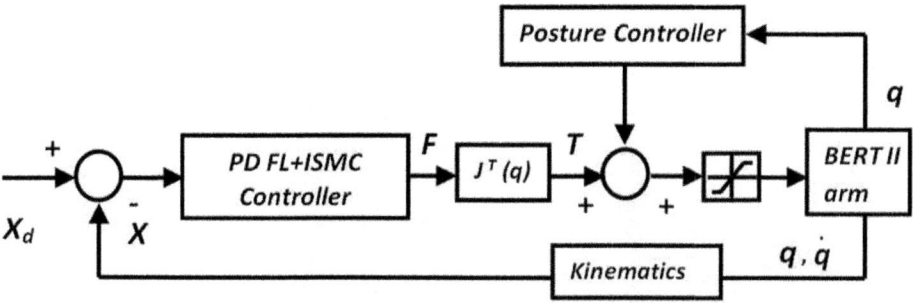

Fig. 2. Block diagram of the control scheme

where I is the identity matrix. $\bar{J}$ is the the inertia weighted pseudo Jacobian inverse:

$$\bar{J} = M^{-1}J^T(JM^{-1}J^T)^{-1} \tag{9}$$

where M is the inertia matrix given by (4). The posture torque, T_p is defined as:

$$T_p = -K_p\frac{\delta U_p}{\delta q} - K_d\dot{q} \tag{10}$$

where, K_p and K_d are proportional and derivative gains respectively. U_p is the 'muscle' effort function defined as :

$$U_p = G^T(K_m)^{-1}G \tag{11}$$

and G is the gravitational vector term from equation (4), and K_m has 4×4 dimensions in our case, actuator activation diagonal matrix (having positive diagonal elements). The diagonal values K_{mi}, define the relative preferential weighting of each actuator $(i = 1..4)$.

$$K_m = \begin{pmatrix} K_{m1} & 0 & 0 & 0 \\ 0 & K_{m2} & 0 & 0 \\ 0 & 0 & K_{m3} & 0 \\ 0 & 0 & 0 & K_{m4} \end{pmatrix} \tag{12}$$

It should be noted that the inertia weighted pseudo Jacobian inverse given by equation (9) uses the inertia matrix M. Since this is done for the posture torque controller, a good model for M in (9) is needed to get robust and acceptable posture control performance. In the Cartesian space, the control law (ISMC+FL) (7) is required for a better tracking accuracy. The overall control scheme is shown in Figure 2, and the applied torques to the robot joints are as given by equation (8):

$$T = J^T F + (I - J^T\bar{J}^T)T_p \tag{13}$$

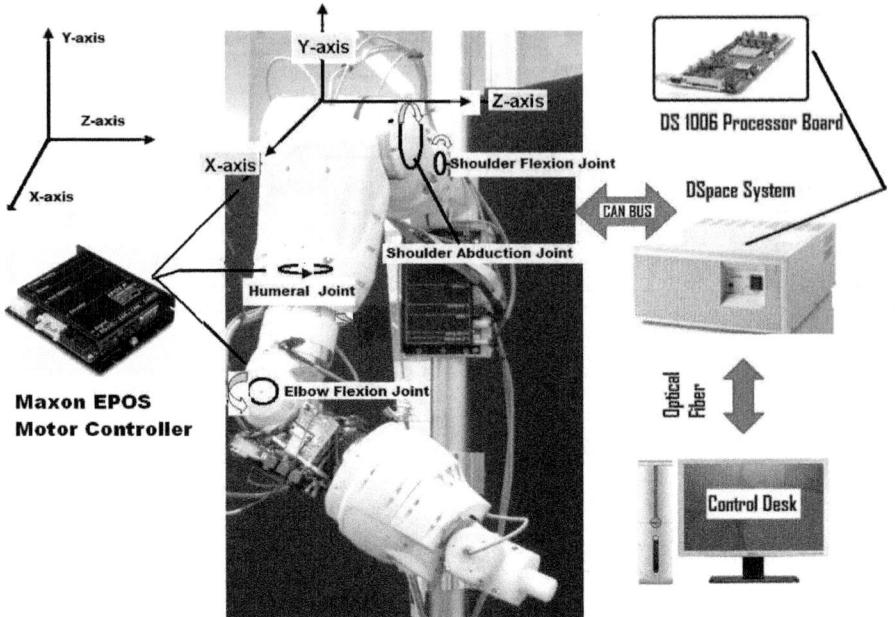

Fig. 3. Main components of the experimental setup including the BERT II robot arm

4   Experimental Setup and Results Discussion

For the real time implementation of the control scheme, a dSPACE DS1106 embedded system is employed. The Simulink model of the scheme is compiled into real-time C code and runs in the dSPACE system. A sampling time of 1 millisecond is used. The BERT II robot arm uses optical encoders for position and velocity measurements as well as torque sensors being attached to each joint. Each joint is equipped with a Maxon EPOS programmable digital motor controller. All the sensors and actuators of each joint are connected to the EPOS positioning controllers, which communicate with dSPACE via a CAN-Bus and provides the low-level current control for torque demands with the brush-less DC motors.

As mentioned, the BERT II robot arm has 7 DOF, however, only 4DOF namely, shoulder flexion, shoulder abduction, humeral rotation and elbow flexion are used as shown in Figure 3. The base coordinate frame is fixed in the shoulder. The end-effector position is specified with respect to the base frame fixed in the shoulder as shown in the Figure 3.

Simulation results are produced for the Cartesian, x, y and z, as shown in Figure 4-6. Where a sinusoidal demand inputs are applied for x & z, while a multi-step demand is applied in the y direction. Cartesian errors in simulation are shown in Figure 7. The following control tuning parameters are used in the simulation: $K_p = 300$, $K_d = 10$, $K_{s1} = K_{s2} = K_{s3} = 8$, $K_{i1} = K_{i2} = K_{i3} = 16$, $\gamma = 3000$ and $\delta = 0.1$.

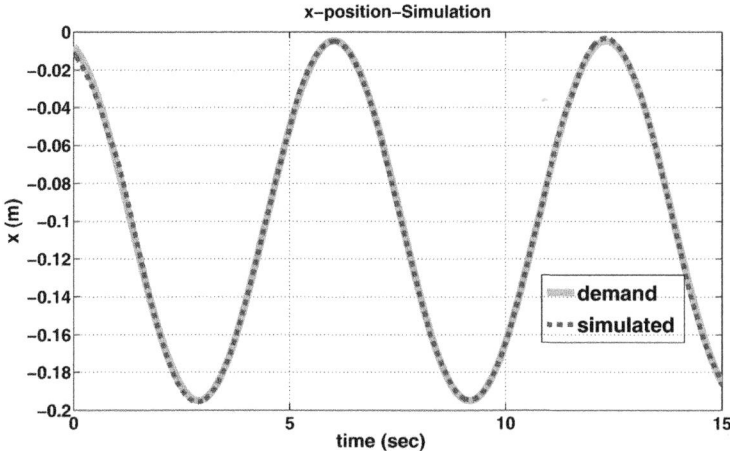

Fig. 4. Cartesian position, x (simulation)

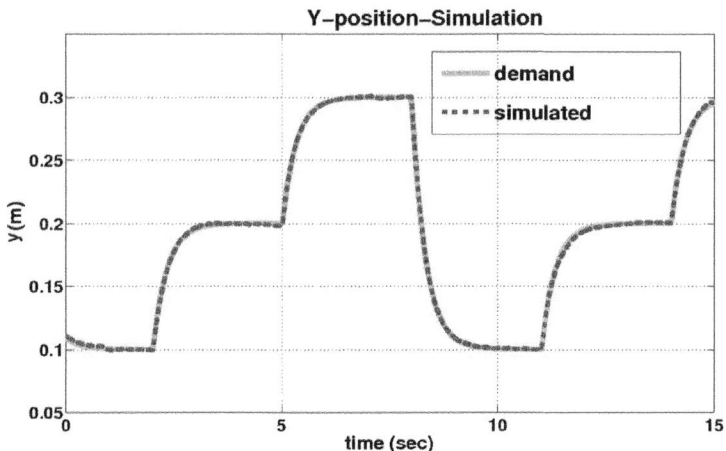

Fig. 5. Cartesian position, y (simulation)

Experimental results are shown in Figures 8-9. In the real robot experiment 1 (Figure 8), a multi-step demand is applied in the x-direction while a constant demand is applied in the y direction. Tracking error is also shown which is very small. The control tuning parameters for experiment 1 are: $K_P = 2000$, $K_D = 10$, $K_{s1} = K_{s2} = 8$, $K_{i1} = K_{i2} = 16$, $\gamma = 50$ and $\delta = 5$.

In the real robot experiment 2 (Figure 9), a sine wave demand is applied in x and a constant demand is applied in the y-direction. The tracking error shown in the Figure 9 for experiment 2. Efficient real time tracking results demonstrate the effectiveness of the proposed control scheme for our BERT II robot arm. The following control tuning parameters are used for experiment 2: $K_P = 2500$, $K_D = 10$, $K_{s1} = K_{s2} = 8$, $K_{i1} = K_{i2} = 16$, $\gamma = 50$ and $\delta = 5$.

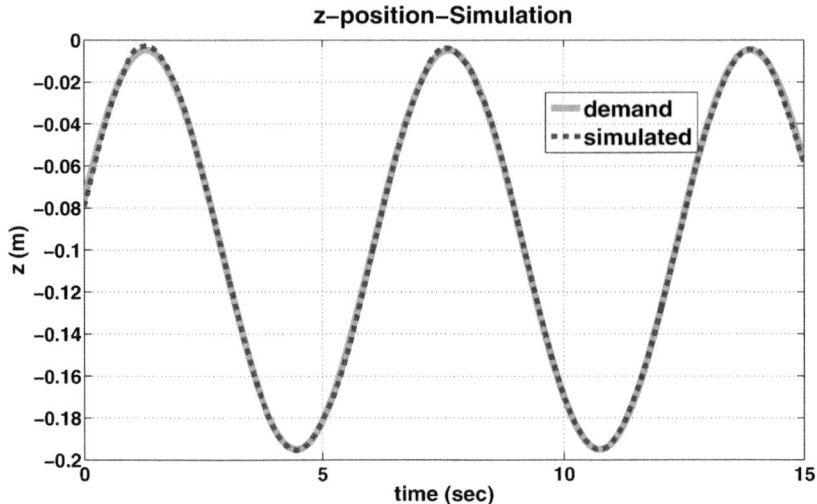

Fig. 6. Cartesian position, z (simulation)

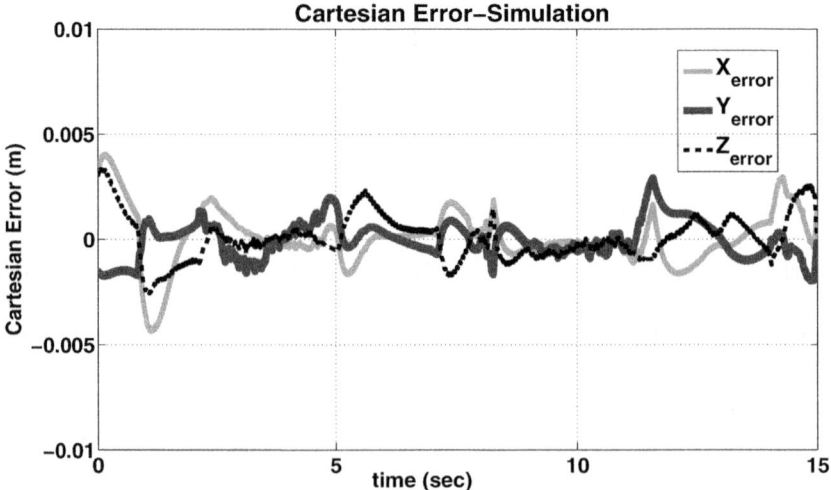

Fig. 7. Cartesian errors (simulation)

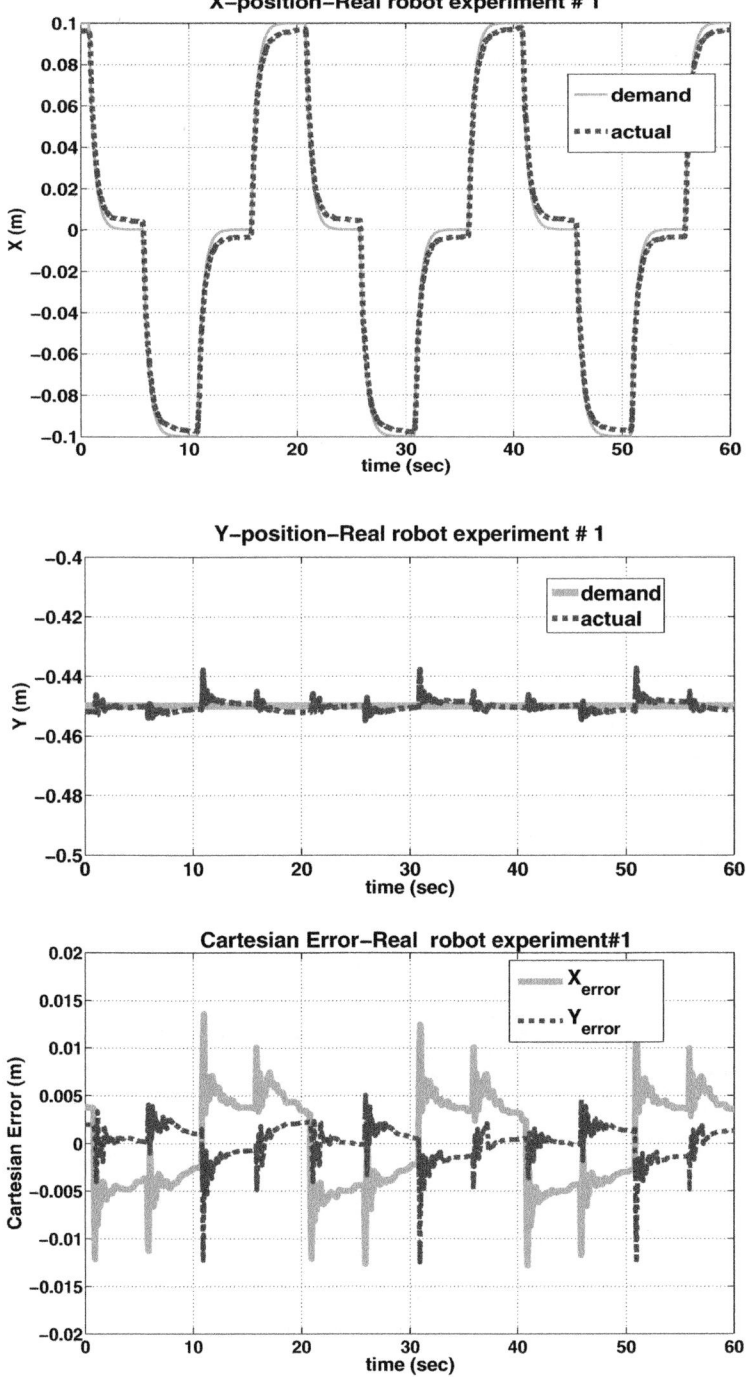

Fig. 8. Cartesian position x & y and Cartesian errors (real robot experiment 1)

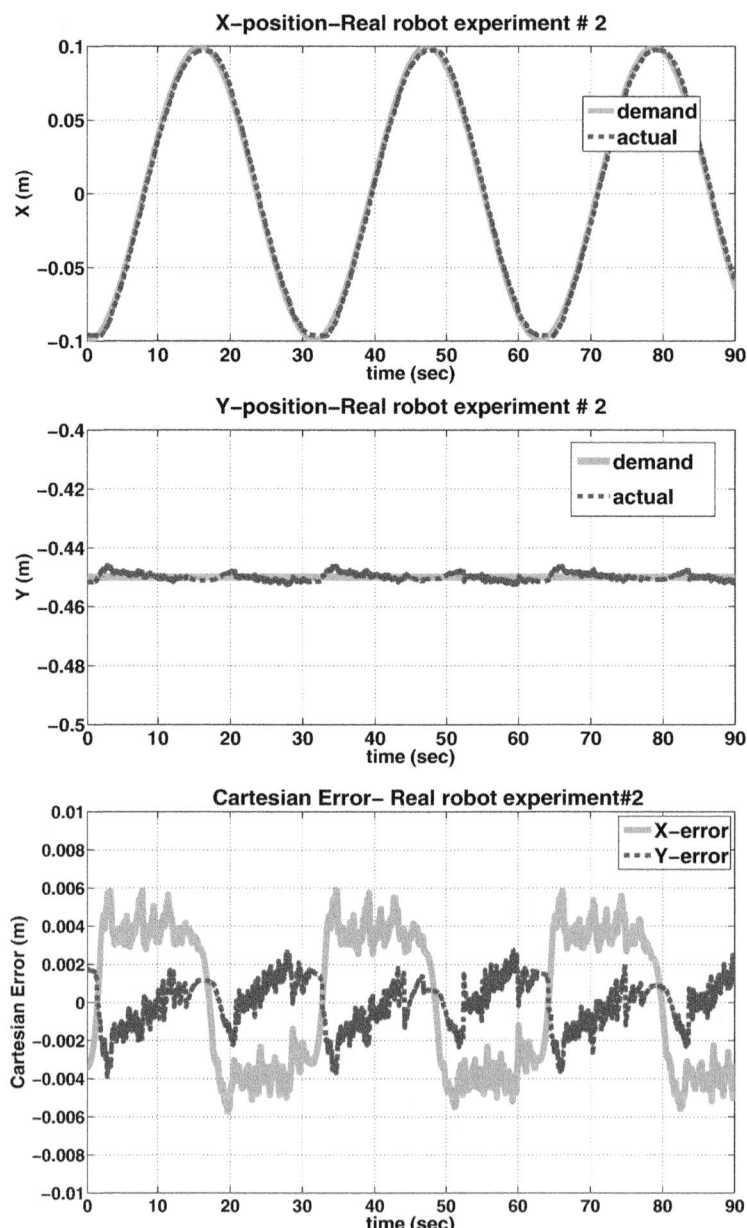

Fig. 9. Cartesian position x & y and Cartesian errors (real robot experiment 2)

The scheme is effectively dealing with the deficiencies and uncertainties in the dynamic model. The ISMC has almost eliminated the chattering effect which normally would occur with traditional sliding mode schemes. The real advantage of this scheme here is to enhance the feedback linearization controller (using the 4DOF dynamic model of the BERT II robot arm) with the ISMC to overcome the nonlinearities and unmodelled dynamics.

5   Conclusion

This paper demonstrates a successful real time implementation of the Cartesian (x & y) integral sliding mode controller for 4DOF (i.e. shoulder flexion, shoulder abduction and humeral rotation and elbow flexion) of the 7DOF humanoid BERT II arm. The redundant DOF are used for an effort-minimizing posture controller to act in a human-like pattern during reaching. Real time results show the effectiveness of the control scheme. Despite parametric uncertainties and unmodelled nonlinearities in the dynamic model (a dynamic model of the BERT II arm for the feedback linearization part and for the posture controller is used), excellent tracking results (approximately accurate to a centimeter or even less) are produced by employing an ISMC controller, while control signal chattering is almost eliminated. Apart from this, the ISMC is simple and very easy to implement. The tuning of the control scheme for the desired performance is also very straightforward as there are not many parameters to be tuned.

Acknowledgements. The research leading to these results has received funding from the European Community's Information and Communication Technologies Seventh Framework Programme [FP7/ 2007.2013] under grant agreement no. [215805], the CHRIS project.

References

1. De Sapio, V., Khatib, O., Delp, S.: Simulating the task level control of human motion: a methodology and framework for implementation. The Visual Computer 21(5), 289–302 (2005)
2. Eker, I., Akinal, S.: Sliding mode control with integral augmented sliding surface: design and experimental application to an electromechanical system. Electrical Engineering (Archiv fur Elektrotechnik) 90, 189–197 (2008)
3. Herrmann, G., Spurgeon, S., Edwards, C.: On robust, multi-input sliding-mode based control with a state-dependent boundary layer. Journal of Optimization Theory and Applications 129, 89–107 (2006)
4. Jalani, J., Herrmann, G., Melhuish, C.: Concept for robust compliance control of robot fingers. In: 11th Conference Towards Autonomous Robotic Systems, Plymouth, UK, pp. 97–102 (2010)
5. Jalani, J., Herrmann, G., Melhuish, C.: Robust trajectory following for underactuated robot fingers. In: UKACC International Conference on Control, Conventry, UK (September 2010)

6. Khan, S., Herrmann, G., Pipe, T., Melhuish, C.: Adaptive multi-dimensional compliance control of a humanoid robotic arm with anti-windup compensation. In: 2010 IEEE/RSJ International Conference on Intelligent Robots and Systems (IROS), pp. 2218–2223 (2010)
7. Khan, S.G., Herrmann, G., Pipe, T., Melhuish, C., Spiers, A.: Safe adaptive compliance control of a humanoid robotic arm with anti-windup compensation and posture control. International Journal of Social Robotics 2, 305–319 (2010)
8. Khatib, O.: A unified approach for motion and force control of robot manipulators: The operational space formulation. IEEE Jounrnal of Robotics and Automation RA3(1), 43–53 (1987)
9. Makoto, Y., Gyu-Nam, K., Masahiko, T.: Integral sliding mode control with anti-windup compensation and its application to a power assist system. Journal of Vibration and Control (2009)
10. Nemec, B., Zlajpah, L.: Null space velocity control with dynamically consistent pseudo-inverse. Robotica 18(1), 513–518 (2000)
11. Shi, J., Liu, H., Bajcinca, N.: Robust control of robotic manipulators based on integral sliding mode. International Journal of Control 81, 1537–1548 (2008)
12. Slotine, J.J.E., Li, W.: Applied Nonlinear Control. Pearson Prentice Hall, Upper Saddle River (1991)
13. Spiers, A., Herrmann, G., Melhuish, C.: Implementing discomfort in operational space: Practical application of a human motion inspired robot controller. In: TAROS Conference: Towards Autonomous Robotic Systems (August 2009)
14. Spiers, A., Herrmann, G., Melhuish, C., Pipe, T., Lenz, A.: Robotic implementation of realistic reaching motion using a sliding mode/operational space controller. In: Edwards, S.H., Kulczycki, G. (eds.) ICSR 2009. LNCS, vol. 5791, pp. 230–238. Springer, Heidelberg (2009)
15. Spurgeon, S.K., Davies, R.: A nonlinear control strategy for robust sliding mode performance in the presence of unmatched uncertainty. International Journal of Control 57(5), 1107–1123 (1993)
16. Utkin, V., Shi, J.: Integral sliding mode in systems operating under uncertainty conditions. In: Proceedings of the 35th IEEE Decision and Control, vol. 4, pp. 4591–4596 (1996)

Towards Autonomous Energy-Wise RObjects

Florian Vaussard[1], Michael Bonani[1], Philippe Rétornaz[1],
Alcherio Martinoli[2], and Francesco Mondada[1]

[1] EPFL – STI – LSRO, Lausanne, Switzerland
[2] EPFL – ENAC – IIE – DISAL, Lausanne, Switzerland
`firstname.lastname@epfl.ch`

Abstract. In this article, the RObject concept is first introduced. This
is followed by a survey of applicable energy scavenging technologies. Energy is a key issue for the large scale deployment of robotics in daily life,
as recharging the batteries places a considerable burden on the end-user
and is a waste of energy which has an overall negative impact on the
limited resources of our planet. We show how the energy obtained from
light, water flow, and human work, could be promising sources of energy
for powering low-duty devices. To assess the feasibility of powering future
RObjects with technologies, tests were conducted on commonly available
robotic vacuum cleaners. These tests established an upper-bound on the
power requirements for RObjects. Finally, based on these results, the
feasibility of powering RObjects using scavenged energy is discussed.

Keywords: Autonomous RObjects, Energy scavenging, Robotic vacuum cleaner, Power awareness.

1 Introduction

Struck by the absence of the so-often promised robotic technology in our daily
life, the project "Robots for daily life" funded by the "NCCR Robotics"[1] has
undertaken a new approach. Instead of creating new robots capable of state-
of-the-art Human Robot Interactions – such as speech or face recognition – the
RObject concept aims at embedding useful robotic technologies within everyday-
life objects. This process relies on the "everyday object" feeling and build upon
existing, natural interactions. In addition, this has the advantage of easing the
overall conception, referred by Pfeifer as "cheap design" [23]. We foresee RObjects
to enhance the long-term interactions between robotic devices and users [12].
The RObject concept has been successfully demonstrated by the creation of
robotic glasses [32]. These glasses are able to move by themselves, for example,
to be refilled when they are empty. The user still keeps the control on them,
using natural interactions, such as lifting the glass. Possible RObjects include,
but are not limited to, tableware, furniture, storage spaces, lightening systems,
decorative plants, or even room dividers.

---

[1] National Competence Center for Research in robotics, a Swiss national coordination
effort in robotics.

R. Groß et al. (Eds.): TAROS 2011, LNAI 6856, pp. 311–322, 2011.

If RObjects prove to be successful, they will spread inside our houses. This article addresses the energetic sustainability of such an approach. In term of autonomy and recharge, RObjects face the same energetic challenges as mobile robots. In addition, they introduce the constraint of their large quantity and diffusion inside the house. This aspect, if not properly addressed, can transform this dream into a nightmare.

In Sec. 2, we will present the available scavenging technologies. Section 3 will discuss further the dream of self-powered objects, by assessing an upper-bound on the power requirements using common robotic vacuum cleaners as a model. Finally, Sec. 4 will discuss the promising opportunities and the challenges which remain in designing useful, energy-wise, RObjects.

2   Energy Scavenging Technologies

Wasted energy is environmentally ubiquitous and represents a potentially cheap source of power. We will present a short survey of available power technologies, with an emphasis on renewable energies in indoor environments. Comprehensive books and state-of-the-art already exist [2,26,29,34], but usually focus on energies that are unavailable to us.

Scavenging the energy required to move a robot has often been considered in previous works, but at present, only a few people have managed to power an autonomous mobile robot using an energy scavenger. This is due to the small power density available. This issue is theoretically addressed for the case of outdoor unmanned vehicles [39], and concludes with the possibility of using solar and kinetic flow energies to power devices in the range of $1 \sim 10\,\mathrm{W}$. Solar power is commonly used in outer space exploration robots [6]. The EcoBot-II is quite original, in the sense that it embeds a microbial fuel cell converting unrefined insect biomass into electricity [19]. This is however just enough energy to allow it to travel a few centimeters.

Most of the envisioned RObjects will have to move around, implying higher energetic needs. But wheels are not always mandatory. By exploiting the interactions already in-place between the humans and those specific objects, the humans can be used as a transportation vector, thus lowering the requirements, by use of the parasitic mobility concept [16]. For example, the above-cited robotic glasses are used like normal glasses most of the time, except for when they need to be refilled. Thus, we could expect a fairly low duty-cycle operation. Another complementary option, as discussed in the Sec. 4, is to focus on RObjects favoring locations where the power can be easily scavenged.

2.1   Heat

Heat is a universal source of energy around us, part of the wasted power generated by thermal engines and exhausts, by poorly insulated buildings and even by the human body. Thermal energy scavenging devices are primarily ruled by the laws of thermodynamic. Their efficiency η is inherently limited by the Carnot cycle [14]

$$\eta_C = \frac{T_H - T_C}{T_H} = \frac{\Delta T}{T_H} \ , \tag{1}$$

where T_H and T_C are the temperatures of the hot and cold sides respectively. The temperature gradient ΔT should be maximized, favoring applications such as in the automotive or electronics cooling industries. Making use of thermal gradients is, however, not well suited for obtaining energy under ambient conditions. For example, when applied to scavenge energy between the human body (T_H=309 K) and the ambient air (T_L=293 K), the maximum achievable efficiency is only 5.2%.

The total efficiency will moreover be lowered by the non-ideal transducer. Thermoelectricity is the most widely used solid-state technology at low temperature. As presented by Min in [2, Chap. 5], ThermoElectric Generators (TEG) are mainly based on the Seebeck effect. A temperature gradient ΔT applied to a junction will generate a voltage V_{ab}, following

$$V_{ab} = \alpha_{ab} \Delta T \ , \tag{2}$$

where α_{ab} is the Seebeck coefficient of the considered junction. Optimizing a TEG is first a matter of material engineering. For ambient temperatures, with Bi_2Te_3 used in the case of the human body, the expected overall efficiency falls around 0.91%.

Seiko was the first company to release a watch powered by the human body heat [13]. Using this strategy, it is also feasible to power wearable sensor nodes, providing about $100\,\mu W$ when placed on the wrist [17]. A TEG half-buried in the soil can provide up-to $1\,W\,m^{-2}$ (peak) to an outdoor sensor node, by exploiting the temperature difference between the buried side and the side exposed to the ambient air [14].

2.2 Light

The sun is a huge source of energy, which can be easily exploited by solar cells, at least during daylight. Indoor, room lighting is another ubiquitous source of energy. The first calculators to rely solely on this source were designed in the 1980's already [35]. There are some discrepancies in the literature regarding the peak power levels available from the solar spectrum. The most widely used values are presented in Table 1 [2,22,34]. The Air Mass (AM) 1.5 model, used as a reference when testing solar panels, considers a value of $1000\,W\,m^{-2}$.

The efficiency of solar cells is quite low and depends on a number of factors: ray angles, illumination level [41] – which depends on the season and time of the day – and spectral source. The optimal technology thus depends on the available light. Single crystal silicon is proposed for outdoor conditions, with common efficiencies around 15 to 20% [10]. Thin film amorphous silicon or cadmium telluride cells are proposed for indoor conditions [34], but they can scavenge a

Table 1. Available and scavenged power densities from light

Raw Power	Scavenged Power	(η)	Conditions
$500 - 1000$ W m^{-2}	200 W m^{-2}	(20%)	Outdoors – Sunny
100 W m^{-2}	15 W m^{-2}	(15%)	Outdoors – Cloudy
$1 - 5$ W m^{-2}	0.1 W m^{-2}	(10%)	Indoors

mere 10% [30]. Expected power densities at the output of the cells are also given in Table 1.

Other technologies, like the dye sensitized cells and the multi-gap cells , have also been studied. This latter technology is promising since efficiency is increased by multiple band gaps which allow for the capture of photons from a wider range of energies making use of more of the solar spectrum.

Exploration robots mainly use the solar power. For example, the Mars rover "Spirit" is powered by triple-junction cells, providing 190 W (150 W m^{-2}) operating under good conditions [6]. The small robot Alice has been powered using only thin film amorphous silicon cells and a 3000 lumens projector, with an achieved power density of 27.8 W m^{-2} [4]. Finally, wireless sensor nodes can use efficiently the power of small solar cells [11,27].

2.3   Ambient Radio Frequencies

Currently, wireless transmissions and various RF emissions are everywhere: TV, radio, cellphones and many others. It would be tempting to exploit part of this radiated power to power RObjects.

The Friis equation [37] can be used to compute the received power P_r in an antenna with gain G_r, assuming a wave of wavelength λ emitted at a distance d by an isotropic source of gain G_t and propagating in free space, far enough from the emitter (far field condition)

$$P_r = G_r G_t P_0 \left(\frac{\lambda}{4\pi d} \right)^2 = G_r \frac{(E\lambda)^2}{4\pi Z_0} , \tag{3}$$

where the radiation impedance of free space, Z_0, equals 377 ohms. P_0 is the emitted power, while E is the field strength. When considering indoor conditions, the received power is closer to $P_r \sim d^{-4}$. Assuming a unity gain and a field strength of 1 V m^{-1} at 2.4 GHz , the scavenged power is around 3.3 μW.

While the scavenged power is too low for our use – apart from actively transmitting energy using microwaves or a laser [7] – this is enough for wireless sensor nodes [24] or RFID tags, even at a distance of a few meters [21].

2.4   Pressure and Temperature Variations

The daily atmospheric changes could also power devices. This technology has been in use since the early 1900's, providing power to clocks [31], and more recently to watches [25], based on a closed dilating volume. However, apart from the watch industry, no other device has been powered by this kind of energy.

In fact, according to [5], a variation of 1 K or 400 Pa provides $3.5 \cdot 10^{-3}$ J of energy. This would be enough to power a small device with a consumption of $50\,\mu$W for only 70 seconds. This is however enough to power the Atmos clock for 2 days, implying a power consumption of only 20 nW.

2.5   Gas and Liquid Flows

Flow has for long been exploited to scavenge energy in wind and watermills. This energy is also available inside houses, hidden in air breezes generated by air-conditioning or heating systems, as well as in water pipes. This principle is, for example, exploited by the self-powered shower handle [1].

According to the Betz' law, for a flow of speed v considered over a surface A, only 59.3% of the total power can be scavenged

$$P_{\max} = C_P \cdot P_{\text{flow}} = C_p \cdot \frac{1}{2}\rho A v^3 \quad \text{where} \quad C_p = \frac{16}{27} \approx 0.59 \; . \tag{4}$$

The power is proportional to the flow density ρ. Air and water have respective densities of 1.2 and $1000\,\text{kg m}^{-3}$. The available power is around 0.35 and $290\,\text{W m}^{-2}$ for a flow speed of $1\,\text{m s}^{-1}$. Water is undoubtedly more powerful.

Even if big mills tend to have an efficiency close to the Betz limit, small devices operating at low speeds suffer from increased mechanical and viscous friction, reducing their efficiency to about 1/6 of the Betz limit [20]. To scavenge energy from the air flow, most researchers are using a custom rotor coupled with a standard DC motor operating in generator mode [8,28], even powering sensor nodes [9,36].

2.6   Mechanical

Mechanical scavenging is a broad topic, with numerous applications [3,40]. However, up until now, no robot has been powered in this manner. Sensor nodes are better suited, by example for monitoring the stress of life-critical materials .

Several technologies are currently being studied. Electromagnetic generators are quite popular, but are difficult to integrate on MEMS devices. Piezoelectric scavengers convert a stress into a voltage, often using a PZT ceramic. Finally, electrostatic devices are based on a variable capacitor, operated either at fixed voltage or fixed charge.

In the home environment, possible mechanical energy sources include machine vibrations such as clothes dryers and microwave ovens, as well as low amplitude vibrations like structural vibrations of the windows and walls [33]. The available power is however low compared to industrial applications (engines, machine tools), on the order of $0.3\,\text{mW cm}^{-3}$.

For our RObject concept, scavenging human power could be an interesting solution. Scavenging the power of the human gait could produce up to a few Watts [38], but scavenging too much energy would result in a disturbance for the user. The first research focused on shoes, producing up to 230 mW [15]. Piezoelectric scavengers implemented in pavements and roads are another option. A prototype [18] is said to produce 3.8 W with a walking person.

As RObjects will closely interact with humans, they could also benefit from this interaction by scavenging some power. Based on the work of the gravity force, the energy involved is around $E = m\,g\,h$. For a lightweight RObject, this would result only in a few Joules, which is not enough for our needs. The work of the friction force, $E = \mu\,m\,g\,d$, is of the same order of magnitude.

3   RObjects Power Requirements

It is clear that scavenging energy for use in powering RObjects is a real possibility. Establishing the energy needs for RObjects and determining an upper bound is crucial to the design of real, autonomous RObjects. This upper-bound has to be representative of an all-purpose mobile robotic system. The previously mentioned moving glasses are not well suited, as they have inefficient stepper motors and perform only small displacements. A more representative device is the robotic vacuum cleaner, now a well established technology. Vacuuming is a typical domestic task, like tidying the house, watering the plants or feeding the cat. All have about the same duty cycle (a few hours per week) and energy requirements.

We have compared several vacuum cleaners with respect to their power consumption, as well as to their performance and efficiency. The tested systems are shown in Fig. 1. The robots differ from the hand-operated vacuum cleaner mainly by their localization and path planning process, but also by their cleaning tools.

(a)   (b)   (c)   (d)

Fig. 1. Tested vacuum cleaners. Hand-operated (Dyson DC05) (a), iRobot Roomba 530 (b), Samsung Navibot (c) and Neato XV-11 (d).

The Roomba has no global localization and relies only on basic displacement patterns (line, circle, wall following), coupled with a simple IR-based detection of the base station. Starting from 2010, two new affordable challengers have appeared. The Navibot is sold by Samsung and relies on an upward-looking camera doing Visual Simultaneous Localization And Mapping (V-SLAM). The XV-11 is sold by Neato Robotics and performs regular SLAM using a cheap laser scanner.

A fake house environment has been built inside our testing facility, as pictured in Fig. 2 (a). The floor is a standard, smooth, industrial grade one, with some minor defects. The efficiency of these vacuums on carpet is not tested since locomotion is not the focus of this article.

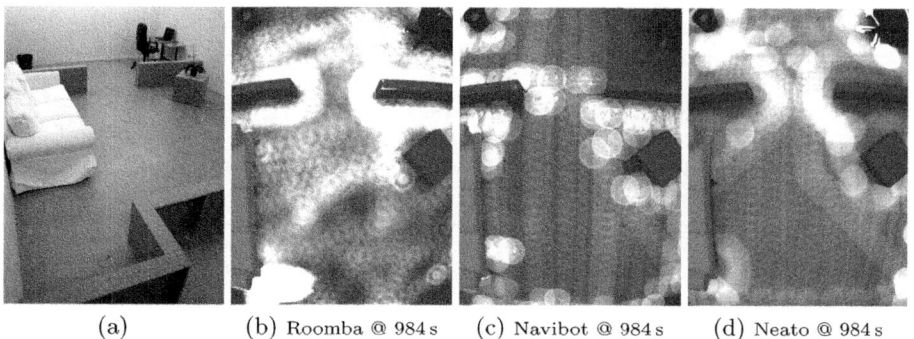

(a)   (b) Roomba @ 984 s   (c) Navibot @ 984 s   (d) Neato @ 984 s

Fig. 2. (a) Testing environment. (b) – (d) Composite images, took with the overhead camera at 2.18 Hz, at completion by the Neato (984 s). The base station is located in the bottom right corner.

Preliminary experiments were conducted in occupied homes and have shown a mean deposition ratio for the dirt of $0.5\,\mathrm{g\,m^{-2}}$ for a whole week. The surface of our arena is roughly $25\,\mathrm{m^2}$. We have used a mixture of 20 g of flour and 10 g of breadcrumb, simulating the result of two busy weeks of activity in a home without vacuuming. The amount of dust was measured for all the experimental steps using a differential methodology coupled with a Metler precision scale, which has a resolution of 0.001 g. The robots were thoroughly cleaned before each run.

The results are shown in Table 2, while composite images are shown in Fig. 2 (b) - (d). The whiter the area, the more often the robot stayed or passed. All three robots have about the same speed. The difference between the Roomba and the two other robots is obvious. The Roomba cleans several times the same spot, while the two others are more systematic. The Neato is slightly more efficient around the obstacles (sofa and alike), as it can plan ahead difficulties with its laser scanner.

The idle power consumptions P_s and P_{sr} are measured using a precision Wattmeter. The figures are really high, especially when the charged robot is connected to its base station (P_{sr}), indicating a huge loss of energy. For each run, the cleaning time was measured directly, while the required energy has been inferred based on the energy consumed during a complete charge. The influence of the idle consumption P_{sr} has been removed beforehand, but the charge losses are still included. The robot operational power is simply computed based on $P_{op} = E/T$.

Table 2. Results for manual and robotic vacuum cleaners

	Idle power		Statistics for each run					
	Station only	Station +robot	Cleaning time		Energy		Robot Power	Collected dust[a]
	P_s	P_{sr}	T		E		P_{op}	
Units	W	W	s	m-m [b]	kJ	m-m^b	W	% m-m^b
Dyson	–	–	390	36	474.0	43.8	1215[c]	97.4 0.005
Roomba	1.14	4.80	3674	482	96.5	12.7	26.3	80.8 3.89
Navibot[d]	1.77	7.75	1264	44	51.9	1.81	41.0	48.9 4.92
Neato	0.33	4.50	930	92	33.8	3.34	36.3	80.4 2.29
Neato (2 times)	0.33	4.50	2067	38	75.1	1.38	36.3	92.8 2.67

[a] This includes both the dust collected in the bin and on other parts of the robot.
[b] $max - min$.
[c] Measured directly using a Wattmeter.
[d] Used in "Edge" mode, for better results.

Several interesting facts should be pointed out. First, cleaning using a robot requires far less energy than doing it by hand, by a factor 5 to 14. Even if a robot takes more time to do its job, its power consumption is nevertheless 30 to 45 times smaller. The cleaning efficiency of the robots is a step behind the classical vacuum cleaner, and is really bad in the case of the Navibot. The Roomba and Neato have a dust efficiency around 80%, while Roomba takes 4 times longer and provides incomplete floor coverage as can be seen in Fig. 2 (e). As shown with the Neato, the efficiency can be improved by executing a second pass. Almost 93% of the dust was collected this way. These results are only a few percent worse than those obtained by the conventional vacuum cleaner. Finally, the Navibot is pretty inefficient overall, but the test conditions were quite harsh for such small robots.

Compared to someone vacuuming once a week, using the Neato twice a week would allow him to spare 80% of the energy, while keeping the dirt level low. It would even be possible to use the Neato everyday and still using only 50% of the power required to vacuum by hand once a week.

The superiority of the Neato undoubtedly stems from its use of the SLAM method coupled with an efficient brush - aspiration system. It has, however, two drawbacks. The most important one is the noise, which is simply unbearable. And the absence of a side brush results in a poor cleaning of the borders and corners. A good compromise would be to combine the SLAM provided by the Neato with the cleaning functions of the Roomba.

4   Conclusion and Discussion

An exhaustive summary of scavenging technologies is given in Table 3. Not surprisingly, water and solar power are the most powerful sources, along with

Table 3. Summary of scavenging technologies. Base units are mW and cm for an easier comparison at the RObject scale.

Device	Theoretic Power	Conditions	Practical Power	Ref.
Light	$100 \ \frac{mW}{cm^2}$	Outdoors; $\eta = 1$	$20 \ \frac{mW}{cm^2}$	[10]
	$0.1 \ \frac{mW}{cm^2}$	Indoors; $\eta = 1$	$0.01 \ \frac{mW}{cm^2}$	[30]
Ambient RF	$0.003 \ mW$	2.4 GHz; $E = 1 \, \mathrm{V \, m^{-1}}$; $G_r = 1$		
Acoustic	$0.96 \cdot 10^{-3} \ \frac{mW}{cm^3}$	$100 dB$		
Thermoelectric	$0.1 \ \frac{mW}{cm^2}$	Body	$0.02 \ \frac{mW}{cm^2}$	[17] [a]
Daily pressure change	$7.8 \cdot 10^{-6} \ \frac{mW}{cm^3}$ [34]	$\Delta p = 677 \, \mathrm{Pa}$		
Daily temperature change	$0.017 \ \frac{mW}{cm^3}$ [34]	$\Delta T = 10 \, \mathrm{K}$	$20 \cdot 10^{-6} \ mW$	[5]
Shoe impact	$8400 \ mW$ [38]		$230 \ mW$	[15]
Vibrations	$0.3 \ \frac{mW}{cm^3}$ [34]		$0.10 \ \frac{mW}{cm^3}$	[40]
Flow (Air)	$0.58 \ \frac{mW}{cm^2}$	$2.54 \, \mathrm{m\,s^{-1}}$ at Betz limit	$0.10 \ \frac{mW}{cm^2}$	[8] [b]
Flow (Water)	$480 \ \frac{mW}{cm^2}$	$2.54 \, \mathrm{m\,s^{-1}}$ at Betz limit	$80.7 \ \frac{mW}{cm^2}$	Hyp.[c]

[a] On the wrist.
[b] $2.54 \, \mathrm{m\,s^{-1}}$.
[c] $2.54 \, \mathrm{m\,s^{-1}}$; 1/6 of Betz limit.

the human gait. Scavenging a water flow could be really interesting, but a free flow is not common inside a house, apart from devices like fountains or shower handles. Indoor lighting is common, but would provide only a small amount of power.

Concerning energy consumption and based on our tests, we can assume as a reasonable case a 30 W RObject, used for two hours per week. This model would represent a consumption of 60 W h per week. Table 4 shows the feasibility of this case, with some of the energy scavenging devices presented in Table 3, mounted on the robot and / or on the base station. An optimized RObject, with good brushless motors and well crafted electronics, would probably consume far less energy than this worst case.

From this analysis, it is clear that powering such a RObject would be feasible. The first choice is a small solar panel operated outdoors. When used indoors, this solar panel should be placed just behind a window in direct sunlight. Even if part of the power will be reflected and absorbed by the glass, a bigger solar panel should still suffice. This will have to be assessed in our future work.

The second choice is to find an application operating near a water source. This would be a good solution for RObjects having a low power consumption, as a reasonable propeller is enough. The last choice is to scavenge mechanical power

Table 4. Possible power sources for a RObject consuming 60 W h per week

Source	Power Density	Duration per Day	Energy per Week	Feasibility Condition	Feasible?
Solar (outdoors)	$10.00 \ \frac{mW}{cm^2}$	8 h	$560.00 \ \frac{mW \, h}{cm^2}$	$1 \, dm^2$	**Yes**
Solar (indoors)	$0.01 \ \frac{mW}{cm^2}$	8 h	$0.56 \ \frac{mW \, h}{cm^2}$	$10 \, m^2$	No
Vibrations	$0.10 \ \frac{mW}{cm^3}$	24 h	$16.80 \ \frac{mW \, h}{cm^3}$	$0.35 \, m^3$	No
Human gait	$200.00 \ mW$	2 h	$2.10 \ W \, h$	**Several humans**	**Maybe**
Air	$0.10 \ \frac{mW}{cm^2}$	24 h	$16.80 \ \frac{mW \, h}{cm^2}$	$35 \, m^2$	No
Water	$80.00 \ \frac{mW}{cm^2}$	1 h	$560.00 \ \frac{mW \, h}{cm^2}$	$1 \, dm^2$	**Maybe**

from human movement, by fitting carpets with mechanical scavenging devices, placed in high-traffic areas. This also sounds interesting, but would require to adapt the human environment.

As a conclusion, this study has provided excellent initial data and analysis. This allows a better understanding of the potential performances and limitations of robotics in term of energy, especially with respect to the long term vision of an intensive use in our daily life. The solar energy is for sure one of the best options to ensure zero-energy robotic systems. We could, for example, design a water-pouring RObject to take care of the houseplants. Or even design moving plants able to take care of themselves, both from the water and sun point of view. Tables and chairs are also often placed in a bright place, and RObjects furniture could be of a great help, for example to increase efficiency when vacuuming.

Low-duty RObjects powered by energy scavenging devices seem imminently feasible. We however have to design them to be energy efficient and cost effective if we want them to reach the market. The robots of Sec. 3 are designed to be cheap, thus their consumption is above state-of-the-art low power robots. This will be one of the main focus of our work on future RObjects, as energy is a scarce resource. Even if the scavenging device shall not suffice by itself to cover all the needs, it will have a great impact in the energetic balance, if it is well designed. Providing such a RObject will be our next step.

Acknowledgments. This research was supported by the Swiss National Science Foundation through the National Centre of Competence in Research Robotics.

References

1. Bean, J.: Energy saving shower head, June 9, US Patent App. 11/148,524 (2005)
2. Beeby, S., White, N.: Energy Harvesting for Autonomous Systems. Artech House Publishers, Boston (2010)
3. Beeby, S., Tudor, M., White, N.: Energy harvesting vibration sources for microsystems applications. Measurement Science and Technology 17, R175 (2006)

4. Boletis, A., Driesen, W., Breguet, J., Brunete, A.: Solar Cell Powering with Integrated Global Positioning System for mm3 Size Robots. In: 2006 IEEE/RSJ International Conference on Intelligent Robots and Systems, pp. 5528–5533. IEEE, Los Alamitos (2007)
5. Callaway, E.: Wireless sensor networks: architectures and protocols. CRC Press, Boca Raton (2004)
6. Crisp, J.A., Adler, M., Matijevic, J.R., Squyres, S.W., Arvidson, R.E., Kass, D.M.: Mars exploration rover mission. Journal of Geophysical Research 108(E12), 8061 (2003)
7. Denninghoff, D., Starman, L., Kladitis, P., Perry, C.: Autonomous power-scavenging MEMS robots. In: 48th Midwest Symposium on Circuits and Systems, pp. 367–370. IEEE, Los Alamitos (2006)
8. Federspiel, C., Chen, J.: Air-powered sensor. In: Proceedings of IEEE Sensors, 2003, vol. 1, pp. 22–25. IEEE, Los Alamitos (2004)
9. Flammini, A., Marioli, D., Sardini, E., Serpelloni, M.: An autonomous sensor with energy harvesting capability for airflow speed measurements. In: 2010 IEEE Instrumentation and Measurement Technology Conference (I2MTC), pp. 892–897. IEEE, Los Alamitos (2010)
10. Green, M., Emery, K., Hishikawa, Y., Warta, W.: Solar cell efficiency tables (version 34). Progress in Photovoltaics: Research and Applications 17(5), 320–326 (2009)
11. Hande, A., Polk, T., Walker, W., Bhatia, D.: Indoor solar energy harvesting for sensor network router nodes. Microprocessors and Microsystems 31(6), 420–432 (2007)
12. Kaplan, F.: Everyday robotics: robots as everyday objects. In: Proceedings of the 2005 Joint Conference on Smart Objects and Ambient Intelligence: Innovative Context-aware Services: Usages and Technologies, pp. 59–64. ACM, New York (2005)
13. Kishi, M., Nemoto, H., Hamao, T., Yamamoto, M., Sudou, S., Mandai, M., Yamamoto, S.: Micro thermoelectric modules and their application to wristwatches as an energy source. In: Eighteenth International Conference on Thermoelectrics, 1999, pp. 301–307. IEEE, Los Alamitos (2002)
14. Knight, C., Davidson, J.: Thermal Energy Harvesting for Wireless Sensor Nodes with Case Studies. Advances in Wireless Sensors and Sensor Networks, 221–242 (2010)
15. Kymissis, J., Kendall, C., Paradiso, J., Gershenfeld, N.: Parasitic power harvesting in shoes. In: Second International Symposium on Wearable Computers, Digest of Papers, pp. 132–139 (1998)
16. Laibowitz, M., Paradiso, J.A.: Parasitic mobility for pervasive sensor networks. Pervasive Computing, 255–278 (2005)
17. Leonov, V., Torfs, T., Fiorini, P., Van Hoof, C.: Thermoelectric converters of human warmth for self-powered wireless sensor nodes. IEEE Sensors Journal 7(5), 650–657 (2007)
18. Mandal, I., Patra, P.: Renewable Energy Source. International Journal of Computer Applications 1(17), 44–53 (2010)
19. Melhuish, C., Ieropoulos, I., Greenman, J., Horsfield, I.: Energetically autonomous robots: Food for thought. Autonomous Robots 21(3), 187–198 (2006)
20. Mitcheson, P., Yeatman, E., Rao, G., Holmes, A., Green, T.: Energy harvesting from human and machine motion for wireless electronic devices. Proceedings of the IEEE 96(9), 1457–1486 (2008)
21. Obrist, B., Hegnauer, S.: A microwave powered data transponder. Sensors and Actuators A: Physical 46(1-3), 244–246 (1995)

22. O'Donnell, T., Wang, W.: Power Management, Energy Conversion and Energy Scavenging for Smart Systems. Ambient Intelligence with Microsystems, 241–266 (2009)
23. Pfeifer, R., Bongard, J., Grand, S.: How the body shapes the way we think: a new view of intelligence. The MIT Press, Cambridge (2007)
24. Philipose, M., Smith, J., Jiang, B., Mamishev, A., Roy, S., Sundara-Rajan, K.: Battery-free wireless identification and sensing. IEEE Pervasive Computing, 37–45 (2005)
25. Phillips, S.: Temperature responsive self winding timepieces, October 1 (2002), US Patent 6,457,856
26. Priya, S., Inman, D.J.: Energy Harvesting Technologies, 1st edn. Springer Publishing Company, Heidelberg (2008) (incorporated)
27. Raghunathan, V., Kansal, A., Hsu, J., Friedman, J., Srivastava, M.: Design considerations for solar energy harvesting wireless embedded systems. In: Fourth International Symposium on Information Processing in Sensor Networks, IPSN 2005, pp. 457–462. IEEE, Los Alamitos (2005)
28. Rancourt, D., Tabesh, A., Fréchette, L.: Evaluation of centimeter-scale micro windmills: aerodynamics and electromagnetic power generation. In: Proc. PowerMEMS 2007, pp. 28–29 (2007)
29. Randall, J.: Designing indoor solar products. Wiley Online Library (2006)
30. Randall, J., Jacot, J.: The performance and modelling of 8 photovoltaic materials under variable light intensity and spectra. In: World Renewable Energy Conference VII Proceedings, Cologne, Germany (2002)
31. Reutter, J.: Horloge à remontage automatique par les variations de température ou de pression atmosphérique, January 15 (1929), Swiss Patent CH130941A
32. Rey, F., Leidi, M., Mondada, F.: Interactive mobile robotic drinking glasses. Distributed Autonomous Robotic Systems 8, 543–551 (2009)
33. Roundy, S., Wright, P., Rabaey, J.: A study of low level vibrations as a power source for wireless sensor nodes. Computer Communications 26(11), 1131–1144 (2003)
34. Roundy, S., Wright, P.K., Rabaey, J.M.: Energy scavenging for wireless sensor networks: with special focus on vibrations. Springer, Netherlands (2004)
35. Sangani, K.: The sun in your pocket. Eng. Technol. 2(8), 36–38 (2007)
36. Sardini, E., Serpelloni, M.: Self-powered wireless sensor for air temperature and velocity measurements with energy harvesting capability. IEEE Transactions on Instrumentation and Measurement PP(99), 1–7 (2010)
37. Smith, A.: Radio Frequency Principles & Applications. Universities Press (1998)
38. Starner, T.: Human-powered wearable computing. IBM Systems Journal 35(3&4) (1996)
39. Thomas, J.P., Qidwai, M.A., Kellogg, J.C.: Energy scavenging for small-scale unmanned systems. Journal of Power Sources 159(2), 1494–1509 (2006)
40. Op het Veld, B., Hohlfeld, D., Pop, V.: Harvesting mechanical energy for ambient intelligent devices. Information Systems Frontiers 11(1), 7–18 (2009)
41. Virtuani, A., Lotter, E., Powalla, M.: Influence of the light source on the low-irradiance performance of Cu (In, Ga) Se2 solar cells. Solar Energy Materials and Solar Cells 90(14), 2141–2149 (2006)

Towards Safe Human-Robot Interaction

Elena Corina Grigore[1], Kerstin Eder[1,2], Alexander Lenz[2],
Sergey Skachek[2], Anthony G. Pipe[2], and Chris Melhuish[2]

[1] Computer Science Department, University of Bristol, Bristol, UK
Kerstin.Eder@bristol.ac.uk
[2] Bristol Robotics Laboratory, Bristol, UK
{Alex.Lenz,Sergey.Skachek,Tony.Pipe,Chris.Melhuish}@brl.ac.uk

Abstract. The development of human-assistive robots challenges engineering and introduces new ethical and legal issues. One fundamental concern is whether human-assistive robots can be trusted. Essential components of trustworthiness are usefulness and safety; both have to be demonstrated before such robots could stand a chance of passing product certification. This paper describes the setup of an environment to investigate safety and liveness aspects in the context of human-robot interaction. We present first insights into setting up and testing a human-robot interaction system in which the role of the robot is that of serving drinks to a human. More specifically, we use this system to investigate when the right time is for the robot to release the drink such that the action is both safe and useful. We briefly outline follow-on research that uses the safety and liveness properties of this scenario as specification.

1 Introduction

Human-assistive robots are machines designed to improve our quality of life by helping us to achieve tasks, e.g. a personal care robot might help us during accident recovery. Such robots will perform physical tasks within our personal space, including shared manipulation of objects and even direct contact. The capability to dynamically adapt to different situations is a prerequisite for these robots to be genuinely useful in practice. To be effective they may also need to be powerful. The combination of both these capabilities makes them potentially dangerous and capable of inflicting significant harm to humans or damage to surrounding objects.

The development of human-assistive robots challenges engineering and introduces new ethical and legal issues that need to be addressed to unlock the clear potential for the huge social and financial benefit that is expected to result from large-scale commercial exploitation. One fundamental concern is whether human-assistive robots can be trusted. Essential components of trustworthiness are usefulness[1] and safety[2]. Demonstrable trustworthiness will be a prerequisite

---

[1] Aka liveness, i.e. robots don't fail to do things they are supposed to do - cf. the definition of a *liveness property* "something good must eventually happen" [7].

[2] Robots operate within their specification and within safe limits - cf. the definition of a *safety property* "something bad does never happen" [7].

R. Groß et al. (Eds.): TAROS 2011, LNAI 6856, pp. 323–335, 2011.

for such robots to pass product certification. This, in turn, will be a fundamental requirement for any large-scale commercialisation of robots that operate in close proximity to humans. Development of a method for undertaking such certification is currently a wide open research question.

This paper describes the setup of an environment to investigate safety and liveness aspects in the context of close Human-Robot Interaction (HRI), i.e. HRI demanding co-location in the same physical space and collaborative interaction with physical objects. We present first insights into setting up and testing a HRI system in which the role of the robot is that of serving drinks to a human. The setting includes a humanoid robot head capable of gaze-tracking, a voice system that allows the user to interact directly with the robot, and a robot hand which grabs a cup from an initial position and takes it to a serving position; the robot must decide whether to release the cup or not.

To ensure safety and liveness of this hand-over we propose safety and liveness properties. We have conducted initial experiments to find safe and useful settings for the control parameters in the system, e.g. the amount of pressure to be applied for the robot to release the drink. The tests performed on this particular system can be a helpful exemplar for other HRI tasks such as handling of objects (e.g. robot teaching/playing with a child), or even more direct interaction such as shaking hands (this is where the pressure sensors would come in, as well as identifying the position of the human's hand). We have identified research questions for challenging follow-on research projects that introduce machine learning to enable the robot to adapt its behaviour to different situations.

2   Related Work

One corner stone for successful verification is a well defined specification. This is particularly important for adaptive systems because the system changes over time adapting to different situations. Verification is the process used to demonstrate the correctness of a design with respect to the specification. Formal methods have been used to establish the correctness of adaptive systems either by verification or by systematic design. An example is the application of model checking [3] to prove that a specification in the form of a set of safety properties is satisfied by a multi-agent control system that adapts by learning from experience [8]. Other work has focused on ensuring safety by construction [2]. Systems such as the LAAS Architecture for Autonomous System, which is based on the BIP (Behaviors Interactions Priorities) framework, support a systematic component-based design, construction and verification approach which can enforce online safety properties [1]. The importance of formal methods has recently been recognized at the ICRA workshop "Formal Methods in Robotics and Automation" [9]. As systems become more and more adaptive, however, the verification challenge increases and new scalable solutions need to be found.

The work presented here prepares the ground for research into finding a compromise between adaptability and verifiability, so as to identify those situations where the emphasis should be put more on one, or the other. We introduce a set of safety and liveness properties acting as specification. Even an apparently simple scenario, such as offering a drink to a person whilst being simultaneously safe and useful, is a challenge because every user is different and will react in a slightly different way to the robot. As the choices increase, and the robot's behavior becomes more adaptive, the more difficult it is to verify it. On the other hand, offering only a small fixed set of behaviors would seriously limit the usefulness of the robot. Clearly, the specification of safety and liveness properties plays a key role in making advances towards verifying adaptive systems.

3   System and Human-Robot Interface

The context of the project is a "drink serving" robot, that interacts with a human to pass over a drink. The user is offered two options, water or coffee, and, dependent on this choice, the robot will need to satisfy different safety constraints. Water is the less dangerous choice, while coffee is more dangerous, because it can stain or burn the user if spilled. The number of choices was limited to two for this initial explorative study.

Our implementation included various components. We used the humanoid head from [5] as shown in Figure 1. A hand/wrist device, monitored by a VICON motion capture suite, was used for tracking the position of the user's hand. A simple robot arm was used to move the cup from a set initial position to the serving position. Finally, a voice system allows for bidirectional vocalization.

For the basic scenario, the robot would use all these cues to identify whether the user is looking at the cup and holding it, so that it can be released in a safe manner. The robot can, however, never be entirely sure if the user is holding the cup or not just by using the position of the hand/wrist. Thus, we advanced the scenario by adding pressure sensors to the robot's hand. Although pressure sensors can introduce new problems themselves, depending on how each person applies pressure to the cup, the introduction of pressure sensors was expected to provide a safer more reliable solution. We compare the two scenarios in Section 7.

BERT2 [6] uses the VICON motion capture (MoCap) system to detect and localise interaction objects and the human's body parts in 3D space. Using a MoCap system as the main vision facility for a humanoid robot may seem far fetched (as it is an external system that cannot be fitted to the robot). However, this allows us to divert away from the machine vision challenges typically encountered when using robot mounted vision systems. Instead, we can directly focus on the subject of HRI. The VICON software stores information about body and object marker topology.

4   Software Architecture

An overview of the software architecture is presented in Figure 2. The following briefly describes each module and inter-module interactions.

The *Main Module* communicates the state of the system at essential points in time to other modules and gathers results about the individual components

Fig. 1. BERT2's expressive head. An embedded LCD screen generates gestures. Animated eyes communicate the direction of attention. The stereo vision system faceLAB from Seeing Machines^{TM} is mounted on the side of the head for head and gaze tracking. A webcam is mounted centrally in the forehead for a wide angle view of the scene.

at that moment. It triggers other modules based on its synchronization with the Voice System. For example, after the Voice System tells the user *"Please hold the cup."* the Main Module sends triggers to the Gaze Tracking and the Hand Wrist Modules to start checking whether or not the user is looking at and holding the cup. When the two modules have the results, they send them to the Main Module, which in turn controls the Voice System.

The *Voice System* is utilized to ask the user for the drink of their choice, and for the robot to utter a small set of instructive sentences. This spoken language interface is based on the Festival speech synthesis and Sphinx-II speech recognition systems. The *EgoSphere* acts as object position and orientation storage. Software modules in the architecture are interconnected using *YARP* [4], an open source library that provides an intercommunication layer to allow data exchange between processes running on different machines. Further information about the above four components can be found in [5] and [6].

The *Gaze Tracking Module* estimates the human's focus of attention using faceLAB^{TM}, an integrated head and gaze tracking system [5].

When triggered, the *Hand Wrist Module* starts to check, within a three second interval, the position of the user's hand. The positions of both the cup and

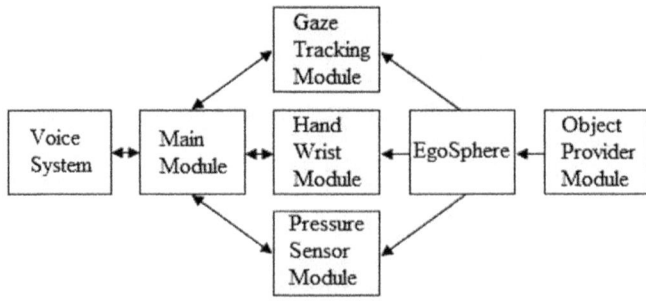

Fig. 2. Software Architecture. Arrows indicate information flow via YARP.

the hand/wrist are always up-to-date because the Object Provider module runs in the background and constantly updates the EgoSphere. The module works by verifying if the user's hand/wrist is within the region defined by the circle $C(O(a,b),r)$, where the centre O of the circle is the centre of the cup (a and b are the coordinates of this point) and the radius r is the distance between the centre of the cup and the hand/wrist. This radius obviously changes depending on the user's hand size and thus would require calibration to ensure that, if the radius is set too large, the cup will not be released when the user is not actually holding it. Alongside this circle region, the algorithm also checks whether the position of the user's hand is at the same height as the cup (within a threshold of measurement error). If, within the time interval, these constraints are satisfied relative to a determined threshold, the module will return a *true* value; otherwise it will return *false*.

When triggered, the *Pressure Sensor Module* first computes the value indicated by the sensor at that moment, averaging over 10 samples. At this point, the robot arm is holding the cup in the serving position and this is the ideal moment for the module to acquire the control value. If, over the next three seconds, the value acquired by the pressure sensor varies by at least 10%-20%, the module will return the *true* value; otherwise it will return *false*. The percentage by which the value should vary depends on how the user holds the cup and this is investigated in Section 7.

5   How the System Works

The core of the algorithm checks for the main cues (hand/wrist position, gaze and, optionally, applied pressure), regardless of what choice the user has made. It then checks, depending on the choice, certain pre- or post-conditions. The conditions are checked within a three seconds time interval, during which the user needs to satisfy the requirements, e.g. looking at the cup, holding it, or applying enough pressure. Figure 3 shows the state diagram of the entire system; it is further explained below.

The basic option is considered to be *Water* and, if the user makes this choice, the system will enter the *Condition* state after it exits the *Choice* state. If all the conditions are satisfied, i.e. the user is simultaneously looking at, holding and applying pressure to the cup, the system will enter the *Release Cup* state, then wait for a period of 25 seconds, after which it will enter the *Another Cup* state. The algorithm loops while the user answers *yes* or *no*. When the user's answer is *exit*, the algorithm will stop. If at least one of the conditions is not satisfied, the algorithm will loop back to the *Condition* state until all conditions are satisfied, looping a maximum number of three times. If the algorithm has looped three times with no satisfactory results, the system will not release the cup; instead it will retract and enter the *25 seconds wait* state.

The alternative option, *Coffee*, will cause the system to enter the *Coffee Precondition* state. Upon satisfaction of the pre-condition, i.e. the user is mostly looking at the cup in the required time interval, the system will enter the core

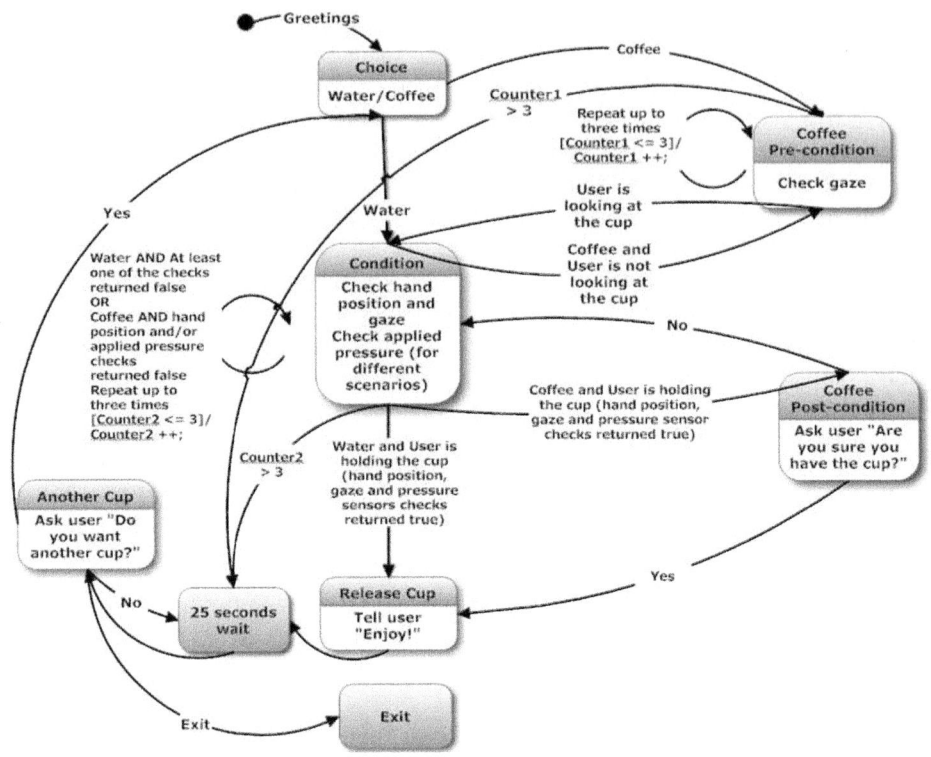

Fig. 3. State Diagram of the entire system

Condition state. The difference in behavior when choosing coffee is that the algorithm will loop back to the *Condition* state if the user is not holding the cup or not applying enough pressure, but continues to look at the cup (i.e. maintains the pre-condition). This will cause the robot to repeat to the user *"Please hold the cup."* If, however, the gaze check pre-condition is falsified in the *Condition* state, then the algorithm will loop back to the *Pre-condition* state and the robot will tell the user *"Please pay attention to the cup."*

This makes the robot pro-active in that if the pre-condition is subsequently falsified, meaning the user no longer looks at the cup, the robot explicitly asks the user to do so. The fact that the user then looks at the cup will have a stronger meaning than if the robot had not requested this.

When the pre-condition and the core conditions are met, the system will enter the *Coffee Post-condition* state. It asks the user: *"Are you sure you have the cup?"* This is a safety measure for the coffee option. It ensures the robot receives verbal confirmation that the human is ready to take the cup. If this post-condition is met, the system will behave in the same way as described above for water. For the *Coffee Pre-condition* state, the same maximum number of loops applies before the algorithm retracts and enters the *25 seconds wait* state without releasing the cup.

6   Trustworthiness

Humans need to trust a robot to be comfortable using it. Trustworthiness is delicate territory with fuzzy boundaries in practice. We analyze a robot's trustworthiness by analyzing its safety and liveness properties. These properties are key to specifying what constitutes safe and useful behaviour and should be maintained by any behavioural adaptations the robot might learn.

6.1   Safety Properties

(S1) A cup of water is never released unless the user is looking at the cup, the hand/wrist is within the circle area, and (when using hand pressure sensors) the user is applying enough pressure on the cup. All these conditions must be fulfilled simultaneously in order for the cup to be released.

(S2) A cup of coffee is never released unless the user is initially looking at the cup (pre-condition), after which the user is required to continue looking at the cup while his/her hand/wrist is within the circle area, and the user is applying enough pressure on the cup (core condition), after which the robot receives verbal confirmation that the user is holding the cup (post-condition). If any one of these conditions is not met, the system loops back to the state where it failed. If the maximum number of times the system can loop is reached, then the cup is not released and another drink is offered, i.e.:

(S2-a) When the choice is coffee, the robot never asks the user *"Please grab/hold the cup."* unless the user previously paid attention to the cup.

(S2-b) When the choice is coffee, the robot never asks the user *"Are you sure you have the cup?"* unless the user previously paid attention to the cup (S2-a), the hand/wrist was within the circle area, and the user applied enough pressure on the cup (implying first property is satisfied).

(S3) The system does not ask if it should offer another cup unless either all the conditions for releasing the current cup are satisfied and the cup is released or else, if the conditions are not satisfied, the cup is not released and the robot arm returns to the initial position.

6.2   Liveness Properties

(L1) Provided the user complies with the hand-over protocol (this ensures the safety aspect and is hence a safety constraint on the robot's liveness), then the drink will eventually be released. This ensures that the algorithm will, indeed, at some point release the cup.

(L2) When the pre-conditions, conditions, or post-conditions are not satisfied, the robot repeats those steps a maximum number of three times and eventually either releases the drink (more likely if water) or not (less likely if coffee) but, importantly for liveness, then asks the user: *"Do you want another cup?"*

The constraint of permitting three repeats reflects the fact that a user would, in normal circumstances, be able to perform the required steps in one or two tries. When considering external factors, such as the users' lack of attention or some external event happening that prevents the user from performing the steps, these would usually allow the user to finish the process within three tries. If something is happening that is not allowing the user to finish the process, then the user has the option of stopping and starting again. Situations in which the user is impaired, e.g. blind or deaf, can be accommodated but were not the focus of this project.

7 Experiments

This project was a short-term initial investigation into finding a working set of safety and liveness properties on which further research to determine the tradeoffs between adaptability and verifiability could be based. The experiments conducted are hence of a "pilot" nature.

Although the number of participants in the experiment was only five, each session was one hour long and covered different scenarios. This allowed us to test how familiarization with the system impacts on user performance and, most importantly, how this impacts on the safety and liveness properties of the system.

7.1 Experimental Setting

The users were seated in a chair in front of the robot and they were wearing the hand/wrist device that enables the VICON system to track the hand position. They were given an information sheet detailing the experiment, as well as a participation consent form. Before starting the experiment, the voice and gaze tracking systems were calibrated.

The users directly interacted with the robot via the Voice System. Details of an interaction depended on the user's choices. A dialogue could proceed according to this example:

- Robot: "Hello, I am BERT2. Do you feel like a drink?"
- *Human: "Yes."*
- Robot: "Water or coffee?"
- *Human: "Water."*
- Robot: "Preparing drink."
  (Pause while the robot is preparing the drink. When ready the robot hand moves upwards with the cup to the serving position.)
- Robot: "Please hold/grab the cup."
  (The human should now hold/grab the cup.)
- Robot: "Enjoy!"
  (The robot releases the cup and the user can have the drink.)

The above script was for the water option. For the coffee option, the robot will also say *"Please pay attention to the cup."* and *"Are you sure you have the cup?"* as these two lines represent the pre- and post-condition, respectively as detailed in Section 5.

7.2   Description of Experiments

Each experiment consisted of three validation scenarios. The first one was without pressure sensor. The second one used the pressure sensor with a low threshold that allowed the cup to be released when the user holds the cup without having to pull it (the robot states *"Please hold the cup."*). The third scenario used the pressure sensor with a slightly higher threshold requiring the user to apply some force for the cup to be released (the robot states *"Please grab the cup."*)

The motivation for having two types of scenarios, with and without the pressure sensor, was to improve upon the first scenario by making the system safer. We expected to observe more failures in experiments without the pressure sensor.

The two pressure sensor settings are used to investigate user reactions. If the person is asked to hold the cup, we expect that less force is applied and the user's hand will remain on the cup until release. If the person is asked to grab the cup, we expect that more force is applied and the user will actively try to pull the cup from the robot's hand. Thus, we intend to investigate whether the cup hand-over is successful more often when the user just holds the cup or when the user actively has to pull it from the robot's hand.

For each validation scenario, we assigned three types of tests:

- *Standard test:* The user is asked to act as if this were a real-world situation, when trying to buy a cup of water/coffee from a robot.
- *Interruption test:* As above but the researcher calls the participant's name at the point in the experiment when the robot says *"Please pay attention to the cup."* or *"Please hold/grab the cup."*
- *Conversation test:* The user is engaged in a conversation and asked to interact with the robot at the same time.

7.3   Expected Outcomes

For the first scenario, without the pressure sensor, the robot action was expected not to be safe, i.e. the cup would be released when the user was not actually holding it or the robot would fail to release it when it should (because of the varying distance between the cup and the hand/wrist).

The second and third scenarios include the pressure sensor which was expected to make the release of the cup safer. While this eliminated the need to calibrate the system with respect to each user's hand length, it required finding an appropriate threshold for the release pressure; we investigated two settings.

Each type of test was to be performed three times and the test type order within each scenario was random. This would enable users to perform better each time and the variation of the results between users for the last sets of tests was expected to be lower than at the start.

The methods of data collection included logging the threshold used for the distance from the centre of the cup to the hand/wrist, the thresholds for the pressure sensors[3] (the threshold for the low pressure scenario varied around 0.01 and the one for the higher pressure scenario varied around 0.2), the number of times the values were between the required thresholds and the number of system failures. The testing sessions were also video recorded.

8   Results

Experiments concentrated on evaluating our system with respect to the set of safety and liveness properties from Section 6. The first scenario focused on the distance from the centre of the cup to the hand/wrist and, thus, the threshold which needed to be determined in this situation. During testing and debug, this threshold was set to 3 cm which fitted the people testing the system. For the experiments the threshold was changed to 5 cm in order to start with a less limiting value with a view to further constrain (i.e. lower) it if the system failed because of it. This threshold actually proved to be quite robust, as it was loose enough to give the impression that the users' hand positions were within the intended circle region. Of course, this threshold would have needed to be modified in the case where the system released the cup when the user's hand was close to the cup, but not actually holding it. This was the very reason for designing the interruption type of test, which would check if the robot would release the cup in the case where the user's hand might, or might not, stay close to the cup when being called.

In preliminary testing, when the hand was intentionally held within the circle delimited by the threshold but the hand was not actually touching the cup, then the system would incorrectly release the cup because it classified this situation as the cup was being touched. This motivated the introduction of the interruption tests. When tested with subjects who were not previously exposed to the system, however, the tests revealed the following results: if the users had already moved their arms towards the cup for more than half of the distance, when called, they would touch the cup, but not look at it; if the users had not moved their arms that far towards the cup, they did not continue the movement. Thus, there was no moment, in general, when the hand was close to the cup, but without actually holding it. The few times when the Hand Wrist Module incorrectly detected the user as holding the cup, the system remained safe because the user was not looking at the cup.

Nonetheless, we cannot foresee all situations in which this can happen, e.g. the user might accidentally hit the cup, and so the hand/wrist would be in the circle area. This is why the pressure sensor scenarios proved to be safer. We used two thresholds one for each pressure sensor scenario. For the first one, the threshold was low, to permit the system to detect when the user was holding the cup without applying too much pressure. In this situation, the robot said

---

[3] These thresholds represent the percentages by which the pressure sensor values should minimally vary from the point when the control value is read (when the robot hand reaches the release position) until the three second time interval for checking whether or not the user is holding the cup ends.

"Please hold the cup." A few adjustments were made along the way to find the right threshold, but overall there were few failures. The difficulty was in tuning the system to release the cup at the right moment. Out of the 15 tests for this scenario during two tests the threshold was too low. The robot released the cup too soon which increased the likelihood of the cup being dropped (although that did not actually happen). During a further two tests the threshold was set too high, resulting in the cup being dropped because users took away their hand assuming the robot would not release the cup.

The higher threshold, used when the robot said *"Please grab the cup."* actually affected the liveness of the system. Some users stopped holding the cup because they thought that the robot would not release it. In addition, because each user grabbed the cup in a slightly different way, the threshold needed to constantly be adjusted. Although the users learned that they should apply more pressure in order for the robot to release the cup after 4-5 attempts (on average), the lower threshold seemed to be the right solution to reach a compromise between the safety and liveness properties of the system. For our "drink serving" robot, the tests revealed that the lower pressure sensor threshold keeps the system safe and live at the same time. However, if the robot were to handle some very dangerous chemicals, it would surely be more important for the system to be designed safer and maintaining liveness becomes a secondary concern.

It was concluded that the safety and liveness properties were appropriate. They provided a good balance between safety and usefulness for this setup and hence form a good basis for further research into methods to ensure safe HRI.

9   Conclusion and Follow-On Research Directions

Clearly, *trust* is a broad category and many elements should be taken into account when analysing the trustworthiness of a robot (e.g. aesthetics, ergonomy). This paper, nevertheless, is focusing on two important factors derived from the field of Design Verification, namely safety and liveness. The scenario discussed in this paper relates to a robot bartender and the trust elements which have been presented are specific to this context. The aim of this investigation was to develop an environment to study trust of HRI in terms of safety and liveness aspects. We designed a safe, yet useful system and presented a state diagram to better illustrate the core algorithm. We extracted the safety and liveness properties of the system and experimentally evaluated these. We concluded that this set of properties provides a good balance between safety and usefulness of our "drinks serving" robot and a good basis for further research.

Our next steps include the development of a formal description of the system including the associated safety and liveness properties with the aim to formally verify that indeed these properties hold. In our view, formal verification complements experimental validation in real-world scenarios. It helps developers to obtain greater confidence in the system's functional correctness.

Another direction for further research is to enable the robot to self-adjust its thresholds based on evaluating failed and successful user interactions. We are currently conducting two follow-on pilot projects that introduce machine learning into the robot bartender setup. The properties established in this paper serve as specification for both these projects. One project is focused on extending reinforcement learning techniques to work within the constraints determined by the safety and liveness properties, thus ensuring that any adaptations maintain the specified properties and hence stay within safe bounds. The other project investigates the use of a requirements-based testing methodology similar to what is used for the certification of avionics software [10] to qualify any proposed adaptations produced by the learning system. By deriving tests from the requirements specification we expect to obtain a set of tests that are "immune" to valid adaptations, i.e. adaptations that do not violate the specification. Adaptations that violate the existing specification may be used to prompt designers/users to investigate whether or not the specification should be modified to permit such adaptations in the future.

In the context of the above two projects we are also investigating the added value gained from introducing *soft* and *hard* constraints. For example, a soft constraint might be that the usual serving distance is about an arm's reach from the human's body, while a hard constraint might be that the robot must not touch or push the human. Hence, for someone with an injured shoulder bone, the serving position would be much closer than usual which would require the soft constraint to be overwritten by an adaptation while the hard constraint must not be violated.

Ultimately, we are working towards finding a compromise between adaptability and verifiability. Our longer term research objectives are to develop design and verification methods that ensure humans can trust adaptive robots by demonstrating that all adaptations remain useful and within safe bounds. This is a significant challenge because the powerful and well-developed methods of Design Verification used in application domains such as the aerospace industries require complete a priori knowledge of all circumstances and states that a system could enter during operation. This is hard enough for, say, a jet-engine control unit. The internal and environmental complexity of a robot operating in unstructured human-inhabited settings, however, renders these approaches completely useless in their current forms and requires significant innovation that can only be achieved through cross-disciplinary collaboration.

Acknowledgements. This work was supported in part by the European Union under the project "Cooperative Human Robot Interaction Systems (CHRIS)"; Grant Agreement No: 215805. The authors acknowledge the help received from Minshuai Zhao, an MSc student collaborating on this project. We also thank Armando Tacchella for constructive feedback.

References

1. Basu, A., Gallien, M., Lesire, C., Nguyen, T.H., Bensalem, S., Ingrand, F., Sifakis, J.: Incremental component-based construction and verification of a robotic system. In: European Conference on Artificial Intelligence, pp. 631–635. IOS Press, Amsterdam (2008)
2. Bensalem, S., Gallien, M., Ingrand, F., Kahloul, I., Nguyen, T.H.: Toward a More Dependable Software Architecture for Autonomous Robots. IEEE Robotics and Automation Magazine 16(1), 67–77 (2009)
3. Clarke, E.M., Grumberg, O., Peled, D.: Model checking. MIT Press, Cambridge (1999)
4. Fitzpatrick, P., Metta, G., Natale, L.: Towards long-lived robot genes. Robotics and Autonomous Systems 56(1), 29–45 (2008)
5. Lallée, S., Lemaignan, S., Lenz, A., Melhuish, C., Natale, L., Zant, T.V.D., Warneken, F., Dominey, P.F.: Towards a Platform-Independent Cooperative Human-Robot Interaction System: I. Perception. In: International Conference on Intelligent Robots and Systems (IROS 2010), pp. 4444–4451. IEEE, Los Alamitos (2010)
6. Lenz, A., Skacheck, S., Hamann, K., Steinwender, J., Pipe, A., Melhuish, C.: The BERT2 infrastructure: An integrated system for the study of human-robot interaction. In: International Conference on Humanoid Robots. IEEE, Los Alamitos (2010)
7. Magee, J., Kramer, J.: Concurrency: State Models & Java Programs. Wiley, Chichester (2006)
8. Metta, G., Natale, L., Pathak, S., Pulina, L., Tacchella, A.: Safe and Effective Learning: A Case Study. In: ICRA 2010, pp. 4809–4814 (2010)
9. Pappas, G., Kress-Gazit, H. (eds.): ICRA Workshop on Formal Methods in Robotics and Automation (2009)
10. RTCA: DO178B Software Considerations in Airborne Systems and Equipment Certification (1992)

Towards Temporal Verification of Emergent Behaviours in Swarm Robotic Systems

Clare Dixon[1], Alan Winfield[2], and Michael Fisher[1]

[1] Department of Computer Science, University of Liverpool, Liverpool, UK
{cldixon,mfisher}@liverpool.ac.uk
[2] Bristol Robotics Laboratory, University of the West of England, Bristol, UK
Alan.Winfield@uwe.ac.uk

Abstract. A robot swarm is a collection of simple robots designed to work together to carry out some task. Such swarms rely on: the simplicity of the individual robots; the fault tolerance inherent in having a large population of often identical robots; and the self-organised behaviour of the swarm as a whole. Although robot swarms are being deployed in increasingly sophisticated areas, designing individual control algorithms that can *guarantee* the required global behaviour is difficult. In this paper we apply and assess the use of *formal verification* techniques, in particular that of model checking, for analysing the emergent behaviours of robotic swarms. These techniques, based on the automated analysis of systems using *temporal logics*, allow us to analyse all possible behaviours and so identify potential problems with the robot swarm conforming to some required global behaviour. To show this approach we target a particular swarm control algorithm, and show how automated temporal analysis can help to refine and analyse such an algorithm.

1 Introduction

The use of autonomous robots has become increasing appealing in areas which are hostile to humans such as underwater environments, contaminated areas, or space [24,2,21]. Rather than deploying one or two, often large and expensive, robots a significant focus is now on the development of *swarms* of robots.

A robot swarm is a collection of simple (and usually identical) robots working together to carry out some task [3,18]. Each robot has a relatively small set of behaviours and is typically able to interact with other (nearby) robots and with its environment. Robot swarms are particularly appealing in comparison to one or two more complex robots, in that it may be possible to design a swarm so that the failure of some of the robots will not jeopardize the overall mission, i.e. the swarm is *fault tolerant*. Such swarms are also advantageous from a financial point of view since each robot is very simple and mass production can significantly reduce the fabrication costs.

Despite the advantages of deploying swarms in practice, it is non-trivial for designers to formulate individual robot behaviours so that the emergent behaviour of the swarm as a whole is guaranteed to achieve the global task of the swarm,

R. Groß et al. (Eds.): TAROS 2011, LNAI 6856, pp. 336–347, 2011.

and that the swarm does not exhibit any other, undesirable, behaviours [19]. Specifically, it is often difficult to predict the overall behaviour of the swarm just given the local robot control algorithms. This is, of course, essential if swarm designers are to be able to effectively and confidently develop reliable swarms. So, we require some mechanism for analysing what the swarm can do, given the behaviour of individual robots and a description of their possible interactions both with each other and with the environment. Using this mechanism, the designer can assess to what extent the swarm behaves as required and, where necessary, redesign to avoid unwanted outcomes.

Currently, the analysis of swarm behaviour is typically carried out by experimenting with real robot swarms or by simulating robot swarms and testing various scenarios (eg see [20,17]). In both these cases any errors found will only be relevant to the particular scenarios constructed; neither provides a comprehensive analysis of the swarm behaviour in a wide range of possible circumstances. Specifically, neither approach can detect a problem where undesirable behaviour occurs in some untested situation. We note that approaches to modelling robot swarms have been proposed in, for example, [15,16,10].

A well-known alternative to simulation and testing is to use *formal verification*, and particularly the technique called *model-checking* [6]. Here a mathematical model of *all* the possible behaviours of the system is constructed and then all possible executions through this model are assessed against a required logical formula representing a desired property of the system. In the case of systems, such as robot swarms, the mathematical model usually represents an abstraction of the real control system, while the logical formula assessed is usually a *temporal* formula representing the presence of a desirable property or the absence of an undesirable property on all paths; see, for example [9].

In this paper we will explore the use of temporal verification for robot swarms in an effort to formally verify whether such swarms do indeed exhibit the required global behaviour. The structure of this paper is as follows. In Section 2 we give details of the temporal logic in which logical properties we wish to show are given and provide an overview to model checking. In Section 3 we describe our use of formal verification to assess swarm algorithms and introduce one existing algorithm, namely Nembrini's *alpha algorithm* [17], that we will analyse. In Section 4 we describe the abstractions we use in more detail. In Section 5 we give verification results from using the temporal model checker and the impact of these results both on the original algorithm and the abstractions. In Section 6 we discuss the results and we provide concluding remarks in Section 7.

2   Temporal Logic and Model Checking

The logic we consider is propositional linear-time temporal logic, called PTL where the underlying model of time is isomorphic to the Natural Numbers, $\mathbb{N}$. A model for PTL formulae can be characterised as a sequence of *states* of the form: $\sigma = s_0, s_1, s_2, s_3, \ldots$ where each state, s_i, is a set of proposition symbols, representing those propositions which are satisfied in the ith moment in time.

In this paper we will only make use two of temporal operators '$\Diamond$' (*sometime in the future*), '$\Box$' (*always in the future*) in our temporal formulae. The notation $(\sigma, i) \models A$ denotes the truth of formula A in the model σ at state index $i \in \mathbb{N}$ defined as follows where PROP is a set of propositional symbols.

$$(\sigma, i) \models p \quad \text{iff } p \in s_i \text{ where } p \in \text{PROP}$$
$$(\sigma, i) \models \Diamond A \text{ iff } \exists k \in \mathbb{N}. \ (k \geqslant i) \text{ and } (\sigma, k) \models A$$
$$(\sigma, i) \models \Box A \text{ iff } \forall k \in \mathbb{N}. \ (k \geqslant i) \text{ and } (\sigma, k) \models A$$

Model checking [6] is a popular technique for verifying the temporal properties of systems. Input to the model checker is a model of the paths through a system (a finite-state transition system) and a formula to be checked on that model. The language of the formula to be checked is usually some form of temporal logic for example the linear-time temporal logic, PTL, or the branching-time temporal logic, CTL. A number of model checkers have been developed but we will use NuSMV [5] which allows properties expressed in both CTL and PTL.

Essentially, we construct a set of finite-state transition systems, corresponding to each of the robots in the swarm, and then model-check a PTL formula against the concurrent composition of these transition systems. The key element of model-checkers is that, if there are execution paths of the system that *do not* satisfy the required temporal formula, then at least one such "failing" path will be returned as a counter-example. If no such counter-examples are produced then all paths through the system indeed satisfy the prescribed temporal formula.

3   Analysing Swarm Algorithms

Our long term aim is to develop, deploy and extend formal verification techniques for use in swarm robotics and so show that formal verification is viable in this context. Yet, even within swarm robotics, the application of formal verification is difficult. The continuous aspects of both the robotic control system and the robots' movements and environment do not sit well with the discrete and finite nature of model-checking. Fortunately, in developing simple robotic control algorithms, engineers typically use finite-state machines as part of their behavioural design. Thus, we can base our verification on such finite-state machines provided by roboticists. There remains the problem of the use of continuous functions/variables in such state machines. As in the verification of *hybrid systems*, we must provide abstractions to simplify such continuous values so that model-checking can be carried out [11]. Finally we note that even with such abstractions, representing the location and movement for a number of robots will generate a huge state space so initially we must focus on small grid sizes and numbers or robots. Thus, in summary, our approach is as follows.

1. Take the design of a swarm control algorithm for an individual robot, as represented in the form of a finite-state machine.
2. Describe an abstraction that tackles the continuous nature of the domain, the potentially large number of robots, and the nature of concurrent activity and communication within the swarm.

3. Carry out (automatic) model checking to assess the temporal behaviour of the model from (2). If model-checking succeeds, then return to (2) refining the abstraction to make it increasingly realistic. If model-checking fails, returning a scenario in which the temporal requirement is not achieved, then analyse (by hand) how the algorithm in (1) *should* cope with this scenario. Either there is a problem with the original algorithm, so this must be revised, or the algorithm is correct for this scenario and so the abstraction in (2) must be revisited and expanded to capture this behaviour.

This process is continued until no errors are found in (3) and the abstraction in (2) is sufficiently close to the physical scenario to be convincing. While this cycle is clearly not (and cannot be) fully automatic, the results from model-checking help direct us in refining the algorithm and/or abstractions used in the design. This approach follows the spirit of the "counter-example guided abstraction refinement" method initiated in [8] and applied to hybrid systems in [1].

3.1   The 'Alpha' Algorithm

As a case study we consider algorithms for robot swarms which make use of local wireless connectivity information alone to achieve swarm aggregation. Specifically, we examine the simplest (alpha) algorithm described in [17,22]. Here, each robot has range-limited wireless communication which, for simplicity, we model as covering a finite distance in all directions from the robot's location. Beyond this boundary, robots are out of detection range. The basic alpha algorithm is very simple:

- The default behaviour of a robot is forward motion.
- While moving each robot periodically sends an "Are you there?" message. It will receive "Yes, I am here" messages only from those robots that are in range, namely its neighbours.
- If the number of a robot's neighbours should fall below the threshold α then it assumes it is moving *out* of the swarm and will execute a 180° turn.
- When the number of neighbours rises above α (when the swarm is regained) the robot then executes a random turn. This is to avoid the swarm simply collapsing in on itself.

Thus, we assume that each robot has three basic behaviours: move forward (default); avoidance (triggered by the collision sensor); and coherence (triggered by the number of neighbours falling below α).

4   Abstraction

In the following we explain, in more detail, the abstractions that are used to build an appropriate model to be checked.

Spatial Aspects. We consider a number of identical robots moving about a square grid and assume that the grid is divided into squares with at most one robot in each square. We assume a step size of one grid square and that a robot can detect other robots for purposes of avoidance in the adjacent squares. The robot has a direction it is moving in. Rather than taking a bearing, we describe this as one of *North, South, East* or *West*. To make the problem finite we limit the grid size to be an $n \times n$ square which wraps round, i.e. a movement North from the upper edge of the grid will result is a move to the lower edge of the grid. Initially the robots may have any direction but are placed on the grid where they are connected but in different grid squares. Initially any robot may move first.

Connectivity. Regarding connectivity, this is calculated from the robots' relative positions. Initially we assume that each robot can detect other robots in the eight squares surrounding it. Hence if a robot is in square $(1,1)$ it can detect robots in squares $(0,0)$, $(0,1)$, $(0,2)$, $(1,0)$, $(1,2)$, $(2,0)$, $(2,1)$, and $(2,2)$. Thus it has a wireless range of one square in all directions. Initially we set the value of $\alpha = 1$ i.e. a robot is connected if it can detect at least one other robot. In our verification, we aim to show that for all i, $\Box \Diamond con_i$ follows from our specification, i.e. each robot stays connected infinitely often. Thus, the swarm need not *always* be connected but, if it becomes disconnected, should eventually reconnect. In particular, no specific robot will remain disconnected forever.

Motion Modes. Avoidance is dealt with as follows. If a robot is moving in some direction and the square ahead is occupied: move to the right or left; if these are both occupied move backwards; else stay in the current position. The original direction the robot was moving in is maintained. Each robot can be in one of two motion modes: *forward* or *coherence*. The connectivity of each robot can also be in one of two modes: *connected* or *not connected*. The combination of motion and connectivity give us four possible alternatives.

- In the forward mode, when connected, move forward and the motion mode remains 'forward'.
- In the forward mode, but not connected, turn $180°$ and change the motion mode to 'coherent'.
- In the coherent mode, but not connected, move forward and the motion mode remains as 'coherent'.
- In the coherent mode, when connected, perform a $90°$ turn (i.e. either $90°$ left or $90°$ right) and change the motion mode to 'forward'.

Concurrency. An important variety of abstraction concerns the representation of concurrent activity within the swarm of robots. Thus, in modelling this aspect we must ask questions about whether all robots run simultaneously, whether some robots run faster than others, etc. Thus, there is a wide variety of different abstractions we might use, which in turn correspond to different mechanisms for concurrently composing the robot transition systems/models.

- *synchrony* — all robots execute at the same time and with the same clock.
- *(strict) turn taking* — execution of the robots is essentially interleaved, but the robots must execute in a certain order, e.g. $r_1, r_2, r_3, r_1, r_2, r_3$, etc.
- *(non-strict) turn taking* — execution of the robots is again interleaved but for m robots in every cycle of m steps each robot moves once, so we can now have a situation where a robot executes two steps consecutively, e.g. $r_1, r_2, r_3, r_3, r_2, r_1$, etc.
- *(fair) asynchrony* — robots execute at the same time, yet some robots are faster than others. However the *fair* aspect ensures that a robot can only take a finite number of steps before all other robots have finished their step.

It is important to note that the particular view of concurrency taken can significantly affect the results.

5 Verification Using Model Checking

We model Nembrini's alpha algorithm and aim to verify $\Box\Diamond con_i$ for each of the i robots. The model is a finite-state transition system written in NuSMV's input language representing the algorithm in Section 4. Due to the large state spaces involved we assume initially that the grid is 5×5 and there are two robots and increase these values. We appreciate both the grid size and number of robots is small but this will increase while allowing the grid to wrap round means that the robots can, for example, move North forever. As well as considering different number of robots and grid sizes we will change the abstraction we use relating to concurrency, α parameter and wireless range.

5.1 Results: Concurrency

First we consider the different concurrency options: fair asynchrony, non-strict turn taking, strict turn taking and then synchrony. In the table below we show some results of running NuSMV on the input files for a number of different sized input grids with two or three robots. Connectedness is a global property calculated from the robots' positions. Note that a response of '*true*' from the model checker means that every path from every initial state of the model satisfies this property. A response of '*false*' means not all paths satisfy the property and NuSMV outputs a failing trace.

Problem	NuSMV Output			
	Fair Async.	Non-Strict	Strict	Sync.
5×5 grid, 2 robots	false	false	false	true
6×6 grid, 2 robots	false	false	false	true
7×7 grid, 2 robots	false	false	false	true
8×8 grid, 2 robots	false	false	false	true
5×5 grid, 3 robots	false	false	false	false
6×6 grid, 3 robots	false	false	false	false

First considering fair asynchrony we can obtain failing traces for all grid sizes and number of robots we tried. Considering the failing trace for the grids just with two robots, after some initial moves the two robots both get in the coherent non-connected mode, moving at right angles to each other and remain disconnected for ever.

Similarly for strict and non-strict turn taking the table shows that NuSMV can show that the property does not occur on all paths from every initial state, size of grid and number of robots we have experimented with. Looking at the failing traces for non-strict turn taking, robot one makes a move resulting in the robots losing connectedness which they never regain. This may be due to a number of reasons. Firstly robot one has to wait two steps before it can take any action relating to this loss of connectedness (recall that the other robot can move twice). This might be avoided by using a strict turn taking abstraction. However, as can be seen in the table, experiments with strict turn taking also give a result of *false* in the same cases as previously. Similar to the above, the failing traces show a loss of connection that is never regained. These results show that changing the computational abstraction (from non-strict to strict) does not, in itself, solve our failure.

However changing the concurrency mode to synchronous we now obtain '*true*' results for *all* the two robot cases but again for all the three robot cases tried we obtain '*false*'. Observing the successful traces of the two robot cases, if the robots are both moving in the same direction originally they continue in this same direction in the forward connected mode forever. If they are in different directions they move away from each other, correct this in the coherent mode by moving back together become re-connected and again repeat one of the above two patterns again. This leads us to believe that synchrony is probably more appropriate for modelling the alpha algorithm.

State	1.1	1.2	1.3	1.4	1.5	1.6	1.7	1.8	1.9	1.10	1.11
Loc r_1	(0,0)	(4,0)	(3,0)	(2,0)	(1,0)	(0,0)	(4,0)	(3,0)	(2,0)	(1,0)	(0,0)
Dir r_1	W	W	W	W	W	W	W	W	W	W	W
Mode r_1	FC	FC	FC	FC	FC	FC	FC	FC	FC	FC	FC
Loc r_2	(1,1)	(0,1)	(4,1)	(3,1)	(2,1)	(1,1)	(0,1)	(4,1)	(3,1)	(2,1)	(1,1)
Dir r_2	W	W	W	W	W	W	W	W	W	W	W
Mode r_2	FC	FC	FC	FC	FC	FC	FC	FC	FC	FC	FC
Loc r_3	(4,1)	(4,2)	(4,3)	(4,2)	(0,2)	(4,2)	(3,2)	(2,2)	(1,2)	(0,2)	(4,2)
Dir r_3	N	N	N	S	E	W	W	W	W	W	W
Mode r_3	FC	FC	FNC	CC	FNC	CNC	CNC	CNC	CNC	CNC	CNC

Above we provide a failing trace for the synchronous 5×5 grid output for three robots produced by NuSMV where r_i denotes robot i. In that trace we use the following abbreviations FC-forward connected; FNC-forward and not connected; CC-coherent connected; CNC-coherent not connected; Loc-location and Dir-Direction. From state 1.6 onwards robots one and two remain in the forward connected mode travelling West whereas robot three is in the coherent non-connected mode also travelling West. State 1.11 is the same as 1.6 showing that the pair of robots can loop round the states 1.6 to 1.11 forever with robot three remaining disconnected.

It could be argued that the reasons for this are that in the three robot case we need a bigger α parameter, or that the wireless range is too small potentially resulting in frequent loss of connection (or both). Hence we next attempt to consider these cases specifically by increasing the α parameter, i.e. in the three robot case each robot needs to remain connected to two others and then by requiring the wireless range to be larger than the step size.

5.2   Results: Changing the Alpha Parameter and Wireless Ranges

We now increase the α parameter from one to two, meaning that each robot must be able to detect two others, and also consider increasing the wireless range. In the following we assume synchronous concurrency. The results are presented in the table below. A dash in the table means that this experiment makes no sense for either an increase in the α parameter or wireless range.

Problem	NuSMV Output	
	$\alpha = 2$	Range $= 2$
6×6 grid, 2 robots	–	true
7×7 grid, 2 robots	–	true
8×8 grid, 2 robots	–	true
9×9 grid, 2 robots	–	true
10×10 grid, 2 robots	–	true
5×5 grid, 3 robots	false	–
6×6 grid, 3 robots	false	false
7×7 grid, 3 robots	false	false
8×8 grid, 3 robots	false	false

Regarding the increase in the α parameter we need to have at least three robots. Hence we only consider three robot cases. Regarding wireless range this was previously set at one square from a robot. We now increase this to be two squares from each robot, i.e. a robot can detect others in the 5×5 squares centred on the robot. Obviously as we have increased the detection area, in grid sizes of 5×5 and smaller all robots will always be connected at all moments. Hence we only consider grid sizes of 6×6 and above.

We note that now, with this modified alpha algorithm (i.e. with $\alpha = 2$) we still obtain results of '*false*' for all the three robot cases tried. Considering the failing traces for three robots we reach a cycle where all the robots are moving in the same direction in the coherent non-connected mode, i.e. there aren't two other robots in range. Something similar happens with the increased wireless range, i.e. we reach a cycle where all the robots move in the same direction, two within wireless range in the forward connected mode and one not within wireless range in the coherent non-connected mode. We note that we can also find a failing trace for both $\alpha = 2$ and wireless range $= 2$ for an 8×8 grid.

We can continue this until either the abstractions are realistic enough, or until our verification attempts take too much time/space. We will discuss our results in the next section.

6   Discussion and Related Work

The alpha algorithm has been well studied and, to date, validated with simulated and real robots [17], and using a probabilistic mathematical modelling
approach [22]. However, neither simulation, real robot experiments, nor mathematical modelling provide formal proof of the algorithm's correctness. Consider
simulation (or real robot experiments, which may be regarded as 'embodied'
simulations). Given that simulated (or real) robots move in a real-valued, not
grid-, world with typical swarm sizes of 40 robots and α values of 5, 10 or 15, for
10,000 seconds [22], we have a practically infinite state-space and each simulation
run tests only a tiny number of paths through that state-space.

As outlined in Section 4, the model checking approach within this paper
attempts to formally establish correctness by reducing the state-space to a
tractable size so that *every* path through that state-space can be tested. That
reduction is achieved by means of introducing, firstly, a number of simplifying
assumptions and, secondly, by running very small swarm sizes (2 or 3 robots)
and very small grid-worlds (5×5 up to 8×8). We thus need to ask ourselves
whether the failure to verify the correctness of the algorithm in some cases of
Section 5 is because (a) the algorithm is flawed or (b) the simplifying assumptions (necessary for tractability) are so severe that they go beyond the bounds
within which the algorithm should be expected to work.

First we consider the method of concurrency used to model the problem. We
believe that fair asynchrony, non-strict and strict turn taking are unsuitable for
modelling this problem because in the first two cases one robot may be able to
make two moves before one of the others makes a move and in the latter case a
robot has to wait for the others to make moves before it can react to a loss in
connectedness. This is confirmed by the *'false'* results obtained in Section 5.1, i.e.
such failing traces would not be represented by the actual runs of the algorithm.
Hence we believe that synchrony is the only reasonable abstraction because
for the timing of the real robots we assume that any two robots that become
disconnected each notice the disconnection at approximately the same time.

However, we note that with the three robot synchronous cases we can still
obtain failing traces. Some of these are of the form presented below. This illustrates a possible scenario in which robot A has become disconnected from the
swarm and turns round but never catches up with B and C. Because robots B
and C remain connected to each other, with $\alpha = 1$, B and C do not react to the
loss of robot A and A remains disconnected.

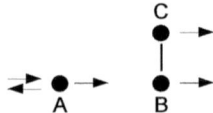

We consider both increasing the α parameter and wireless range but still
obtain failing traces similar to the above form. These results appear to confirm
a known problem with the alpha algorithm, when a robot or group of robots is

linked to the rest of the swarm by a single link (known as a bridge or cutvertex). This problem was discussed by Nembrini [17] and overcome by the improved beta algorithm [17].

Regarding the *wrap round* grid abstraction we note that potentially some failing traces could not appear in actual runs of the algorithm. Take a small grid size, for example 4×4, and a disconnected robot travelling in some direction, for example West. The wrap round nature of the grid may mean the robot becomes re-connected to robots that are some distance East of the robot. For this reason we have excluded any results from 4×4 grids and have checked that the failing traces listed above do not suffer from this problem. Let the term *grid independent* mean that the robot movement in a failing trace for a wrap round grid can be translated into an infinite grid obtaining the same connectedness values for each robot. We conjecture that, in the case of synchrony, once we have found a *grid independent trace* for an $n \times n$ grid be can extend this to a *grid independent trace* for an $m \times m$ grid where $m > n$. Hence when we find one such trace for a number of synchronous robots and some grid size we do not have to try any larger sized grids.

Obviously we would like to consider larger number of robots and, even with the above observation, may need to consider larger grid sizes but we are faced by the well known state explosion problem [7]. Even with the simplifications we use here, the state space explored is huge. Modelling a robot's position on an $n \times n$ grid, with 4 directions, and two motion modes for r robots requires of the order of $(n \times n \times 4 \times 2)^r$ states to be explored. The state space increases further with variables to deal with turn taking etc. For example, in the case of three robots, using non-strict turn taking within a grid of 7×7, we have more than a thousand million states. Because the underlying model used by the model checker is the product of each robot transition system we will not be able to scale this up to larger number of robots and grid sizes. To combat this we will have to apply and develop some of the work that has been carried out in the model checking field using more sophisticated abstractions, clever representations, and reduction techniques such as slicing or symmetry, in order to limit the state space.

The model checking approach we use here provides the swarm algorithm designer with a systematic way of analysing robot swarms. Whilst we are limited by the state explosion problem, the failing traces obtained in this analysis provide an invaluable starting point for further investigation either to reject as not possible due to the abstraction used or for further consideration by swarm designers to guide simulation or experimentation. Formal verification is a useful tool for deeper analysis of swarm algorithms, and finds more potential faults than simulation or real-robot tests.

Related Work. The application of temporal logics to robots swarms has not been widely used. In [23] temporal logic is used to provide a high level specification of robot swarms however there the focus is on specification rather than verification. We have applied probabilistic model checking to foraging robots, using a counting abstraction, in [14]. In that paper we focus on the mode the robot is in without dealing with either their location or connectedness. In [13] a model checking approach is adopted to considering the motion of robot swarms. A hierarchical framework is suggested to abstract away from the many details of the problem including the location of the individual robots. Further, the paper assumes a centralised communication architecture which is not assumed here. A related paper is [9] which again considers model checking robot motion but doesn't discuss robot swarms.

7   Conclusions and Future Work

In this paper we have shown how formal verification can be used as part of the development of reliable robot swarm algorithms. Although our examples involving two or three robots might be considered too small to constitute a swarm, they do represent a valid test-case for self-organised flocking or aggregation algorithms, such as the case study of this paper. Indeed swarms of two or three robots are useful in that they test the lower limit bounds of swarm size and are therefore more demanding of the algorithm than larger swarm sizes with higher values of α, as tested in simulation studies. Further the production of failing traces provides a focus for further consideration by the swarm designer.

We clearly need further work to tackle the state explosion problem. This will involve development of abstractions and reduction techniques relevant to swarm algorithms as well as the application of techniques such as symmetry detection to allow the analysis of larger grids and swarms sizes. Additionally we could use probabilities to represent uncertainty such as the unreliability of the robot sensor near to the maximum range leading to the use of probabilistic model checkers such as PRISM [12]. Future work involves not only further analysis of the alpha algorithm but also studying more complex swarm algorithms such as Nembrini's Beta algorithm [17] and the Omega-algorithm developed by Bjerknes [4]. Other properties apart from connectedness such as emergent swarm taxis toward a beacon (i.e. an Infra-Red light source) can also be considered.

References

1. Alur, R., Dang, T., Ivančić, F.: Counterexample-Guided Predicate Abstraction of Hybrid Systems. Theoretical Computer Science 354(2), 250–271 (2006)
2. Arai, T., Pagello, E., Parker, L.: Editorial: Advances in Multi-Robot Systems. IEEE Trans. Robotics and Automation 18(5), 655–661 (2002)
3. Beni, G.: From Swarm Intelligence to Swarm Robotics. In: Şahin, E., Spears, W.M. (eds.) Swarm Robotics 2004. LNCS, vol. 3342, pp. 1–9. Springer, Heidelberg (2005)
4. Bjerknes, J.D.: Scaling and Fault Tolerance in Self-organised Swarms of Mobile Robots. Ph.D. thesis, University of the West of England (2010)
5. Cimatti, A., Clarke, E.M., Giunchiglia, E., Giunchiglia, F., Pistore, M., Roveri, M., Sebastiani, R., Tacchella, A.: NuSMV 2: An OpenSource Tool for Symbolic Model Checking. In: Brinksma, E., Larsen, K.G. (eds.) CAV 2002. LNCS, vol. 2404, p. 359. Springer, Heidelberg (2002)
6. Clarke, E., Grumberg, O., Peled, D.A.: Model Checking. MIT Press, Cambridge (2000)

7. Clarke, E.M., Grumberg, O.: Avoiding The State Explosion Problem in Temporal Logic Model Checking. In: ACM Symposium on Principles of Distributed Computing (PODC), pp. 294–303 (1987)
8. Clarke, E.M., Grumberg, O., Jha, S., Lu, Y., Veith, H.: Counterexample-Guided Abstraction Refinement for Symbolic Model Checking. J. ACM 50(5), 752–794 (2003)
9. Fainekos, G., Kress-Gazit, H., Pappas, G.: Temporal Logic Motion Planning for Mobile Robots. In: Proceedings of the IEEE International Conference on Robotics and Automation (ICRA), pp. 2020–2025 (2005)
10. Hamann, H.: Space-Time Continuous Models of Swarm Robotic Systems: Supporting Global-to-Local Programming. Cognitive Systems Monographs. Springer, Heidelberg (2010)
11. Henzinger, T., Ho, P.H., Wong-Toi, H.: HYTECH: A Model Checker for Hybrid Systems. International Journal on Software Tools for Technology Transfer 1(1-2), 110–122 (1997)
12. Hinton, A., Kwiatkowska, M., Norman, G., Parker, D.: PRISM: A Tool for Automatic Verification of Probabilistic Systems. In: Hermanns, H. (ed.) TACAS 2006. LNCS, vol. 3920, pp. 441–444. Springer, Heidelberg (2006)
13. Kloetzer, M., Belta, C.: Temporal Logic Planning and Control of Robotic Swarms by Hierarchical Abstractions. IEEE Transactions on Robotics 23, 320–330 (2007)
14. Konur, S., Dixon, C., Fisher, M.: Formal Verification of Probabilistic Swarm Behaviours. In: Dorigo, M., Birattari, M., Di Caro, G.A., Doursat, R., Engelbrecht, A.P., Floreano, D., Gambardella, L.M., Groß, R., Şahin, E., Sayama, H., Stützle, T. (eds.) ANTS 2010. LNCS, vol. 6234, pp. 440–447. Springer, Heidelberg (2010)
15. Martinoli, A., Easton, K., Agassounon, W.: Modeling Swarm Robotic Systems: A Case Study in Collaborative Distributed Manipulation. International Journal of Robotics Research 23(4), 415–436 (2004)
16. Milutinovic, D., Lima, P.: Cells and Robots: Modeling and Control of Large-Size Agent Populations. Tracts in Advanced Robotics, vol. 32. Springer, Heidelberg (2007)
17. Nembrini, J.: Minimalist Coherent Swarming of Wireless Networked Autonomous Mobile Robots. Ph.D. thesis, University of the West of England (2005)
18. Sahin, E., Winfield, A.F.T.: Special issue on Swarm Robotics. Swarm Intelligence 2(2-4), 69–72 (2008)
19. Spears, W.M., Spears, D.F., Hamann, J.C., Heil, R.: Distributed, Physics-Based Control of Swarms of Vehicles. Autonomous Robots 17(2-3), 137–162 (2004)
20. Støy, K.: Using situated communication in distributed autonomous mobile robotics. In: SCAI 2001: Proceedings of the Seventh Scandinavian Conference on Artificial Intelligence, pp. 44–52. IOS Press, Amsterdam (2001)
21. Truszkowski, W., Hallock, H., Rouff, C., Karlin, J., Rash, J., Hinchey, M., Sterritt, R.: Swarms in Space Missions. In: Autonomous and Autonomic Systems: With Applications to NASA Intelligent Spacecraft Operations and Exploration Systems. NASA Monographs in Systems and Software Eng., pp. 207–221. Springer, Heidelberg (2009)
22. Winfield, A., Liu, W., Nembrini, J., Martinoli, A.: Modelling a wireless connected swarm of mobile robots. Swarm Intelligence 2(2-4), 241–266 (2008)
23. Winfield, A., Sa, J., Fernández-Gago, M.C., Dixon, C., Fisher, M.: On Formal Specification of Emergent Behaviours in Swarm Robotic Systems. International Journal of Advanced Robotic Systems 2(4), 363–370 (2005)
24. Yuh, J.: Design and Control of Autonomous Underwater Robots: A Survey. Autonomous Robots 8(1), 7–24 (2000)

Walking Rover Trafficability - Presenting a Comprehensive Analysis and Prediction Tool

Brian Yeomans and Chakravathini M. Saaj

Surrey Space Centre, University of Surrey, Guildford, UK
{B.Yeomans,C.Saaj}@surrey.ac.uk

Abstract. Although walking rovers perform well in rocky terrain, their performance over sands and other deformable materials has not been well studied. A better understanding of walking rover terramechanics will be essential if they are to be actually deployed on a space mission.

This paper presents a comprehensive walking rover terramechanics model incorporating slip and sinkage dependencies. In addition to quantifying the leg / soil forces, the superior trafficability potential of a walking rover in deformable terrain is demonstrated, and a control approach is described which can reduce the risk inherent in traversing soils with unknown physical parameters. This work enhances the state of the art of legged rover trafficability and highlights some potential benefits from deploying micro-legged rovers for future surface exploration missions.

Keywords: walking rover, terrain, trafficability, terramechanics, control.

1 State of the Art

All missions to date to other planetary bodies have employed wheeled rovers, probably because the wheel as a means of locomotion is quite energy efficient, wheeled vehicle technology is well known and extensively studied, the wheel itself is a mechanically simple device even if the full vehicle is very complex, and as a consequence there is extensive Space heritage.

Despite this sucessful track record, there are some challenges. A wheeled rover cannot climb very steep slopes or traverse extremely high relief regions. Also soft, sandy terrain presents a significant risk of complete mission failure; both Mars Exploration Rovers have been stuck at some time, Spirit probably permanently so - see Fig. 1. It is also very likely that the performance expectations of rovers deployed on future missions will increase; mission planners will be aware that many of the most interesting scientific sites are in hard to reach locations high up in the cliffs and gullies of Mars, and in the rugged and often permanently shadowed cratered areas of the Lunar polar regions. Accessing these sites will demand excellent rover trafficability over many types of terrain, and there must be some question whether wheeled vehicle technology can adequately meet the greater challenge.

Walking rovers may offer a highly effective alternative solution. Walking vehicle capability on steep, rocky terrain has been studied on a number of occasions,

R. Groß et al. (Eds.): TAROS 2011, LNAI 6856, pp. 348–359, 2011.

Fig. 1. Spirit Rover showing front left wheel stuck in soft soil in location 'Troy'(Source: NASA)

and impressive results have been achieved using reflexive behaviours to negotiate unstructured environments - for example, the "Big Dog" project [11]. Combining the long distance capabilities and energy efficiency of a wheeled rover with the agility of a walking rover has been explored in [4] which utilised the SCORPION eight legged vehicle as a scout adjunct to a wheeled rover.

Walking rovers are complex machines, and this could mean increased risk of failure; however designed-in redundancy can help. A six - legged rover not only permits speed, efficiency and climbing ability trade-offs through gait pattern variation, but also incorporates redundancy given that only four legs are required for a statically stable gait. Finally, whilst there is no history of successful walking rover operation in Space, the individual components from which they are assembled do have history, as was demonstrated in [1].

Although a walking rover's superior agility may not be in doubt, performance over all types of terrain, and particularly the soft deformable areas which have proved such a hazard on Mars, has not been well studied. The work presented in this paper addresses this gap through the development of a detailed analytical model of leg / terrain interaction for a walking vehicle.

2   The LPTPT Tool

2.1   Introduction

The Legged Performance and Traction Prediction Tool (LPTPT) comprises a comprehensive model of the interaction between deformable terrain and the legs of a walking rover [15].

LPTPT uses the MATLAB computation engine for its caculations and can produce both numerical and graphical output as required. The model contains a database of reference vehicles and physical data on a range of planetary soil types, both real and simulated; currently interactions between ten vehicle and twenty five soil types can be assessed. The model analyses the forces arising between the vehicle and soil, and predicts the trafficability performance. Leg loading, and the effect of gait modification can be varied. The model's force predictions have been validated by experiment using a test rig comprising a manipulator arm moving a representative leg / foot assembly through simulated planetary soils.

2.2   LPTPT Model - Phase 1

In Phase 1, LPTPT develops the basic force calculations. Fig. 2 identifies the forces arising on a leg stepping into soil.

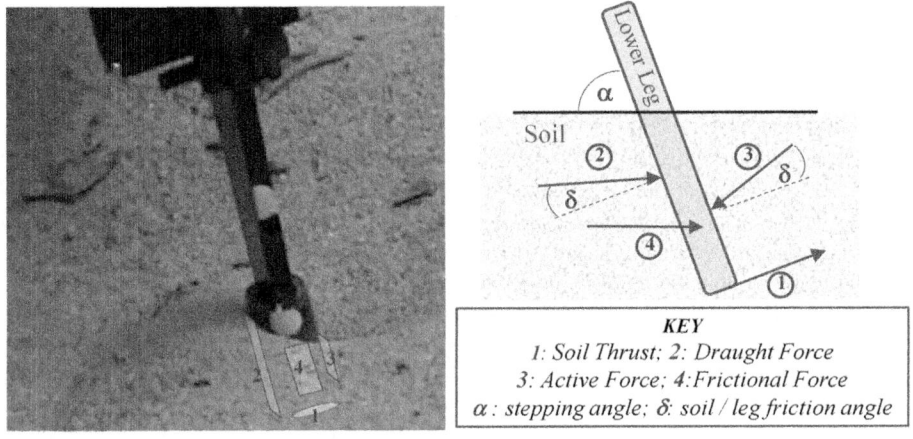

Fig. 2. Leg / Soil Forces - location (left image) and direction (right image)

Four force types are described and quantified by LPTPT:

(i) *Soil Thrust* H_o, is the shear force acting on the foot / soil interface, providing forward traction.

The soil thrust at the foot / soil shear interface is based on the Mohr - Coulomb equation [2]. The *maximum* shear stress $\tau_{\max}$ arising is:

$$\tau_{\max} = C_o + \sigma \tan\phi, \tag{1}$$

where C_o and ϕ are the soil physical properties of cohesion and friction angle, and σ is the normal stress on the soil / foot interface. Soil Thrust can be derived from this equation by integrating the foot / soil shear stress values over the area of the contact patch.

(ii) *Draught Force* F_d, is the force between the soil and the leg / foot assembly cutting through the soil. This force provides additional traction for the vehicle unless the leg is stepping "forward" into the soil, in which case it will act to resist forward motion.

The draught force derivation in LPTPT is based on the application of tillage theory, the study of the mechanics of tool / soil interaction [10]. Terzaghi's Universal Earthmoving Equation [16] as further developed by Reece [12] is used, and is expressed as follows:

$$F = \gamma g z^3 N_\gamma + C_o z^2 N_c + q z^2 N_q + C_a z^2 N_{ca} \tag{2}$$

where γ = unit weight of soil, g is acceleration due to gravity, z is the sinkage, $N_{c,q,a,ca}$ are Terzaghi's four dimensionless soil bearing capacity factors, q is the soil surcharge pressure, and C_a is soil adhesion.
In order to compute the forces arising, the four soil bearing capacity factors must be determined. Many methods have been devised to compute the factors, based on various interpretation's of Terzaghi's equation; McKyes [10] summarises the 2-D and 3-D models which have been used. LPTPT can compute these factors using several alternative models and selects the most appropriate approach based on the scenario, the principal criterion being the ratio of leg / foot width to soil sinkage.

The model predicts the angle of the failure plane of the soil in front of the moving leg / foot and then solves for the soil bearing capacity factors. An approach based on the study of narrow tools by Grisso et al [7] has been found to give good agreement with experimental results when applied to the narrow legs typically seen on walking vehicles.

(iii) *Active Force* F_a, which arises as soil falls back into the trench created by the leg as it moves through the soil. This force acts to assist the leg moving through the soil and so reduces traction.

The active force derivation is also based on Terzaghi's analysis [16]. It is not described in detail here as at low to moderate levels of sinkage, active force has a negligible effect on the total forces arising.

(iv) *Frictional Force* F_f, the effect of friction between the soil and the foot / leg. Where the stepping angle α is high ($\alpha \approx 90°$) , this force will provide further resistance to the leg moving back through the soil and so provides additional forward traction.

Frictional force is modelled in LPTPT following the same principles applied to determining the shear force at the foot / soil interface. At high values of stepping angle ($\alpha \approx 90°$), friction derives principally from the sides of the leg / foot assembly and depends on adhesion between the soil and leg assembly, the geometry of the leg / foot, and the sinkage depth.

(v) *The Effect of Gravity.* Unlike a wheeled vehicle, the leg attitude is decoupled from the attitude of the vehicle as a whole and therefore gravity does not directly affect the forces on the leg / soil interface; indirectly there is an effect however through alteration of the direction and amount of the loading on each leg due to the vehicle's weight.

(vi) *Changes in leg loading.* The complex locomotion patterns of a legged vehicle will have a direct effect on the leg loading and thus on all the forces
computed; the load per leg will be directly affected by the gait employed,
and in the case of more complex gait patterns, the load per leg is likely
to vary during the gait cycle. LPTPT allows "per leg" computations to be
made and summed to derive overall Drawbar Pull (*DP*).

Results and Forces Summary. The results are illustrated in Fig. 3 which
shows each of the forces arising, plotted against sinkage for an example 20kg
hexapod walking in a low density granular soil. In this and subsequent plots,
forces are resolved in the horizontal direction to derive the component which
aids or hinders forward motion of the rover. The vehicle is modelled as adopting
the slow but highly stable wave gait locomotion pattern under which only one
leg is lifted at any time, the other five remaining in ground contact. The soil
data is based on the in-house developed simulant Surrey Space Centre Simulant
2 *(SSC-2)*, a fine grained garnet sand with a particle distribution profile similar
to many lunar soils [3]. The figure highlights a number of significant features:

 (i) The above analysis shows that each of the four forces modelled other than
the Active Force act to increase *DP* per leg as they increase.
(ii) All forces other than the Soil Thrust increase as sinkage increases; in the
case of the Draught Force markedly so.

It follows that unlike a wheeled vehicle, where sinkage has a negative effect
through increased compaction, bulldozing and other resistances, a degree of sinkage, provided this is not so large as to overwhelm the vehicle, is of assistance to a
legged vehicle as the force available to generate forward movement is increased.

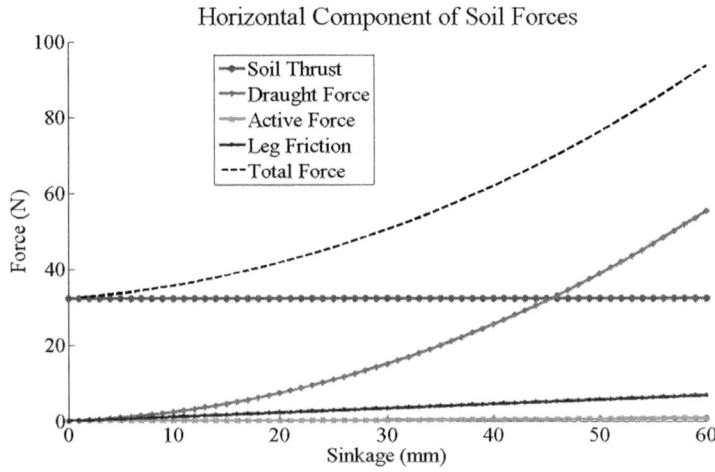

Fig. 3. Predicted Rover soil /leg interaction

2.3   LPTPT Model - Phase 2

LPTPT - 1 develops the basic force analysis and derives the maximum horizontal force available from all sources at the soil interface, given a known level of sinkage. Whilst this information enables a view to be formed of the potential of the rover / terrain combination to deliver thrust, it does not describe the impact of slip at the soil interface, and the consequent effect on sinkage and so on the forces generated.

LPTPT - 2 incorporates slip and sinkage modelling to enable it to directly show the effect on forces as these parameters vary. Additionally, sinkage and slip are vehicle operational parameters that can be measured or estimated, enabling terrain interaction predictions to be directly linked with vehicle performance.

Stress / Slip Relationship. Equation (1) gives the maximum available shear stress; however the actual stress at the interface will depend on the amount of shear displacement at that interface, which is in turn dependent on the amount of slip.

The stress / shear relationship for sands can typically be characterised by one of two types of exponential function [17]. In both cases the shear stress tends to a constant residual level; in one case, characteristic of compacted sands [17], the curve shows a pronounced peak, whereas in the other case, characteristic of loose sands, there is no peak. Given that the degree of compaction may not be known, and to avoid overstating the forces available at the soil interface, LPTPT models the stress / shear relationship on the basis of a curve with no peak, using this relationship [8]:

$$\tau = \tau_{\max} \left(1 - e^{(-j/K)} \right), \tag{3}$$

where j is the shear displacement at the relevant point in the interface, and K is a further physical property of the soil, the *shear deformation parameter*; K can be considered as a measure of the shear displacement required to develop the maximum shear stress [17].

Slip Ratio. The slip ratio measures the extent to which the forward traction theoretically available at the leg / soil interface fails to be converted to actual forward motion. The slip ratio i can be defined as [17]:

$$i = 1 - \frac{V}{V_{t,}} \tag{4}$$

where V is the actual forward speed and V_t is the theoretical forward speed with perfect traction.

The force developed at the foot / soil interface is computed by substituting the Mohr - Coulomb relation from equation (1) for $\tau_{\max}$ and integrating the resulting stress values over the contact patch area. The result will depend on the geometry of the foot, and therefore variations in foot design will directly affect the forces available. In the simple case of a flat, rectangular foot, of width b, length l and area A, the shear displacement j at a point under the foot is related to the slip ratio i and the distance from the front of the foot x as $j = ix$ [17]. In this simple case, the shear stress increases linearly across the length of the contact area l and the force can be derived analytically as:

$$F = b \int_0^l \left(C_o + \frac{W}{bl} tan\phi \right) \left(1 - e^{(-ix/K)} \right) dx \tag{5}$$

$$F = (AC_o + W tan\phi) \left[1 - \frac{K}{il} \left(1 - e^{(-il/K)} \right) \right] \tag{6}$$

The impact of variations in K. K is required to compute the above relation; however it may well not be known for the particular soil type and conditions applicable. Some commentators suggest K is constant; for example, Wong [17] quotes values for K of between 1cm and 2.5cm for sandy terrain. However, our test results gave values which were not constant and were much smaller than those quoted by Wong. LPTPT adopts a model proposed by Godbole and Alcock [6] to compute K derived from known laboratory reference values. This approach was found to be well supported by experimental results, and derives K for the conditions applicable from values of K measured in the laboratory using the following relationship:

$$(K_1/K_2) = \sqrt{(A_1/A_2)}, \tag{7}$$

where K_1 is the value of K sought, K_2 is the laboratory measured reference value, A_1 is the actual contact patch area for the given conditions and A_2 is the contact patch area applicable to the laboratory measurements.

Sinkage and Slip Sinkage. Static sinkage is modelled using the Bernstein-Bekker methodology relating pressure p and sinkage z [2]:

$$p = \left(\frac{k_c}{b} + k_\phi \right) z^n \tag{8}$$

where k_c is the cohesive modulus of soil deformation, k_ϕ the frictional modulus of soil deformation, n is an experimentally determined exponent (typically between 0.7 and 1.3), and b is the smaller dimension of the contact patch (the radius, in the case of a circular contact area).

With respect to slip sinkage, a number of methodologies have been developed, beginning with Bekker [2] who proposed a linear relationship between slip and slip sinkage:

$$z_{\text{total}} = z_o + z_j \tag{9}$$

where z_o = static sinkage, and $z_j = 2hi$, where i is the slip ratio and h equals the boundary layer of soil being sheared; equal to 1.2 times the height of grousers.

Reece in [13] proposed the following relationship, which also depends on grouser height:

$$z_j = \frac{h_{\text{gr}} i}{(1 - i)} \tag{10}$$

One concern with this relation is that sinkage $\to \infty$ at high slip levels.

Richter [14] proposes a non-linear relationship, whilst retaining the link with grouser height:

$$z_{\mathrm{j}} = 2h \left(i - \frac{i^2}{2} \right) \tag{11}$$

Lyasko [9] proposed the following relationship,which has been verified against experimental results:

$$z_{\mathrm{total}} = K_{ss}.z_o \tag{12}$$

where

$$K_{ss} = \left(\frac{1 + i}{1 - 0.5i} \right) \tag{13}$$

This relationship does not depend on grouser height, and does not $\to \infty$ at high slip levels. Fig. 4 plots the respective slip and sinkage relationships, equating grouser height and initial sinkage for comparison purposes. It can be seen that the three methods give similar results at lower levels of slip, diverging at higher levels. LPTPT adopts the Lyasko model, because:

(i) The curve does not trend to infinity at high slip levels;
(ii) Grousers may or may not be fitted, depending on the foot design, and this model does not depend on grouser height;
(iii) It is non linear, in accordance with observed results, and accords well with observed data.

LPTPT - 1 has been thoroughly tested using a lower leg segment moved through soil by a robotic arm manipulator. LPTPT - 2 testing is currently in progress, and employs a new test facility. The prior approach cannot be employed to test

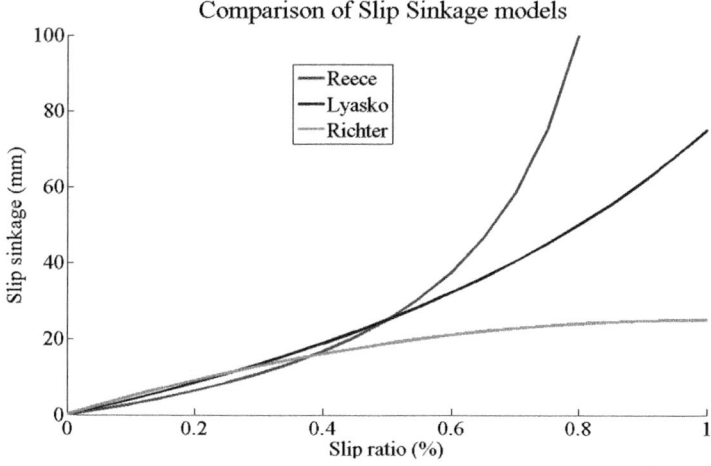

Fig. 4. Slip - Sinkage profiles

the slip dependency as the manipulator base and the soil are fixed relative to each other, and so only zero or 100% slip ratios can be achieved. A specially developed single leg testbed is used, comprising a suspended carriage to which the test leg is attached, which replicates the kinematics of the moving leg on the vehicle.

3   LPTPT Model Predictions

3.1   Wheeled Rover Comparison

Whilst *DP* is not the only criterion to comparatively evaluate rover performance, and typically a walking rover will operate at a lower level of energy efficiency that a comparable wheeled vehicle, there may be situations where maximising *DP* is paramount,and so it is useful to evaluate the maximum *DP* available to a vehicle of comparable mass of each type. Ellery in [5] computed the *DP* available to *Sojourner*, the 11.5kg vehicle used in the Mars Pathfinder mission, as 6.88N net of bulldozing, compaction and other resistances. In contrast, LPTPT computes *DP* of a 11.5kg legged vehicle, walking with a wave gait in soil with the characteristics of that at the Viking lander 2 site, as 31.5N, a very substantial increase. This comparison provides an illustration of how significant are the resistances encountered by a wheeled rover; whilst in this example the force available for thrust at the soil interface in both cases is of a similar amount, the resistances to motion of a legged rover typically do not act to impede forward motion as it can simply pick up its legs and step across intervening obstacles. Additionally given a high ($\approx 90°$) stepping angle, the principal forces reinforce rather than degrade *DP*, as described in Section 2.2.

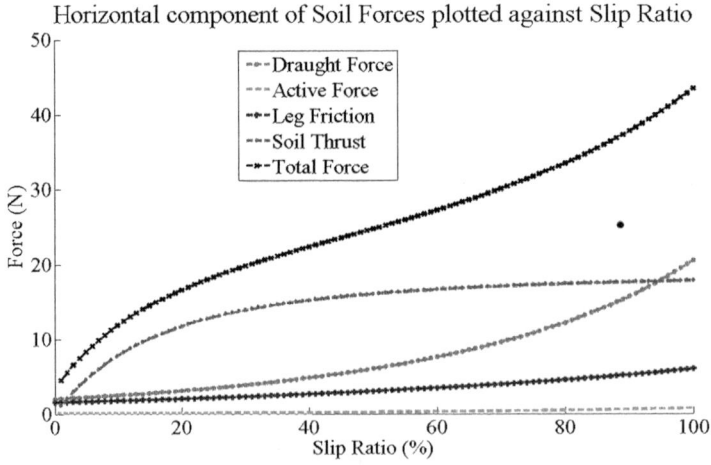

Fig. 5. Force / Slip relation

3.2   Slip Dependency

LPTPT - 2 enables the Forces / Slip relation to be plotted, giving insights into the dynamics of legged vehicle mobility, and illustrated in Fig. 5.

The plot demonstrates how total force available at the soil interface increases strongly with slip ratio. Unlike a wheeled rover, available traction will increase with slip provided sinkage is not so large as to overwhelm the rover.

3.3   Trafficability Prediction on Unknown Soils

In many cases in the field, precise information on the soil material's *in situ* physical properties will not be available; for example, whilst it may be possible to identify the soil (for example) as a coarse sand, its packing ratio may not be known or may vary as a result of wind and thermal action. Despite the lack of complete information, it may be possible to reliably predict performance using prior data sourced across a class of terrain materials.

Figure 6 illustrates how a minimum trafficability performance could be determined, as follows:

(i) The characteristics of the type of soil identified are assessed (sandy, coarse or fine grained) , and the upper and lower bounds of the total force available plotted for all previously tested examples of that soil type; there should therefore be a high likelihood that the unknown soil example in the field has properties which fall within this range;

(ii) the lower bound position is assessed in the light of the demands on the vehicle - can forward progress still be made, allowing for any climbing required,

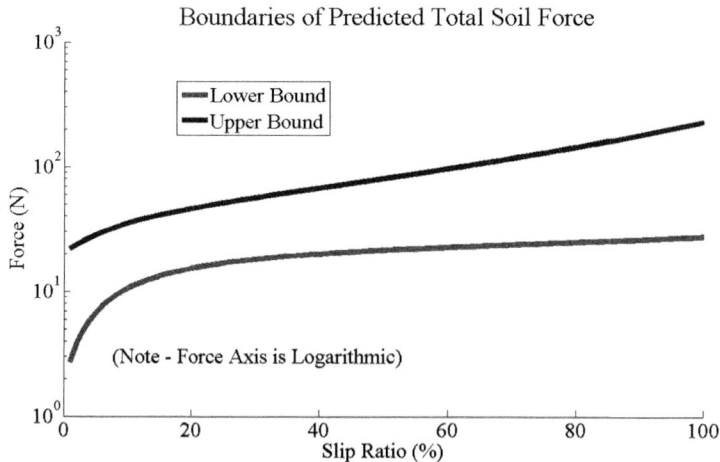

Fig. 6. Force Boundaries

at a moderate level of slip, by the application of forces at the leg / soil interface of an amount equal to or less than the minimum? This process may need to iterate to evaluate the effect of vehicle geometry changes (varied leg positioning) and gait modification;

(iii) If the answer is yes (subject to an appropriate safety margin), then an *in principle* assurance can be achieved that the vehicle will not become stuck, and can proceed with caution.

4   Conclusions and Future Work

LPTPT is presented as a comprehensive tool to predict, analyse and quantify the forces available at the leg / soil interface of a walking rover. In LPTPT - 1 form, it enables maximum performance predictions to be made of a wide variety of vehicles operating on many soil types, and suggests that a legged rover, in addition to demonstrating superior agility over rough, rocky terrain, can also be an effective vehicle to traverse soft sands and other types of deformable materials.

Incorporation of the Slip / Sinkage analysis in LPTPT - 2 enables the dynamics of the Force / Slip relationship to be modelled, and shows that slip and associated sinkage, rather than being a disadvantage, can positively aid legged vehicle traction.

LPTPT can reduce the risk that incorrect "stop / go" decisions are made in challenging terrain scenarios, by increasing confidence that a traverse is feasible despit incomplete information on terrain physical characteristics.

Further improvements are envisaged using LPTPT - 2's explicit derivation of force, slip, and sinkage relations to reduce the impact of unknown terrain parameters and enable improved performance to be achieved. Following a valid "go" decision, the vehicle would proceed with limited initial thrust application and employ on - line estimation of slip and sinkage to compare actual and predicted slip and sinkage. The terrain parameters used in the model will be refined on - line based on this comparison, deriving a more accurate view of the terrain characteristics and enabling increased rover performance.

References

1. Bartsch, S., Birnschein, T., Cordes, F., Kuehn, D., Kampmann, P., Hilljegerdes, J., Planthaber, S., Roemmermann, M., Kirchner, F.: SpaceClimber: Development of a Six-legged Climbing Robot for Space Exploration. In: International Symposium on Robotics (2010)
2. Bekker, M.G.: Introduction to Terrain Vehicle Systems, Part 1 - The Terrain & Part 2 - The Vehicle. University of Michigan Press, Ann Arbor (1959)
3. Brunskill, C., Lappas, V.: The effect of relative soil density on microrover trafficability under low ground pressure conditions. In: 11th European Regional Conference of the International Society for Terrain-Vehicle Systems, Bremen, Germany (2009)
4. Colombano, S., Kirchner, F., Spenneberg, D., Hanratty, J.: Exploration of Planetary Terrains with a Legged Robot as a Scout Adjunct to a Rover. In: American Institute of Aeronautics and Astronautics, Space 2004 Conference, San Diego, California (2004)

5. Ellery, A.: Environment-robot interaction - the basis for mobility in planetary micro-rovers. Robotics and Autonomous Systems 51(1), 29–39 (2005)
6. Godbole, R., Alcock, R.: A Device for the in situ Determination of Soil Deformation Modulus. Journal of Terramechanics 32(4), 199–204 (1995)
7. Grisso, R.D., Perumpral, J.V., Desai, C.S.: A soil tool interaction model for narrow tillage tools. In: American Society of Agricultural Engineers (ASAE) Annual Meeting, pp. 80–1518 (1980)
8. Janosi, Z., Hanamoto, B.: The analytical determination of drawbar pull as a function of slip for tracked vehicles in deformable soils. In: ISTVS 1st Int. Conf. on Mechanics of Soil-Vehicle Systems, Turin, Italy (1961)
9. Lyasko, M.: Slip sinkage effect in soil-vehicle mechanics. Journal of Terramechanics 47(1), 21–31 (2010)
10. McKyes, E.: Soil Cutting and Tillage. Elsevier Science Publishers, Amsterdam (1985)
11. Raibert, M., Blankespoor, K., Nelson, G., Playter, R.: the BigDog Team: BigDog, the Rough-Terrain Quadruped Robot. In: Proceedings of the 17th World Congress, The International Federation of Automatic Control, Seoul, Korea (July 2008)
12. Reece, A.R.: The fundamental equation of earth-moving mechanics. In: Proceedings of the Symposium on Earthmoving Machinery, vol. 179, pp. 16–22 (1965)
13. Reece, A.R.: Problems of soil-vehicle mechanics. Tech. rep., US Army Land Locomotion Lab, US Army Tank - Automotive Center, Warren, Michigan, AD Number 450151 (1964)
14. Richter, L., Ellery, A., Gao, Y., Michaud, S., Schmitz, N., Weiss, S.: A predictive Wheel-Soil Interaction Model for Planetary Rovers Validated in testbeds and Against MER Mars Rover Performance Data. In: The International Society for Terrain - Vehicle Systems 10th Europ. Conference, Budapest, Hungary (2006)
15. Scott, G.P., Saaj, C.M.: Measuring and Simulating the Effect of Variations in Soil Properties on Microrover Trafficability. In: American Institute of Aeronautics and Astronautics, SPACE 2009 Conference, Pasadena, CA (2009)
16. Terzaghi, K.: Theoretical Soil Mechanics, 3rd edn. John Wiley and Sons, Inc., London (1943)
17. Wong, J.Y.: Theory of Ground Vehicles, 4th edn. John Wiley & Sons Inc., New York (2008)

What Can a Personal Robot Do for You?

Guido Bugmann and Simon N. Copleston

School of Computing and Mathematics,
University of Plymouth,
Plymouth, UK
gbugmann@plymouth.ac.uk

Abstract. This is a report on the expectations of future users of personal robots based on a survey of 358 respondents with a median age of 22 years. A questionnaire was designed using a "text open ended" approach. The questions along with an introduction and sketch of a humanoid robot in a home were used to "paint the picture" of the respondents having a robot assistant at home. Respondent were then asked what they would ask the robot to do at various times of the day, at weekends and while away on holiday. The task category of "Housework" was the most popular, with 39% of the overall answers. "Food Preparation" and "Personal Service" were the second and third most popular categories with 16% and 14% respectively. Although many of such tasks can also be done by humans, there are potential qualitative benefits in using robots. These results suggest that research should provide solutions for cooking meals, tidying up, general cleaning and the preparation of drinks.

Keywords: Personal robots, Service robot, User expectations, Survey, User survey, Household tasks, Robot tasks.

1   Introduction

The robotic market has been predicted to boom in the near future, especially in the personal and service sectors [1]. For this boom to actually happen, robot engineers need to develop useful application and corresponding robot skills. The present research was undertaken with the aim to quantify the user's expectations and inform on potential applications.

Previous surveys have involved a limited selection of tasks. In one survey [2], respondents were asked to select from a list of 6 applications (Household (vacuuming etc.), Gardening, Guarding the house/family, Looking after children, Entertainment, and an open category "Other". The five named applications appeared to be the users' preferred tasks. In another survey [3], respondents had to select tasks from a list of 28 tasks that the literature usually assigns to assistant robots. Both studies produced results strongly biased by the initial selection of tasks. In contrast, this study uses an

R. Groß et al. (Eds.): TAROS 2011, LNAI 6856, pp. 360–371, 2011.
© Springer-Verlag Berlin Heidelberg 2011

open question format where respondent can mention any task. It is however biased by the humanoid robot shape suggested to the respondents, the pre-defined situations, e.g. time of the day, described in the questionnaire.

In the next two sections, the survey method and process is explained, followed by an explanation of how the open-ended answers are quantified and categorized. The results are presented in section 4, including the frequency of individual task requests and the relative importance of categories such as household work, cooking, etc. In section 5, features of the requested tasks and survey method are discussed. Implications for research and market are proposed.

2 Survey Design

Surveys can be carried out using a variety of methods. Here we need to discover uses of a "really new" product that is not an evolution of an existing product. In such a case, "information acceleration" methods are generally used, bringing the future to the respondent, or projecting the respondent into the future [4]. For that purpose, an in-depth questionnaire (figure 1) was designed to induce the respondent to imagine life with a robot and "day dream" interactions with the robot, and the working day of his or her robot. The design benefited from the advice of specialists in social and market research.

Three data collection methods were used for this questionnaire, personal interviews, paper questionnaires and internet survey. In total 358 respondents completed the questionnaire. 260 aged 11 to 17, 87 aged 18 to 60 and 11 aged 61 or above. 11 to 17 year olds were children attending secondary school. 18 to 60 year olds were considered adults and 61 years or above were considered as Retired/Elderly. The data for 15 respondents were collected using the personal interview methods and all fell under the adult and elderly age groups. Data for 21 respondents were collected using the Internet survey method and all were adults. The remaining 322 respondents were asked to fill out a paper questionnaire. The respondents had a median age of 22 years.

The introduction to the questionnaire included a short sentence to describe how the subject has been given a robot and that it is theirs for a year. In addition a sketch of a humanoid robot in a domestic setting was supplied to enhance the subjects "mental picture" before answering the questions. A humanoid shape was selected to avoid suggesting any functional limitation that may bias the answers.

The questions were of type "text open ended" [5], giving the respondent freedom of expression and allowing for multiple answers per question. The level of detail gathered in each question was decided by the respondent himself. Additional multiple-choice questions were used to gather information on the respondent's age, sex and if they have any help from helpers, nannies, cleaners or gardeners.

Personal robot user expectation survey

You have been given a robot as part of a trial program. It is yours to live with for one year.

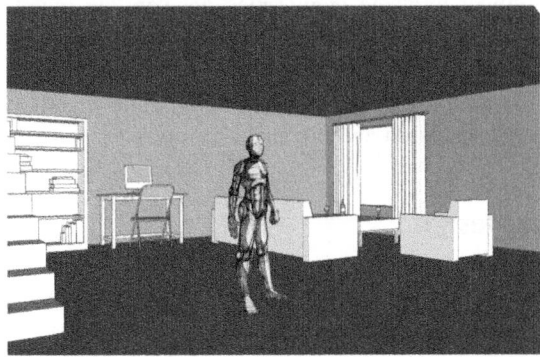

1. You get up and get ready for your day, what will you ask your robot to do today?

2. The evening comes, what will you ask your robot to do during the evening?

3. You are going to bed, what will you ask your robot to do before you go to sleep?

4. It is Sunday morning. What will you ask your robot to do?

5. You have booked two weeks holiday and plan to go away. What will you ask your robot to do while you are gone?

6. You can "upgrade" your robot by teaching it new activities. 6 months have passed since you got your robot. Have you upgraded your robot by teaching it anything?

7. Instead of a robot you have been given a trial of an intelligent appliance. Which appliance would you choose and how would it be intelligent?

Sex: Male [ ] Female [ ]
Age: 18 – 20 [ ] 21 – 30 [ ] 31 – 40 [ ] 41 – 50 [ ] 50 and over [ ]
Nationality:_____
Do you have any of the following?
Cleaner [ ] Helper [ ] Nanny [ ] Gardener [ ]

Fig. 1. The questionnaire form

3   Results Analysis

Due to using the open-ended question method it was important to find the right approach to analysing the results. Here we focus on the overall level of interest for various robot tasks. All tasks mentioned by respondents were recorded and the number of references to that task was counted. When a respondent mentioned the same task several times, e.g. in the responses to different questions, each mention was counted separately and added to the total. This was to reflect the higher interest for such a task. The answers to question 6 on "what task would you teach your robot" were amalgamated with the answers to the other questions on "what tasks would you ask a robot to do", as both answer groups reveal tasks of interest to the user. The answers to question 7 "what kind of intelligent appliance would you like instead of a robot" are only briefly summarized in this paper.

The number of responses was normalized to 100 respondents in each age group, so that the presented results can easily be evaluated by the reader. No further normalisation took place. Note that the small number (11) of respondents in the age group above 60 makes their results statistically less reliable than those of the other groups, but they are still informative. The results presented in a bar chart (figure 2) show the cumulated answers of 3 populations of 100 respondents in three age classes. The pie chart (figure 3) reflects the raw number of answers in each category. This approach should therefore give a clear indication of how popular tasks were.

4   Results

Figure 2 shows the normalized quantitative data recorded from the survey. It contains data from questions 1 to 6 from the three age groups. The chart shows the most popular tasks and the number of times they were mentioned in the survey by respondents in each age group. General cleaning for example was mentioned 63 times by the normalized population of 100 adult respondents (18-60 years old). As mentioned in section 3, this does not necessary mean that 63 separate respondents mentioned general cleaning. Some respondents could have mentioned the tasks more than once in their questionnaire. The bar chart in Figure 2 gives a clear indication of the most popular requested tasks: general cleaning, tidying, prepare tea, (this also means "prepare supper" in the UK), and guard the house.

The tasks included in the chart are varied. We grouped them into categories to give an overview of the areas in which the robot would be asked to perform tasks. The tasks were grouped into the following categories; Housework, Food Preparation, Gardening, Family Help, Pet Care, Security, Stay Quiet, Personal Service, School Work and Play. These categories were derived from the collected data and were not pre-decided at the start of the research. Table 1 below shows which tasks came under which category.

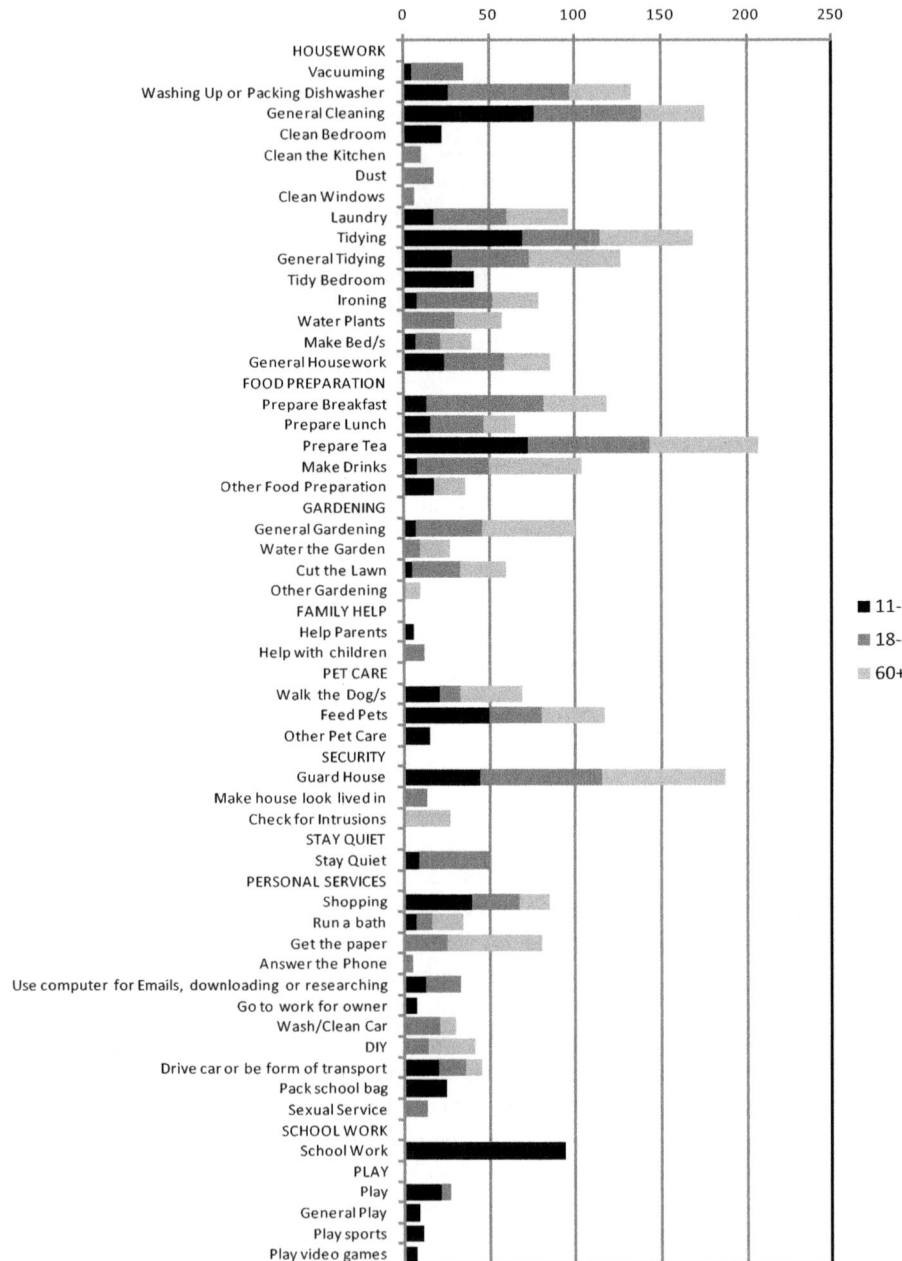

Fig. 2. Quantitative results. Number of requests of the indicated tasks normalized to 100 respondents in three age groups: 11-17, 18-60 and 61 or more. Only tasks mentioned more than 4 times are shown. Task categories are indicated by labels with zero counts, e.g. HOUSEWORK.

Table 1. Tasks Categories and corresponding tasks.

Housework
Vacuuming, washing up or packing dishwasher, general cleaning, clean bedroom, clean the kitchen, clean the bathroom, dust, clean windows, laundry, tidying, ironing, water plants, make bed/s general housework, other housework.
Food Preparation
Prepare breakfast, prepare lunch, prepare tea, make drinks, other food preparation
Gardening
General gardening, water the garden, cut the lawn, other gardening
Family Help
Help parents, help with children, help elderly
Pet Care
Walk the dog/s, feed pets, other pet care
Security
Guard house, make the house look lived in, check for intrusions, protect against fire
Stay Quiet
Stay quiet, stay out of my way
Personal Service
Shopping, run a bath, get the paper, answer the phone, use computer for emails, downloading or research, got to work for owner, wash/clean car, DIY, drive car or be form of transport, pack school bag, Sexual Service
School Work
School work
Play
General play, play sports, play games or with toys, play video games, acrobatics

The popularity of each task category is represented in figure 3 with a pie chart showing the fraction of the actual number of mentions over all 358 respondents. Housework had the largest number of answers with 39%. Food preparation and personal services also had relatively high portions with 16% and 14% respectively. Finally, Pet care, Security and School Work appeared in 7-8% of the requests.

Question 7, concerning the interest for intelligent appliances, was only posed to adults and elderly in its form of figure 1. The replies cover almost all current appliances, with the most mentioned being hoover/cleaner, iron, cooker, television, fridge/freezer and food preparation devices. The latter are an exception in that they do not cover existing appliances. How would appliances be intelligent? Typically, a hoover should be capable of regular daily cleaning, sensing of dirt or spills, opening of doors automatically, emptying itself and to differentiate between types of dirt. An

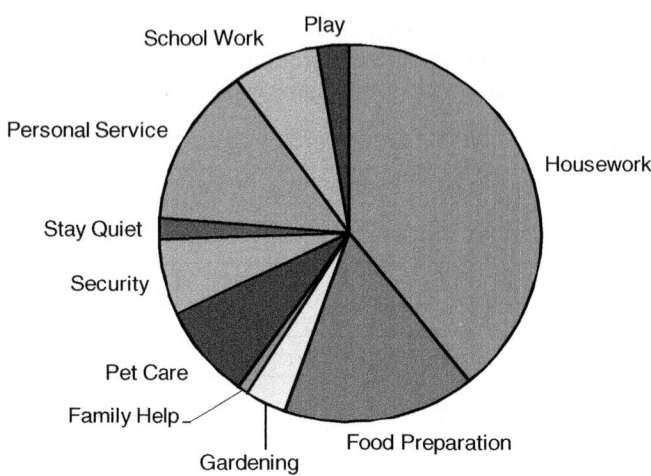

Fig. 3. Popularity of task categories

iron should iron cloth, fold and put away the clothes after ironing, the iron should detect fabric types and automatically adjust its temperature. A cooker should detect when food is burning or boiling over, automatically cook a meal for a specific time and create an entire meal from scratch including preparation and cooking. A television should sense when the user is in the room so that it only switches on when the user is present, fast forwarding through advertisements without prompting and operate by voice control. A fridge should keep an inventory of what is inside it, defrost meat in time for cooking, give the user suggestions of recipes based on its contents, dispose of spoiled food or even be able to prepare meals itself. General food preparation devices were varied and included appliances which would provide healthy meals to the user, tell the user what is in their cupboards and what they need to replenish (all based on a specified budget), a steamer which would wash, peel, cut and steam vegetables and a drinks maker that can be sent text messages from a mobile phone to create drinks.

5   Discussion

5.1   Attitude towards Robots

The results show that people have an overall positive attitude towards living with robots. Although there was no question to ask if the respondents would accept the robot into their homes, it is clear that they were all willing to give answers when asked what they would ask a robot to do for them. A previous survey conducted at a robotics exhibition in Switzerland backs up this view [6]. In that survey of 2000 people, 71% of them answered "yes" to the question "could you imagine to live on a daily basis with robots which relieve you from certain tasks that are too laborious for you?"

The robot in our survey was considered as a servant. It should be noted that the questions never said that the robot was there to do work for you. People just assumed that this is its purpose. This conforms to a distinctly western view on robotics suggested by [7]: If a robot does not do specific tasks, it is not useful. However, tasks mentioned in our survey were not always "work".

5.2 Task Types

There seems to be two types of tasks the robot could do for the owner. Tasks which remove labour from the owner and tasks which enhance the owner's life. Housework, gardening or food preparation all relieve work from the owner whereas playing adds to the person's life through entertainment. Security tasks 'enhance' the user by duplicating his/her presence. The DIY task also represents an enhancement, as not everyone has the required skills. The same probably applies to cooking. Curiously, respondents did not mention collaborative tasks, e.g. help in moving a piece of furniture.

The popularity of the housework category appears to indicate that people are looking for solutions to the "laborious" jobs around the home. Another survey into the top ten most dreaded household chores highlights cleaning the kitchen, cleaning the bathroom and washing the dishes as the three most dreaded chores [8]. In the results in figure 2 cleaning and dish washing are also the task most asked of the robot under the housework category. General cleaning was often mentioned but respondents were not specific to exactly what object or what rooms they wanted cleaning. Some respondents did state that they wanted the kitchen or bathroom to be cleaned however this represented a very small number of replies compared to general cleaning. It is unclear whether the respondents had bathrooms or kitchens in mind when answering "do the cleaning". However, in our list of tasks that a household robot should be able to do, cleaning is of a high priority. The results of our survey seem to back up the attitudes to certain chores. However, preparing tea and preparing breakfast were very popular tasks, while cooking only ranked 9th out of the ten most dreaded chores. There seems to be a contradiction here, as food preparation is a high priority task for household robots even though it is not as dreaded as others. People may not dislike cooking meals, but would nevertheless be happy letting the robot prepare their evening meal. We can speculate that robots will have better cooking skills than the average person, and will be able offer a greater variety of meals. Meal preparation also appears in the replies to the question on intelligent appliances. To highlight another difference between our and the dreaded chores surveys, tidying was not included in the top ten list but appears prominently in the housework category. People do not seem to dread this task very much but would still pass it off onto their robot.

The significant number of responses in the "food preparation" category is one of the interesting findings of this survey. This is an application where robots could make a valuable contribution to the quality of life of the user, especially in view of the limited cooking skills of the general population.

Among other tasks, it is noteworthy that schoolwork was the third most popular tasks overall. This task was only requested by children up to the age of 17. Therefore unlike the majority of tasks, it is not an answer distributed over all of the respondents. It is not surprising that children want to receive help with schoolwork or in fact to have it done for them. These results could be highlighting a need for children to have more support when doing their schoolwork. The form of this support needs further investigation.

There were a few mentions of sexual services by adult respondents. Whether respondents were serious or not, the fact is sex was seen in the results. Sexual robots have already been seen in Hollywood, e.g. in the 2001 film "AI" by S. Spielberg, and it might just be a matter of time before robots made for sex become a reality. There are several mentions on the internet of developments in that area.

Playing with the robot was also seen in the results. Among younger respondents, males were the ones who wished to play sports and video games. Females answered to generally play with the robot or to play games or with toys. Among older respondents, card games were mentioned. Thus, not all the tasks required from the robot were "work".

There were clear differences between age groups in the categories of Homework, Play and Gardening (figure 2). However, there is very little difference in the other categories, reflecting similar needs for services at home.

The survey does not reveal "new" tasks that were not known before, or that could only be realized with or by a robot. Such tasks may have emerged if respondents had the use of a real robot.

5.3 Methodology

It is very likely that the results of the survey were influenced by the introduction and sketch. These were deemed necessary to put the respondent in the "mind set" of having a robot at home. To help respondents imagine their life with the robot, the questions were designed to guide the respondent though their day and during a weekend and holiday.

The use of an image of humanoid robot of human size had the expected effect that no task was limited by the assumption that the robot had limited cognitive capabilities or dexterity. However, it has probably eliminated tasks where superhuman physical or cognitive robot capabilities could be exploited.

There was a definite relationship between the time of day and the tasks answered (data not shown here). Obviously food preparation was already governed by the time of day however other tasks like gardening or washing up were answered in the questions corresponding to periods when the respondent themselves would perform these tasks. For example, gardening was mentioned most often in the weekend question. A robot could perform these tasks at any time but the respondents chose to ask the robot when they would normally do them. This trend could simply be a bias introduced by the way the questions were worded and the questionnaire design but may also reveal an attitude of the respondents towards cohabitation with robots. It may be significant that several respondent wanted the robot to "be quiet" during the night.

A different introduction would most likely have led to different results. For instance, a picture of a robot with two wheels and no manipulators would have reduced the range of tasks requested. This could be confirmed by a future survey to investigate how different introductions and sketches lead to different results. For this survey however, the introduction served its purpose and very few respondents questioned the robot being in their home or its functionality.

Different data collection methods were used. It is unclear if this has affected the response pattern. If it had, this would mainly have affected data for the 61+ age group. As there also is a small number of respondents in that group, its results are only indicative. A follow study would be required for more refined results.

Quantifying the results proved challenging for the open-ended questions used here, as task categories had to be determined after the survey. The use of multiple-choice questions may have been simpler, but it was considered more important not to influence the respondent by giving them answers to choose from. Various normalization schemes could have been implemented, e.g. to mimic a flat distribution of ages, but there was no strong reason for doing so. The current results in figure 3 thus reflect the views of a rather young population. However, figure 2 has shown that in most categories, there is little differences between age classes.

It is a question for future work of whether a more imaginative list of tasks could have been generated with a different questionnaire design. This survey probably reflects correctly what users would ask from their robot after taking delivery. However, they are likely to start inventing new tasks soon after, which may point to the need for user-programmable robots.

5.4 Market Considerations

The demands of respondents were in great majority for tasks that they could do themselves, raising the question of whether the main benefit of personal robots is additional free time. In many cases the robot could theoretically do a task better than its user, e.g. cleaning, DIY and cooking. Thus, personal robots could not only provide free time to the user, but add value through better execution of tasks. This however requires advanced robots.

Certain tasks are services that users cannot provide themselves, such as entertainment, home security while absent and feeding the pets. These benefits could probably also be provided by simpler non-humanoid robots.

Overall, no single task identified in this survey has the flavour of a life-changing "killer application". It is tempting to speculate however that, if robots were flexible enough, user would find new applications of their robot, e.g. to expand the user's sensing and action capabilities, or to reduce life's uncertainties, etc.

Regarding the commercialization of early versions of personal robots of the humanoid type considered here, it is anticipated that these will be very expensive, with a price probably similar to that of a luxury sport car. Thus the initial market appears limited, with only the very rich being able to afford a robot. Such users can also afford human servants who can probably provide a higher quality of services than early robots. One of the added benefits of robots, however, would be a greater privacy. Another possible market is the assistance of the elderly, where a support organization could bear the costs of the robot and supply it as and when required. Recent robots for this market, however, have not been well received [9].

One big limitation to market growth is the lack of user-centred applications. Users expect robots to work, and much research and development is still needed to develop the required robot skills. The development of assistant robots requires applied research in areas such as handling of fabric, cloth and bed sheets, preparing food and drinks, and cleaning surfaces and awkward places. The specification of corresponding research problems would require a more detailed analysis of the demands of the users. For instance, what do respondents actually mean by "tidying" or "cleaning"? Other tasks such as "prepare tea" or "feed the pet" probably pose more clear challenges. The implementation of these tasks could benefit from decomposition into more elementary operations in defined domestic environments[1]. This decomposition is likely to reveal gaps in the knowledge needed for the development of specific applications.

Another issue is the certification and development of safety standards [10], as personal robots will operate in close proximity of children, elderly and pets. Once the questions of functionality and safety are resolved, more social aspects of robot behaviour will gain importance.

6 Conclusion

This research has produced unique quantitative data on peoples' expectations from robots in the home. The study focused on humanoid robots but also inquired about future intelligent appliances.

The robot was seen mainly as a servant and, to a lesser extent, a play partner for children. People expect robots to take care of the usual chore, and provide services in the area of homework and security. There are strong demands in the categories of housework (mainly cleaning), food and drink preparation, and in a range of personal services, e.g. shopping. Although many of such tasks can also be done by humans, there are potential qualitative benefits in using robots.

These results suggest that research should provide solutions for cooking meals, tidying up, general cleaning and the preparation of drinks. Adaptable designs would enable users to develop their own applications.

Acknowledgements. The following people are acknowledged for their contributions, informative discussions, support and advice during the preparation of the survey and analysis of the data: Andy Phippen, Phil Megicks, Peter White and Liz Hodgkinson.

References

1. Dan, K.: Sizing and Seizing the Robotics Opportunity. RoboNexus, Robotics Trends Inc. (2003), http://www.robonexus.com/roboticsmarket.htm (accessed on January 17, 2007)
2. Dautenhahn, K., Woods, S., Kaouri, C., Walters, M.L., Koay, K.L., Werry, I.: What is a robot companion - friend, assistant or butler. In: IEEE/RSJ Int. Conf. on Intelligent Robots and Systems (IROS 2005), pp. 1192–1197 (2005)

---

[1] You can see and contribute to analyses examples at vrcpersonalrobotics.org.

3. Oestreicher, L., Eklundh, K.S.: User Expectations on Human-Robot Co-operation. In: Proceedings of the 15th IEEE International Symposium on Robot and Human Interactive Communication (ROMAN 2006), pp. 91–96 (2006)
4. Glen, U., Bruce, W., John, H.: Premarket Forecasting of Really New Products. Journal of Marketing 60(1), 47 (1996)
5. Creative Research Systems Corporation Web Site (2007), The Survey System – Survey Design, http://www.surveysystem.com/sdesign.htm (accessed on September 19, 2007)
6. Arras, K.O., Cerqui, D.: Do we want to share our lives and bodies with robots? A 2000-people survey., Technical Report EPFL-ASL-TR-0605-001, Autonomous Systems Lab, Swiss Federal Institute of Technology Lausanne (June 2005)
7. Matthew, S.: Japanese Robotics. IET Computing & Control Engineering 17(5), 12–13 (2006)
8. Consumer Electronics Association (CEA) Market Research (2007), http://www.ce.org/ (accessed on January 17, 2007) (Technologies to Watch, Robotics 5, Winter 2006)
9. Fitzpatrick, M.: No, robot: Japan's elderly fail to welcome their robot overlords (2011), BBC website http://www.bbc.co.uk/news/business-12347219 (accessed on March 24, 2011)
10. Virk, G.S., Moon, S., Gelin, R.: ISO standards for service robots. In: Advances in Mobile Robotics — Proc. 11th Int. Conf. on Climbing and Walking Robots and the Support Technologies for Mobile Machines, Coimbra, September 8-10, pp. 133–138 (2008)

A Systems Integration Approach to Creating Embodied Biomimetic Models of Active Vision

Alex Cope, Jon Chambers, and Kevin Gurney

Department of Psychology, University of Sheffield, Sheffield, UK
`k.gurney@sheffield.ac.uk`

Navigating the visual world is a challenging problem for autonomous agents, which must be flexible, robust, and preferably easily extensible in order to meet changing task demands. Here, we outline the rationale for an approach to constructing such agents *biomimetically*, even if they appear 'over engineered' at first glance, using the problem of gaze redirection in an attentional task.

Many attempts to create methods for analysing the visual world and redirecting gaze, notably the work of Itti and Koch [2], use biologically *inspired* solutions where simple features (e.g. colours, orientations) are used to construct a visual *salience map*. The problem with such models is that they require significant redesign to correctly choose eye movement targets where, for example, the same scene is used with different task instructions (as in the free viewing experiments of Yarbus [3]).

This is a typical failing of systems which, though *bioinspired*, are 'just sufficiently' engineered for a specific task. In contrast, we advocate attempting to more precisely recreate the function of relevant biological systems, even if there appears to be more complexity than is strictly necessary for the particular task in hand, with the extra complexity being exactly that required to allow the system to perform a wider range of tasks, and be integrated with other subsystems.

In previous work we created a model of primate oculomotor control [1], which was capable of reorienting gaze to simple isoluminant point stimuli. The architecture of the model is shown in Fig. 1 (right hand side). This diagram is shown only to emphasise that the complexity of the model is well in excess of the demands of the task, clearly much simpler solutions could be used.

We have extended this model with the ability to recognise objects and then redirect gaze to these based on the current task demands. These extended parts of the model (left side of Fig. 1) were integrated with few changes to the existing architecture, demonstrating the flexibility that is found in biological solutions.

In cases where there is insufficient constraint from the biological literature we advocate the use of phenomenological, 'engineered', solutions as 'black boxes', with a described function and fixed input and output interfaces, to be replaced when a biologically derived solution is possible, as the replacement will provide a more flexible and robust foundation for further extension.

We therefore created an agent capable of more advanced behaviour than the original model by using the extended model as a 'biomimetic core', and integrating engineered competencies as an 'engineered surround' adding additional functionality. We exercised this agent on a simple ethological task where

R. Groß et al. (Eds.): TAROS 2011, LNAI 6856, pp. 372–373, 2011.
© Springer-Verlag Berlin Heidelberg 2011

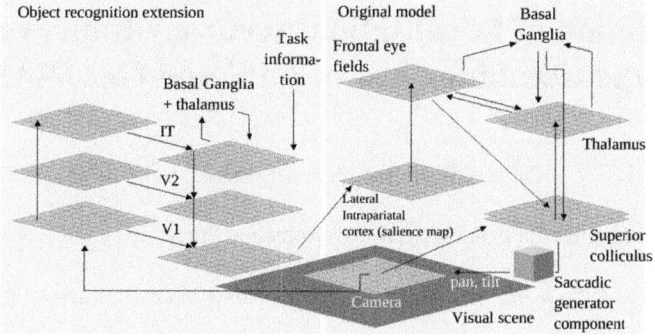

Fig. 1. A diagram of the extended model, with the original oculomotor model the right hand panel

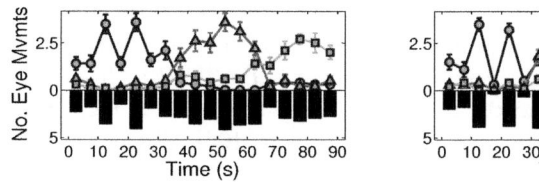

Fig. 2. Agent behaviour. Each symbol represents a different object class. The histogram shows the total number of eye movements in a time period. Left: with task irrelevant phasic distraction. Right: with task relevant phasic distraction.

directing gaze to certain types of object elicited a reward for the agent, leading to the direction of gaze to rewarding objects more often, thus maximising reward. Throughout each trial the type of object eliciting the reward was altered, and the agent adapted to the new circumstances (See Fig. 2).

It is clear this behaviour could be achieved in a simpler way, however the aim of this work is to demonstrate the flexibility and extensibility of our approach. Thus, we have shown an agent which has been extended to with more competencies, without requiring the redesign of the gaze control model. We believe that by continuing this approach it can be used to further extend the competencies of the agent, with each advance building on a robust and flexible base.

References

1. Chambers, J.M.: Deciding where to look: A study of action selection in the oculomotor system. Ph.D. thesis, Sheffield: University of Sheffield (2007)
2. Itti, L., Koch, C., Niebur, E.: A model of saliency-based visual attention for rapid scene analysis. IEEE Transactions on Pattern Analysis and Machine Intelligence 20(11), 1254–1259 (1998)
3. Yarbus, A.L.: Eye Movements and vision. Plenum Press, New York (1967)

A Validation of Localisation Accuracy Improvements by the Combined Use of GPS and GLONASS

Dennis Wildermuth and Frank E. Schneider

Fraunhofer Institute for Communication, Information Processing and Ergonomics (FKIE),
Wachtberg, Germany
{dennis.wildermuth,frank.schneider}@fkie.fraunhofer.de

For autonomous navigation in outdoor environments, robust and reliable positioning is an indispensable prerequisite. Looking at unstructured or a priori unknown surroundings the use of global navigation satellite systems (GNSS) is a reasonable approach [1]. The Global Positioning System (GPS) is definitely the most popular GNSS. There are several efforts to build competing GNSS, from which only the Russian GLONASS is nearly operational. Nowadays, even recreation-grade GPS receivers often achieve an accuracy of less than 10m. But, since this is still not enough for many localisation and navigation tasks, several techniques have been developed to improve the positioning accuracy. In principle, all these methods use differential data coming from a base station at a well-known position. The GPS receiver applies the differential information in order to eliminate error sources like signal delays or inaccurate satellite orbits. Depending on the used method, this is called Code Differential GPS (DGPS) or Real Time Kinematics (RTK). Using sufficiently sophisticated receivers, with DGPS accuracy in the metre range can be reached. For RTK systems centimetre or even millimetre ranges are achievable.

A GPS receiver can obtain differential data via various communication means. One approach is the use of satellite-based augmentation systems (SBAS), which send correction data over additional satellites. Additionally, many countries have set up ground-based radio beacons, which transmit DGPS corrections at frequencies around 300kHz. For surveying and other high precision applications, normally a dedicated base station is positioned at a known position. The base station sends differential data through some radio link directly to the GPS receiver. A rather new approach is to retrieve corrections over the Internet via the standardised Ntrip protocol. This is becoming increasingly popular due to rapid technical development and decreasing traffic costs in the mobile GSM networks.

Several studies address the performance improvements caused by the various augmentation systems, for example [2] for DGPS and [3] for RTK. Mostly, GPS accuracy is evaluated under optimal conditions – static measurements with an open sky view – thus providing some kind of ground truth for realistic outdoor scenarios. Realistic environments include nearby buildings and forests and – as documented for instance in [4, 5, 6] – generally lead to significant deviations and sudden jumps in the GPS positioning. Furthermore, in [5] and [6] the GPS receiver was put on a moving platform, leading to results directly transferable to the field of outdoor robotics.

R. Groß et al. (Eds.): TAROS 2011, LNAI 6856, pp. 374–375, 2011.

A major drawback of the cited literature concerning accuracy of GNSS is the exclusive covering of GPS. The additional use of GLONASS should be considered, since the system is planned to be fully operational at the end of 2011 and all differential augmentation systems are, in principle, applicable to GLONASS as well. According to the manufacturers state-of-the-art GPS receivers are equally able to cope with GPS and GLONASS signals. Thus, the authors are currently conducting an evaluation of the combined use of GPS and GLONASS together with all freely available augmentation services. Long-term datasets of more than 24 hours have been recorded for different setups. Thereby, receiver hardware ranges from a simple Navilock GPS mouse over more sophisticated Ashtec GPS boards up to surveying-grade Topcon receivers, namely a Legacy-E$^+$ and a GRS-1 handheld receiver. Since the latter are capable of receiving GLONASS signals as well, one of the major aims of this study is to evaluate possible benefits for mobile robot localisation using GPS and GLONASS. We have already acquired more than 20 full day datasets with open sky view and nearly perfect GPS reception, which are meant as some kind of ground truth data. Currently, data is gathered under more realistic conditions, e.g. near (tall) buildings and under tree foliage. However, already the preliminary results from the open sky data are rather disappointing. Neither do all augmentation systems really work with GLONASS, nor seems the integration of GLONASS at least into the tested GPS receivers fully completed. To give only two examples, the slightly older Legacy-E$^+$ took considerably longer to acquire an RTK fix when using both GPS and GLONASS. Because accuracy was nearly the same, the overall precision for this setup was even worse than with the use of GPS alone. As a second example, none of the Topcon receivers was able to integrate the German radio beacon signals into DGPS corrections for GLONASS satellites. We are currently in contact with the manufacturer to examine this faulty behaviour. The poster will give a detailed overview of all so far obtained results.

References

1. Cooper, S., Durrant-Whyte, H.: A Kalman Filter for GPS Navigation of Land Vehicles. In: Proceedings of the IEEE/RSJ International Conference on Intelligent Robots and Systems (IROS), Munich, pp. 157–163 (1994)
2. Kuter, N., Kuter, S.: Accuracy comparison between GPS and DGPS: A field study at METU campus. Italian Journal of Remote Sensing 42(3), 3–14 (2010)
3. Dammalage, T.L., Samarakoon, L.: Test Results of RTK and Real-Time DGPS Corrected Observations Based on Ntrip Protocol. The International Archives of the Photogrammetry, Remote Sensing and Spatial Information Sciences 37(B2), 1119–1123 (2008)
4. Danskin, S.D., Bettinger, P., Jordan, T.R., Cieszewski, C.: A Comparison of GPS Performance in a Southern Hardwood Forest: Exploring Low-Cost Solutions for Forestry Applications. Southern Journal of Applied Forestry 33(1), 9–16 (2009)
5. Morales, Y., Tsubouchi, T.: GPS moving performance on open sky and forested paths. In: Proceedings of the IEEE/RSJ International Conference on Intelligent Robots and Systems (IROS), San Diego, pp. 3180–3185 (2007)
6. Ohno, K., Tsubouchi, T., Shigematsu, B., Yuta, S.: Outdoor Navigation of a Mobile Robot between Buildings based on DGPS and Odometry Data Fusion. In: Proceedings of the IEEE International Conference on Robotics and Automation (ICRA), Taipei, pp. 1978–1984 (2003)

Adaptive Particle Filter for Fault Detection and Isolation of Mobile Robots

Michał Zając

Institute of Control and Computation Engineering,
University of Zielona Góra, Zielona Góra, Poland
M.Zajac@weit.uz.zgora.pl

Particle filters have recently gained major attention as a powerful diagnostic tool. Their severe drawback is the computational burden closely related to the number of particles used. Therefore, it is often necessary to work out a compromise between computation time and the quality of results, especially in the case of systems with limited computational resources such as mobile robots. This work outlines the concept of a fault detection and isolation (FDI) system for a mobile robot which is based on a bank of adaptive particle filters and accounts for the aforementioned problems.

In this paper the model-based fault diagnosis problem of mobile robots is considered [1,3]. The main task of the proposed FDI system is to detect any kind of fault which may occur and to isolate the faults related to wheels (e. g. flat tires). It is assumed that the mobile robot is powered by a differential drive which motion is described by the velocity motion model [7], but the approach can be implemented for other wheeled robots as well.

The key element of the proposed FDI system is the likelihood-based decision function suggested by Li and Kadirkamanathan [4,2,5], as well as the mechanism of activation of particle filters based on a set of heuristic rules which is similar in its basic idea to the mixed-abstraction particle filter proposed by Plagemann et al. [6]. Further details can be found in the paper of Zając and Uciński [8].

The proposed FDI system is composed of a bank of particle filters. The bank consists of a nominal filter and adaptive filters (their number depends on the number of faults to be isolated). The fault detection task is based on thresholding the likelihood-based decision function and fault isolation is achieved by choosing the filter which currently yields the highest value of the function. In the nominal filter f_0 the state vector is simply the robot pose $x_0 = [x, y, \theta]'$, and an m-element vector of constant (nominal) parameters – $\Psi_0 = [p_1, ..., p_m]'$ describes the robot structure and their changes form a basis for fault detection. In the adaptive filters f_k, where $k = 1, ..., n$, the state vector is augmented with an additional parameter p_j, where $j \in 1, ..., m$, is at the same time the j-th element of the parameter vector Ψ_k and is of the form $x_0 = [x, y, \theta, p_j]'$. For a robot in a nominal state, each of filter would be able to estimate the state of the system. Hence, to decrease the computational resources demand, a filter activation block was introduced. The fault isolation filters can only be activated in cases when a fault is being detected, what, when combined with sample adaptation for each filter, considerably lowers the computational power consumption of the system.

R. Groß et al. (Eds.): TAROS 2011, LNAI 6856, pp. 376–377, 2011.

To verify the proposed solution, a number of simulations in MATLAB and tests with data from a real robot Pioneer 3-AT were performed. The considered faults were flat tires on each side of the robot. The proposed FDI scheme with its sample size and filter activation mechanism allows to decrease the computational burden of fault diagnosis when compared to systems with constant particle number and without filter activation, while still providing satisfactory parameter estimates.

Acknowledgment. The author is a scholar within Sub-measure 8.2.2 Regional Innovation Strategies, Measure 8.2 Transfer of knowledge, Priority VIII Regional human resources for the economy Human Capital Operational Programme co-financed by the European Social Fund and the Polish state budget.

References

1. Chen, J., Patton, R.: Robust model-based fault diagnosis for dynamic systems. Kluwer, Boston (1999)
2. Kadirkamanathan, V., Li, P., Jaward, M.H., Fabri, S.G.: Particle filtering-based fault detection in non-linear stochastic systems. International Journal of Systems Science 33, 259–265 (2002)
3. Korbicz, J., Kościelny, J., Kowalczuk, Z., Cholewa, W.: Fault Diagnosis: Models, Artificial Intelligence, Applications. Springer, Berlin (2004)
4. Li, P., Kadirkamanathan, V.: Particle filtering based likelihood ratio approach to fault diagnosis in nonlinear stochastic systems. IEEE Transactions on Systems, Man and Cybernetics - Part C: Applications and Reviews 31, 337–343 (2001)
5. Li, P., Kadirkamanathan, V.: Fault detection and isolation in non-linear stochastic systems–a combined adaptive monte carlo filtering and likelihood ratio approach. International Journal of Control 77, 1101–1114 (2004)
6. Plagemann, C., Stachniss, C., Burgard, W.: Efficient failure dtection for mobile robots using mixed abstraction particle filters. In: Proceedings of European Robotics Symposium (2006)
7. Thrun, S., Burgard, W., Fox, D.: Probabilistic Robotics. The MIT Press, Cambridge (2005)
8. Zając, M., Uciński, D.: Adaptacyjny filtr cząsteczkowy w detekcji i lokalizacji uszkodzeń robota mobilnego. Prace Naukowe Politechniki Warszawskiej. Elektronika 175, 635–644 (2010)

An Approach to Improving Attitude Estimation Based on Low-Cost MEMS-IMU for Mobile Robot Navigation

Lu Lou[1,2], Mark Neal[1], Frédéric Labrosse[1], and Juan Cao[1,2]

[1] Department of Computer Science, Aberystwyth University, Ceredigion, UK
{lul1,mjn,ffl,juc3}@aber.ac.uk
[2] College of Information Science and Engineering,
Chongqing Jiaotong University Chongqing, People's Republic of China

An inertial measurement unit (IMU) is an electronic device to measure vehicle states like attitude, orientation, velocity, and position. Recently, many low-cost micro electro mechanical systems (MEMS) IMUs have emerged for only several hundred US dollars [1]. These MEMS-IMUs usually consist of three-axis accelerometers,gyros and magnetometers. In comparison to high-end IMUs (usually used in aerospacecrafts, missiles, rockets and artificial satellites), an entire Inertial Navigation System (INS) can be implemented with smaller size/volume, lower weight and costs. In exchange, they have a rather low accuracy performance due to their large systematic errors such as biases, scale factors and drifts, which are strongly dependent on disturbance and temperature. Hence, the raw signal output of a low cost IMU must be processed to reconstruct smoothed attitude estimates. For many of the mobile robot navigation considered the algorithms need to run on embedded processors with low memory and processing resources.

No sensor is perfect and no sensor suits all the applications. However these sensors have themselves distinctive characters, for example, although the gyro is not free from noise, it is less sensitive to linear mechanical movements because it measures rotation. The drift problem is a fatal weakness of gyro. However, the accelerometer does not drift. Therefore, by 'averaging' data that comes from accelerometer and gyro we can obtain a relatively better estimate of the vehicle than we would obtain by using the accelerometer or gyro data alone. The estimation accuracy of IMUs highly depends on the sensor fusion algorithm.

Unlike other related works where the COTS MEMS-IMUs were used as a testbed, in our project of mobile robot navigation we select a cheap custom build MEMS-IMU which costs only about 100 US dollars and incorporates one cheap MCU (ATmega328, 8Mhz) and three-axis gyros, accelerometers and magnetometers. We use a modified fusing Direction Cosine Matrix (DCM) algorithm for orientation estimation. In short, this algorithm is to continuously update the 3×3 matrix that defines the relative orientation of the robot and ground reference frames, and uses gyro information as the primary link between the two reference frames and uses magnetometer and accelerometer information to correct for gyro drift [2]. The description of the algorithm is shown as Fig.1(a).

This fusing DCM algorithm works well with accuracy and computational efficiency at a rate of 50Hz in our low cost IMU. Accelerometers are used for

R. Groß et al. (Eds.): TAROS 2011, LNAI 6856, pp. 378–379, 2011.

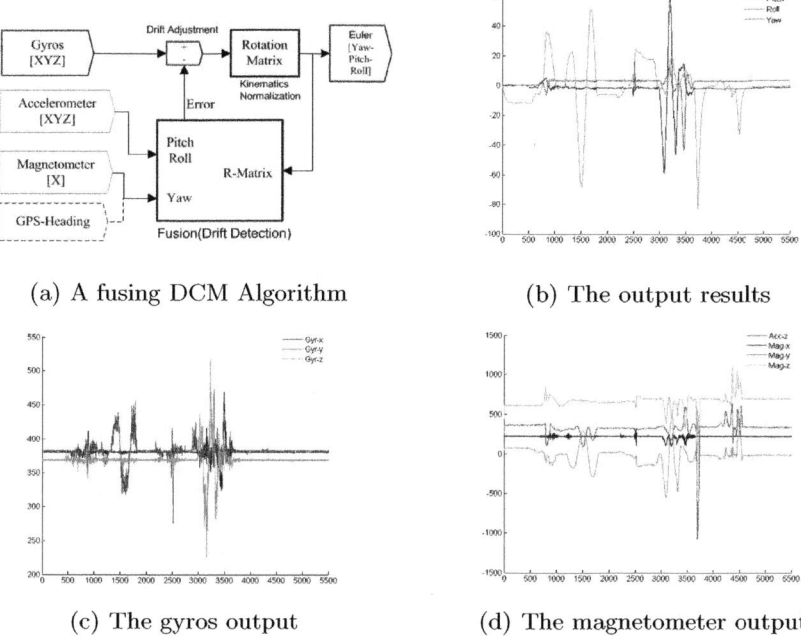

(a) A fusing DCM Algorithm

(b) The output results

(c) The gyros output

(d) The magnetometer output

Fig. 1. The fusing DCM Algorithm and results

roll-pitch drift correction because they have zero drift. Magnetometers or GPS (outdoor) have to be used for yaw drift correction because accelerometers don't up to this task (due to the gravity). However, in some cases, such as local magnetic disturbance, static or moving slowly, the measurement error of yaw from magnetometers or GPS occur and decrease the accuracy of this DCM algorithm. Therefore, we have developed a simplified kalman filter to minimize these errors.

We evaluate the errors of this fusing algorithm by comparing the difference with traditional DCM algorithm (without drift correction). The experimental results are shown as Fig.1(b)(c)(d). These results indicate that the fusing DCM algorithm can improve attitude estimation, especially for roll and pitch by means of the roll-pitch drift correction with accelerometers, but the error of yaw cannot be fully eliminated even though magnetometers and filter were added.

We will present result comparing these algorithms with ground truth obtained using a motion tracking system. Further, we will also consider using other sensors, such as camera, to develop the more robust fusing algorithm.

References

1. Munoz, J., Premerlani, B.: Arduimu open source project (2011),
   http://diydrones.com/profiles/blogs/arduimu-v2-demo-video
2. Phuong, N., Kang, H.J., Suh, Y.S., Ro, Y.S.: A DCM based orientation estimation algorithm with an inertial measurement unit and a magnetic compass. Journal of Universal Computer Science 15(4), 859–876 (2009)

Cooperative Multi-robot Box Pushing
Inspired by Human Behaviour

Jianing Chen and Roderich Groß

Natural Robotics Lab, Department of Automatic Control and Systems Engineering,
The University of Sheffield, Sheffield, UK
{j.n.chen,r.gross}@sheffield.ac.uk

This paper investigates mechanisms underlying cooperative behaviour in a group of miniature mobile robots around the problem of coordinating a group of robots to push collectively a heavy object. Numerous solutions to this problem have been proposed [5,7,4]. The performance of these however typically deteriorates as the number of robots increases to more than a dozen. The cause of this is often said to be robot interference—there are many robots but insufficient space to manipulate the object effectively. The situation is particular difficult when the object itself occludes the view of robots [3]. In this case, robots can benefit from division of labour (e.g., see [2]). Here, we take inspiration of the division of labour in teams of humans pushing a large object towards a target location: persons who can see the target push the box only when the transporting direction needs to be corrected, while all other persons simply push the box forward. The two roles in this cooperation are indicated in Fig. 1(a).

We have developed a simulation framework based on the Bullet Physics and Enki Libraries[1], see Fig. 1(a). The simulated robot model is the e-puck robot [8]. It has two differential wheels, a directional colour camera pointing forward, and eight proximity sensors distributed around the robot's perimeter. Each robot executes an identical finite-state machine. Transitions between states are triggered based on perceptual cues [6]. The robots reside in a bounded arena of $5\,\text{m} \times 5\,\text{m}$ dimensions. The object to be transported is a blue cube which is tall enough to occlude the target location. It is placed in the centre of the arena. The target location, a red cylinder, is located at a fixed distance. The robots start from random locations within a starting zone. Further details can be found in the on-line supplementary material [1].

Fig. 1(b) shows the results of a series of computational experiments. A trial was considered successful, if the object reached the target location (5 cm tolerance) within 1200 s. The figure shows that there was a benefit from increasing the number of robots up to 64. Visual inspection indicated that the controller succeeded in allocating the different roles to the robots as intended (for video recordings, see [1]). The transport behaviour was reliable as long as there was a sufficient number of robots involved. However, in most cases, the box moved on average by about 1 cm/s only. By comparison, the robots can move with up to 12.8 cm/s when having no load. In the future, the finite state machine model will be refined and the role allocation will be studied in more detail.

---

[1] http://bulletphysics.org/ and http://home.gna.org/enki/

R. Groß et al. (Eds.): TAROS 2011, LNAI 6856, pp. 380–381, 2011.
© Springer-Verlag Berlin Heidelberg 2011

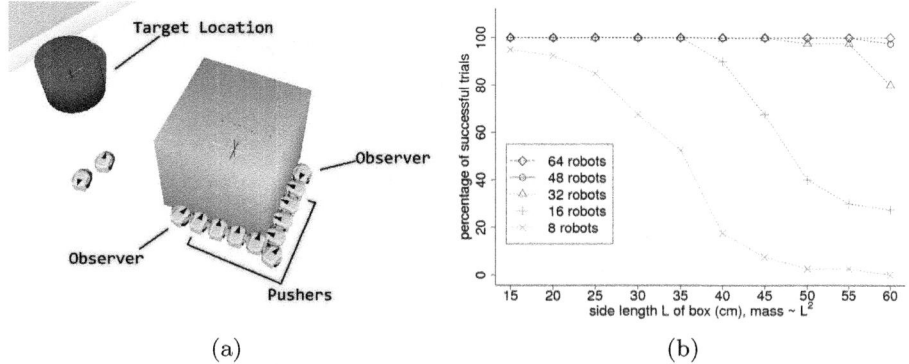

Fig. 1. (a) Robots that have pushed a cuboid object near to the target location. (b) Percentage of successful trials for 8, 16, 32, 48 and 64 robots transporting objects of different size/mass (40 trials per data point).

Acknowledgements. This research was supported by a Marie Curie European Reintegration Grant within the 7th European Community Framework Programme (grant no. PERG07-GA-2010-268354).

References

1. Chen, J., Groß, R.: Online supplementary material,
   http://naturalrobotics.group.shef.ac.uk/supp/2011-002
2. Gerkey, B.P., Matarić, M.J.: Sold!: Auction methods for multirobot coordination. IEEE Trans. Robot. Autom. 18(5), 758–768 (2002)
3. Groß, R., Dorigo, M.: Cooperative transport of objects of different shapes and sizes. In: Dorigo, M., Birattari, M., Blum, C., Gambardella, L.M., Mondada, F., Stützle, T. (eds.) ANTS 2004. LNCS, vol. 3172, pp. 106–117. Springer, Heidelberg (2004)
4. Groß, R., Dorigo, M.: Evolving a cooperative transport behavior for two simple robots. In: Liardet, P., Collet, P., Fonlupt, C., Lutton, E., Schoenauer, M. (eds.) EA 2003. LNCS, vol. 2936, pp. 305–316. Springer, Heidelberg (2004)
5. Kube, C.R., Zhang, H.: Collective robotics: From social insects to robots. Adapt. Behav. 2(2), 189–218 (1993)
6. Kube, C.R., Zhang, H.: Task modelling in collective robotics. Auton. Robot. 4(1), 53–72 (1997)
7. Matarić, M.J., Nilsson, M., Simsarian, K.T.: Cooperative multi-robot box-pushing. In: Proc. of the 1995 IEEE/RSJ Int. Conf. on Intelligent Robots and Systems, vol. 3, pp. 556–561. IEEE Computer Society Press, Los Alamitos (1995)
8. Mondada, F., Bonani, M., Raemy, X., Pugh, J., Cianci, C., Klaptocz, A., Magnenat, S., Zufferey, J.C., Floreano, D., Martinoli, A.: The e-puck, a robot designed for education in engineering. In: 9th Conf. on Autonomous Robot Systems and Competitions, vol. 1, pp. 59–65. IPCB: Instituto Politécnico de Castelo Branco (2009)

Cooperative Navigation and Integration of a Human into Multi-robot System

Joan Saez-Pose[1], Amir M. Naghsh[1], and Leo Nomdedeu[2]

[1] Sheffield Hallam University, Sheffield, UK
[2] Universitat Jaume, Castello, Spain
a.naghsh@sheffield.ac.uk

In the recent years, considerable research efforts have been conducted towards the control of a group of autonomous robots. A human moving cooperatively with a group of robots could as well broadly benefit in several applications (e.g. search and rescue operations). The key focus of this study is to investigate the inclusion of a human in the multi-robot system and consequently the robot motion coordination.

The multi-robot system used in this study [5] has taken inspiration and useful insight for cooperative movement of human with a group of robots from the extensive literature in this field [4, 6]. This system considers that a robot of the group is capable of detecting and identifying the relative positions of other *robot mates* and the *obstacles* of the environment. The social potential function definitions and the commutation of individual robot's motion (direction and speed) are described in details at [5]. The laser range finder information is used to detect and track a human walking in front of a mobile platform. Given the initial position of the human within the field of view of the robot, the method searches in a specific area for an object of a certain width. This object represents the legs of the human for the robot to follow. A more detailed explanation of detection/following algorithms can be found in [3]

The multi-robot system uses a distributed architecture in which the system is able to share the robots and human poses through the whole team, allowing grouping techniques to take advantage of these essential data. In cases where the robots identify specific safe routes they provide information for the fire fighter to employ and act on at their discretion. Based on recommended direction, the fire fighter's pose and direction is calculated and presented to him using a prototype of Light Array Visor (LAV) which was developed following good design practice of user centered design and was conducted with fire fighters [2].

To test the performance of the control strategy, a simulation was carried out using the simulation software Player/Stage. Models of the real ERA-MOBI platform and the Hokuyo range finder device were used and the human was simulated as an independent agent that may move freely in the environment. In total five different trials were performed, both in simulation and in the real-world. The experiments consisted in placing the mobile robots at different starting positions and situating the human at the same starting point for all the different trials. At start of each trial some time was given to the robots to reach their stable formation and became motionless. Once the formation was stable then the human started to move and follow the instructions through the LAV interface. The interface indicated when the human had reached its final point (on average 7 meters of distance was travelled) and at this point the trial was considered as over.

R. Groß et al. (Eds.): TAROS 2011, LNAI 6856, pp. 382–383, 2011.
© Springer-Verlag Berlin Heidelberg 2011

In order to compare the results some typical metrics had to be chosen [1]. *(A) Percentage of time out of formation, %tof. to evaluate the stability performance of the robot system. (B) System Efficiency, SE. to evaluate the average distance travelled by the robots divided by the distance travelled by the human.*

Figure 1 (right) show the explanatory metrics of the simulation and real-world experiments. The simulation results show an average of time out of formation of 7.108% and system efficiency of 1.02 which indicates that the human and the robots are mostly in formation and both travel the same amount of distance. In the real-world experiments the results are satisfactory although the percentage of time out of formation increased to 22.32% while the system efficiency decreases to 0.96. The overall results in figure 1 indicate that the real-world trials performed satisfactory close to the simulated experiment.

	Simulation		Real-World	
	◇	□	◇	□
Trial 1	0.93	8.21%	0.87	20.09%
Trial 2	1.09	6.95%	1.01	17.69%
Trial 3	1.08	7.93%	0.99	36.59%
Trial 4	0.97	5.52%	0.93	16.50%
Trial 5	1.04	6.93%	1.04	20.86%
Mean	1.022	7.108%	0.968	22.332%
std.dev.	0.069	1.057%	0.068	8.165%

◇ SE   □ %tof

Fig. 1. Experimental snapshot (left) Trail results (right)

This study shows that no global positioning system is required for inclusion of the human user in a multi-robot system. However, further work is required to investigate the effectiveness and robustness of this approach over longer distances.

References

1. Balch, T., Arkin, R.: Behavior-based formation control for multirobot teams. IEEE Transactionson Robotics and Automation 14(6), 926–939 (1998)
2. Naghsh, A., Roast, C.: User interfaces for robots swarm assistance in emergency settings. In: BCS HCI 2009: Proceedings of the 2009 British Computer Society Conference on Human- Computer Interaction, UK, pp. 324–328 (2009)
3. Nomdedeu, L., Sales, J., Cervera, E., Alemany, J., Sebastia, R., Penders, J., Gazi, V.: An experiment on squad navigation of human and robots. In: 10th International Conference on Control, Automation, Robotics and Vision, ICARCV 2008, pp. 1212–1218 (2008)
4. Reif, J., Wang, H.: Social potential fields: A distributed behavioral control for autonomous robots. Robotics and Autonomous Systems 27, 171–194 (1999)
5. Saez-Pons, J., Alboul, L., Penders, J., Nomdedeu, L.: Multi-robot team formation control in the guardians project. Industrial Robot: An International Journal 37(4), 372–383 (2010)
6. Yamaguchi, H., Arai, T.: A distributed navigation strategy for multiple mobile robots to make group formations adapting to geometrical constraints. JSME International Journal Series C Mechanical Systems, Machine Elements and Manufacturing 45(3), 758–766 (2002)

Coordination in Multi-tiered Robotic Search

Paul Ward and Stephen Cameron

Oxford University, Computer Science Department, Oxford, UK
{Paul.Ward,Stephen.Cameron}@cs.ox.ac.uk

1 Introduction

Over recent years robot vehicle technology has matured to allow groups of mobile robots to be simultaneously deployed as *Multi-Robot Systems* (MRS), whose advantages include robustness to failures, reduced completion time and the ability to tackle more complex goals. A major design consideration for heterogeneous groups is effective coordination; often handled using task allocation procedures.

Task allocation in MRS has been studied for some time; the work on area exploration is of particular relevance to our own [3,5]; other examples include [7,1,6]. Our work has also been heavily influenced by the post-disaster scenario found in the RoboCup Rescue Virtual Robots Competition [2].

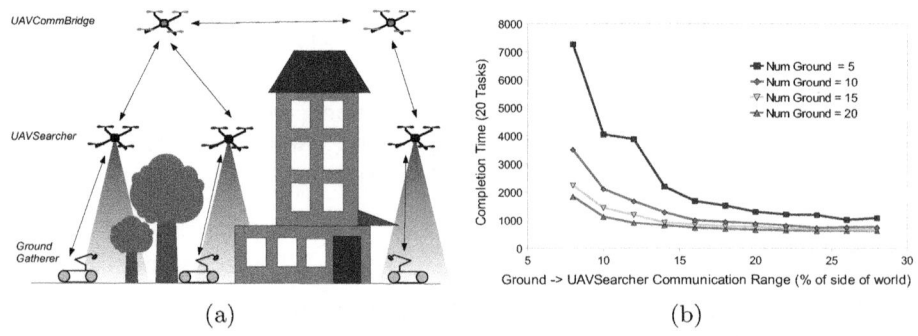

| (a) | (b) |

Fig. 1. (a) Three Tier Multi-Robot Search, comprising 2 tiers of rotorcraft *Unmanned Aerial Vehicles* (UAVs) and 1 tier of ground-based robots, (b) Average time to complete 20 tasks for a variety of robot group sizes with varying communication ranges

We propose describing a MRS in terms of the number of classes of robots, or *tiers*, it contains. For example, the hypothetical searching system sketched in Figure 1(a) contains 3 tiers of (i) aerial communication bridges; (ii) aerial searchers; and (iii) ground-based investigators. This designation looks from a system design perspective, where the participants are designed to perform one of a set of specific functions within a multi-agent scenario, each of which will be labeled a tier, and coordination mechanisms can be described in terms of tiers rather than individuals. For clarity, this way of categorising systems is not the same as general agent heterogeneity, which implies an arbitrary variation

R. Groß et al. (Eds.): TAROS 2011, LNAI 6856, pp. 384–385, 2011.

amongst the population. In the simplest system containing a single type of agent all with identical purpose, for example a group of identical ground vehicles all performing the same search behaviour, coordination is often just division of work. A common two tier system consists of a search or exploration tier and an exploitation tier, for example the site preparation problem [4] requires firstly the detection of obstacles within an area and subsequent clearance; two tasks that require markedly different capability. A search-based 3 tier system with minimal autonomy is presented in [8] and comprises a ground vehicle, low altitude rotor-craft UAV and fixed wing UAV.

2   Simulation Experiments

A three tier robot system along the lines of Figure 1(a) was described using a custom-built discrete-event simulation of a robotic search scenario. Results from these simulations include the importance of inter-robot communication and how its reduction, from environmental interference for example, can adversely affect overall system performance; an example graph is shown in Figure 1(b).

Acknowledgements. This work was supported by the EPSRC-funded SUAAVE project, grant EP/F064179/1. Further information can be found at http://www.suaave.org.

References

1. Dias, M.B., Dias, M.B., Zlot, R., Zinck, M., Stentz, A., Gonzalez, J.: A versatile implementation of the Traderbots approach for multirobot coordination. In: Proc. of the Int. Conf. on Intelligent Autonomous Systems, IAS (2004)
2. de Hoog, J., Cameron, S., Visser, A.: Dynamic team hierarchies in communication-limited multi-robot exploration. In: IEEE Int. Workshop on Safety, Security & Rescue Robotics (SSRR 2010), Bremen (July 2010)
3. Howard, A., Mataric, M.J., Sukhatme, G.S.: An incremental deployment algorithm for mobile robot teams. In: Proc. of the IEEE/RSJ Int. Conf. on Intelligent Robots and Systems (IROS), pp. 2849–2854 (2002)
4. Parker, L.E., Guo, Y., Jung, D.: Cooperative robot teams applied to the site preparation task. In: Proc. of the 10th Int. Conf. on Advanced Robotics, pp. 71–77 (2001)
5. Rooker, M.N., Birk, A.: Multi-robot exploration under the constraints of wireless networking. Control Engineering Practice 15(4), 435–445 (2007)
6. Rossi, C., Aldama, L., Barrientos, A.: Simultaneous task subdivision and allocation for teams of heterogeneous robots. In: ICRA, pp. 946–951 (2009)
7. Sheng, W., Yang, Q., Tan, J., Xi, N.: Distributed multi-robot coordination in area exploration. Robotics and Autonomous Systems 54(12), 945–955 (2006)
8. Wahren, K., Cowling, I., Patel, Y., Smith, P., Breckon, T.: Development of a two-tier unmanned air system for the MoD grand challenge. In: 24th Int. Unmanned Air Vehicle Systems Conf., UAVS (2009)

Covert Robotics: Improving Covertness with Escapability and Non-Line-of-Sight Sensing

Tom Moore, Richard Ratmansky, Bob Chevalier,
David Sharp, Vincent Baker, and Brian Satterfield

Lockheed Martin Advanced Technology Laboratories, Cherry Hill, NJ, USA
{richard.ratmansky,bob.chevalier,david.sharp,
vincent.baker,brian.satterfield}@lmco.com

Covert robotics is an application that presents many challenges in perception, modeling, and planning due to the need to minimize visibility within the environment while perceiving as much as possible and reacting to appropriate events. We have modified a Pioneer 3-AT robot for night operation and incorporated a microphone array for acoustic sensing of sentries. Extending previous research [1] [2] [3], we have developed a novel method for considering the value of goals with respect to escapability: a measure of the degree that a location affords escape if detection is imminent (Fig. 1). Our simulation analysis demonstrates that using an escapability value for goal selection results in a 25 percent decrease in the time a robot is visible to a sentry.

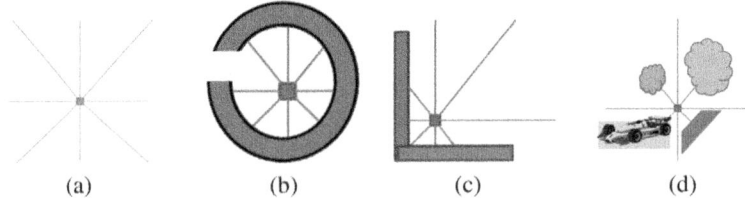

| | | | |
| (a) | (b) | (c) | (d) |

Fig. 1. Scenarios showing (a) poor intervisibility, excellent escapability; (b) excellent intervisibility, poor escapability; (c) good intervisibility, poor escapability; and (d) good escapability, good intervisibility

Our software architecture employs a finite state machine for behaviors such as goal seeking, observing, hiding and waiting. These behaviors are enabled by the assessment of a world map represented as a standard 2D cost grid. The cost map represents all perceptions and prior knowledge of the robot in a multi-layered fashion and is combined by applying a heuristic method of weighting for each map (terrain, light, intervisibility, desirability and escapability) to produce a single cost map that represents the world model. We perform a least cost search through this space using A* to find the optimal goal and path.

To test the effectiveness of the escapability parameter, we conducted experiments in simulation and in the field (Fig. 2). We captured the results from forty-five simulated missions, starting the robot at the same location for each trial and varying the weights for stealth, escapability and desirability. In each trial, the weights were

R. Groß et al. (Eds.): TAROS 2011, LNAI 6856, pp. 386–387, 2011.
© Springer-Verlag Berlin Heidelberg 2011

some multiple of 0.1, and all weights had to sum to 1. When the robot reached a simulated goal location, a sentry was automatically placed in a cell adjacent to it, and the amount of time that the robot was visible to the sentry before escaping was recorded. When the robot had escaped and began its hiding behavior, the sentry was removed and the robot continued its traverse. Our results suggest that the best weighting scheme is 80% stealth, 10% escapability, and 10% desirability (Fig. 3). This follows our intuition that some combination of stealth and escapability are required to minimize the chances of detection. Relying only on escapability results in very poor performance, because robots only interested in choosing very escapable regions typically select wide-open spaces that are far from any hiding spots.

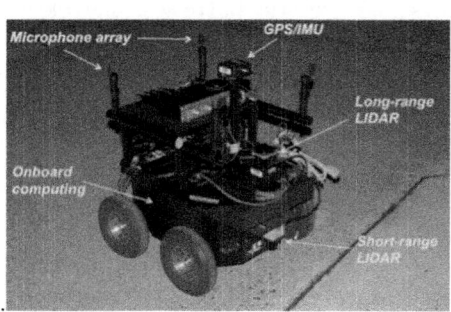

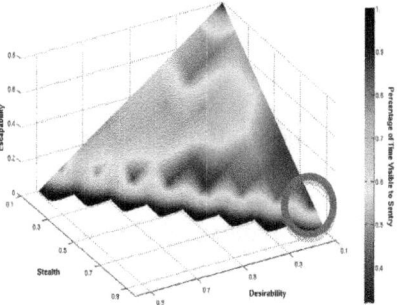

Fig. 2. We modified a Pioneer 3-AT with non-reflective paint and integrated LIDAR, GPS/IMU and audio sensors

Fig. 3. Analysis shows the importance of escapability as a cost factor. The circled area in the figure represents the optimal distribution of weights and results in the least time visible to sentries.

Future work would include modifying the robot's sensor configuration to add seismic or vision capabilities to allow the robot to make better choice in goal selection by considering terrain, preventing silhouettes against backlit areas and detecting dynamic light sources.

References

1. Tews, A.D., Sukhatme, G.S., Mataric, M.J.: A Multi-Robot Approach to Stealthy Navigation in the Presence of an Observer. In: Proc. IEEE Int. Conf. Robotics and Automation, pp. 2379–2385. IEEE, Los Alamitos (2004)
2. Marzouqi, M.S., Jarvis, R.A.: Covert Robotics: Covert Path Planning in Unknown Environments. In: Proc. of the Australasian Conference on Robotics and Automation (2004)
3. Marzouqi, M.S., Jarvis, R.A.: Covert Robotics: Hiding in Known Environments. In: Proc. IEEE Conf. on Robotics, Automation and Mechatronics, pp. 804–809. IEEE, Los Alamitos (2004)

Designing Electric Propulsion Systems for UAVs

Mohamed Kara Mohamed, Sourav Patra, and Alexander Lanzon

Control Systems Centre, EEE, The University of Manchester, Manchester, UK
Mohamed.Karamohamed@postgrad.manchester.ac.uk,
{Sourav.Patra,Alexander.Lanzon}@manchester.ac.uk

1 Introduction

Due to the advantages of electric motors and electric batteries, the electric propulsion systems are widely used in UAVs [1,2]. The UAV's performance and endurance, the payload capacity and the flight time of the vehicle are affected highly by the structure and design of the electric propulsion system. In this paper, a new systematic design methodology for electric propulsion system (propeller, electric motor and battery pack) is proposed based on the design specifications given in terms of the required thrust, permissible propulsion system's weight and required flight time.

For design, a mathematical model of the propeller is considered to quantify the generated thrust and the corresponding mechanical power. Using the momentum theory [3] with certain assumptions, the static thrust T developed by a propeller and the mechanical power P needed to generate this thrust are given respectively as $T = \frac{\rho}{128\pi} N^2 d^2 k_1^2 \omega^2 R^2 ((1 + \frac{64\pi R}{3Ndk_1}\theta)^{0.5} - 1)^2$ and $P = \frac{\rho}{2048\pi^2} N^3 d^3 k_1^3 \omega^3 R^2 ((1 + \frac{64\pi R}{3Ndk_1}\theta)^{0.5} - 1)^3$, where ρ, N and d are respectively the air density, the number of blades and the chord length of the blade of the propeller. k_1 is known as the two-dimensional lift slope factor and is considered as $k_1 = 5.7$ [4]. R, θ and ω are respectively the radius, the pitch angle, and the rotational speed of the propeller.

2 Design Methodology

Inputs to the design algorithm: (i) the maximum allowable radius of the propeller R_{max}, (ii) the total mass requirement of the UAV M_{total} and the allowance for the propulsion system mass M_p, and (iii) the required minimum flight time t_f.

Step 1: Set the required thrust $T_h = \alpha g M_{total}$, where $\alpha > 1$ is a safety factor to be chosen by the designer (e.g. $\alpha = 1.2$).

Step 2: Selecting a set of propellers: **2a)** Choose a set of commercially available propellers $\mathbb{Y}$ whose radii are $R_y \leq R_{max}\ \forall\ y \in \mathbb{Y}$. **2b)** For each propeller $y \in \mathbb{Y}$, calculate the minimum rotational speed ω_y and the corresponding minimum mechanical power P_y (using the expression given in Section 1) necessary to generate the required given thrust T_h. A propeller y is infeasible for the design if $\omega_y > \omega_{max_y}$, where ω_{max_y} is the maximum allowed rotational speed of the propeller y (specified by the manufacturer) and hence the propeller y must be excluded from the design process. Let $\overline{\mathbb{Y}}$ the set of all feasible propellers. 2c) Over

R. Groß et al. (Eds.): TAROS 2011, LNAI 6856, pp. 388–389, 2011.

all propellers $y \in \overline{\mathbb{Y}}$, find the minimum rotational speed and minimum mechanical power necessary to generate the required thrust; i.e., $\omega_{min} = \min_{y \in \overline{\mathbb{Y}}}(\omega_y)$ and $P_{min} = \min_{y \in \overline{\mathbb{Y}}}(P_y)$.

Step 3: Selecting a set of motors: **3a)** Select a set of commercially available BLDC motors $\mathbb{Z}$ such that $\forall \ z \in \mathbb{Z}$ the following conditions are fulfilled: $P_{max_z} \geq P_{min}, \omega_{max_z} \geq \omega_{min}$ and $M_z < M_p$ where $P_{max_z}, \omega_{max_z}$ and M_z are respectively the power rate, the maximum rotational speed and the mass of the motor z. **3b)** Construct motor-propellers groups $G_j, j = 1, 2, .., n(\mathbb{Z})$, where $n(\mathbb{Z})$ denotes the number of the motors in the set $\mathbb{Z}$. The jth group G_j contains a motor $z_j \in \mathbb{Z}$ and a subset of propellers $\mathbb{I}_j \subseteq \overline{\mathbb{Y}}$, where $\mathbb{I}_j := \{y \in \overline{\mathbb{Y}} : \omega_y \leq \omega_{max_{z_j}}, P_y \leq P_{max_{z_j}}\}$. **3c)** For the jth group G_j, calculate V_y^j and $I_y^j \ \forall \ y \in \mathbb{I}_j$, where $V_y^j = \frac{\omega_y}{k_{v_{z_j}}}$ and I_y^j is obtained from the operational chart of the motor z_j. V_y^j and I_y^j are respectively the required voltage and current for the motor z_j to rotate the propeller y at the minimum speed ω_y necessary to generate the required thrust. In group G_j, the pair (z_j , y), where $y \in \mathbb{I}_j$, is feasible for the design if $I_y^j \leq I_{max_{z_j}}$, where $I_{max_{z_j}}$ is the maximum allowed continuous current of the motor z_j (specified by the manufacturer). Select all feasible pairs in G_j, $j = 1, 2, ..., n(\mathbb{Z})$.

Step 4: Selecting a set of batteries: **4a)** For each feasible pair $(z_j , y) \in G_j$, select a set of commercially available battery packs $\mathbb{B}_y^j$ such that $\forall \ b \in \mathbb{B}_y^j$ the following conditions are fulfilled: $V_b \geq V_y^j$, $I_{max_b} \geq I_y^j$ and $M_b \leq M_p - M_{z_j}$, where V_b and I_{max_b} are respectively the effective voltage and maximum continuous discharging current of the battery $b \in \mathbb{B}_y^j$, and M_b is the mass of the corresponding battery pack. **4b)** $\forall \ b \in \mathbb{B}_y^j$, calculate the mass of the propulsion system (z_j , y, b) and the full load (i.e. minimum) flight time as: $M_{z_j,y,b} = M_{z_j} + M_b$ and $t_{z_j,y,b} = \frac{I_{0_b}}{I_y^j}$, where $(z_j , y) \in G_j$ is a feasible pair and I_{0_b} is the current rate (A.h) of the battery $b \in \mathbb{B}_y^j$. If $t_{z_j,y,b} < t_f$, the battery pack b cannot provide the required flight time when used with the pair (z_j , y) and must be excluded from $\mathbb{B}_y^j$. **4c)** Calculate $M_{z_j,y,b}$ and $t_{z_j,y,b}$ for all feasible pairs $(z_j , y) \in G_j$, $j = 1, 2, ..., n(\mathbb{Z})$.

Step 5: Based on the values of $M_{z_j,y,b}$ and $t_{z_j,y,b}$ for all feasible pairs, we can choose the best design to achieve maximum flight time (z_t , y_t , b_t) or minimum propulsion system weight (z_w , y_w , b_w) or a design that satisfy a trade off between the two factors.

References

1. Freddi, A., Lanzon, A., Longhi, S.: A feedback linearization approach to fault tolerance in quadrotor vehicles. In: Proceedings of The 2011 IFAC World Congress, Milan, Italy (2011)
2. Moir, I., Seabridge, A.: Aircraft Systems: Mechanical, Electrical and Avionics Subsystems Integration, 3rd edn. Wiley, Chichester (2008)
3. Phillips, W.F.: Mechanics of Flight. Wiley & Sons, Inc., Chichester (2004)
4. Seddon, J., Newman, S.: Basic Helicopter Aerodynamics, 2nd edn. Blackwell Science, Malden (2001)

Enhancing Self-similar Patterns by Asymmetric Artificial Potential Functions in Partially Connected Swarms

Giuliano Punzo, Derek Bennet, and Malcolm Macdonald

Advanced Space Concepts Laboratory, University of Strathclyde, Glasgow, UK
{giuliano.punzo,derek.bennet,
malcolm.macdonald.102}@strath.ac.uk

The control of mobile robotic agents is required to be highly reliable. Artificial potential function (APF) methods have previously been assessed in the literature for providing stable and verifiable control, whilst maintaining a high degree of non-linearity. Further, these methods can, in theory, be characterised by a full analytic treatment. Many examples are available in the literature of the employment of these methods for controlling large ensembles of agents that evolve into minimum energy configurations corresponding in many cases to regular lattices [1-2]. Although regular lattices can present naturally centric symmetry and self-similarity characteristics, more complex formations can also be achieved by several other means. In [3] the equilibrium configuration undergoes bifurcation by changing a parameter belonging to the part of artificial potential that couples the agents to the reference frame. In this work it is shown how the formation shape produced can be controlled in two further ways, resulting in more articulated patterns. Specifically the control applied is to alter the symmetry of interactions amongst agents, and/or by selectively rewiring inter-agent connections. In the first case, the network of connections remains the same, and may be fully connected. In the second some links are rewired with possible changes of APF parameters, this can be better understood considering a group of 5 mobile agents interacting through Morse-like potentials, which C_a, C_r, l_a, l_r to defined the potential shaping parameters and $\left| x_{ij} \right|$ as the inter-agent distance is defined as;

$U_{ij} = -C_a \exp\left(-\left|x_{ij}\right|/la\right) + C_r \exp\left(-\left|x_{ij}\right|/lr\right)$. When all the agents are subjected to the same APF the swarm will relax into the minimum energy configuration, a circle. This configuration can be modified by tuning the attraction or repulsive scale distance, l_a or l_r, in one agent. Consider a change in the l_r parameter for a agent and say that for this agent the parameter takes value $l_{rr}<l_r$, creating an asymmetry in the global potential field. As the difference between the two increases, the minimum energy level corresponding to a circle formation turns to be higher, and hence less favourable, than the one corresponding to a cross formation which the swarm will then naturally evolve into. This can be better understood by looking at Figure 1 where the global artificial potential $U = \sum_{i,j} U_{ij}$ as function of a characteristic distances and of the free parameter l_{rr} is shown. Consider now the case of 5 groups, each as the one just discussed, giving 25 agents in total. Linking the centres of all the groups together with the same attraction-repulsion function coefficients would lead again to 5 groups (each one cross shaped) self arranging into a circle.

R. Groß et al. (Eds.): TAROS 2011, LNAI 6856, pp. 390–391, 2011.

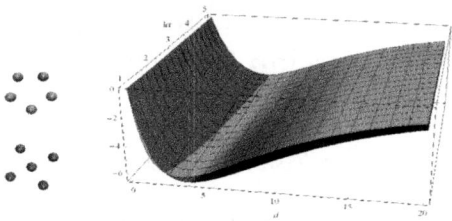

Fig. 1. Global artificial potential as function of the inter-agent characteristic distance and of the free parameter l_r for the circle configuration (green) and the cross configuration (red). The characteristic distance considered is the radius (from the centre to a peripheral agent) for both configurations

On the other hand, once again adjusting the same free parameter as previously shown, one group will move to the centre, surrounded by the other 4, however this will not guarantee that all the groups keep the same relative orientation. It is otherwise possible to organize the 5-agent groups into a cross formation as the agents in the groups spontaneously relax. This can be achieved by adding links between each of the four other agents of one group (that will be the central one) to an agent other than the central one belonging to the other groups. This will guarantee that 5 groups of 5 agents organize amongst them in exactly the same way as the agents which they are composed of do amongst themselves. The final outcome is a self-similar pattern at both agent and group level, as seen in Figure 2. Furthermore, as the second level agents are connected in couples, the separation distance corresponding to the minimum energy configuration can be calculated exactly from the analytic expression of the Morse potential, giving fully verifiable swarm shape and size control.

Fig. 2. 25 agent self-similar formation obtained by linking the central agents of each group amongst them plus one agent for each side group to one "arm" of the central group

References

1. Izzo, D., Pettazzi, L.: Equilibrium shaping: Distributed motion planning for satellite swarm. European Space Agency, Special Report ESA SP (Compendex), pp. 727–736 (2005)
2. Gazi, V., Passino, K.M.: Stability analysis of swarms. IEEE Transactions on Automatic Control 48(Compendex), 692–697 (2003)
3. Bennet, D.J., McInnes, C.R.: Pattern transition in spacecraft formation flying via the artificial potential field method and bifurcation theory. In: Proceedings of 3rd International Symposium on Formation Flying, Missions and Technologies, ESA-ESTEC, Noordwijk, Netherlands (2008)

Evolving Modularity in Robot Behaviour Using Gene Expression Programming

Jonathan Mwaura and Ed Keedwell

College of Engineering, Mathematics and Physical Sciences,
University of Exeter, Exeter, UK
{jm329,e.c.keedwell}@ex.ac.uk

Incremental learning [3] and layered learning [4] have been proposed as suitable approaches to improve evolutionary robotic (ER) algorithms by subdividing the required behaviour into simpler tasks. However, incremental learning does not divide the controller to unique task modules and although layered learning subdivides the problem into modules it does not offer continuous learning for the various sub-behaviours. Moreover, both methods involve the modification of the fitness function in every module thus increasing computational overhead.

In this paper we use gene expression programming (GEP) [1] to evolve multiple genes where each gene corresponds to a layer in the layered architecture. Results are presented on a number of wall following tasks using unigenic GEP (ugGEP) and multigenic GEP (mgGEP) and a comparison is made with similar behaviour evolved using standard genetic programming (GP) [2].

The main aims of the experiments were as follows; firstly, to compare GEP and GP in solving a wall following problem with increasing complexity, and secondly to determine whether the multigenic nature of GEP can be used to evolve a layered controller architecture. The experimentations and parameters used are based on [2].

The algorithm was run 50 times for both algorithms and each room. In mgGEP, a logical IF was used as the linker with three arguments. The linker arbitrates between the two genes by checking the reading of the front sensor. If F= 0, i.e. no obstacle, then gene one is used to effect the motor output, however if F =1, meaning an obstacle is detected, then gene 2 is executed.

It is clear from Table 2 that GEP in general performs better than GP on this task, resulting in far fewer evaluations on Rooms 2 & 4 and narrowly incurring more evaluations on Room 5. Figure 1 shows that mgGEP performed better on average than ugGEP in solving the problem. The success of mgGEP in this domain could be explained by two reasons; Firstly, the increased diversity in its chromosome due to gene transposition and gene recombination [1]. Secondly, mgGEP is able to divide the task into its two constituent elements, obstacle avoidance and wall following.

Table 1. Comparison of average evaluations (best result in bold)

	Room 2	Room 3	Room 4	Room 5
ugGEP	30616.67	**19871.43**	19425	15825.58
mgGEP	**27611.11**	21652.78	**18089.74**	19885.42
Lazarus[2]	90000	20050	50000	**14000**

R. Groß et al. (Eds.): TAROS 2011, LNAI 6856, pp. 392–393, 2011.
© Springer-Verlag Berlin Heidelberg 2011

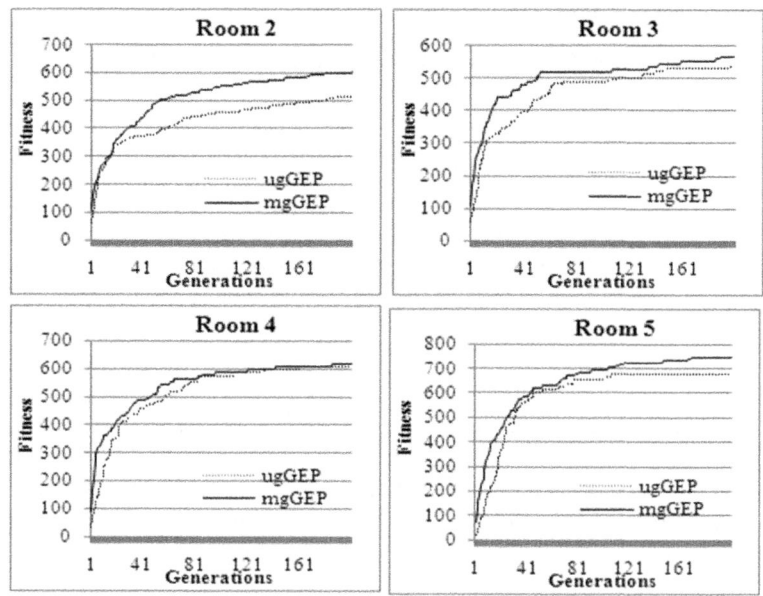

Fig. 1. Comparison of ugGEP and mgGEP performances across rooms 2-5

This paper shows that mgGEP performed better than both GP and ugGEP in solving the wall following problem. The technique evolved is one of manually dividing the problem and then allowing each gene to solve specific tasks. The overall behaviour then emerges due to the interaction of the two genomes and the environment.

References

[1] Ferreira, C.: Gene Expression programming: A new Adaptive Algorithm for Solving Problems. Complex Systems 13(2), 87–129 (2001)
[2] Lazarus, C., Hu, H.: Using Genetic Programming to Evolve Robot Behaviours. In: Proceedings of the 3rd British Conference on Autonomous Mobile Robotics and Autonomous Systems, Manchester, UK (2001)
[3] Nolfi, S., Floreano, D.: Evolutionary Robotics: The Biology, Intelligence, and Technology of Self-Organizing Machines. The MIT Press, Cambridge (2000)
[4] Togelius, J.: Evolution of Subsumption Architecture Neurocontroller. J. Intelligent Fuzzy System 15, 15–20 (2004)

Forming Nested 3D Structures Based on the Brazil Nut Effect

Stephen Foster and Roderich Groß

Natural Robotics Lab,
Department of Automatic Control and Systems Engineering,
The University of Sheffield, Sheffield, UK
stephen@sdfoster.co.uk, r.gross@sheffield.ac.uk

This study investigates the formation of nested structures in swarms of intelligent agents that can freely move in three dimensions. The underlying segregation mechanism is inspired by the Brazil nut effect, which occurs when granular mixtures are subjected to vibrations [6,1]. Similar effects were reported for brood items sorted by ants [3]. This sorting behaviour was validated with swarms of mobile robots [7,5]. Different from these studies we are concerned with sorting the agents themselves. In [4], we proposed a controller based on the Brazil nut effect that was capable of segregating groups of simulated e-puck robots reliably in two dimensions. In the present study, we investigate a 3D particle system implemented in NetLogo[1]. The agents mimic the behaviour of particles of distinct sizes. The motion of each agent is determined by three types of vectors [4]: (i) a repulsion vector for every agent that intrudes the particle's virtual body, (ii) a random vector simulating vibrations, (iii) and a "gravitational" vector that points to a "centre" location. The agents do not communicate. However, the repulsion behaviour requires them to sense each others' relative positions within their particle range. The segregation quality is measured as the percentage of pairs of particles from different groups that are segregated correctly (based on distance to centre) [4]. A value of 100% corresponds to perfect structures with all large agents surrounding the small agents, 50% corresponds to purely random structures, whereas 0% corresponds to perfect but inverted structures.

The structure formed by 2 groups of 50 agents was investigated for different particle size ratios [see Fig. 1(a)]. For ratio 1.00, all agents behave identical and hence no segregation occurred. For ratios near 0, structures achieved consistently a segregation quality of about 100%. The study was expanded to include a third group with results that matched the ones obtained with two groups [see Fig. 1(b)]. We observed the formation of asymmetric structures when using group specific centres of gravity [see Fig. 1(c)]. Video recordings are available in [2].

The formation of 3D structures could be useful in medical, military and space applications. Future work will have to investigate a physical implementation.

---

[1] http://ccl.northwestern.edu/netlogo

R. Groß et al. (Eds.): TAROS 2011, LNAI 6856, pp. 394–395, 2011.
© Springer-Verlag Berlin Heidelberg 2011

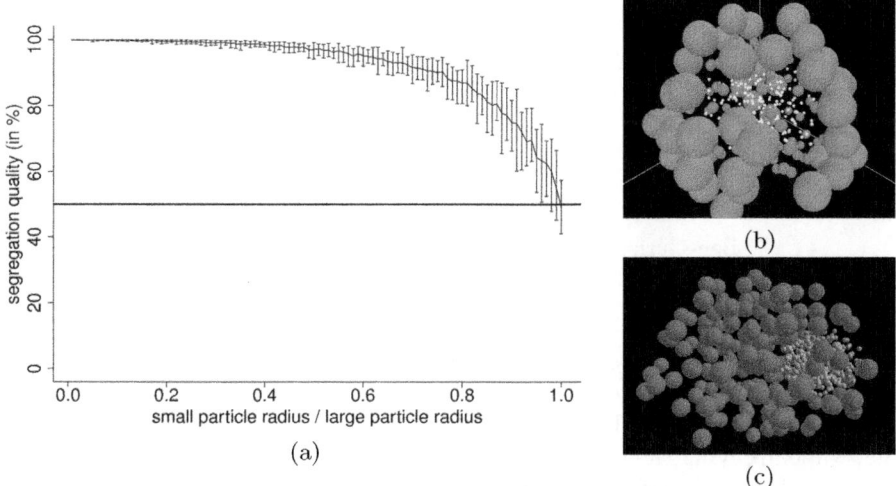

(a)

(b)

(c)

Fig. 1. (a) Mean segregation quality as observed for two groups of particles (30 trials per particle size ratio). Arrows stretch from the smallest to the largest observation. (b) Three groups of particles organise into a nested structure. The smallest particles form a sphere located in the centre. The sphere is surrounded by a layer of medium sized particles, which in turn is surrounded by a layer of large particles. (c) Asymmetric structures formed by two groups of particles with different gravitational centres.

Acknowledgements. This research was supported by a Marie Curie European Reintegration Grant within the 7th European Community Framework Programme (grant no. PERG07-GA-2010-268354).

References

1. Barker, G., Grimson, M.: The physics of muesli. New Scientist 126(1718), 37–40 (1990)
2. Foster, S., Groß, R.: Online supplementary material (2011),
   http://naturalrobotics.group.shef.ac.uk/supp/2011-003
3. Franks, N.R., Sendova-Franks, A.B.: Brood sorting by ants: Distributing the workload over the work-surface. Behav. Ecol. Sociobiol. 30(2), 109–123 (1992)
4. Groß, R., Magnenat, S., Mondada, F.: Segregation in swarms of mobile robots based on the Brazil nut effect. In: Proc. of the 2009 IEEE/RSJ Int. Conf. on Intelligent Robots and Systems, IROS 2009, pp. 4349–4356. IEEE Computer Society Press, Los Alamitos (2009)
5. Melhuish, C., Sendova-Franks, A.B., Scholes, S., Horsfield, I., Welsby, F.: Ant-inspired sorting by robots: The importance of initial clustering. J. Roy. Soc. Interface 3(7), 235–242 (2006)
6. Rosato, A., Strandburg, K.J., Prinz, F., Swendsen, R.H.: Why the Brazil nuts are on top: Size segregation of particulate matter by shaking. Phys. Rev. Lett. 58(10), 1038–1040 (1987)
7. Wilson, M., Melhuish, C., Sendova-Franks, A.B., Scholes, S.: Algorithms for building annular structures with minimalist robots inspired by brood sorting in ant colonies. Auton. Robot. 17(2-3), 115–136 (2004)

Grasping of Deformable Objects Applied to Organic Produce

Alon Ohev-Zion and Amir Shapiro

Department of Mechanical Engineering,
Ben-Gurion University of the Negev, Be'er-Sheva, Israel
{alonohev,ashapiro}@bgu.ac.il

Grasping mechanisms (*grippers*) are used for a wide range of applications including fixturing arrangements, industrial, agricultural, and service robotics for medical and home use. The gripper must immobilize the object it is manipulating, while applying the minimal necessary grasping force, in order to prevent the grasped object's bruising. Performing grasp analysis and synthesis, requires the development of a grasp model.

The grasp model includes kinematic, dynamic and contact models, where the last define the coupling between the contact forces and their deformations [4]. All grasp models greatly depend on the system's parameters, such as the number of contact points, the object's geometry, the gripper's structure, the desired object's manipulation, and the mechanical properties of the object and gripper [3].

Contact models were developed in order to assess and predict the reaction forces and local deformations. They vary from considering rigid components and sharp fingertips with no friction at the contact point, to soft finger models – assuming compliant components, which imply a contact area with a force and torque distribution [1,5]. Note that the contact's location is referred as a *contact point*, although geometrically it is a *contact patch* in some of the discussed compliant contact models.

The goal of this work is to provide the set of grasp forces and torques that can be applied to an object, without losing equilibrium and stability constraints. The grasp planner, provides an optimal set of positions for n contact points, which are optimal in the sense of minimizing contact force and torque at each contact point. The planning is independent in the selection of a specific gripper, as oppose to more traditional techniques [2]. Under the constraints of maintaining grasp's stability and not bruising the object, two grasp – quality criteria are currently in consideration: 1) The minimal required grasping force and torque. 2) The maximal allowed grasping force and torque.

In order to achieve this goal, the following assumptions and definitions are made: consider a virtual gripper, whose n hemisphere fingertips can be positioned at every point in space, and actuated at any desired direction. The fingertips and the object are considered as compliant (quasi-rigid) [1,6]. The object's B geometry and mechanical properties are known, and it may be subjected to a known set of external forces and torques (*loads*) $\tilde{F}$. The object can be grasped by a set of grasp configurations $\tilde{K}$, where each configuration is defined by the locations of the n contact points on B's circumference.

R. Groß et al. (Eds.): TAROS 2011, LNAI 6856, pp. 396–397, 2011.

The optimal grasp – planing objective is described by (1), and as follows: Given an object B, grasped by n contact points positioned on its circumference in a specific locations – that is a "grasp configuration", loaded by a specific external load in $\tilde{F}$, search B's circumference for an optimal grasp – quality measure. The search is comprise two stages. First, evaluate a grasp – quality measure for the current grasp configuration, which is follows by the evaluation of a grasp – quality measure for the current grasp configuration, for all loads in $\tilde{F}$. This algorithm can be analytically described as,

$$\text{Optimal Grasp Configuration} = \arg\min_{\tilde{\mathbf{K}}} \left(\min_{\tilde{\mathbf{F}}} \left(\mathbf{J}\left(\tilde{\mathbf{F}}\right)\right)\right), \qquad (1)$$

where J is the grasp – quality evaluation function.

The search is performed in the eight dimensional space, whose parameters are the object's surface geometry coordinates (two), the applied external spatial forces vectors (three) and torques (three). The algorithm output is the locations of the n contact points on the object's circumference that defines the optimal grasp configuration, in the sense of the the grasp quality criterion.

Currently, a virtual environment that enables simulation of the object's dynamics under the integrated contact model is constructed. Both non – linear contact models by Elata [1] and Xydas's [6] are studied for their properties and differences. An experimental system is being built, in order to validate the grasp's analysis and synthesis.

Acknowledgements. Supported by the European Commission in the 7th Framework Programme (CROPS GA no 246252).

References

1. Elata, D., Berryman, J.G.: Contact force-displacement laws and the mechanical behavior of random packs of identical spheres. Journal of Mechanics of Materials 24(3), 229–240 (1996)
2. Goldfeder, C., Allen, P., Lackner, C., Pelossof, R.: Grasp planning via decomposition trees. In: Proceedings of the International Conference on Robotics and Automation, pp. 4679–4684. IEEE, Los Alamitos (2007)
3. Miller, A., Allen, P.: Graspit! a versatile simulator for robotic grasping. IEEE Robotics Automation Magazine 11(4), 110–122 (2004)
4. Shapiro, A., Rimon, E., Burdick, J.W.: On the mechanics of natural compliance in frictional contacts and its effect on grasp stiffness and stability. In: Proceedings of the International Conference on Robotics and Automation, vol. 2, pp. 1264–1269. IEEE, Los Alamitos (2004)
5. Xiong, C., Wang, M., Tang, Y., Xiong, Y.: Compliant grasping with passive forces. Journal of Robotic Systems 22(5), 271–285 (2005)
6. Xydas, N., Kao, I.: Modeling of contact mechanics and friction limit surfaces for soft fingers in robotics, with experimental results. The International Journal of Robotics Research 18(8), 941–950 (1999)

Learning to Grasp Information
with Your Own Hands

Dimitri Ognibene[1], Nicola Catenacci Volpi[3], and Giovanni Pezzulo[2],[*]

[1] Intelligent System Networks, Imperial College London, London, UK
[2] Istituto di Linguistica Computazionale "Antonio Zampolli", CNR, Italy
[3] IMT Institute for Advanced Studies, Lucca, Italy
d.ognibene@imperial.ac.uk

Autonomous robots immersed in a complex world can seldom directly access relevant parts of the environment by only using their sensors. Indeed, finding relevant information for a task can require the execution of actions that explicitly aim at unveiling previously hidden information. Informativeness of an action depends strongly on the current environment and task beyond the architecture of the agent. An autonomous adaptive agent has to learn to exploit the epistemic (e.g., information-gathering) implications of actions that are not architecturally designed to acquire information (e.g. orientation of sensors). The selection of these actions cannot be hardwired as general-purpose information-gathering actions, because differently from sensor control actions they can have effects on the environment and can affect the task execution. In robotics information-gathering actions have been used in navigation [7]; in active vision [4]; and in manipulation [3]. In all these works the informative value of each action was known and exploited at design time while the problem of actively facing un-predicted state uncertainty has not received much attention[1].

We propose a definition of epistemic actions for POMDP that allows to formally frame the problem of task-aware use of information-gathering actions. In classical planning literature Herzig and colleagues [2] characterize epistemic actions by *informativeness* and *non-intrusiveness*. We translate them in POMDP formalism. Cassandra et al. [1] define *action entropy* $AH(b)$ as the entropy of the distribution of the optimal action $w_a(b)$ computed considering $Q^*(s,a)$, the optimal Q-value function associated to the MDP underlying the POMDP. The *expected action entropy* for action a' in belief state b is: $EAH(b,a') = \sum_{b'} \tau(b,a',b')H(w_a(b'))$. The information gain ΔI of executing action a in belief state b can be defined as: $\Delta I(b,a) = AH(b) - EAH(b,a)$. If $\Delta I(b,a) > 0$ then the action a is informative in current belief state b.

We define an action e to be *non-intrusive* in the belief state b if $E[Q(b,e)] = \sum_{b'} \tau(b,e,b') \sum_s b'(s) \max_a Q^*(s,a)$ is equal to the value of the current belief state b $V^*(b) = \sum_s b(s) \max_a Q^*(s,a)$. This formalizes the concept that the execution of a non-intrusive action doesn't change the reward that the agent can

---

[*] This research was funded by the European Projects *HUMANOBS*, contract no FP7-ICT-STREP-231453.
[1] See [5] in which however the information-gathering actions, i.e. directing gaze, are wired to motor actions so that they were selected for their executive value.

R. Groß et al. (Eds.): TAROS 2011, LNAI 6856, pp. 398–399, 2011.
© Springer-Verlag Berlin Heidelberg 2011

receive[2]. Non-intrusiveness allows to disregard long term effects of executing an epistemic action thus permitting to manage more easily the trade-off between the values of information gathered and the cost of executing the epistemic action. This consideration can be more useful if there exists a policy π_e allowing to reduce the action entropy to zero after a finite number of epistemic actions for any belief state. We can name this kind of POMDP *MDP-reducible-POMDP*. They can be solved combining the policy π_e with the optimal policy π^*_{MDP}. Solving this problem instead of solving the POMDP will give a solution which will be less efficient only at most for the time spent and the cost given by π_e. Active vision tasks like those proposed in [8] are elements of the MDP-reducible-POMDP and so are many other active vision tasks where the task relevant uncertainty can be removed by proper selection of actions.

Using this definition, the implications of epistemic actions on learning have been tested in an integrated arm-eye control neural architecture [6] based on the attention-for-action principle and reinforcement learning framework in a task involving the use of reaching as an epistemic action to remove occlusions on visual landmarks. Rewarding the top-down attention controller only when the target is reached produces a strong integration of the attention process with the learning process. This integration allows to learn the task even though the attention for action principle constrains the gazing actions to be executed to support reaching actions while the opposite is required by the task. Still, the experiment showed that the agent can develop easily a scanning behaviour instead of an efficient exploration. To develop the latter it is necessary sometimes to allow the agent to access freely the information usually revealed by the epistemic action.

References

1. Cassandra, A.R.: Exact and Approximate Algorithms for Partially Observable Markov Decision Processes. Ph.D. thesis, Brown University (1998)
2. Herzig, A., Lang, J., Marquis, P.: Action representation and partially observable planning in epistemic logic. In: Proc. of IJCAI 2003 (2003)
3. Hsiao, K., Kaelbling, L.P., Lozano-Perez, T.: Task-driven tactile exploration. In: Proceedings of Robotics: Science and Systems, RSS (2010)
4. Kwok, C., Fox, D.: Reinforcement learning for sensing strategies. In: Proceedings of IROS 2004 (2004)
5. McCallum, A.: Efficient exploration in reinforcement learning with hidden state. In: AAAI Fall Symposium on Model-directed Autonomous Systems (1997)
6. Ognibene, D., Pezzulo, G., Baldassarre, G.: How can bottom-up information shape learning of top-down attention control skills? In: ICDL 2010 (2010)
7. Roy, N., Thrun, S.: Coastal navigation with mobile robots. In: Advances in Neural Information Processing Systems, vol. 12 (2000)
8. Whitehead, S., Lin, L.: Reinforcement learning of non-Markov decision processes. Artificial Intelligence 73(1-2), 271–306 (1995)

---

[2] Note that state uncertainty effects on action selection is not taken into account. Note also that to check epistemicity is not necessary to find the optimal policy and the optimal value function for the POMPD but only for the underlying MPD.

Long-Term Experiment Using an Adaptive Appearance-Based Map for Visual Navigation by Mobile Robots

Feras Dayoub, Grzegorz Cielniak, and Tom Duckett

School of Computer Science, University of Lincoln, Lincoln, UK
{fdayoub,gcielniak,tduckett}@lincoln.ac.uk

Building functional and useful mobile service robots means that these robots have to be able to share physical spaces with humans, and to update their internal representation of the world in response to changes in the arrangement of objects and appearance of the environment – changes that may be spontaneous and unpredictable – as a result of human activities. However, almost all past research on robot mapping addresses only the initial learning of an environment, a phase which will only be a short moment in the lifetime of a service robot that may be expected to operate for many years.

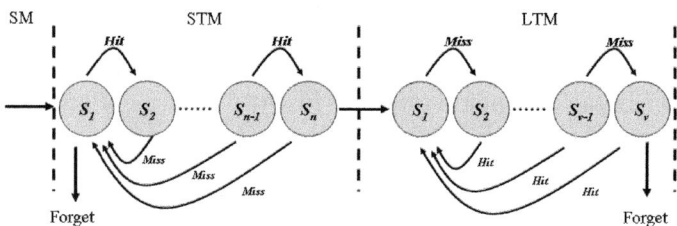

Fig. 1. The proposed multi-store memory model. SM: Sensory memory. STM: Short-term memory. LTM: Long-term memory [1].

Working towards mobile service robots capable of working in changing environment for long periods of time, we proposed in [1] a long-term updating mechanism inspired by the classic modal model of human memory. (See Fig. 1). In further work [2], the memory model was integrated with a hybrid map that represents the global topology and local geometry of the environment. The environment is represented as an adjacency graph of nodes on a topological level, and each node on the metric level of the map represents the 3D location of image features on a sphere. The spherical representation of the nodes creates a connection between the topological and metric level of the map. A group of image features is used as a qualitative descriptor for global localization on the topological level, and the 3D location of these features on the sphere is used for estimating the heading needed for visual navigation between the nodes.

The work presented in [2] focused on how to update the reference views of the map in response to the changes in the appearance of the environment while

R. Groß et al. (Eds.): TAROS 2011, LNAI 6856, pp. 400–401, 2011.

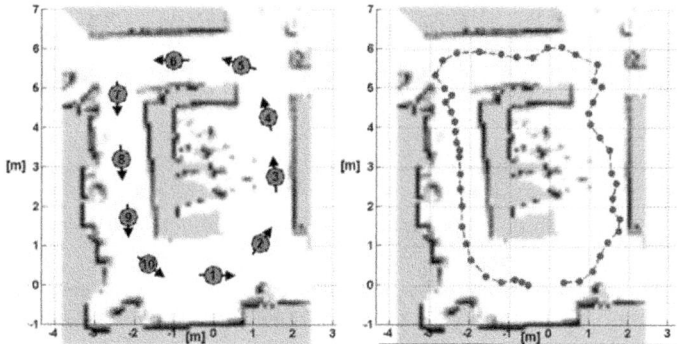

Fig. 2. Left: Occupancy map for the area of the room where the experiments took place along with the position of 10 nodes. Right: The trajectory of the path taken by the robot at day 34 of the experiment.

maintaining the ability to use the map for visual navigation. In this work we extended the approach such that it forms a complete navigation system.

In order to navigate, the robot constructs a path consisting of a sequence of nodes between its current node and the destination node. By using the nodes in the path as way-points, the robot estimates the heading orientation required to move in the direction of the next node in the path, and then rotates accordingly. It follows that by moving forward for a certain distance. The robot repeats these rotations and translations until it reaches the destination.

Fig. 2 shows a map of 10 nodes for an office environment where our long-term experiment took place. This map was used by the robot to perform a daily visual navigation routine from node number 1 to node 10. The 10 nodes route was repeated 38 times over a period of 8 weeks and after each run the robot used the memory model to update the map as explained in [2].

The robot was able to perform the path following task successfully during all runs. We used an image similarity metric as an indicator for the adaptability of the map. The mean number of matched points between the view which has the best number of matching points and the reference views of the map was 170.9 ± 84.6 when the static reference views were used for the map, and 255.7 ± 92.6 when the adaptive map was used. The results show the effect of using our memory model in increasing the similarity of the map to the environment and that the proposed system has a persistent performance in real changing environments.

References

1. Dayoub, F., Duckett, T.: An adaptive appearance-based map for long-term topological localization of mobile robots. In: Proc. of the IEEE/RSJ Intl. Conf. on Intelligent Robots and Systems (IROS), Nice, France, September 22-26 (2008)
2. Dayoub, F., Cielniak, G., Duckett, T.: Long-Term experiments with an adaptive spherical view representation for navigation in changing environments. Robotics and Autonomous Systems 59(5) (2011)

Occupancy Grid-Based SLAM Using a Mobile Robot with a Ring of Eight Sonar Transducers

George Terzakis[1] and Sanja Dogramadzi[2]

[1] School of Computing and Mathematics, University of Plymouth, Plymouth, UK
georgios.terzakis@plymouth.ac.uk
[2] Department of Engineering Design and Mathematics,
University of the West of England, Bristol, UK
sanja.dogramadzi@uwe.ac.uk

The degree of accuracy by which a mobile robot can estimate the properties of its surrounding environment, and the ability to successfully navigate throughout the explored space are the main factors that may well determine its autonomy and efficiency with respect to the goals of the application. This paper focuses on the implementation of a SLAM framework comprising a planner, a percept and a displacement/angular error estimator using a regular occupancy grid spatial memory representation.

The overall process is arbitrated by a finite state machine (FSM) invoking the components of the framework according to the deliberate paradigm, SENSE-PLAN-ACT as described by R. Murphy [1], with the occasional exception of the interstitial invocation of a correction process, based on the estimated angular and displacement error during the execution of the planned tasks. The FSM uses a global task queue in order to execute the tasks that the planner produced during the previous call.

The robot uses a sonar ring of eight equally spaced transducers in its circumference; it utilizes four DC motors connected in pairs to provide differential drive, while the possible moves include forward-backward linear motion by a distance equal to the circumference of the circle to which the robot can be inscribed, and rotations by 90° to the left or to the right. Accordingly, the cells are square-shaped with side length equal to the diameter of the robot's circumcircle.

The map is stored in spatial memory as a 2D array of characters; its entries can be one of the following: 'B' for BLOCK, ' ' for VOID and 'X' for UNKNOWN. The percept decides whether the cell is occupied (BLOCK) or empty (VOID). The UNKNOWN entry is simply used to denote that the cell has not been inspected (the A* heuristic of the planner uses these entries to plan a series of moves towards the nearest unexplored cell).

The perception mechanism perceives the map as 0-order Markov random field [2]. The latter implies that each cell is "tied" to a probability value that is recursively updated using Bayes's rule upon each cell inspection. The robot's impression regarding the state of a neighboring cell is based on whether the sonar readings provide minimal "assurances" that it can fit in that particular cell (VOID). The robot uses the three transducers facing the cell under inspection in quite a similar manner to the one that somebody would use his/her hands to fumble about in the dark in order to decide whether the space in a particular direction is empty. In that aspect, three transducers are used by the percept to obtain an impression as to whether the robot can fit breadth

R. Groß et al. (Eds.): TAROS 2011, LNAI 6856, pp. 402–403, 2011.

wise in the examined area. These conditions are then used to obtain likelihood probability values for the sonar measurements given the state of the cell. The probability of the cell is then updated using Bayes's rule and the new state is determined with respect to whether the new probability of it being BLOCK is greater than the corresponding of it being VOID [3]. The measurement likelihood probability density functions (PDF) are chosen as a increasing (BLOCK) and decreasing (VOID) quadratic in order to depict the increasing/decreasing likelihood of obtaining greater/less measurements given that the cell is VOID/BLOCK.

During casual motion, the robot performs a set of motion primitives: GO_FORWARD, GO_BACKWARD, ROTATE_LEFT and ROTATE_RIGHT. The rotation primitives involve a 90^0 rotation which, in theory, should not affect the robot's pose and position. In practice, due to several factors (i.e., wheel spinning, gyro integration errors, etc) the robot introduces a slight displacement (up to 5 cm in both axes) and loss of orientation (up to 10^0). To estimate these discrepancies, a Kalman filter [4] was used, in which, displacement corresponds to process noise, while misalignment is seen as measurement noise. The displacement calculation is based on the difference between the state estimates before and after a 90^0 turn. The average angular error is calculated as the average inverse cosine of the fraction of the sum of the state estimate and the robot's radius, divided by the sum of the corresponding raw transducer measurement for each case in which the measurement is greater than the state estimate.

Despite the relatively limited number of sensors and several issues related to the equipment employed (i.e., low budget materials such as serial wireless transceivers, batteries, gyros and motor driver), the robot can successfully perform mapping in relatively wide spaces with respect to its own size without significantly losing track of its position and orientation.

References

1. Murphy, R.: Introduction to AI Robotics, pp. 42–44. MIT Press, Cambridge (2001)
2. Elfes, A.: Using occupancy grids for mobile robot perception and navigation. Carnegie Mellon University, Pittsburgh (1989)
3. Theodoridis, S., Koutroumbas, S.: Pattern Recognition, 4th edn., pp. 14–16. Elsevier Inc., San Diego (2009)
4. Welch, G., Bishop, G.: An introduction to the Kalman filter. University of North Carolina, Chapel Hill (2006)

On the Analysis of Parameter Convergence for Temporal Difference Learning of an Exemplar Balance Problem

Martin Brown and Onder Tutsoy

Control Systems Group, School of Electrical and Electronic Engineering,
The University of Manchester, Manchester, UK
martin.brown@manchester.ac.uk,
Onder.Tutsoy@postgrad.manchester.ac.uk

Bipedal walking/locomotion is a challenging control problem but also an interesting problem for studying learning algorithms. In 1981, Barto and Sutton developed a RL method based on TD which used the concept of learning from failure. Moreover, over the last few years the poor/slow convergence issues has gained more attention by researchers [1]. In this paper, a closed form value function solution for an unstable plant and optimal polynomial basis for the value function are presented. The linear TD(0) algorithm is stated and it is shown that the finite horizon effect which is due to repeatedly simulating the system over a finite horizon introduces a near singularity/bias in the parameter estimation process. A method is proposed to overcome this problem. Finally, the simulation results for the exemplar problem are presented, and the parameter convergence is analyzed.

Description of the System and Closed Form Solution of the Value Function

A 1st order unstable discrete time plant with sampling time 0.1, time constant and gain 1 is given by:

$$x_{k+1} = -u_k + e^{0.1*k}\left(x_k + u_k\right) \tag{1}$$

The optimal control signal which minimizes the time/state error from upright position is a bang-bang control: $u_k = -sign(x_k)$. By symmetry, we can assume that the state is negative and a positive control signal is applied.

$$V(x_k) = \sum_{t=0}^{t^*} \gamma^t \left|x_{k+1}\right| \tag{2}$$

When the pendulum reaches the upright position $x_k = 0$, then a zero torque is applied and all future rewards are zero. This occurs when $t^* = -10*\ln(x_k + 1)$.

$$V(x_k) = \frac{1}{1-\gamma} - \frac{1}{1-\gamma e^{0.1}}(x_k+1) + \left(\frac{\gamma e^{0.1}}{1-\gamma e^{0.1}} - \frac{\gamma}{1-\gamma}\right)(x_k+1)^{-10*\ln(\gamma)} \tag{3}$$

R. Groß et al. (Eds.): TAROS 2011, LNAI 6856, pp. 404–405, 2011.
© Springer-Verlag Berlin Heidelberg 2011

From equation (3) optimal polynomial basis function is constructed.

$$\phi(x_k) = \underbrace{\frac{1}{\text{Basis for Bias Parameter}}} \quad \underbrace{(1+x_k)}_{\substack{\text{Basis for} \\ \text{Linear Parameter}}} \quad \underbrace{(1+x_k)^{-10*\ln(\gamma)}}_{\substack{\text{Basis for} \\ \text{Higher Order Polynomial}}} \qquad (4)$$

Estimated value function is $\hat{V}(x_k, w_k) = w_k^T \phi(x_k)$ and its parameter vector update;

$$w_{k+1} = w_k + \eta \delta_k \left[\phi(x_k) - \gamma\phi(x_{k+1}) \right] \qquad (5)$$

where η is the learning rate and $\delta_k = r(x_k) + \gamma\hat{V}(x_{k+1}, w_k) - \hat{V}(x_k, w_k)$ is the temporal difference error. In order to provide learning at the end of the horizon, a new basis function must be reconstructed $\Phi(x_{k^*}) = \left[(1-\gamma) \quad (1-\gamma) \quad (1-\gamma) \right]$.

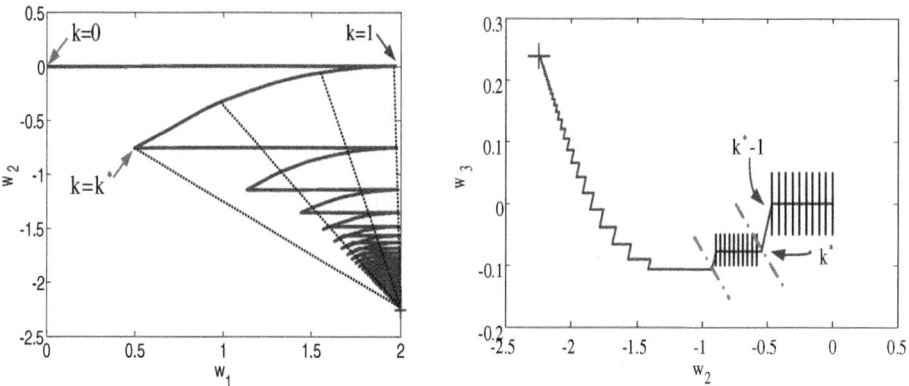

Fig. 1. a) w_1 & w_2 parameter convergence, b) w_2 & w_3 parameter convergence

In Fig. 1, w_1, w_2 & w_3 parameters convergence is shown. The solid red line shows the parameter trajectories and dotted blue line represents the solution lines associated with each training datum (corresponding to the differenced basis functions). As can be seen, there is a repeated zigzag pattern in the bias/linear parameter learning where each zigzag pattern represents a batch consists of 64 samples.

Acknowledgments. This research has been done on the CICADA project which is financed by the EPSRC grant EP/E050441/1 and the University of Manchester.

References

1. Bertsekas, D.P.: Temporal Difference Methods for General Projected Equations. IEEE Trans. on Automat. Contr. (in press)

Online Hazard Analysis for Autonomous Robots

Roger Woodman, Alan F.T. Winfield, Chris Harper, and Mike Fraser

Bristol Robotics Laboratory, University of the West of England, Bristol, UK
roger.woodman@brl.ac.uk, alan.winfield@uwe.ac.uk
cjharper@avian-technologies.co.uk, fraser@compsci.bristol.ac.uk

Robotic systems require rigorous analysis at all stages of development to ensure system safety. The manufacturing industry has developed many of the robotic design methods used today. These methods were adapted from design practices taken from other industrial sectors [3]. Incorporated into the design process are proven techniques such as hazard analysis, failure analysis and testing. In addition a number of robotic safety standards have been developed; most notably ISO 10218-1 [4]. As discussed by [1], these methods are not appropriate for designing safe robots operating in unstructured environments, due to the high complexity associated with a system that must adapt to a changing environment.

Our research is looking at novel approaches for designing autonomous personal robots. The methodology we are developing builds on traditional hazard analysis, which involves identifying and evaluating hazards in a system that may cause or contribute to an unplanned or undesirable event. We have integrated into this process a safety protection system, developed during the hazard analysis stage. This protection system serves dual purposes. Firstly to verify that safety constraints identified during hazard analysis have been implemented appropriately. Secondly as a high-level safety enforcer governing the actions of the robot, preventing the control system from performing unsafe operations. Separating safety processes from control is a relatively new technique and not one which is generally done in commercial systems. However, a report by [1] has identified a number of European robotic manufactures that have recently included software safety modules to monitor the robots working area for any perceivable hazards.

At the core of the safety protection system are safety policies. These policies are treated independently and therefore can be amended, added or removed at any time. The design of the safety policies are such that they are not tied to any specific hardware. This gives more flexibility to the construction of the robot and allows identical policies to be used on different types of robotic system. Each safety policy is constructed as an independent set of rules, which use facts derived from sensor data, to impose restrictions on actuators. These restrictions can be either limitations on output or as total suppression. This idea is based on principles taken from knowledge based system design [5]. The key benefit of which, is the inherent parallelism from structuring rules in the system as separate tasks that can be processed simultaneously. Other benefits include probabilistic reasoning and priority based inference [2]. A safety policy can be thought of as a traditional hardware interlock implemented in software. These software

R. Groß et al. (Eds.): TAROS 2011, LNAI 6856, pp. 406–407, 2011.

interlocks, or safety policies, aim to prevent the robot from generating unsafe actions, by means of intervention between the control system and the actuators.

An example safety policy is shown below. This policy is based on an ISO 10218-1 [4] safety requirement. The requirement states that while operating in reduced speed mode the maximum speed of the robot must not exceed 250 mm/s.

```
IF    robot_state = reduced_speed_control
AND   end_effector_speed > 250mm/s AND confidence_level >= 0.9
THEN  restrict actuators
```

This example shows that it is possible to explicitly represent safety standard requirements, opening up the possibility that standards could not only be used as guidelines, but also as specifications for actual safety constraint implementation.

Navigating unstructured environments safely requires intense processing of a large data set. To handle this data some have suggested using probabilistic models in the form of Bayesian networks [6]. This technique uses Bayes rule to combine data to produce a reasoned output. However, we have chosen to use confidence levels (also referred to as confidence factors), which is a method used in expert systems for dealing with uncertainty [2] [5]. This differs from Bayesian networks as it does not require priori probability to be assigned to each node of the network. Instead it allows us to assign a belief value to sensor readings that can be combined to give an overall confidence level for use in the safety policies.

The protection system aids the hazard analysis in two main areas. Firstly by having a collection of generic safety policies, it is possible to iterate through them and verify that they all exist and adhere to the specification requirements. Secondly metrics can be used in a quantitative assessment of hardware criticality.

Further experimentation is required to determine the effectiveness of the proposed safety techniques. However, preliminary experiments reveal the benefits from organising safety constraints as a set of rules in a knowledge based system.

Acknowledgements. Part of a PhD Studentship in Human Robot Interaction funded by Great Western Research in partnership with Avian Technologies Ltd.

References

1. Desantis, A., Siciliano, B., Deluca, A., Bicchi, A.: An atlas of physical human robot interaction. Mechanism and Machine Theory, 253–270 (2008)
2. Hopgood, A.A.: Intelligent systems for engineers and scientists. CRC Press, Florida (2001)
3. Hgele, M., Nilsson, K., Pires, N.J.: Springer Handbook of Robotics: Industrial Robotics. Springer, Heidelberg (2008)
4. ISO-10218-1: Robots for Industrial Environments - Safety Requirements - Part 1: Robot. ISO, Geneva (2006)
5. Kendal, S., Creen, M.: An Introduction to Knowledge Engineering. Springer, London (2007)
6. Marzwell, N.I., Tso, K.S., Hecht, M.: An integrated fault tolerant robotic control system for high reliability and safety. In: Proceedings of Technology 2004 (1994)

Results of the European Land Robot Trial and Their Usability for Benchmarking Outdoor Robot Systems

Frank E. Schneider and Dennis Wildermuth

Fraunhofer Institute for Communication,
Information Processing and Ergonomics (FKIE),
Wachtberg, Germany
{frank.schneider,dennis.wildermuth}@fkie.fraunhofer.de

It is generally a problematic task to compare different approaches and methods in the field of outdoor robotics [1]. In the majority of cases, results are reported only for a specific robotic system. All tasks are solved in a static and often specially defined environment, making it hard to compare the outcome with results from other research groups, other approaches, and other robots. The commonly used means of "proof by video" or "proof by (one) example" are insufficient for obvious reasons. As one possible solution, robot competitions can be a benchmark for real robot systems [2].

Two of the largest and best-known competitions are the RoboCup with its different leagues [3] and the DARPA Grand Challenges. Another large event with hundreds of competitors and many different categories is the annual RoboGames competition [4]. In the recent years also a variety of smaller contests have been established, among which several are explicitly regarding outdoor robotics. Examples include the Robotour [5] and the International Ground Vehicle Competition (IGVC), the International RoboSub Competition (WRSC) and the International Aerial Robotics Competition (IARC) – all organized by the AUVSI foundation [6].

The European Land Robot Trial (ELROB) was designed to demonstrate and compare the capabilities of unmanned systems in realistic scenarios [7]. It was founded by the European Robotics Group and is organised by the Fraunhofer Institute for Communication, Information Processing, and Ergonomics (FKIE).[1] The ELROB is held annually at changing locations throughout Europe and is conducted with a focus on short-term realisable robot systems. All tasks and scenarios are developed in close collaboration with experienced users from possible application domains.

In contrast to other outdoor contests – like for example the DARPA Grand Challenges – the ELROB defines a variety of scenarios instead of only one single mission. These tasks include, for example, security missions, convoying, or reconnaissance by day and night. For all these different tasks one important topic is, of course, to generate a reasonable ranking. Omitting the details of task design, it is still obvious that many different parameters might have an influence on the overall benchmark for a mission. Taking a convoying scenario as an example, average speed, totally driven distance, or degree of autonomy are possible choices from a wider range of feasible parameters. Each parameter has to be measured in a precise and

---

[1] Major support and funding for the M-ELROB comes from the German Ministry of Defence.

R. Groß et al. (Eds.): TAROS 2011, LNAI 6856, pp. 408–409, 2011.
© Springer-Verlag Berlin Heidelberg 2011

reproducible manner, which often raises serious problems, and afterwards has to be weighted in its influence on the final benchmark.

Due to its relevance for the robotics community, the ELROB organizers decided to take autonomy as the most important benchmarking parameter. A scenario design with long distances in combination with a hilly and woody environment forces the teams to implement a high degree of autonomy for their robots leading to a broad range of possible problems and challenges. In the autonomous navigation scenario of ELROB 2011, for instance, the goal was to follow a path over more three kilometres, defined only by a few intermediate GPS points. The track contained several narrow passages, and the roads mainly consisted of forest paths with no clear distinction from the surroundings.

Apart from autonomy, as additional benchmarking data the totally driven distance, total runtime, and the delivery of relevant mission data (e.g. a digital map or GPS log file) was recorded. In contrast to these factors, for which it is clear how they can be measured, the definition of autonomy has to be explained. We use the ratio of total driving time and the so-called "interaction time", which starts each time someone of the team interacts with the vehicle directly or, for example, via an operation console. It ends as soon as this interaction is over and the vehicle continues its autonomous action. A weighted combination of all benchmarking parameters leads to the final ranking. For a description of the full ELROB trial benchmarks one can look at [8].

The poster addresses the sixth European Land Robot Trial titled "Robotics in security domains, fire brigades, civil protection, and disaster control", which took place from 20th until 24th of June 2011 in Leuven, Belgium. Apart from the already mentioned autonomous navigation task, the participating teams could choose among three more challenges. In the reconnaissance scenario special markers had to be found and located. During the "mule" mission, a person guided the robot to a goal point and, afterwards, the robot had to drive the path repeatedly on its own. Finally, for "camp security" a defined area had to be monitored autonomously, thereby detecting specially marked human intruders. The authors, who belong to the ELROB organising team, will briefly picture the ongoing development of ranking system and scenario design, interface standardisation and, of course, the results of the participants.

References

1. Del Pobil, A.P.: Why do we need benchmarks in robotics research? In: Proceedings of the IROS 2006 Workshop on Benchmarks in Robotics Research, Beijing (2006)
2. Behnke, S.: Robot competitions – Ideal benchmarks for robotics research. In: Proceedings of the IROS 2006 Workshop on Benchmarks in Robotics Research, Beijing (2006)
3. The RoboCup© Federation, http://www.robocup.org
4. The official RoboGames website, http://robogames.net
5. Robotika.cz outdoor delivery challenge (Robotour),
   http://robotika.cz/competitions/robotour/en
6. The Association for Unmanned Vehicle Systems International (AUVSI) Foundation,
   http://www.auvsifoundation.org
7. The European Land Robot Trial (ELROB), http://www.elrob.org
8. Schneider, F.E., Wildermuth, D., Brüggemann, B., Röhling, T.: European Land Robot Trial (ELROB) – Towards a Realistic Benchmark for Outdoor Robotics. In: Proceedings of the 1st International Conference on Robotics in Education (RIE), Bratislava (2010)

Solutions for a Variable Compliance Gripper Design

Maria Elena Giannaccini, Sanja Dogramadzi, and Tony Pipe

Bristol Robotics Laboratory, Bristol, UK
{maria.elena.giannaccini,sanja.dogramadzi,Tony.Pipe}@brl.ac.uk

Grippers used in industry and robotics as peripheral tools to grasp objects are often limited to only one type of object surface shape. On the other hand, grippers with the capability to adaptively conform to disparate objects surfaces could be used for handling multi-surfaced workparts [1]. Furthermore, a gripper with a compliant grasp allows pressure-sensitive objects to be grasped without damage. In addition, a low inertia end-effector improves the energy efficiency of the overall system and can be easily moved in a highly dynamical manner [2]. The first concern of this ongoing research has been to prove that, notwithstanding their variable compliance, the grippers we designed are able to firmly hold objects.

A good example of a successful lightweight and flexible gripper is the concept implemented by Festo on their BionicTripod with FinGripper [1]. Similarly, compliant grasping is implemented on the universal gripper of Chicago University which smartly exploits jamming together of granular material using vacuum [3].

The aforementioned devices have inspired two gripping solutions reported on here which share the same working principle but have different physical design. The working principle is the one described in [3].

In order to perform the grasping, the University of Chicago gripper is pressed against an objects surface. However, such technique does not work with certain very pliable objects, for example a used napkin on a table, since it flattens them onto the surface. For this reason, in our first prototype, we have inserted two granular material filled pockets into the internal part of the two hard rubber jaws of a traditional gripper. The hard rubber jaws enable a rigid interaction between the object and the gripper so that even very pliable objects can be picked up.

The second gripper has a biomimetic shape: it is inspired by an octopus arm. The device proposed here is similar to the solution developed in [4]. Our adaptation of this concept consists of a roughly conic balloon filled with ground coffee. The device grips the object by wrapping around it. This action is obtained by pulling a wire fixed on the external side of the gripper. Once the object is fully encircled by the gripper, a vacuum is applied to create a hard structure that grips the object.

Experiments have been performed with each prototype to assess their functionality. In the experiment with the first prototype, the two jaws close to grasp the 0.003 kg object that creates a contact with the compliant pockets. The vacuum pump is then activated and the pressure is maintained at -33.86 kPa. Consequently, the gripper and the object are lifted from the ground. An additional weight of 0.17 kg is attached to the object while the suction is still activated. This additional weight does not loosen the grip and the object is being held for another minute. As soon as the suction is released, the object and the weight drop instantly. This simple experiment proves that the suction caused by the vacuum pump is necessary to maintain a firm grip on the

R. Groß et al. (Eds.): TAROS 2011, LNAI 6856, pp. 410–411, 2011.

object. Furthermore, this device is able to grasp very pliable objects such as unfolded napkins, clothes, plastic bags, provided they do not lie completely flat on the ground.

In the experiment with the second device, the octopus arm gripper grasps a 0.086 kg object. The granular material inside the gripper is jammed and the artifact and the object are firmly held in place. As soon as the vacuum is released, though, the object drops. Conversely, even without the use of vacuum, a 0.029 kg object has been successfully grasped and held in place stably, probably because of its lower weight. Further trials with different object weights are necessary to establish a characterization of the capabilities of the gripper regarding different weight lifting.

In conclusion, the experiments proved that both devices can successfully grip and hold both rigid and very pliable objects.

Furthermore, while its filling material is still malleable, the compliant nature of the octopus arm-like gripper causes less harm to a human, incidentally colliding with such a device, than a conventional stiff industrial-like robotic grasping tool. This is due to the fact that a compliant gripper can decrease the object velocity such as a human hand colliding with the gripper tool. Simultaneously, this reduces the impact energy (E_k), $E_k = (1/2)mv^2$.

The proportionality between impact energy and mass implies that the 0.055 kg octopus arm gripper causes less collision damage than heavier grippers. Compared to other lightweight grippers such as the Festo gripper, (0.08 kg) [1], or the Lynxmotion arm little gripper (0.06 kg) [5], it has a similar or smaller weight. Moreover, a lightweight gripper is only a first but important step towards a gripper-manipulator system safer than traditional ones. In fact, a lightweight gripper can be supported by a lightweight robotic arm which is inherently safer than heavy traditional manipulators [6].

Acknowledgments. This work is part of the INTRO (INTeractive RObotics Research Network) project, in the Marie Curie Initial Training Networks (ITN) framework, grant agreement no.: 238486, funded by the European Commission. We would like to thank Ian Horsfield and Dr Ioannis Ieropoulos for their excellent hardware support.

References

1. Festo, http://www.festo.com/rep/en_corp/assets/pdf/Tripod_en.pdf
2. Festo, http://www.festo.com/net/SupportPortal/Downloads/46268/Brosch_Tripod_3_en_RZ_110311_lo_einzel.pdf
3. Brown, E., Rodenberg, N., Amend, J., Mozeika, A., Steltz, E., Zakin, M., Lipson, H., Jaeger, H.: Universal robotic gripper based on the jamming of granular material. Proceedings of the National Academy of Sciences 107, 18743–19132 (2010)
4. Calisti, M., Arienti, A., Giannaccini, M., Follador, M., Giorelli, M., Cianchetti, M., Mazzolai, B., Laschi, C., Dario, P.: Study and fabrication of bioinspired octopus arm mockups tested on a multipurpose platform. In: IEEE/RAS-EMBS Int. Conf. on Biomedical Robotics and Biomechatronics (BioRob 2010), Tokyo, Japan (2010)
5. Lynxmotion, http://www.lynxmotion.com/p-161-little-grip-kit-no-servos.aspx
6. Haddadin, S., Albu-Schäffer, A., Hirzinger, G.: Safe Physical Human-Robot Interaction: Measurements, Analysis & New Insights. In: International Symposium on Robotics Research (ISRR 2007), Hiroshima, Japan, pp. 439–450 (2007)

Study of Routing Algorithms Considering Real Time Restrictions Using a Connectivity Function

Magali Arellano-Vázquez[1], Héctor Benítez-Pérez[2],
and Jorge L. Ortega-Arjona[3]

[1] Posgrado en Ciencia e Ingeniería de la computación
[2] Departamento de Ingeniería de Sistemas Computacionales y Automatización
[3] Departamento de Matemáticas, Facultad de Ciencias, UNAM, México D.F., Mexico
arellano_m@uxmcc2.iimas.unam.mx, hector@uxdea4.iimas.unam.mx,
jloa@ciencias.unam.mx

The paper focuses on the study of a mobile distributed system that is characterized by frequently changing topology. The routing algorithms [4,5,1] for such a system should be, in general, fully adaptive. Additionally, it is important to know the state of the task scheduler of each node in order to determine whether it acts as a router. Traditionally, existing routing algorithms [3,2] resort to the path discovery process for each modification in topology.

This paper introduces an adaptive routing algorithm based on a connectivity function that evaluates the state of the node's task scheduler as well as the general conditions of the network. The connectivity function assesses the status of the current node and the connection states of its neighboring nodes, thus obtaining the overall state of the system through local data. It is necessary to quantify the cost of the routes, as there may be more than one, considering hops from node to node as a measure. This measure can be bounded, i.e. the number g of hops in a route from node A to node B belongs to the set $g = 0, 1, \ldots, n-1$, where n is the number of nodes.

The construction of the path is performed by using the following five steps:

1. Calculating node's availability.
2. Evaluating connection state of neighboring nodes.
3. Calculating the connectivity function per node.
4. Calculating or updating the adjacency matrix.
5. Calculating node $k+1$ using Floyd-Warshall's algorithm.

The connectivity function is defined as follows: $f(x) = e^{-\frac{s_d{}^2 + \delta n^2 + hops^2 + C_{i,j}^2}{\sigma}}$, where s_d is the space needed to transmit, δn is the data loss in the channel, $hops$ is the number of nodes the message passed and $C_{i,j}$ is the load in the data channel between node i and node j.

Let A be a symmetric matrix, the element $A_{i,j}$ indicates whether a node i is connected to the node j. Each row or column of the matrix A represents the node's connections. Therefore, any node can be used as a router, if a node has enough idle time then it will be available for a routing service, this service will be a low priority process, so this node will first complete its own services and then it will respond to external requests. The proposed algorithm does not attempt

R. Groß et al. (Eds.): TAROS 2011, LNAI 6856, pp. 412–413, 2011.

to reserve a channel communication, it assures that the message arrives to its destination despite frequent changes in the network's topology, because the route is rebuilt at each step.

The worst case scenario is that the network topology changes faster than routing itself. This would be reflected in the fact that for the n units time interval we would have n different adjacency matrices. To avoid a common error in adaptive routing algorithms, like RIP [6], which is falling into a cycle (this happens when the routing information is not updated), the proposed algorithm evaluates periodically the conditions of the network, so the matrices are independent and they do not keep a relation to the previous state of the network.

To obtain the full state of the network through local data, the adjacency matrix has been implemented, so that each node possesses information of the nodes to which it connects. Since the adjacency matrix is a mathematical representation of the network's connectivity, we can calculate the cost of the paths between all nodes, adding the restriction mentioned above on the number of hops. To calculate the cost of the paths from each node to the other $n-1$ nodes, the Floyd-Warshall algorithm is used.

Implementation results of our algorithm show that the obtained route is optimal in every transitory state. However, it is possible to get a transitory state in which the destination is unreachable if the node's scheduler is too busy to transmit a data package. In this case a constraint can be added so that upon arrival of a data package enough space is assigned to the node's scheduler. This idea has not yet been implemented and is left for future work. In most cases the routes are not reversible. The run-time results also show that the time is oscillating and not converging to any particular value. A future challenge is to improve the execution time so the time will converge.

References

1. Jacquet, P., Mühlethaler, P., Clausen, T., Laouiti, A., Qayyum, A., Viennot, L.: Optimized link state routing protocol for ad hoc networks. In: Proc. IEEE INMIC 2001, pp. 62–68. IEEE, New York (2001)
2. Liu, Z., Kim, J., Lee, B., Kim, C.: A routing protocol based on adjacency matrix in ad hoc mobile networks. In: Proc. 7th Int. Conf. on Advanced Language Processing and Web Information Technology, pp. 430–436. IEEE Computer Soc., Los Alamitos (2008)
3. Medhi, D., Ramasamy, K.: Network Routing: Algorithms, Protocols, and Architectures. Morgan Kaufmann, San Francisco (2007)
4. Park, V.D., Corson, M.S.: A highly adaptive distributed routing algorithm for mobile wireless networks. In: Proc. IEEE INFOCOM 1997, vol. 3, pp. 1405–1413. IEEE Computer Soc., Los Alamitos (1997)
5. Perkins, C., Royer, E.: Ad-hoc on-demand distance vector routing. In: Proc. of the 2nd IEEE Workshop on Mobile Computing Systems and Applications, pp. 90–100. IEEE Computer Soc., Los Alamitos (1999)
6. Peterson, L.L., Davie, B.S.: Computer Networks: A Systems Approach, 4th edn. Morgan Kaufmann, San Francisco (2007)

Systematic Design of Flexible Magnetic Wall and Ceiling Climbing Robot for Cargo Screening

Yuanming Zhang and Tony Dodd

Department of Automatic Control and Systems Engineering,
University of Sheffield, Sheffield, UK
{t.j.dodd,yuanming.zhang}@sheffield.ac.uk

1 Introduction

Cargo screening is an important process to inspect illegal contraband, such as drugs, nuclear materials, weapons and explosives at seaports and airports. A great deal of research has been carried out to address the problem of cargo screening to explore new methods and solutions to improve detection accuracy and rates [1]. In some cases, human or sniffer dogs are required to enter the internal cargo to inspect unidentified substances. However, this is time consuming and potentially dangerous. These problems with correct systems have motivated the development of an innovative solution for cargo inspection in this study. It is known that robotic inspection techniques have been applied to many industries in recent years. However, so far, very little work has been done to develop a cargo screening robotic system. According to the literature, only one related piece of work was found to be proposed by Siegel *et al* [2], who pointed out that robotic systems could be used to detect illegal contraband in cargo. However, the approach was suggested to use sensors for external cargo detection rather than for internal cargo screening. Consequently, this study aims to demonstrate a systematic design approach to develop an autonomous robotic vehicle to enter the cargo container to detect the contraband for cargo screening process. Fundamentally, several types of robotic vehicles might be applied to enter a cargo container, such as miniature unmanned aerial vehicle (UAV), long robotic arm, flexible snake robot and climbing robot equipped with advanced sensing and navigating system. However, UAVs would need more working space whilst long robotic arm might lack mobility and snake robots require complex mechanics. In these cases, one possible solution is to design a wall and ceiling climbing robot (WCCR) to climb along the wall and ceiling of the cargo container to implement the subsequent inspections.

2 Systematic Design of Flexible Magnetic WCCR

Systematic design involves issues of process analysis, robot platform design, electrical system, user interface, system evaluation, prototyping with practical testing for cargo screening. A homing beacon and relay station device is proposed to set up in the entrance of cargo container, which acts as a matchmaker to provide wireless signal receiving and transmitting between remote device and mobile robot, whilst the beacon acts as a reference navigation point for home position. Infrared light source

R. Groß et al. (Eds.): TAROS 2011, LNAI 6856, pp. 414–415, 2011.

and reflector are located on the two sides of the entrance of the cargo container, which constitutes a safety light curtain as a virtual wall to prevent robot moving outside of cargo container. A miniature single board computer Gumstix with an embedded Linux system was chosen to WCCR, which can be easily extended to connect with other interface boards such as Wifistix and Robostix with advantages such as open source software, wireless communication and low power consumption [3].

There are various flexible mechanisms which can potentially be used in the WCCR platform. In this study, spring hinge joints with suspension device are used to implement wheel bending, whereas novel passive rubber-spring mechanisms are explored to connect the three sections longitudinally to implement the body bending. In addition, the magnetic wheel is a key component to produce a suitable adhesive force. Finite element method is applied to analyse the magnetic field and estimate the magnetic force to achieve wheel optimisation. The flexible WCCR platform with six magnetic wheels is developed and the prototype is built as shown in Fig. 1, which provides both wheel and body bending to achieve an adaptive climbing capability to follow the wave-shaped surface of steel cargo container. The specification of the robot is weight 5 kg, size 400×340×140(mm) and power requirement approx. 20W.

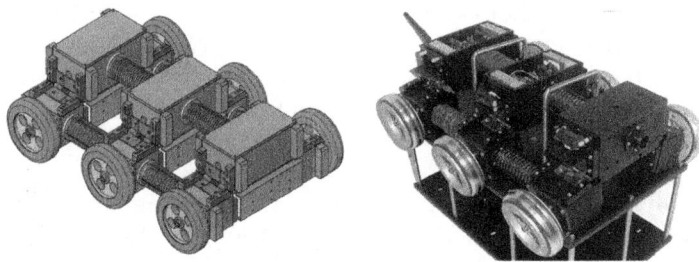

Fig. 1. Development of magnetic robot platform with 3D model (left) and prototype (right)

3   Conclusions and Further Work

A magnetic WCCR platform has been developed with highly-flexible mechanisms. The compact single board embedded system has been proposed to use in the current system. An experimental study will be carried out to verify the current design.

References

1. Verbinski, V.V., Orphan, V.J.: Vehicle and cargo container inspection system for drugs. Application of Accelerators in Research and Industry, Pts 1 and 2 475, 682–686 (1999)
2. Siegel, M.W., Guzman, A.M., Kaufman, W.M.: Robotic systems for deploying sensors to detect contraband in cargo. In: Brandenstein, A.E. (ed.) Proceedings of the Washington DC Meeting, pp. 345–352 (October 1992)
3. Gumstix (2010), http://www.gumstix.com

Tactile Afferent Simulation from Pressure Arrays

Rosana Matuk Herrera

Department of Computer Science, Facultad de Ciencias Exactas y Naturales,
Universidad de Buenos Aires, Buenos Aires, Argentina
rmatuk@dc.uba.ar

At present, autonomous, robotic dexterous manipulation in unknown environments still eludes us. Children only a few years old lift and manipulate unfamiliar objects more dexterously than today's robots. Thus, robotics researchers increasingly agree that ideas from biology can strongly benefit the design of autonomous robots.

In this article, different models for the simulation of human tactile afferents from pressure arrays are proposed and compared. The modeling of robotic tactile perception from pressure arrays is relevant, because currently most of the tactile array devices used in robotics are pressure sensors.

A bio-inspired model of robotic tactile sensing can be useful to build models of robotic tactile perception, based on the neurophysiology of human tactile perception. Besides, tactile signals are essential in human dexterous manipulation, and thus, their modeling is crucial to implement bio-inspired algorithms of dexterous manipulation that use tactile feedback.

There are four types of tactile afferents in the glabrous skin of the human hand: FA-I, SA-I, SA-II and FA-II. The SA I and SA II units are referred to as slowly adapting afferents, which means that they show a sustained discharge in the presence of ongoing stimuli. In contrast, the fast adapting afferents, FA I and FA II, fire rapidly when a stimulus is presented (or removed) and then fall silent when the stimulation cease to change. The skin area where responses can be evoked from a receptor is called its receptive field (RF). The RFs of type I units typically have sharply delineated borders and show several small zones of maximal sensitivity. In contrast, the type II units have large RFs [2].

Neurophysiologists have studied how humans lift small objects, and have identified different phases in a lifting task [3]. Short-lasting, specific patterns of tactile sensory activity seem to trigger the transition between the different phases of a lifting task [4,2]. Besides, humans use their tactile signals to estimate crucial parameters in a manipulative task, like the coefficient of friction and the incipient slips.

Israelsson [1] proposed a model for the simulation of human tactile afferent signals from 3D force signals. However, the commercially-available tactile devices are all pressure mapping systems, i.e. they only measure forces that are perpendicular to the surface. Thus, in this article, the simulation of tactile afferent signals from 3D force signals of the work of Israelsson will be adapted, to obtain afferent signals only from normal forces. Besides, alternative models for the simulation of human tactile afferents were designed and tested, to compare their performance.

R. Groß et al. (Eds.): TAROS 2011, LNAI 6856, pp. 416–417, 2011.

The Tekscan Grip is used to test the performance of the different functions in the simulation of human tactile afferent signals, from a real pressure array. The tactile sensing device TekScan Grip is a system specifically designed to acquire the pressures applied by the different regions of a human hand (fingers, thumb, palm) during the execution of tasks which require grasp movements. The Tekscan Grip is integrated by a tactile device (Tekscan sensor model 4256E), which is a flexible thin film with embedded pressure sensors. The device has 349 pressure sensels distributed as squared arrays, with a density of 40 sensels/in^2. The size of each sensel is of 15mm x 15mm, and it uses resistive technology.

In our model the receptive fields of type I afferents correspond to 1 sensel. For the FA II, the RF corresponds to the whole sensor array. The SA IIs RF is calculated using the contact point (cp) in which the center of the SA II is located and from there calculates the total force in the adjacent cps. Each SA II has a receptive field covering 9 cps.

Experiments consisting of a human subject lifting a cube of weight 340 g. with the hand covered with the Tekscan Grip were performed. To change the friction between the hand and the object, lifting experiments were performed covering the object with a very slippery material, and with a very rough material. The tactile device was set to its highest sensitivity (range 0 - 12.5 PSI measured as a raw number in a scale between 0 and 255). The experiments were recorded at 750 Hz. Afferent signals were generated from the pressure recordings using the different models for the simulation of tactile afferents, and their performance compared.

Resistance-based force distribution sensors, such as those made by Tekscan are appealing for biomechanics research because they are thin and flexible and offer high resolution and straightforward data acquisition. They also provide realtime dynamic feedback. However, limitations in their accuracy have been identified by several authors in a range of biomechanical applications. Thus, in order to improve the results, currently these factors are being analyzed.

References

1. Israelsson, A.: Simulation of responses in afferents from the glabrous skin during human manipulation. Master's thesis, Master thesis in Cognitive Science, Umeå University, Umeå, Sweden (2002)
2. Johansson, R.: Sensory and memory information in the control of dexterous manipulation. In: Lacquaniti, F., Viviani, P. (eds.) Neural Bases of Motor Behaviour, pp. 205–260. Kluwer Academic Publishers, Dordrecht (1996)
3. Johansson, R., Westling, G.: Signals in tactile afferents from the fingers eliciting adaptive motor responses during precision grip. Exp. Brain Res. 66, 141–154 (1987)
4. Johansson, R., Edin, B.: Mechanisms for grasp control. Restoration of Walking for Paraplegics Recent Advancements and Trends, 57–63 (1992)

The Interaction between Vortices and a Biomimetic Flexible Fin

Jennifer Brown[1], Lily Chambers[1], Keri M. Collins[1], Otar Akanyeti[2], Francesco Visentin[2], Ryan Ladd[1], Paolo Fiorini[2], and William Megill[1]

[1] Ocean Technology Laboratory, Department of Mechanical Engineering, University of Bath, Bath, UK
j.c.brown@bath.ac.uk
[2] Department of Computer Science, University of Verona, Verona, Italy

The fluid-structure interaction of flexible bodies in steady and unsteady flow is a key area of interest for the development of underwater vehicles. In the design of marine vehicles the flow can often be seen as an obstacle to overcome, whilst in nature a fish interacts with the flow and is capable of achieving a high level of efficiency. Therefore by understanding how fish – or flexible bodies – interact with the flow we may be able to achieve a better level of co-operation between our vehicles and their environment, potentially attaining a better efficiency in design.

A large part of the interaction between a fish and the flow is its sensing capabilities. A fish is able to sense the flow around it and react to the fluid structures that are present, such as turbulence or vortices; one of its senses that may provide information for this is a sensory array called the lateral line. If an underwater vehicle had a similar hydrodynamic sensing system it may be able to react to the fluid flow autonomously. With the emergent technology of MEMs sensors [1] it will be possible to sense the flow in ways other than using visual or sonar technology.

There is a gap in the literature that investigates the fluid flow over the body, especially in turbulent flows. Focus has been on the flow around a body in steady flow [2], where drag reduction by the motion of the body and therefore the boundary layer behaviour was investigated; the boundary layer of an oscillating body stays attached, corroborating swimming fish experiments [3]. Studies into the wake behind foils [4-7] have described and quantified the different wake patterns that are produced dependent on the frequency and amplitude of the foil, or on the position of the foil with respect to an oncoming vortex street. The different patterns that can occur at the leading edge of a rigid aerofoil when interacting with a vortex street have been classified [8]. Presented here is a study on the fluid-structure interaction of a flexible biomimetic fin in a Kármán vortex street. The effect of the fin on the wake behind a cylinder is considered; specifically changes in vortex speed, strength and size. This provides insights on the features available to sense in the flow and how flow is reconfigured due to the presence of a flexible body.

A silicone rubber fin with a biomimetic stiffness profile [9], of chord length 120 mm and an aspect ratio of 1 was placed in a flow channel (1300 x 300 x 240 mm). A 45 mm diameter circular cylinder was placed 150 mm upstream of the fin. The cylinder was used to create a Kármán vortex street, a fluid phenomenon that has been extensively studied. It is a common occurrence in natural water-ways,

R. Groß et al. (Eds.): TAROS 2011, LNAI 6856, pp. 418–419, 2011.
© Springer-Verlag Berlin Heidelberg 2011

caused by the fluid flowing around bluff bodies. Two-dimensional digital particle image velocimetry (DPIV) was used to visualise the flow. A high-speed CCD camera was used to capture the images of the seeded particles in the flow (Vestosint 1301, mean diameter 100 µm), illuminated by a solid state laser of wavelength 532 nm. Insight 3G software was used for capture and processing of the images (field of view 308 x 247mm).

Initial results show that vortices in the wake of the cylinder, with no other obstacle introduced, travel downstream at a higher speed than that of the freestream: 0.42 m/s as opposed to 0.3 m/s. With the biomimetic fin placed in the Kármán vortex street the velocity of the vortices travelling downstream from the cylinder towards the fin was much lower: 0.18 m/s. By tracking a vortex down the length of the fin it was calculated that it accelerated; by the trailing edge the vortex was travelling at approximately the freestream velocity and in the wake the average vortex velocity was 0.33 m/s. In addition to studying the vortices' downstream velocity, the variations in their strength and size (before, along and after the fin) have been determined.

Incident vortices have been shown to deform before the leading edge of a foil [8] suggesting the presence of a bow wake, which may contribute to their slowing. We believe that the shape of the fin causes a positive pressure gradient near the nose, which slows the flow down. The flow then accelerates around the fin as the effect of this positive pressure gradient is diminished and a favourable pressure gradient is encountered. This set of experiments is a first step towards better understanding the fluid flow interaction with a flexible body in steady and unsteady flow. The knowledge gained from these experiments could allow for better modelling and design of biomimetic submersibles, particularly if they intend to use flow detection as a means for locomotion and control optimisation.

References

1. Qualtieri, A., Rizzi, F., Todaro, M.T., Passaseo, A., Cingolani, R., De Vittorio, M.: Stress-Driven AlN Cantilever Based Flow Sensor for Fish Lateral Line System. Microelectronic Engineering (2011), doi:10.1016/j.mee.2011.02.091
2. Kunze, S., Brucker, C.: Flow Control over an Undulating Membrane. Experiments in Fluids 50, 747–759 (2011)
3. Anderson, E.J., McGillis, W.R., Grosenbaugh, M.A.: The Boundary Layer of Swimming Fish. Journal of Experimental Biology 204, 81–102 (2001)
4. Godoy-Diana, R., Marais, C., Aider, J., Wesfried, J.E.: A Model for the Symmetry Breaking of the Reverse Bernard-von Kármán Vortex Street Produced by a Flapping Foil. Journal of Fluid Mechanics 622, 23–32 (2009)
5. Schnipper, T., Andersen, A., Bohr, T.: Vortex Wakes of a Flapping Foil. Journal of Fluid Mechanics 633, 411–423 (2009)
6. Muijres, F.T., Lentink, D.: Wake Visualization of a Heaving and Pitching Foil in a Soap Film. Experiments in Fluids 43, 665–673 (2007)
7. Wang, S., Jia, L., Yin, X.: Kinematics and Forces of a Flexible Body in Kármán Vortex Street. Chinese Science Bulletin 54, 556–561 (2009)
8. Gursul, I., Rockwell, D.: Vortex Street Impinging upon an Elliptical Leading Edge. Journal of Fluid Mechanics 211, 211–242 (1990)
9. Riggs, P., Bowyer, A., Vincent, J.: Advantages of a Biomimetic Stiffness Profile in Pitching Flexible Fin Propulsion. Journal of Bionic Engineering 7, 113–119 (2010)

Toward an Ecological Approach to Interface Design for Teaching Robots[*]

Guillaume Doisy, Joachim Meyer, and Yael Edan

Ben-Gurion University of the Neguev, Be'er Sheva, Israel
doisyg@bgu.ac.il

With robots implementing learning capabilities, a whole new range of issues regarding the human interaction with the learning process arises. Among these issues, the question of the nature of the interaction while the human is teaching the robot is critical. To solve the problem of interface design for teaching robots, we propose a new approach based on Gibson's theory of visual perception [4].

We state that there are two main domains of application that must be distinguished.

The first domain is the one which uses robot learning only during the development process of the robot. By using learning techniques the capabilities of the robot are extended before it is used by the end-user. The learning or training is done by the developers of the robot, hence they have an extensive understanding of what the robot is learning and how. Therefore, users will not experience interaction with the learning process and there is no need to consider additional issues than for a regular robot.

However, the situation is different for the second domain which corresponds to the cases where the robot is learning from the end-user while performing tasks. Here learning takes place in the real world, and is used on-line to continuously increase the autonomous and adaptive capabilities of the robot. This raises several issues about how the users should interact with the learning process.

Among these issues the question of what should be the interface while the robot is learning a task is critical. Indeed it has been shown [8] that when a user is teaching a robot it is unlikely that he properly understands what the robot is learning, even for simple tasks. This is caused by the lack of feedback available during the learning process: the user can compare his expectation and the result of his teaching only after the teaching is complete and applied a first time. Thus, as pointed out in [3], the feedback must form proper user's expectations that should be provided continuously by the interface. The interface should be able to communicate to the user sufficient information for him to understand the constraints of the learning process. For instance, in the case of teaching by demonstration a robot to navigate, the user should understand what the robot is using to estimate its position. If it uses landmarks, the user should understand that when the robot is not in the range of a landmark, it is not able to learn the path the user is teaching it.

This need for feedback from the interface in order for the user to properly teach the robot has been also observed experimentally by Rouanet et al. [7]. Additionally, they showed that the efficiency of an interface to provide feedback to the user is strongly

---

[*] This work was supported by the EU FP7-INTRO project.

R. Groß et al. (Eds.): TAROS 2011, LNAI 6856, pp. 420–421, 2011.
© Springer-Verlag Berlin Heidelberg 2011

dependent on its nature. More precisely, the human-like interface tested was found to be the least usable. More generally, we argue that if robots are considered as tools or cognitive artifacts in Norman's sense [6], human-like interfaces are not the optimal way of interacting with robots. This is due to the ambiguity and the inconsistency inherent to human interactions. As pointed by Rouanet et al. [7], the interface should be as transparent as possible and provide feedback about what the robot perceives. The work of Crick et al. [2] goes in the same direction: they showed that in a robot learning situation, the user should be able to see the world through the eyes of the robot to understand the learning process.

To reach this goal, the ecological theory of visual perception of Gibson [4] can provide a valuable framework. Basically, Gibson states that humans do not construct a representation of the world but perceive directly *affordances*. Affordances are defined as the link between the environment and the possible actions. This theory led to the development of original approaches to interface design which demonstrated successful applications in the industry [1] and more recently in the robotic field [5].

We argue that this ecological approach can be extended to interface design for teaching robots. The idea is to consider that the learning process consists of possible actions that can be taken by the user and which are constrained by the robot environment. Thus, it is possible to define *learning affordances* which are the link between the environment and what the robot can possibly learn. This approach consists then of allowing the user to perceive these learning affordances through the interface. Similarly to the Gibson's theory which suppresses the need for an internal representation of the world, we think this approach can avoid the need for the users to develop an internal model of the robot learning behavior, and hence to overcome the expectation and feedback issues which arise when a user is teaching a task to a robot.

References

1. Burns, C., Jamieson, G., Skraaning, G., et al.: Supporting Situation Awareness through Ecological Interface Design. Cognitive Engineering and Decision Making 5, 205–209 (2007)
2. Crick, C., Osentoski, S., Jay, G., et al.: Human and Robot Perception in Large-Scale Learning from Demonstration. In: Proceedings of 2011 HRI International Conference on Human-robot Interaction, pp. 339–346. ACM, New York (2011)
3. Doisy, G., Meyer, J.: Expectations regarding the Interaction with a Learning Robotic System. In: Proceedings of the Workshop of the 2011 HRI Conference on Expectations in Intuitive Human-robot Interaction, pp. 41–44 (2011)
4. Gibson, J.J.: The Ecological Approach to Visual Perception. Houghton Mifflin, New York (1979)
5. Nielsen, C.W., Goodrich, M.A., Ricks, R.W.: Ecological Interfaces for Improving Mobile Robot Teleoperation. IEEE Transactions on Robotics 23, 927–941 (2007)
6. Norman, D.A.: Cognitive artifacts. Dept. of Cognitive Science, University of California, San Diego (1990)
7. Rouanet, P., Danieau, F., Oudeyer, P.Y.: A Robotic Game to Evaluate Interfaces used to show and Teach Visual Objects to a Robot in Real World Condition. In: Proceedings of 2011 HRI International Conference on Human-robot Interaction, pp. 313–320. ACM, New York (2011)
8. Saunders, J., Otero, N., Nehaniv, C.L.: Issues in human/robot Task Structuring and Teaching. In: Proceedings of the RO-MAN 2007 International Symposium on Robot and Human Interactive Communication, pp. 708–713 (2007)

Towards Adaptive Robotic Green Plants

Janine Stocker, Aline Veillat, Stéphane Magnenat,
Francis Colas, and Roland Siegwart

Autonomous Systems Lab, ETH Zurich, Zürich, Switzerland
janine.stocker@gmail.com, firstname.lastname@mavt.ethz.ch,
aveillat@hotmail.com

Humans often use green plants as furniture in their built environments. More generally, the status of plants as living beings is highly dependent on their lack of mobility. The goal of this project, which is a collaboration between artists and roboticists, is to lead the general public to a questioning about the role of plants in the society through an artistic installation. We do so by endowing a green plant with motion, perception, learning and adaptation capabilities while retaining the emotional characteristic of a plant, such as calm and gently random behaviours.

Fig. 1. Prototype of the cyborg, the final version will feature a casing hiding the electronics

Several existing works involve green plants and robotics. For instance, [6] builds a "fake" plant mimicking the behaviour of a real one. In [2], a group of robots based on the iRobot Create manage a garden, caring for the plants. More into the artistic direction, [5] demonstrates a robotic flower pot moving imperceptibly slowly, and thus mimicking this particular attribute of green plants, that is, the slow reaction to the environmental changes. In this paper, the plant adapts its behaviour using a genetic algorithm. In this direction, previous works have emphasized the importance of adaptation [4] and the implied "play" behaviour [1] to create interesting robots from an artistic point of view. At the level of interaction, works have shown that an artistic robot should have its own life, avoiding or ignoring humans [3].

In this work, we propose a cyborg, that is, an agent combining biological and technological elements. This cyborg consists of an iRobot Create, a computer running Linux, a normal plant and additional sensors (Fig. 1). The final version will also feature a casing hiding the electronics. The cyborg lives its own life, following its internal needs of *water*, *sunlight* and *electrical energy*. It satisfies these needs by performing actions to fetch water, to find sunny locations and to recharge the robot's battery. To do so, and in contrast with existing works, the robot employs a well-founded probabilistic planning algorithm. This algorithm takes as input the current state of the cyborg needs, and produces as output a sequence of actions.

R. Groß et al. (Eds.): TAROS 2011, LNAI 6856, pp. 422–423, 2011.

The planning consists of optimizing the expected fitness value of the cyborg's needs at a certain time in the future, given a probabilistic model of the effect of actions on needs. To do so, the algorithm enumerates the possible action sequences of a certain length, and for each estimates the final fitness value. If a sequence would lead the cyborg into starvation (an empty battery), the sequence is discarded. Then, the sequence resulting in the best value is selected. The cyborg adapts by online learning the model of the effect of actions on needs.

Currently, the prototype of the cyborg is performing its tasks over several battery cycles. The cyborg avoids obstacles using ultrasonic sensors, finds the best light spot using light sensors and goes to a recharge and to a mock-up water station using iRobot's infrared sensor. We have conducted experiments by implementing the probabilistic model using histograms: for each type of action, we track the probability distribution of the state change. As needs are independent, we have one histogram per need per action. This model allowed the robot to run continuously for durations up to 28 hours, while alternating behaviours in an interesting way from an artistic point of view. In the future, we will include other actions, for instance in the context of group behaviours with multiple cyborgs. Indeed, one of the artistic goals is to run several cyborgs at an art exhibition.

An adaptive robotic green plant provides a simple application scenario for the development of simultaneous planning and learning algorithms. The implementation of such an algorithm allows a robotic plant to perform autonomously during a long period of time and enables its deployment in an art exhibition. On the artistic level, by endowing green plants with animal-level autonomy while keeping plant-level goals and drives, we expect interesting reactions from the general public. Ultimately, we hope to awake questions about the role of plants in the society, and to highlight their presence as actors in the world.

References

1. Arata, L.: Can your autonomous robot come out and play? In: International Conference on Integration of Knowledge Intensive Multi-Agent Systems, pp. 14–18. IEEE Press, Los Alamitos (2003)
2. Correll, N., Arechiga, N., Bolger, A., Bollini, M., Charrow, B., Clayton, A., Dominguez, F., Donahue, K., Dyar, S., Johnson, L., et al.: Building a distributed robot garden. In: IEEE/RSJ International Conference on Intelligent Robots and Systems, IROS 2009, pp. 1509–1516. IEEE Press, Los Alamitos (2009)
3. Fujita, M., Kotani, K., Kawaguchi, Y.: Artistic concept for negative-style interaction robotics. In: International Conference on Artificial Reality and Telexistence, pp. 240–245. IEEE Computer Society, Los Alamitos (2006)
4. García, R.P., Aróstegui, J.M.M.: A cooperative robotic platform for adaptive and immersive artistic installations. Computers & Graphics 31(6), 809–817 (2007)
5. Mondada, F., Legon, S.: Interactions between art and mobile robotic system engineering. Evolutionary Robotics. From Intelligent Robotics to Artificial Life, 121–137 (2001)
6. Park, H., Jung, S., Choi, J., Park, S., Yoon, C., Park, J.: A study on the moving mechanism for flower robot. In: International Conference on Control, Automation and Systems, ICCAS 2007, pp. 2514–2518. IEEE Press, Los Alamitos (2007)

Using Image Depth Information for Fast Face Detection

Sasa Bodiroza

Cognitive Robotics Group, Institut für Informatik,
Humboldt-Universität zu Berlin, Berlin, Germany
bodiroza@informatik.hu-berlin.de

Human-robot interaction relies heavily on the ability to locate and track agents in real-time. For joint attention [1] participating agents must be able to locate one another. Therefore, face detection (FD) and consequent tracking represents an area of great importance for interaction.

Current approaches for FD, such as the Viola-Jones (VJ) method [3] present a robust, yet relatively slow method. Other approaches based on it use depth information to determine the possible search area for every pixel in the image, or in the post-detection step to eliminate false positives [4]. In this paper, we investigate if depth-based image preprocessing can significantly reduce the search area, lowering the number of false positives while shortening the execution time.

Depth and RGB images of both sitting and standing people are recorded and then processed offline using the algorithm presented in this article. Images are acquired using Microsoft Kinect. The user is kept at a fixed distance from the camera (between 1 and 1.5 meters). However, with slight modifications, the proposed algorithm can also be used with people at various distances.

The processing is performed on a laptop with Intel T5750 processor and a Kinect. A sample output from RGB and IR cameras is shown in Fig. 1.

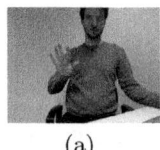 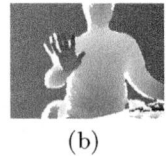 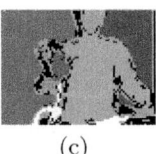

(a)　　　　　　　(b)　　　　　　　(c)

Fig. 1. (a) and (b): Sample RGB and depth images; (c) Sample segmented image

The method consists of two stages: image segmentation (IS) and FD.

IS is executed to enable finding the regions which are at different depth levels (Fig. 1(c)). Note that in the sample image head and body regions are separate regions, which have similar average depth. The black pixels represent regions which are too small to contain a face and white pixels are invalid since they were too close to the device to get depth information. The image is scaled down by factor 8 in order to increase the speed of execution, since high-level details are not needed for finding region borders. The depth neighbors of each pixel are then

R. Groß et al. (Eds.): TAROS 2011, LNAI 6856, pp. 424–425, 2011.
© Springer-Verlag Berlin Heidelberg 2011

found. A depth neighbor is a neighbor pixel which depth is within the threshold range of the depth of the current pixel. The threshold value lies between the maximum difference of two neighboring pixels in a face region and the minimum difference of pixels from a face and a neighboring region (i.e. the neck). The scale factor is chosen such that the face is still easily separable when the image is sc! aled down. The pixel and its neighbor are stored as pairs in an $n \times 2$ matrix, where n is the number of neighbor pairs. Using the matrix we construct an adjacency matrix. Dulmage-Mendelsohn decomposition [2] is used to analyze the graph, represented with the adjacency matrix and locate all connected components which represent separate regions. Every region is characterized with five parameters: coordinates of top left and bottom right corners, and its average depth level. Among the remaining regions, certain regions are discarded based on the low probability of locating a face of a certain size at a certain depth level. Selected regions are then analyzed with the VJ algorithm to locate faces.

The algorithm was implemented in Matlab and tested on a set of 21 images with one person on each image, containing 2 different persons. The mean search area, compared to the original was 4.63%, the mean execution time was $0.1427\,s$ for the image segmentation and $3.2723\,s$ for the Viola-Jones method for selected regions. This results in a mean execution time of $3.415\,s$. Using the same setup, it took $16.1359\,s$ using the Viola-Jones method, finding also two false positives in one image. These results present a considerable decrease in the execution time, resulting in only 21.16% of the original time. However, for five images the IS was not successful. This happened when the head was turned left or right, featuring smooth transition between the face and the neck, which resulted in detection of a region which included both the face and the torso.

In conclusion, we have presented the use of depth information in a human FD process. Our approach presents a significant speed-up compared to traditional only vision-based FD methods.

Acknowledgements. This research has been supported by the EU funded Initial Training Network in the Marie-Curie People Program (FP7): INTRO (INTeractive RObotics research network), grant agreement no.: 238486. The author would like to thank Verena Hafner and Guido Schillaci for useful discussions.

References

1. Kaplan, F., Hafner, V.V.: The challenges of joint attention. Interaction Studies 7(2), 135–169 (2006)
2. Pothen, A., Fan, C.-J.: Computing the block triangular form of a sparse matrix. ACM Trans. Math. Softw. 16(4), 303–324 (1990)
3. Viola, P., Jones, M.J.: Robust real-time face detection. Int. J. Comput. Vision 57(2), 137–154 (2004)
4. Wu, H., Suzuki, K., Wada, T., Chen, Q.: Accelerating face detection by using depth information. In: Proceedings of the 3rd Pacific Rim Symposium on Advances in Image and Video Technology, pp. 657–667. Springer, Heidelberg (2008)

Using Sequences of Knots as a Random Search

C.A. Pina-Garcia and Dongbing Gu

School of Computer Science and Electronic Engineering,
University of Essex, Colchester, UK
{capina,dgu}@essex.ac.uk

1   Introduction

Fink and Mao define a knot as "a sequence of moves creating an aesthetic structure or topology, where its properties are preserved under continuous deformations" [1]. Thus, it is possible to emulate a random search behavior [5], using a set of steps that represents a knot. However, a single knot is not enough to cover a specific area, due to this lack of coverage, we suggest link several knots in order to increase the searching scope.

2   Formal Basis for Designing Knots

Every knot can be described by a polygonal curve, due to this, a sequence of points $(p_1, p_2, ..., p_n)$ can be joined with the aim to build edges $[p_0, p_1]$, $[p_1, p_2]$ and $[p_{n-1}, p_n]$, where every set of ordered pairs can be grouped in a single knot $K = \{[p_0, p_1], [p_1, p_2], ..., [p_{n-1}, p_n]\}$. Consequently $K_0, K_1, ..., K_n$ might be seen as:

$$\bigcup_{i=0}^{n} K_i = \sum_{i=0}^{n} K_i \quad \text{where} \sum_{i=0}^{n} K_i \quad \text{denotes their connected sum} \quad K_0 \# K_1 \# ... K_n \quad (1)$$

Schematically, the connected sum [2] of several knots can be seen as a knot diagram on the Euclidean space 2D (Fig. 1).

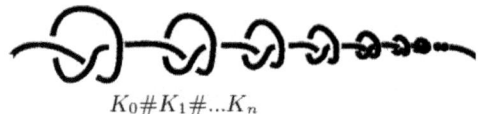

$$K_0 \# K_1 \# ... K_n$$

Fig. 1. The connected sum of several knots $K_0 \# K_1 \# ... K_n$

3   Selecting Knots According Size and Shape

The purpose of this research is to use the notion of a knot as a set of moves that can be achieved by a robot device. Initially, this set of moves must be modelled from a well defined knowledge base. With this in mind, in [1] Thomas Fink

R. Groß et al. (Eds.): TAROS 2011, LNAI 6856, pp. 426–427, 2011.

and Yong Mao suggest six states: $R_\odot, R_\otimes, C_\odot, C_\otimes, L_\odot$ and $L_\otimes$. These states are equally considered as moves, hence, a consecutive array of well structured steps on a square lattice represents a knot. Given these facts, our knowledge base is described as follows: R for a right move, L for a left move and C for a centre move. $\odot$ and $\otimes$ denote the directions of the active end as viewed from in front, that is to say, out of the page and into the page, respectively [1,3].

4 Results

Preliminary results suggest that the shape of the knot is related to search efficiency. the best running was carried out by knot labelled with number 13 (see [3]) obtaining a mean search efficiency of 4.60×10^{-5} due to the correlation between number of targets found (7.6 ± 1.42) and travelled distance $(1.70 \times 10^5 \pm 1.99 \times 10^4)$.

5 Conclusions

Knots were experimentally derived with the help of a simulator [4], in order to compare performance between every knot. Thus, our preliminary experiments show that it is possible to develop an acceptable random search behavior [5], from a sequence of steps corresponding to a knot. We have shown that the shape of the knot is closely related with search efficiency in every knot. In conclusion, here we present a new sort of family of trajectories with the aim to emulate a random search.

Acknowledgements. The authors thank the reviewers of this paper for their useful comments. Mr. Pina-Garcia has been partially supported by the Mexican National Council of Science and Technology (CONACYT), through the program "Beca para estudios de posgrado en el extranjero" (no. 213550).

References

1. Fink, T., Mao, Y.: Tie knots, random walks and topology. Physica A 276, 109–121 (2000)
2. Livingston, C.: Knot theory. The Mathematical Association of America (1993)
3. Fink, T.M., Mao, Y.: Designing Tie Knots by Random Walks. Nature 398, 31 (1999)
4. Wilensky, U.: Netlogo, center for connected learning and computer-based modeling (1999), http://ccl.northwestern.edu/netlogo
5. Berg, H.C.: Random walks in biology. Princeton University Press, Princeton (1993)

Vision-Based Segregation Behaviours
in a Swarm of Autonomous Robots

Michael J. Price and Roderich Groß

Natural Robotics Lab, Department of Automatic Control and Systems Engineering,
The University of Sheffield, Sheffield, UK
michaelprice@theiet.org, r.gross@sheffield.ac.uk

The accelerating development and usage of autonomous robots has led to increased interest in decentralised systems and the cooperation of individual elements within a swarm, in particular the ability to self-organise and form patterns or self-segregate. Such segregation is readily exhibited in nature, as highlighted by prior work looking at the collective sorting of brood items by ant colonies [1,3]. This work focuses on the ability of robots to spatially segregate themselves. In particular, it takes inspiration primarily from the Brazil nut effect [6], which is responsible for the emergent striped patterns observed when a container holding particles of different sizes is agitated.

Previous work has concentrated on simulating recognized elemental segregation behaviours in synthetic environments. This study extends a specific series of simulation work [2] in which a relatively simple algorithm involving three basic behaviours and no communication between agents was used to create annular and stripe patterns in groups of virtual mobile robots where each robot mimics a particle of a certain size. The aim is to validate the results already obtained by implementing the identified constituent behaviours on the e-puck desktop mobile robot [4].

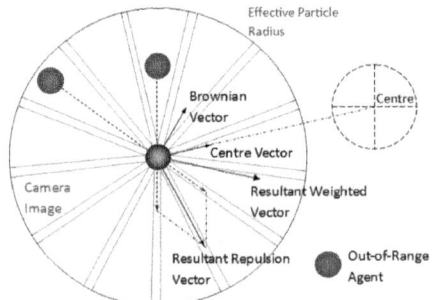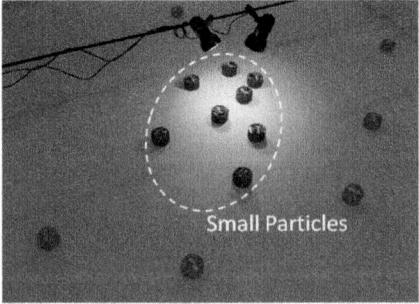

Fig. 1. Left: Brazil nut sub-behaviour integration. Right: Typical final segregation pattern for 14 robots assigned to 2 distinct groups of particles (image post-processed to improve visibility).

Each robot calculates a new vector to follow once every control cycle based on a weighted sum of the unit vectors produced by each constituent behaviour (see Figure 1, Left). The first behaviour simulates a particle's random movement in a container subjected to vibration (similar to *Brownian* motion). It was implemented around Marsaglia's MWC algorithm [7]. The second behaviour simulates a particle's attraction

R. Groß et al. (Eds.): TAROS 2011, LNAI 6856, pp. 428–429, 2011.

towards a gravitational *centre*. It was implemented using the e-puck's passive IR sensors to locate a light source. The third behaviour simulates a particle's *repulsion* from other particles upon collision. The e-puck detects (virtual) particle collisions by taking a series of thirteen camera images while turning on the spot.

Testing of the integrated behaviour involved placing a group of identical robots inside an empty arena measuring 3.0mx2.5m, with the initial positions and bearings for each experimental run being randomised beforehand. The environment is predominantly white, with the robots being made to appear as a constant diameter, matte-black object to improve image contrast, and with an IR-intensive light source being placed in the middle of the arena to define the centre of the gravitational field. The resulting patterns produced at the end of each run were then measured based on the final positions of the robots, where ideally the robots representing small particles should group together in the centre of the simulated gravitational field with a small inter-robot spacing, while the robots representing large particles should form an annular group around the central cluster with a larger inter-robot spacing.

We conducted 45 runs of 600s in length using two groups of robots (14-19 robots in total) representing particles of radius 0.20m and $0.20 \times \beta$ m respectively (β=1.0, 1.2, 1.4, 1.6, 1.8, 2.0, 2.25, 2.5 and 3.0, five trials per setup). A unity size ratio (i.e., β=1.0) produced patterns where no segregation was present (control runs). For higher size ratios, two distinct annular groups were consistently formed. The average segregation errors were 14.5% and 16.6% for ratios β=1.2 and 1.4 respectively and below 3.7% for all ratios β>1.4. A typical arrangement is shown in Figure 1, Right. Full results, including video material, can be found at [5].

Acknowledgements. This research was supported by a Marie Curie European Reintegration Grant within the 7th European Community Framework Programme (grant no. PERG07-GA-2010-268354).

References

1. Franks, N.R., Sendova-Franks, A.B.: Brood sorting by ants: Distributing the workload over the work-surface. Behav. Ecol. Sociobiol. 30(2), 109–123 (1992)
2. Groß, R., Magnenat, S., Mondada, F.: Segregation in swarms of mobile robots based on the Brazil nut effect. In: Proc. 2009 IEEE/RSJ Int. Conf. on Intelligent Robots and Systems, IROS 2009, pp. 4349–4356. IEEE Computer Society Press, Los Alamitos (2009)
3. Melhuish, C., Sendova-Franks, A.B., Scholes, S., Horsfield, I., Welsby, F.: Ant-inspired sorting by robots: The importance of initial clustering. J.R. Soc. Int. 3(7), 235–242 (2006)
4. Mondada, F., Bonani, M., Raemy, X., Pugh, J., Cianci, C., Klaptocz, A., et al.: The e-puck, a Robot Designed for Education in Engineering. In: Proc. 9th Conference on Autonomous Robot Systems and Competitions, Robotica 2009, Portugal, pp. 59–65 (2009)
5. Price, M., Groß, R.: Segregation in Autonomous Robots (2011), Supplementary information available online at
   http://naturalrobotics.group.shef.ac.uk/supp/2011-001
6. Rosato, A., Strandburg, K.J., Prinz, F., Swendsen, R.H.: Why the Brazil nuts are on top: Size segregation of particulate matter by shaking. Phys. Rev. Lett. 58(10), 1038–1040 (1987)
7. Marsaglia, G.: Simple Random Number Generation (2011), The Code Project Web site:
   http://www.codeproject.com/KB/recipes/SimpleRNG.aspx
   (retrieved January 05, 2011)

Visual-Inertial Motion Priors
for Robust Monocular SLAM

Usman Qayyum and Jonghyuk Kim

School of Engineering, Australian National University, Canberra, Australia
{usman.qayyum,jonghyuk.kim}@anu.edu.au

Monocular visual SLAM approaches are mostly constrained in their performance due to general motion model and availability of true scale information. We proposed an approach which improves the motion prediction step of visual SLAM and results in better estimation of map scale. The approach utilizes the short term accuracy of inertial velocity with visual orientation to estimate refined motion priors. These motion priors are fused with sparse number of 3D map features to constraint the positional drift of moving platform. Experimental results are presented on large scale outdoor environment, yielding robust performance and better observability of map scale by monocular SLAM.

Visual and inertial sensors are natural candidates for SLAM research whereas availability of GPS depends upon operating environment. We have presented an approach to estimate robust motion priors from visual orientation and accelerometer based velocity to improve the EKF-SLAM filter consistency with observable scale. The recent work to provide efficient priors to SLAM filter with sparse set of features is conducted by [1], in which stereo based system is used whereas the focus of our research is monocular cameras. Various visual-inertial SLAM approaches proposed so far, work in limited scenarios i.e. known feature locations or stereo cameras having limited working environment due to its predetermined baseline. In monocular-inertial SLAM, the work of [2] presented their results on small scale hand-held environment while maintaining full Inertial Measurement Unit (IMU) states whereas our approach is directed to large scale environment relying on relative inertial velocity and visual orientation as motion priors to EKF-SLAM filter.

The prime limitation with the dead-reckoning sensors on long term navigation is the unconstrained drift in position and orientation due to integration errors whereas the short term accuracy and high throughput makes them an ideal choice for integrated localisation applications with other sensors. Inertial sensors are comprised of tri-axial gyros and accelerometers, arranged orthogonally to each other and provide 6DOF measurements of the attached platform/body. We have used the relative velocity information of accelerometer instead of integrated position to generate the translational magnitude in navigation frame (an internal IMU complimentary filter estimates the on line biases whereas gyro compensation is applied with subtraction of gravity vector). The inertial velocity magnitude provides the translation offset of the vehicle in metric space whereas angular orientation between the consecutive images is provided by adopting the

R. Groß et al. (Eds.): TAROS 2011, LNAI 6856, pp. 430–431, 2011.
© Springer-Verlag Berlin Heidelberg 2011

work of [3], in which a dense motion estimation algorithm compares the consecutive image arrays by Average Absolute intensity Difference (AAD). The motion estimates from accelerometer are preferred over the visual estimates as visual approaches reveal the translation in arbitrary scale. The visual-inertial relative motion priors provide the predictive stage to the EKF SLAM-filter which maintains 3D features to constrainst the long-term inertial drift.

Current research on sequential monocular SLAM has shown that it performs the drift free tracking by matching the sparse set of map features to the image features. The visual-inertial motion priors estimated earlier are provided to sequential; EKF-SLAM filter [4], which helps in better prediction of feature locations and results in efficient generation of map. The feature map consists of n number of inverse depth features and visual matching is based upon patch correlation. Each observed/re-observed feature imposes the constraint on camera location by using the correspondence of image-to-map feature. Each observed feature imposes the constraint on vehicle/camera location by using the correspondence of image to map features. The innovation vector between the predicted feature position and visually matched observation is used to calculate the innovation vector, for EKF-SLAM filter updation.

An experimental evaluation is presented on a dataset collected at campus by placing the sensors (IMU/camera) on testing Van for a traveled distance of approx. 1.0 Km. To fully evaluate the proposed approach, a comparison is performed with IMU integrated-only, visual odometry [3] and Monocular SLAM [4] approaches. The result of the presented approach shows improved and more stable performance by observing the metric scale whereas inertial drift is minimized by inverse depth features maintained in EKF-SLAM filter.

The proposed work presents the use of short term relative velocity magnitude with visual orientation to provide refine priors to EKF-SLAM approach. The use of inverse depth feature provide the positional constraint on the traveled distance. The proposed approach is evaluated on large scale outdoor sequence showing a more stable and consistent performance of monocular SLAM in observing the map scale.

References

1. Alcantarilla, P., Bergasa, L., Dellaert, F.: Visual odometry priors for robust EKF-SLAM. In: Proc. of IEEE Int. Conf. on Robotics and Automation, pp. 3501–3506. IEEE, Los Alamitos (2010)
2. Pinies, P., Lupton, T., Sukkarieh, S., Tardos, J.D.: Inertial Aiding of Inverse Depth SLAM using a Monocular Camera. In: Proc. of IEEE Int. Conf. on Robotics and Automation, pp. 2797–2802. IEEE, Los Alamitos (2010)
3. Milford, M., Wyeth, G.: Single Camera Vision-Only SLAM on a Suburban Road Network. In: Proc. of IEEE Int. Conf. on Robotics and Automation, pp. 3684–3689. IEEE, Los Alamitos (2008)
4. Civera, J., Davison, A.J., Montiel, J.: Inverse Depth Parametrization for Monocular SLAM. IEEE Trans. on Robotics 24(5), 932–945 (2008)

Author Index

Akanyeti, Otar   418
Alonso, Antonio   24
Antero, Unai   24
Aouf, Nabil   36
Arellano-Vázquez, Magali   412
Armesto, Leopoldo   195, 277
Asthenidis, Alexandros   1
Astiz, Mikel   24
Azkune, Gorka   24

Baiboun, Nadir   90
Baker, Vincent   386
Barkana, Duygun Erol   125
Benítez-Pérez, Héctor   412
Bennet, Derek   390
Biggs, James   207
Birattari, Mauro   90
Bodiroza, Sasa   424
Bonani, Michael   311
Brooker, Graham   265
Brown, Jennifer   418
Brown, Martin   404
Brutschy, Arne   90
Bugmann, Guido   360

Cameron, Stephen   384
Cao, Juan   378
Chambers, Jon   372
Chambers, Lily   418
Chen, Jianing   380
Chevalier, Bob   386
Chorley, Craig   114
Cielniak, Grzegorz   400
Colas, Francis   422
Collins, Keri M.   418
Cope, Alex   372
Copleston, Simon N.   360

Dayoub, Feras   400
Dehghani, Abbas   173
Di Caro, Gianni A.   137
Dixon, Clare   336
Dodd, Tony   414
Dogramadzi, Sanja   402, 410
Doisy, Guillaume   420

Dorigo, Marco   90
Drury, David   114
Ducatelle, Frederick   137
Duckett, Tom   400

Edan, Yael   420
Eder, Kerstin   323
Esnaola, Urko   24
Espinosa, Felipe   241
Evans, Mathew H.   13, 102

Fiorini, Paolo   418
Fisher, Michael   336
Foster, Stephen   394
Fox, Charles W.   13, 102, 183, 253
Fraser, Mike   406
Frison, Marco   90

Gambardella, Luca M.   137
Gasteratos, Antonios   289
Giannaccini, Maria Elena   410
Girbés, Vicent   195, 277
Grigore, Elena Corina   323
Groß, Roderich   380, 394, 428
Gu, Dongbing   426
Gurney, Kevin   13, 372

Ham, Andy   102
Han, Weicheng   54
Harper, Chris   406
Herrmann, Guido   299
Hirata, Shinnosuke   46

Iglesias, Roberto   241
Imran, Saad Ali   36

Jalani, Jamaludin   299
Jayne, David G.   173

Keedwell, Ed   392
Khan, Said Ghani   299
Khazravi, Mojtaba   173
Kim, Jonghyuk   430
Kostavelis, Ioannis   289
Kurosawa, Minoru Kuribayashi   46
Kyriacou, Theocharis   66

Labrosse, Frédéric   378
Ladd, Ryan   418
Lanzon, Alexander   388
Lenz, Alexander   323
Lepora, Nathan F.   13, 102, 253
Li, Liyuan   54
Liskiewicz, Tomasz   173
López, Alfonso Montellano   173
Lou, Lu   378

Macdonald, Malcolm   390
Magnenat, Stéphane   422
Martinoli, Alcherio   311
Matuk Herrera, Rosana   416
Megill, William   418
Melhuish, Chris   114, 299, 323
Meyer, Joachim   420
Mitchinson, Ben   13
Mohamed, Mohamed Kara   388
Mondada, Francesco   311
Moore, Tom   386
Morina, Ardian   173
Motiwala, Asma   13, 253
Mwaura, Jonathan   392

Naghsh, Amir M.   382
Nalpantidis, Lazaros   289
Navarro, Nicolás   231
Neal, Mark   378
Neville, Anne   173
Nomdedeu, Leo   382

Ognibene, Dimitri   398
Ohev-Zion, Alon   396
Ohkura, Kazuhiro   161
Orino, Yuichiro   46
Ortega-Arjona, Jorge L.   412
Ozkul, Fatih   125

Patra, Sourav   388
Pearson, Martin J.   13, 102
Petrou, Georgios   1
Pezzulo, Giovanni   398
Pina-Garcia, C.A.   426
Pini, Giovanni   90
Pipe, Anthony G.   13, 114, 299, 323, 410
Prescott, Tony J.   13, 102, 183, 253
Price, Michael J.   428
Punzo, Giuliano   390

Qayyum, Usman   430
Quintia, Pablo   241

Ratmansky, Richard   386
Reina, Andreagiovanni   137
Reina, Giulio   265
Rétornaz, Philippe   311
Richardson, Robert   173
Rodríguez, Miguel A.   241
Roke, Calum   114
Roli, Andrea   90
Roshan, Rupesh   173

Saaj, Chakravathini M.   348
Saez-Pose, Joan   382
Saito, Shinya   46
Santos, Carlos   241
Santos, Paulo E.   219
Satterfield, Brian   386
Schneider, Frank E.   149, 374, 408
Shapiro, Amir   396
Sharkey, Amanda J.C.   78
Sharp, David   386
Shi, Ji Yu   54
Siegwart, Roland   422
Skachek, Sergey   323
Sogorb, Jose Vicente   24
Solanes, J. Ernesto   195, 277
Souza, Carlos R.C.   219
Stocker, Janine   422
Sullivan, J. Charlie   13

Terzakis, George   402
Tornero, Josep   195, 277
Tran, Nam-Luc   90
Tutsoy, Onder   404

Underwood, James   265

Valdes, Fernando   241
Vaussard, Florian   311
Veillat, Aline   422
Visentin, Francesco   418
Volpi, Nicola Catenacci   398

Wada, Motohiro   161
Wang, Gang   54
Ward, Paul   384

Webb, Barbara   1
Weber, Cornelius   231
Welsby, Jason   13
Wermter, Stefan   231
Wessnitzer, Jan   1
Wildermuth, Dennis   149, 374, 408
Winfield, Alan F.T.   336, 406
Woodman, Roger   406

Yasuda, Toshiyuki   161
Yeomans, Brian   348
Yu, Xinguo   54

Zając, Michał   376
Zhang, Jiaming   78
Zhang, Yuanming   414